Engineering Design
An Introduction
2nd Edition

John R. Karsnitz, Ph.D.
Professor, Technological Design Strand
Chair, Department of Technological Studies
School of Engineering
The College of New Jersey

Stephen O'Brien, Ph.D.
Associate Professor, Electrical Engineering Design Strand
Department of Technological Studies
School of Engineering
The College of New Jersey

John P. Hutchinson, Ph.D.
Professor, Technological Design Strand and Professional Studies
Department of Technological Studies
School of Engineering
The College of New Jersey

Pat Hutchinson, Ph.D.   **Donna Matteson, M.Ed.**   **Karen Schertz, M.Ed.**

Australia • Brazil • Japan • Korea • Mexico • Singapore • Spain • United Kingdom • United States

Engineering Design: An Introduction
Second Edition
John Karsnitz, Stephen O'Brien, and John Hutchinson

Vice President, Careers & Computing: Dave Garza

Director of Learning Solutions: Sandy Clark

Senior Acquisitions Editor: James DeVoe

Managing Editor: Larry Main

Senior Product Manager: Mary Clyne

Editorial Assistant: Aviva Ariel

Vice President, Marketing: Jennifer Ann Baker

Marketing Director: Deborah Yarnell

Senior Marketing Manager: Erin Brennan

Associate Marketing Manager: Jillian Borden

Senior Production Director: Wendy A. Troeger

Production Manager: Mark Bernard

Content Project Manager: David Barnes

Art Director: Casey Kirchmayer

Media Editor: Deborah Bordeaux

Cover Image: © Shutterstock Images, LLC/MC_PP

© 2013 Delmar, Cengage Learning

ALL RIGHTS RESERVED. No part of this work covered by the copyright herein may be reproduced, transmitted, stored, or used in any form or by any means graphic, electronic, or mechanical, including but not limited to photocopying, recording, scanning, digitizing, taping, Web distribution, information networks, or information storage and retrieval systems, except as permitted under Section 107 or 108 of the 1976 United States Copyright Act, without the prior written permission of the publisher.

> For product information and technology assistance, contact us at
> **Cengage Learning Customer & Sales Support, 1-800-354-9706**
> For permission to use material from this text or product,
> submit all requests online at **www.cengage.com/permissions.**
> Further permissions questions can be e-mailed to
> **permissionrequest@cengage.com**

Library of Congress Control Number: 2011942183

ISBN-13: 978-1-111-64582-3

ISBN-10: 1-111-64582-5

Delmar
5 Maxwell Drive
Clifton Park, NY 12065-2919
USA

Cengage Learning is a leading provider of customized learning solutions with office locations around the globe, including Singapore, the United Kingdom, Australia, Mexico, Brazil, and Japan. Locate your local office at: **international.cengage.com/region**

Cengage Learning products are represented in Canada by Nelson Education, Ltd.

To learn more about Delmar, visit **www.cengage.com/delmar**

Purchase any of our products at your local college store or at our preferred online store **www.cengagebrain.com**

Notice to the Reader
Publisher does not warrant or guarantee any of the products described herein or perform any independent analysis in connection with any of the product information contained herein. Publisher does not assume, and expressly disclaims, any obligation to obtain and include information other than that provided to it by the manufacturer. The reader is expressly warned to consider and adopt all safety precautions that might be indicated by the activities described herein and to avoid all potential hazards. By following the instructions contained herein, the reader willingly assumes all risks in connection with such instructions. The publisher makes no representations or warranties of any kind, including but not limited to, the warranties of fitness for particular purpose or merchantability, nor are any such representations implied with respect to the material set forth herein, and the publisher takes no responsibility with respect to such material. The publisher shall not be liable for any special, consequential, or exemplary damages resulting, in whole or part, from the readers' use of, or reliance upon, this material.

Printed in the United States of America
4 5 6 7 8 9 10 22 21 20 19 18

Brief Contents

PREFACE   X

PART 1   THE ENGINEERING DESIGN PROCESS

Chapter 1	Technology: The Human-Designed World	2
Chapter 2	The Process of Design	30
Chapter 3	Development of the Team	60
Chapter 4	Generating and Developing Ideas	86
Chapter 5	Drawing to Develop Design Ideas	106
Chapter 6	Reverse Engineering	156
Chapter 7	Investigation and Research for Design Development	184
Chapter 8	Technical Drawing	202
Chapter 9	Testing and Evaluating	244
Chapter 10	Manufacturing	262

PART 2   RESOURCES FOR ENGINEERING DESIGN

Chapter 11	Designing Structural Systems	290
Chapter 12	Designing Mechanical Systems	320
Chapter 13	Designing Electrical Systems	346
Chapter 14	Designing Pneumatic Systems	386
Chapter 15	Human Factors in Design and Engineering	404
Chapter 16	Math and Science Applications	428
Chapter 17	Design Styles	470
Chapter 18	Graphics and Presentation	504

GLOSSARY   537
INDEX   543

Table of Contents

PREFACE X

PART 1 THE ENGINEERING DESIGN PROCESS

Chapter 1
Technology: The Human-Designed World 2

Introduction 3
Design Professionals 4
   The Design Continuum 5
Design and The Industrial Revolution (1750–1850) 8
   The American System 11
Engineering Societies 12
Standardization 14
Greatest Engineering Achievements of the Twentieth Century 16
   Electricity in the Twentieth Century 16
Engineering Careers 19
Science and Technology 19
Impacts of Technology 20
   Inputs 20, Processes 20, Outputs 21, Product Life Cycle 23
Technology and The Earth's Resources 26
Ethics and Design 27
Summary 28
Bring It Home: Observation/Analysis/Synthesis 29
Extra Mile: Engineering Design Analysis Challenge 29

Chapter 2
The Process of Design 30

Introduction 31
Polya's Four Steps to Effective Problem Solving 32
   Planning 32, Order 33, Iteration 33, Managing a Project 34
Design Process 36
Creativity and Innovation in the Design Process 54

Design Limitations 56
Summary 58
Bring It Home: Observation/Analysis/Synthesis 58
Extra Mile: Engineering Design Analysis Challenge 59

Chapter 3
Development of the Team 60

Introduction 61
Utilization of Teams in Industry Today 61
   Engineering Professionals 62
   How Would You Answer These Questions? 63
      1. Would You Like to Make Something? 63,
      2. Would You Like to Save the World? 64,
      3. Would You Like to Play All Day? 65,
         Engineering Education 65
   Involving Everyone in the Design Process 65,
   Virtual Teams and Their Place in Engineering Design 67, Benefits and Challenges of Virtual Teams 68
      Benefits 68, Challenges 69, Best Practices: Building a Virtual Team for Success 69
   Development of the Team 70, Team Success 70, Group Norms 71, The Cycles of Team Maturity 75, Most Teams Go Through Five Stages of Development 75
Team Leadership and Team Control 76
Information Sharing as It Relates to Teams 78
   Making Team Communications Work 78, Reaching Consensus 79, Using Feedback for Furthering Open Communications 79, Team Communications Should Be Organized 81, Listening Skills Complete the Communication Loop 81, Effective Techniques for Active Listening 82

Summary 84
Bring It Home: Observation/Analysis/Synthesis 84
Extra Mile: Engineering Design Analysis
    Challenge 85

Chapter 4
Generating and Developing Ideas 86

Introduction 87
Creative Thinking 87
Generating Design Ideas 87
    Lateral Thinking 88, Analogies 89,
    Brainstorming 89
        Mindmapping 90, Breaking the Rules 90,
        The Nominal Group Technique 91
    Synectics 92, Sketching and Doodling 93,
    Incubation Period 94
Development Work 95
    Construction Kits 95, Computer-Aided
    Design 96, Appearance Development 97,
    Visual Design Elements 99
        Line 99, Color 99, Form/Shape 100, Space 101,
        Texture 102, Value 102
Choosing the Best Solution 102
    Criteria Selection 103
Summary 104
Bring It Home: Observation/Analysis/Synthesis 105
Extra Mile: Engineering Design Challenge 105

Chapter 5
Drawing to Develop Design Ideas 106

Introduction 107
The Many Uses of Drawing 107
    Exploring the Visible World 107, Developing
    Ideas 107, Documenting the Process 108
        Design Portfolio 109, Engineer's Notebook 110,
        Student Design Portfolio 111
Communicating Through Drawing 114
    Whole-Brain Drawing 114, Warm-Up
    Exercises 114
        Exercise 1: Drawing Mirror Images 115,
        Exercise 2: Turn It Upside Down 115,
        Exercise 3: Blind Contour Drawing 117,
        Exercise 4: Positive and Negative Shapes 118
Drawing Basics 119
    Line 119, Shape/Form 120, Value 122
        Light Source and Shading 122, Highlights 123
    Color 124

Hue, Chroma, and Value 124, Primary
    Colors 124, Secondary and Tertiary Colors 125
Texture 126, Space 128
Sketching and Drawing Techniques 129
    Perspective Drawing 129
        Linear Perspective 130, Point Perspective 130,
        Exercise 5: Cube in One-Point Perspective 131,
        Interior Views 132, Two-Point Perspective 134,
        Exercise 6: Horizon Line and Vanishing
        Points 135, Understanding Perspective
        Visually 137, Exercise 7: Using a Grid to Capture
        the Whole Scene 138, Isometric Drawing 139,
        Exercise 8: Isometric Drawing 140
Other Drawing Conventions 142
    Crating 142, Sighting for Proportion 143,
    Outlining 143, Adding a Background 144,
    Colored Pencil Techniques 144, Color Marker
    Techniques 144
Using Drawings in the Design Process 147
    Preliminary Sketches 147, Annotated
    Sketches 148, Developmental Sketches and
    Drawings 148, Production Drawings 149
Developing an Engineer's Notebook and Design
    Portfolio 150
Portfolio 151
    Title Page, Page Numbering, and Table of
    Contents 151, Page Orientation 151, Logo 151
    Binding 151, Page Content 153, Portfolio Page
    Layout 154
Summary 155
Bring It Home: Observation/Analysis/Synthesis 155

Chapter 6
Reverse Engineering 156

Introduction 157
Reverse Engineering Leads to a New
    Understanding About Products 158
    Reverse Engineering Decreases Product
    Waste 158
Reverse Engineering and Patents 160
Reverse Engineering: The Big Picture 162
    Reasons for Reverse Engineering 162
Reverse Engineering: Step By Step 163
    Functional Analysis 168, Structural
    Analysis 169, Materials Analysis 171,
    Manufacturing Analysis 172
Summary 181
Bring It Home: Observation/Analysis/
    Synthesis 182
Extra Mile: Engineering Design Analysis Challenge 183

Chapter 7
Investigation and Research for Design Development 184

Introduction 185
Asking Questions 188
Using Market Research 189
Using Primary Sources 189
  Consumer-Based Information 190, Patent Information 192, Trademark and Copyright Protection 192, Patent Searches and Independent Inventor Resources 193, Visiting Stores 193, Laboratory Studies 194
Using Secondary Sources 194
  Library Homepages 194, Web Portals 195, Encyclopedias 195, Internet 196, Magazines, Trade Journals, and Newspapers 197, Saving Information and Citing Sources 198, Using Human Factors Information 199, The Myth of the Average Person 199, Human Scale 199
Summary 200
Bring It Home: Observation/Analysis/Synthesis 201
Extra Mile: Engineering Design Analysis Challenge 201

Chapter 8
Technical Drawing 202

Introduction 203
Technical Drawings 203
Measurement for Engineering Design 208
Architectural Scale 209
Measurement: The Real World of Variability 210
  Accuracy and Precision 212, Simple Rules for Variability of Measurement 213, Who Uses Technical Drawings? 213, Technical Drawings and the Consumer 216, Technical Drawing Standards 216, Technical Drawings and the Engineering Process 218
Drafting and Cad 219
Isometric and Oblique Pictorial Drawings 220
  Isometric Grid Paper 221, Oblique Views 222
Orthographic Drawing and Sketching 224
  Arrangement of Views 224, Envisioning an Object in Three Views 224, Spacing between Views 227, Scale 227
Line Conventions 228
  Object Lines 228, Construction Lines 228
    Hidden Lines 228
  Centerlines 228, Extension Lines 229, Dimension Lines 229, Section Views 230, Auxiliary Views 232
Dimensioning 232
  Dimension Precision 234, Dimension Tolerances 234, Dimensioning Features 235
Computer-Aided Design 237
  Creating Sketches in Solid Modeling Software 239
Summary 240
Bring It Home: Observation/Analysis/Synthesis 241
Extra Mile: Detailed Orthographic Drawing 243

Chapter 9
Testing and Evaluating 244

Introduction 245
Developing Appropriate Tests 246
  Testing a Toy 246
Testing an Engineering Solution 248
  Materials Testing 249
    Stress versus Strain 249, Tensile Testing 249
    Hooke's Law 250, Fatigue 251, Hardness
    Testing 252, Other Engineering Tests 252
Testing and Evaluating Your Own Design Work 252
  Aesthetics 252, Ergonomics 253, Performance/Functionality 253, Durability/Reliability 253, Cost 254, Impacts 254
Presenting Test Results 256
  Descriptions 256, Numbers 256, Checklists 256, Testimonials 259
Evaluating Your Design Skills 259
Summary 260
Bring It Home: Observation/Analysis/Synthesis 260
Extra Mile 261

Chapter 10
Manufacturing 262

Introduction 263
The Industrial Revolution 263
  The Agrarian Revolution 264, The Steam Engine 264, Principles of the Steam Engine 265, Interchangeable Parts 266
The Assembly Line 267
Material Processing 267
  Material Production Cycle 268, The Importance of Materials 269, Forming Materials 269
    Bending 269, Pressing 271, Forging 272,
    Extruding 273, Drawing 273, Casting 273, Types of Molds 274
  Separating Materials 275
    Mechanical Separating Techniques 276

Combining Materials 278
    Mechanical Fastening 278, Chemical Fastening 279
Organizing for Production 280, *Computer-Integrated Manufacturing* 280
    Just-In-Time 281, CAD/CAM 281, Flexible Manufacturing System (FMS) 281

Designing for Manufacture 282
    Rapid Prototyping 282
Future Impacts 286
Summary 287
Bring It Home: Observation/Analysis/Synthesis 287
Extra Mile: Mass-Production Line 288

PART 2   RESOURCES FOR ENGINEERING DESIGN

Chapter 11
Designing Structural Systems 290

Introduction 291
Structural Systems 292
Technological Structures 294
    Structural Failure 294, *Safety Factor* 297
Newtonian Mechanics 297
Structural Loads and Forces 299
    Equilibrium 299, *Stress* 300, *Strain* 300
Common Forces 301
    Compression 302, *Tension* 302, *Bending* 302, *Shear* 304, *Torsion* 304
Structural Components 305
    Beams 305
Bridge Structures 305
    Beam Bridges 307, *Truss Bridges* 308, *Arch Bridges* 309, *Suspension Bridges* 309
Calculating Loads on Structures 310
    Simple Testing of Loads 310, *Using Mathematics to Calculate Loads* 312, *Using Graphical Analysis* 313, *Bow's Notation* 313, *Interpreting the Graphical Analysis* 315
Summary 317
Bring It Home: Observation/Analysis/Synthesis 318
Extra Mile: Engineering Design Analysis Challenge 319

Chapter 12
Designing Mechanical Systems 320

Introduction 321
Mechanisms and Machines 323
Kinematics 324
    Reuleaux's Six Mechanical Elements 325, *Kinematic Diagrams* 326, *Kinematic Members* 327
        Motion 327, Momentum 327, Work 327, Power 327, Energy 328

Creating Motion in a Mechanism 329
    Linear Motion 330, Reciprocal Motion 330, Rotary Motion 330, Oscillating Motion 330
Levers and Linkages 331
    Linkages 333
        Bell Crank and Reversing Linkages 333, Parallel Linkages 334, Treadle Linkage 334, Toggle Linkages 334
Rotary Mechanisms 335
    Gears 336
        Other Rotary Systems 338
    Cam and Crank Slider Mechanisms 339
        Cam Motion 339
    Ratchet Mechanisms 340, *Clutches and Brakes* 340
Modeling Mechanical Designs 341
Summary 343
Bring It Home: Observation/Analysis/Synthesis 344
Extra Mile: Engineering Design Analysis Challenge 345

Chapter 13
Designing Electrical Systems 346

Introduction 347
The Science of Electricity 350
    Electrical Conductors and Insulators 351, *Electrical Resistance* 354, *Ohm's Law* 355
        Current 355, Resistance 357, Voltage 357
    Kirchhoff's Laws 358, *Magnetism* 360
Major Circuit Components 361
    Resistors 361, *Power Consumption of Resistors* 363, *Capacitor* 364, *Inductor* 365
Electronic Systems Design 366
    Input 366, Process 366, Output 366
    System Inputs 367
        Light 367, Heat 368, Sound 369, Position 370
    System Outputs 370
        Light-Emitting Diodes 370, Actuators 371

System Processors 374, Analog Processors 374
    Transistors 374, 555-Timer Chips 375,
    Operational Amplifiers 376
Digital Processors 376
    Coding 377, Logic Operation 378,
    Binary Numbers and Arithmetic 378,
    Microprocessors 380
Summary 384
Bring It Home: Observation/Analysis/Synthesis 385
Extra Mile: Engineering Design Analysis
    Challenge 385

Chapter 14
Designing Pneumatic Systems 386

Introduction 387
Characteristics of Fluids 388
Pneumatics versus Hydraulics 389
Principles of Pneumatics 390
Pneumatic-System Components 392
    Cylinders and Control Valves 393
        Single-Acting Cylinders 393, Three-Port
        Valve 394, Double-Acting Cylinders 394,
        Five-Port Valve 394, Pressure-Operated
        Five-Port Valve 395
    Other Components 395
        Flow Regulator 395, Shuttle Valve 396, Solenoid
        Valve 396
Calculating Forces in Fluidic Systems 396
Basic Pneumatic Circuits 399
Safety in Fluidic Systems 401
Summary 402
Bring It Home: Observation/Analysis/Synthesis 402
Extra Mile: Engineering Design Analysis
    Challenge 403

Chapter 15
Human Factors in Design and Engineering 404

Introduction 405
Human Scale 406
    *The Myth of the Average Person 406, Picking
    the Right Numbers 407, Not All Measures Are
    Equal 409, Reach and Clearance 409*
Human Behavior 410
    Compatibility 411
Posture and Movement 412
    *Capability (Abilities and Limitations) 412,
    Range of Motion 412, The Hand 413,
    Lifting 414*

Universal Design 415
    *Assistive or Adaptive Technology 415, Elderly and
    Physically Challenged 416*
Designing with Anthropometric Data 418
    Designing a Chair 419
Evaluating Design for Human Factors 422
    Computer Workstation 422
        Establish a Neutral Body Position 422,
        Establishing Component Placement 423
Summary 425
Bring It Home: Observation/Analysis/
    Synthesis 426
Extra Mile: Engineering Design Analysis Challenge 427

Chapter 16
Math and Science Applications 428

Introduction 429
Measurement: The Real World of Variability 429
    Units Analysis 433
        Graphing: Absolute, Relative, and Polar 434
    *Absolute versus Relative Graphs (Using
    Cartesian Plotting) 434, Polar (Non-Cartesian)
    Graphing 436, Spreadsheets and Structured
    Programming 437*
        Organization of Data/Information 438,
        Calculations 439, Graphing 443,
        Statistics 445
    *Newton's Three Laws of Motion 448, Statics and
    Vectors 450, Dynamics 452*
        Acceleration 452, Free-Falling Objects 453,
        Projectile Motion 454, Rotational Motion 454
Swinging Doors 456
    *Springs: Nature's Trigonometric Function 460,
    Exponential Functions 462, Probability/Statistics:
    Applications 464*
Summary 468
Bring It Home: Observation/Analysis/Synthesis 469
Extra Mile: Engineering Design Analysis
    Challenge 469

Chapter 17
Design Styles 470

Introduction 471
Architectural Design 473
    *Frank Lloyd Wright 473, Architectural
    History 475*
        Greek Architecture 475, Roman Architecture 476

What Gives a Building Style? 477
   Floor Plan 477, Elevations 479, Shape of
   Roof 479, Shape of the Eaves 479, Window and
   Door Styles 481, Exterior Walls 483
Architectural Styles 484
   *Colonial Style (1600–1800) 485, Georgian Style
   (1700–1780) 485, Federal Style (1780–1840) 486,
   Greek Revival Style (1825–1860) 486, Victorian
   Style (1860–1900) 486, Twentieth-Century
   Styles 487, Prairie Style (1900–1920) 488,
   Craftsman Style (1905–1930) 488, Art Deco
   (1920–1940) 489, International Style
   (1925–present) 490, Modern Style
   (1945–present) 490*
Postmodern Architects 491
Industrial Design 494
   *Raymond Loewy, the Father of Industrial
   Design 495, Norman Bel Geddes and
   Streamlining 496, Henry Dreyfuss and Human
   Factors 497, Louis Comfort Tiffany and Art
   Nouveau 498, William Morris and Arts and
   Crafts 499, Walter Gropius and Bauhaus 499,
   Postmodern Industrial Designers 499*
Summary 502
Bring It Home: Observation/Analysis/Synthesis 503
Extra Mile: Engineering Design Analysis
   Challenge 503

Chapter 18
Graphics and
Presentation 504

Introduction 505
The Graphic Design Process 506
   *Collecting Information for the Design 506,
   Creating Possible Solutions 507,*

Producing a Graphic Image 511
   What Specifications Are Needed When Placing
   an Order? 511, What Software Should Be
   Used to Create the Designs? 511, What Image
   Resolution Is Needed for the Job? 511, What
   Is the Difference Between the RGB CMYK
   Color? 513, Will the Color on the Finished
   Job Match the Original on the Computer? 514,
   Additional Considerations 514, What Are the
   Legal Considerations? 514
Creating Effective Graphic Designs Using Design
   Principles 515
   *Unity 516, Balance 517, Rhythm 518,
   Emphasis 518, Proportion and Scale 520*
Using Typography Effectively 521
   Understanding Type Specifications 522
   Type Size 522, Line Spacing 523, Typeface
   or Font 523, Line Length 524, Format 525,
   Pagination 525
Using Photographs 527
   *Technical Qualities 527, Composition 528,
   Editing Photographs for Design and
   Production 529*
Using Illustrations, Symbols, and Logos 529
Web Design 530
   *Web Page Development 530, Brochure
   Design 531*
Making Presentation Using PowerPoint 532,
   Planning a Presentation 532
Summary 535
Bring It Home: Observation/Analysis/
   Synthesis 536
Extra Mile: Engineering Design Analysis
   Challenge 536

GLOSSARY 537
INDEX 543

Preface

Learning to design is a journey. Whether you want to design energy-efficient homes or urban megaprojects, whether you want to invent a more effective toothbrush or a new surgical robot, whether you want to make passenger cars safer and more energy efficient or design sports equipment, you will journey through a process that helps you understand that design is a unique form of human thinking and doing that results in new solutions to human problems. This text provides the roadmap that engineers and other design professionals follow as they work to find new solutions to help improve the lives of people.

This new edition of *Engineering Design: An Introduction* has been revised and expanded for teachers who want to inspire their students to explore career pathways in engineering and technology. By presenting the learned principles and concepts that engineers and design professionals use to shape our modern, human-designed world, *Engineering Design: An Introduction* will help students develop the problem-solving skills and technological literacy they need to complete the journey.

ENGINEERING DESIGN AND PROJECT LEAD THE WAY, INC.

This text resulted from a partnership forged between Delmar Cengage Learning and Project Lead The Way, Inc. in February 2006. As a nonprofit foundation that develops curriculum for engineering, Project Lead The Way, Inc. provides students with the rigorous, relevant, reality-based knowledge they need to pursue engineering or engineering technology programs in college.

The Project Lead The Way® curriculum developers strive to make math and science relevant for students by building hands-on, real-world projects in each course. To support Project Lead The Way's® curriculum goals, and to support all teachers who want to develop project/problem-based programs in engineering and engineering technology, Delmar Cengage Learning is developing a complete series of texts to complement all of Project Lead The Way's® nine courses:

- ▶ Gateway to Technology
- ▶ Introduction to Engineering Design
- ▶ Principles of Engineering
- ▶ Digital Electronics
- ▶ Aerospace Engineering
- ▶ Biotechnical Engineering
- ▶ Civil Engineering and Architecture

- Computer Integrated Manufacturing
- Engineering Design and Development

To learn more about Project Lead The Way's® ongoing initiatives in middle school and high school, please visit www.pltw.org.

HOW THIS TEXT WAS DEVELOPED

This book's development began with a focus group that brought together experienced teachers and curriculum developers from a broad range of engineering disciplines. Two important themes emerged from that discussion: (1) teachers need a single resource that fits the way they teach engineering design today, and (2) teachers want an engaging, interactive resource to support project/problem-based learning.

For years, teachers have struggled to fit conventional textbooks to STEM-based curricula for engineering design. Most high school teachers have collected stacks of thick textbooks in their search for the right mix of materials to teach design concepts. *Engineering Design: An Introduction* finally answers that need with an interactive text organized around the principles and applications of engineering design. For the first time, teachers will be able to choose a single text that addresses the challenges and individuality of project/problem-based learning while presenting sound coverage of the essential concepts and techniques used in the engineering design process.

This book is unique in that it was written by a team of authors with pedagogy and professional engineering expertise to ensure a textbook that students can understand and one that contains valid engineering content. *Engineering Design: An Introduction* supports project/problem-based learning by:

- Creating an unconventional, show-don't-tell pedagogy that is driven by engineering *concepts*, not traditional textbook content. Broad concepts are mapped at the beginning of each chapter, and clearly identified as students navigate the chapter.

- Reinforcing major concepts and learned principles with Applications, Projects, and Problems based on real-world engineering discipline–focused problems.

- Providing a text rich in features designed to bring the design process to life in the real world. Case studies, Career Profiles, related STEM principles, Boxed Articles highlighting human achievements, and resources for extended learning will show students how engineers develop career pathways, work through product failures, and innovate to continuously improve product success and quality of life.

- Reinforcing the text's interactivity with an exciting design that invites students to participate in a journey through the engineering design process.

Organization

Engineering Design presents concepts and applications in a flexible, two-part format. Part I describes the process of engineering and technology product design. Students will learn the story of engineering by studying the great engineering achievements of the twentieth century. They will also learn how to identify problems and opportunities, how to organize a design team including the role of engineering and other design professionals, how to conduct research and document information about a problem, and how to work toward innovative solutions using drawing and other design techniques and practices.

Part II develops the specific skill sets and competencies needed to apply and participate in the process. In these skill-building resource chapters, students can learn how to design structural, mechanical, electrical, or pneumatic systems. They can explore human factors, and they can study how math and science principles—the cornerstone of good engineering design—are applied to real-world problems. They can learn how the visual form or product style is important to the success of the final product. Finally, students will discover how they can use graphic design principles to enhance product design and make more effective personal presentations.

New to This Edition

The second edition reflects a two-year effort to improve this text for students and teachers alike. The authors have added information on current engineering careers, including the unique characteristics of many engineering sub-disciplines and career profiles. Special attention has been given to acknowledging the importance of diversity in engineering and other design fields, all in an effort to help students explore the range of possibilities within engineering fields today.

Additional information on technical drawing strengthens the text's emphasis on this essential practical skill, including information on measuring and commonly used measuring tools and systems. The chapter on drawing has been moved to Part I, focusing on the importance of drawing to the design process and design solutions.

The text includes extensive updates to provide current coverage of the latest research tools, making it easier to find information for design assignments and gain familiarity with resources and concepts used by modern industry professionals. Book features such as STEM elements and Your Turn were redesigned to improve the flow of the text and readability for the students.

A new chapter on design styles helps to complete the story of engineering design by providing information and related visuals that show classic architectural and product designs. Case studies highlight two important designers, one who is working to improve the lives of handicapped individuals and the other the designer of the iPad® and other popular Apple® products. The chapter reinforces the fact that the visual form of any design is a critical factor in the commercial success of that design.

An all-new CourseMate supplement for engineering and design has been developed to support the second edition. Students will find a rich array of online learning resources to help them succeed, while instructors gain a suite of assessment and classroom management tools. Please see the full description following for instructions on how to access CourseMate through www.cengagebrain.com.

Features

Teachers want an interactive text that keeps students interested in the story behind the engineered products and solutions that shape our world. This text delivers that story with plentiful boxed articles and illustrations of current technology. Here are some examples of how this text is designed to keep students engaged in a journey through the design process.

▶ **Case Studies** let students explore the process design teams use to create new technologies.

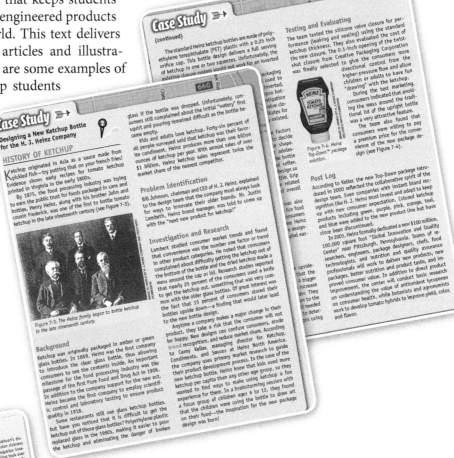

▶ **Boxed Articles** highlight fun facts and points of interest on the road to new and better products.

Your Turn
Logo and Package Design

Design World Learning would like to create a new curriculum to promote environmental awareness. They are looking for a new logo and package design that conveys a universally understood message to support environmental awareness. Create at least six thumbnails for a new environmental awareness logo using uniquely created universal symbols.

▶ **Your Turn** activities reinforce text concepts with skill-building activities.

xiv   Preface

OFF-ROAD EXPLORATION

Career information about all design professions can be found at the U.S. Department of Labor, Bureau of Labor Statistics website, using the Occupational Outlook Handbook at http://www.bls.gov.

▶ **Off-Road Exploration** provides links to extended learning with resources for additional reading and research.

sustainable design: Also known as green design, the process of producing products, systems, or environments that are environmentally friendly because they reduce the use of nonrenewable resources and minimize any negative impact on the environment.

▶ **Key Terms** are defined throughout the text to help students develop a reliable lexicon for the study of engineering.

▶ **STEM** connections show examples of how science and math principles are used to solve problems in engineering and technology.

Mathematics can help you understand financial questions. In 1854, Elias Howe won a $2-million settlement. What is the present-day equivalent of this amount? Is it considered substantial? Here are some suggestions to help you compare:

1. A loaf of bread in the year 1854 cost approximately 7 cents.
2. Go to http://measuringworth.com/. Under Calculators, select "Relative Values – US$" to calculate the current value for the patent infringement award of $2,000,000 granted in 1854. Do not use the $ sign or commas when entering the amount in the calculator. Read *Six Ways to Compute the Relative Value of a U.S. Dollar Amount, 1790–2005* to better understand economic factors that impact relative value.

Careers in the Designed World

KRISTIN WEARY: ELECTRICAL SYSTEMS ENGINEER, LOCKHEED MARTIN

▶ Design a solution in the software
▶ Test/debug the software on a computer simulator and with a test box
▶ Merge the new design with the rest of the equipment and computer model

Documentation: A Link from the Past to the Future

Kristin Weary sees her position in electrical engineering as a critical link to the future. As a systems engineer for Lockheed Martin, Kristin often works on equipment and systems that were designed many years ago. Well-documented systems allow technology upgrades to be implemented more easily. Kristin says, "Documentation of your work is crucial, because twenty years from now you won't be around to explain what you did, and someone will want to know why you did it the way you did."

On the Job

Kristin supports the design, documentation, and troubleshooting of complex electrical systems in remote locations. "I write documentation for very sophisticated electrical systems," explains Kristin. She starts with complicated information and writes it in a simple manner so that someone without an engineering degree can comprehend and safely operate the systems.

"Most of the time I can't see the equipment when issues come up, and detailed information is limited," says Kristin. "When there is a problem with a system, I work with a team to develop strategies to resolve the problem."

A computer model is used to simulate the actual complex systems Kristin works with. She works in a lab where she has designed a simulator to model a smaller system that has been integrated with the computer model. Here are the steps that Kristin used to set up and use these computer models:

▶ Understand the problem
▶ Understand the overall system
▶ Conceptually frame out a solution

Kristin uses communication technology such as email to communicate with off-site team members and customers.

Inspirations

Kristin loves to solve logic problems and come up with creative solutions. She realized early on that an engineering career would let her apply this skill in real life.

For Kristin, fun is being able to debug a complex system by breaking it into smaller areas to find the problem. Being able to identify and fix problems provides satisfaction for Kristin.

Education

Kristin earned a bachelor's of science degree in electrical engineering from Virginia Tech. In addition, she completed two summer internships with General Electric Corporation. During one of the internships, she was challenged to fix a problem with a complex database that someone else designed. This experience set the stage for Kristin's future employment as a systems engineer at Lockheed Martin.

Advice for Students

Kristin says that students who plan to study engineering already know that math and science are important parts of their preparation. She also advises students to develop their writing skills. "A lot of engineering is the documentation of the design and also the ability to communicate in writing about the design to other people," says Kristin.

"Don't be afraid to ask for help," adds Kristin. "If you can't solve a problem or find the answer and you have put an honest effort into doing so, ask for help, because most likely there is someone who knows the answer and is more than willing to mentor you."

▶ **Career Profiles** provide role models and inspiration for students to explore career pathways in engineering.

▶ **Design Briefs** and **Engineers' Notebooks** are included where relevant to model effective communication techniques.

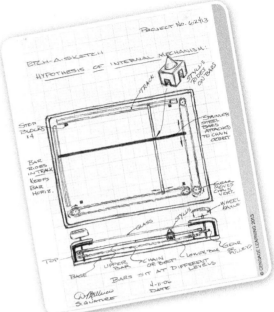

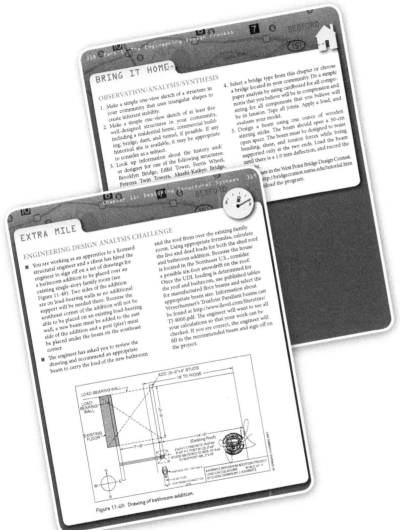

▶ **Bring it Home:** *Observation/Analysis/Synthesis* activities are provided at the end of each chapter. The activities progress in rigor from simple, directed exercises and problems to more open-ended projects.

▶ **Extra Mile:** *Engineering Design Analysis Challenge* at the end of each chapter provides extended learning opportunities for students who want an additional challenge.

Supplements

A complete supplements package accompanies this text to help instructors implement 21st-century strategies for teaching engineering design:

▶ A **Student Workbook** reinforces text concepts with practice exercises and hands-on activities, including an extended golf club design project.

▶ An **Instructor's e-resource** includes solutions to text and workbook problems, instructional outlines and helpful teaching hints, a STEM mapping guide, PowerPoint presentations, and computerized testing options.

▶ An **Activities Manual** contains 30 projects that explore virtual and physical modeling for engineering design.

COURSEMATE

The second edition of *Engineering Design: An Introduction* includes an all-new Technology and Engineering CourseMate to help students make the grade.

Technology and Engineering CourseMate for Engineering Design includes:

▶ An interactive eBook, with highlighting, note taking, and search capabilities

▶ Interactive learning tools including:

✓ Quizzes

✓ Flashcards

✓ Games

✓ PowerPoint Lecture slides

✓ and more!

Instructors will be able to use CourseMate to access Instructor Resources and other classroom-management tools.

To access these supplemental materials, please visit www.cengagebrain.com. At the cengagebrain.com homepage, search for the ISBN for your title (from the back cover of your book) using the search box at the top of the page. This will take you to the product page where these resources can be found.

HOW *ENGINEERING DESIGN* SUPPORTS STEM EDUCATION

Understanding math and science are critical to the design process and are part of the languages we use to communicate ideas about engineering and technology. It would be difficult to find even a single paragraph in this text that does not discuss the importance of understanding the integration of Science, Technology, Engineering, or Mathematics (iSTEM). The authors of this text have taken the extra step of showing the links that bind math and science to engineering and technology. The STEM icon shown here highlights passages throughout this text that explains how engineers use math and science principles to support successful designs in engineering and technology. In addition, the Instructor's e-resource contains a STEM mapping guide to this career cluster.

ACKNOWLEDGMENTS

A text like *Engineering Design: An Introduction* could not be produced without the patient support of family and friends and the valuable contributions of a dedicated educational community. The authors wish to acknowledge several individuals for their support and patience throughout this process.

Early in this book's development, the authors met Christopher Gibson and Daryn Mount from Watchung Hills Regional High School, and Kenneth Mathis from Williamstown High School. All three professionals are certified Project Lead The Way, Inc. teachers and graduates of the Department of Technological Studies at The College of New Jersey. Their practical classroom experience provided valuable insight into the Introduction to Engineering Design course. We also want to thank Kevin Schauer for his technical support, and other undergraduate students in our technology education/pre-engineering and M/S/T programs who read and commented on chapter drafts. We are very grateful to Brett Handley for his attention to detail in providing thoughtful feedback on early chapter drafts.

Many people contributed to the textbook beyond the named authors. Special thanks must be given to Donna Matteson of the State University of New York at Oswego, who identified elements of the book and contributed Chapter 6, Reverse Engineering. We were also fortunate to have the expert contributions of Karen Schertz and Terry Whitney, who wrote about the importance of team development in Chapter 3. Mr. Whitney's discussion on virtual teams is informed by his 12 years experience teaching in engineering graphics. He is currently the Senior Training and Development Manager, Securitas USA.

Dr. Patricia Hutchinson, who wrote Chapter 5 on Drawing to Develop Design Ideas, is nationally recognized for her work in technological design. She also shared her creative talents by providing special pieces of art illustrating important technological concepts. We are sure you will quickly recognize her work as it adds a special and unique dimension to the book. Suzanne Karsnitz, a school librarian, contributed to the development of Chapter 7, Investigating and Researching for Design Development. Ms. Karsnitz has more than 30 years of experience working with teachers and students in design project research and was one of the first librarians to fully computerize her library in central New Jersey. Jennifer Markarian, a chemical engineer and technical writing expert, reviewed chapters for both engineering content and for written clarity.

The publisher wishes to acknowledge the invaluable wisdom and experience brought to this project by our focus group and review panel.

Focus Group

Connie Bertucci, Victor High School, Victor, NY
Omar Garcia, Kearny High School, San Diego, CA
Brett Handley, Wheatland-Chili Middle/High School, Scottsville, NY
Donna Matteson, State University of New York at Oswego
Curt Reichwein, North Penn High School, Lansdale, PA
George Reluzco, Mohonasen High School, Rotterdam, NY
Mark Schroll, Program Coordinator, The Kern Family Foundation
Lynne Williams, Coronado High School, Colorado Springs, CO

Review Panel

Brian Benton, Walton High School, Marietta, GA
David Boe, Avoyelles Public Charter High School, Mansura, LA
Mike Braceland, Westerly High School, Westbrook, CT
Ted Branoff, North Carolina State University, Raleigh, NC
Andrew Bucci, Greece Athena High School, Rochester, NY
Brenda Campbell, Centerville High School, Centerville, IN
Todd Dischinger, Liverpool High School, Liverpool, NY
Brett Handley, Wheatland-Chili Middle/High School, Scottsville, NY
John Mativo, Ohio Northern University, Ada, OH
Terry Nagy, Shenendehowa High School, Clifton Park, NY
Michael G. Osterman, Middleton High School, Tampa, FL

Curt Reichwein, North Penn High School, Lansdale, PA
Robert Schmeling, South Division High School, Milwaukee, WI
Thomas Singer, Sinclair Community College, Dayton, OH

We also wish to thank our special consultant for this series:
Aaron Clark, North Carolina State University, Raleigh, NC

In addition, Project Lead The Way's® curriculum directors Sam Cox and Wes Terrell provided feedback on the previous edition and advice on curriculum updates and content changes for this edition.

ABOUT THE AUTHORS

John Karsnitz is Professor and Chair of the Department of Technological Studies in the School of Engineering at The College of New Jersey. He holds a Ph.D. from The Ohio State University and has authored books on *Graphic Communication, Society Ethics and Technology,* and *Design and Problem Solving in Technology*. Dr. Karsnitz has taught at the middle school and high school level and has taught at the college level for more than 30 years. He is active in the International Technology and Engineering Educators Association, the New Jersey Technology and Engineering Educators Association, and serves on the NJ Engineering Education Commission. Dr. Karsnitz helped lead the establishment of Technology as a core content area (NJDOE Standard 8) in New Jersey. The Department was one of the first teacher preparation programs nationally to include pre-engineering content as part of its core program and to create an integrated STEM-based major for elementary education teachers. Dr. Karsnitz serves on the board of directors for the Center for Excellence in STEM Education at The College of New Jersey.

Steve O'Brien holds bachelor's degrees in mathematics and physics from Western Washington University, and a master's and Ph.D. in electrical engineering from Cornell University. Currently, Dr. O'Brien is an associate professor in the Department of Technological Studies in the School of Engineering at The College of New Jersey. Dr. O'Brien has more than 20 years of engineering experience in industry as well as over six years of teaching experience. His area of technical expertise is the design and application of semiconductor lasers, especially high-power and high-speed lasers. Prior to his focus on education, Dr. O'Brien co-founded T-Networks, a company specializing in high-speed lasers for telecommunications, and accumulated more than 40 publications and 13 patents. Currently, Dr. O'Brien's area of focus is curriculum design for pre-engineering courses in K-12, with an emphasis on effectively integrating math and science topics. Dr. O'Brien has shared these results through presentations and papers at regional and national conferences and advises the Technology Education Society, the Department's student professional organization.

John Hutchinson is an Emeritus Professor in the Department of Technological Studies at The College of New Jersey. He holds a Ph.D. from The Pennsylvania State University and has authored books on *Design and Problem Solving in Technology* and *Designing with Pro/DESKTOP*. Dr. Hutchinson has taught at the high school, community college, and college level for more than 35 years. During his tenure, Dr. Hutchinson worked closely with colleagues from the United States, United Kingdom, Germany, and other countries in design and technology curriculum development. He served on the Editorial Board of *Design and Technology Education: An International Journal*.

Pat Hutchinson received a B.A. in Art at Gettysburg College in 1970 and an M.F.A in Painting from the Pennsylvania State University in 1972. She painted, showed her work and taught art at Trenton State College, Mercer County Community College and Ocean County College for ten years before beginning a doctoral program in Design and Technology at New York University. Dr. Hutchinson spent a year as a Fulbright Scholar doing doctoral research on design-based education at Oxford University, before receiving her Ph.D. in 1987. She spent five years at Drexel University, where she co-founded and served as Editor-in-Chief of *TIES Magazine* before moving to The College of New Jersey to direct the NSF-funded Children Designing and Engineering™ Project.

Donna M. Matteson is an Associate Professor in the Department of Education at the State University at Oswego. Her area of expertise is Architecture and Computer Aided Design with AutoCAD certification through Autodesk Inc. Ms. Matteson has a total of 27 years of teaching experience and 8 years of experience in business and industry. During the summer months Ms. Matteson is a Master Teacher for Project Lead the Way, and has provided graduate level summer institutes at ten different universities. For more than a decade, Ms. Matteson has worked with Project Lead the Way to develop strategic plans, curriculum, and course-end exams. Ms. Matteson was the project director for the Introduction to Engineering Design curriculum released in 2008. Ms. Matteson is the lead author of *Civil Engineering and Architecture*.

Karen Schertz is a thirty-year veteran in the field of technical education. She has taught in high school, community college, and at the university level in the areas of drafting, CAD, and design. Ms. Schertz served for five years as Director of the Advanced Technology division at Front Range Community College, Fort Collins, Colorado, and received a Faculty Member of the Year award from the Colorado State Board of Community Colleges. Ms. Schertz holds a Bachelor of Science degree in Education from the State University of New York at Buffalo and a Master's degree in Education from Penn State University. Ms. Schertz is the co-author of *Design Tools for Engineering Teams, an Integrated Approach*.

PART I
The Engineering Design Process

Chapter 1 Technology: The Human-Designed World
Chapter 2 The Process of Design
Chapter 3 Development of the Team
Chapter 4 Generating and Developing Ideas
Chapter 5 Drawing to Develop Design Ideas
Chapter 6 Reverse Engineering
Chapter 7 Investigating and Researching for Design Development
Chapter 8 Technical Drawing
Chapter 9 Testing and Evaluating
Chapter 10 Manufacturing

CHAPTER 1
Technology: The Human-Designed World

Menu

 Before You Begin
Think about these questions as you study the concepts in this chapter:

1. Why do humans design?
2. Who are design professionals and how do they contribute to society?
3. How did the Industrial Revolution shape the world as we know it?
4. How were the early engineering societies established?
5. Why did engineering societies develop standards?
6. What are the greatest engineering achievements of the twentieth century?
7. What are the modern engineering professions?
8. What is the relationship between science and technology?
9. How does technological design impact the world?
10. What is an ethical design dilemma?

INTRODUCTION

Our natural world is filled with human-designed objects. We study about the natural world in science but it is equally important that we understand our human-designed world—technology. All people, young and old, male and female, need to understand how technology affects individuals, society, and our environment. We call this understanding technological literacy. Technologically literate citizens are able to *use*, *manage*, and *evaluate* technology in their lives.

The term *technology* can mean different things to different people. For example, some people use it to refer to computer hardware. In this book, we will use the term *technology* very broadly to include:

- The process of design, that is, solving problems with criteria and constraints
- The product, system, or environment that results from the design
- The new understanding or knowledge gained through the design and production of a product

Early humans lived in a natural world and developed quite a few tools to help them adapt. Today, we live in a technological world, one that we have adapted extensively. Even when we venture into the natural world, such as when we hike in a wilderness area, we carry such things as specially designed clothing, backpacks, GPS navigation, or freeze-dried food.

A technologically dependent society needs:

- Technology professionals who can understand design processes and create new products
- Technology experts who can help consumers use and manage those products
- Citizens who can make intelligent technology choices

Informed citizens can make smart decisions about what to buy, for whom to vote, and where they fall on the many technology issues in our society. For example, as young adults, you will need to intelligently consider the technological component of many important current issues: the role of nuclear generation as a renewable energy source, how to regulate the manufacture of products that are environmentally friendly and safe for human use, the impact of technological activity on global warming, genetic technologies and its applicability on humans, and many other issues that could affect important aspects of our lives today and for generations to come. Growing in your understanding of technology will give you a framework to be able to develop well-informed opinions about technological questions. Before you can make a good decision about an issue, you need to understand the issue.

As a first step to becoming a technology professional, you must study the objects in your daily life—your microwave oven, your running shoes, your bicycle—and begin to ask yourself how these objects have been *conceived*, *designed*, and *made* by others. Engineers, architects, graphic designers, fashion designers, and industrial designers are just a few of the more common design professionals who bring new products to the marketplace. They are responsible for everything from the packages that contain your breakfast cereal, to the clothing that you wear, to the water supply system that lets you wash in the morning. They also let you play music on your iPod, text your friends, and enjoy the comforts of climate control. Figure 1-2 shows some examples of the technology that surrounds us. Each photo lists some of the concepts that the design team had to understand or apply for the product to be designed and produced. Common to all the products was the need to understand the human user, materials, manufacturing processes, budgeting, and marketing.

Figure 1-1: Our natural world is filled with human-designed objects.

> **design:** An iterative or repeating decision-making process that results in a plan to produce a new product.

INTRODUCTION

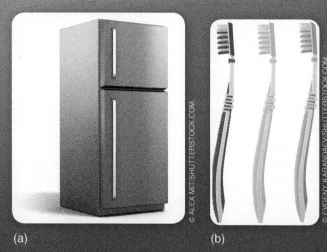

Figure 1-2: *Everyday items that exist in our "designed world":*
- (a) In the home (application of thermodynamics).
- (b) For personal use (biomechanics and human factors).
- (c) For play (plastic injection molding, robotics).
- (d) To entertain (microelectronics and aesthetics).
- (e) For exercise (advanced composites and biomechanics).
- (f) To transport (hybrid engines, or "green technology").

DESIGN PROFESSIONALS

Design professionals may work individually or as part of a design team, but always *design under constraint*. This means that they have limits on the time and money allowed to do their work. They usually are employed by a company that provides the resources and organization needed to produce their successful designs.

Most products require teams of design professionals who have varied backgrounds and expertise. Consider the cell phone in Figure 1-3. Why would someone want to spend money on this product? Is it because it looks nice, works well, fits comfortably in

> **OFF-ROAD EXPLORATION**
>
> Career information about all design professions can be found at the U.S. Department of Labor, Bureau of Labor Statistics website, using the Occupational Outlook Handbook at http://www.bls.gov.

your hand, is environmentally friendly, or has a good price? If you buy the phone for its looks, then you have been influenced by the design professionals who apply aesthetic principles to make the phone attractive. Aesthetic principles are rules used to explain how people respond to what they see. Aesthetic judgments are therefore subjective. In other words, "beauty is in the eye of the beholder."

If, on the other hand, you are impressed by the phone's sound quality and its ability to obtain or send a signal, then you have been influenced by the design professional who applied math and science principles to create a more effective electronic circuit—in this instance, an electrical engineer. Other design professionals made sure the phone is comfortable to use and can be produced and sold at a competitive price. Someone else in the organization may have addressed the impact that the phone's production and use will have on the environment. Many skilled design professionals are needed to create a successful product.

The Design Continuum

All design professionals apply their understanding of aesthetic and math/science principles to create successful designs. Design professionals also should apply ethical principles to their work by being concerned about the impact of their work on the environment and society. The design continuum in Figure 1-4 shows the importance of aesthetic and math/science principles to the work of different design professionals. For example, architects who design residential and commercial structures must consider all the sets of principles shown in Figure 1-5. In addition, architects must understand and meet all governmental requirements such as building codes and other regulations. Architects fit very nicely near the middle of the design continuum.

Figure 1-3: A modern cell phone employs sophisticated design for both its visual appearance and operation.

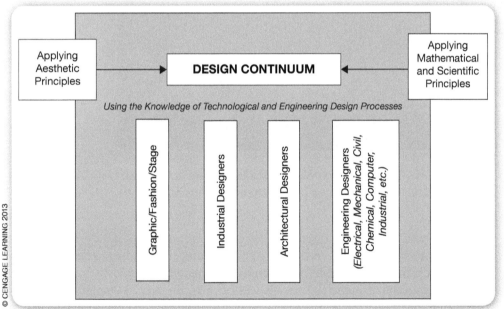

Figure 1-4: The designed world requires many different skill sets as shown on a design continuum.

Architects Consider	What That Means
Visual principles	How the building looks
Functional principles	How the building serves the people housed in it
Structural principles	How the building reacts to natural forces
Ethical principles	How the building will affect the environment and the society in which it is placed

Figure 1-5: Architects fit in the middle of the design continuum.

OFF-ROAD EXPLORATION

Architectural career information and schools offering academic programs can be found on the American Institute of Architects website at *http://archcareers.org*.

Some design professionals, such as graphic designers and fashion designers, are most concerned with aesthetic principles, but still need to understand the technology that helps them produce their products. For example, graphic designers use software packages to lay out pages and manipulate photos and artwork. They also estimate costs for paper and ink, and write specifications for producing their final products. *Stage designers* must consider the unique design constraints set by the physical space in a theater and the seated audience (see Figure 1-6).

Figure 1-6: Two examples of how a stage designer can merge objects and light to create a unique theatrical space.

Industrial design is an important design profession. Industrial designers work to improve the function, value, and appearance of the product. These improvements help both the consumer and producer (manufacturer).

The concept of industrial design was born in the early twentieth century as competition grew among producers. Wider newspaper and magazine circulation made it easier for people to see and compare print images of products. This meant that sales often hinged on how a product looked.

In the United States, these new industrial designers, including Raymond Loewy and Henry Dreyfuss, had trained as graphic, fashion, and stage designers. Industrial designers are similar to architects as they apply similar principles to their work. Most new product development teams today include an industrial designer as a key member.

Figure 1-7: ResQTec G2 Cutter. This rescue tool was an Industrial Designers Society of America industrial design excellence award winner. It is a hydraulic rescue device designed to extricate people trapped in a vehicle. It won the award not only for its appearance but also for reduced weight, increased efficiency, and improved ergonomics.

Businesses know that a well-designed product will increase sales. The Industrial Designers Society of America (IDSA; http://www.idsa.org) gives awards annually for successful new product designs (see Figure 1-7).

As we move toward the mathematics and science end of the design continuum, we find the more traditional engineering professions. **Engineers** use their detailed knowledge of physical principles to design and produce new products, systems, or environments. Engineers need strong math and science understanding, and gain additional knowledge in the specialties they choose. An aerospace engineer might determine the shape and size of an aircraft wing. A civil engineer might calculate the forces on a concrete foundation pillar. An electrical engineer might analyze the current-carrying capacity of high-tension cable. The U.S. Department of Labor reported that about 1,600,000 engineers were working in the United States in 2008, making engineering one of the largest professions in the country.

OFF-ROAD EXPLORATION

Industrial design career information and schools offering academic programs can be found at *http://www.idsa.org/education*.

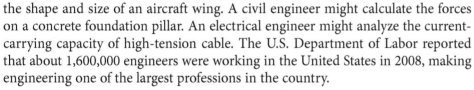

Your Turn

Consider getting involved in your next school play by joining the stage-design team. This team will use stage-design principles to design and make the staging that is appropriate for the play and the school auditorium constraints.

There are 1.6 million engineers in the United States. Is this a relatively small or large number? Mathematics can help us answer this question. For example, in 2008, the U.S. population was around 300 million. If 1.6 million people were engineers, then we can find that engineers made up 0.5 percent of the U.S. population:

$$(1.6/300)\ 100 = 0.533\%$$

Has the percentage of engineers in the U.S. population increased or decreased over time? How does this fractional population compare to other countries? How does this population of engineers compare to other well-known professions in the United States, such as medical doctors, nurses, or teachers?

DESIGN AND THE INDUSTRIAL REVOLUTION (1750–1850)

Humans have always modified their natural environment. They created shelters and clothing to shield them from cold temperatures, scorching sun, and harsh weather. They created tools to help build new useful objects (archaeologists call these objects artifacts), grow crops, or provide protection. Most early human design was done by people using craft techniques to make goods for themselves or their families. Each crafter designed and made each product. Products made in a craft system take longer to produce and are expensive to sell. In a craft-based society, only a small number of people could afford high-quality products.

Before the twentieth century, technology was used mostly to provide food, shelter, clothing, and other basic needs. Because more than 50 percent of the population worked on farms, historians call this period the Agricultural Age (see Figure 1-8).

More people began to move to cities and work in industries that produced nonagricultural products. For the first time in the United States, more people were involved in producing goods than in producing food. This period is known as the Industrial Age (1907–1957), and it was a significant change from previous human history.

By the mid-twentieth century, another big change had occurred. In the United States and in other developed countries, the number of white-collar professional jobs outnumbered the blue-collar industrial jobs. This meant that companies employed more workers to manage information about their products than they employed to *produce* the products. The United States has been in the Information Age or Technological Age since 1957.

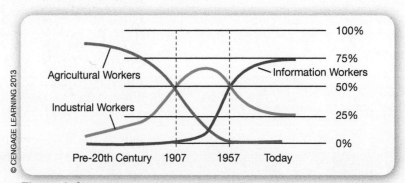

Figure 1-8: **Historical perspectives on technological development.**

Careers in the Designed World

Designing the Future

Randy Rausch is rarely satisfied with the view from his desk. As a computer engineer for General Electric Corporation, thousands of current technologies lie at his fingertips. However, Randy constantly looks beyond them to identify emerging computing technologies and match them to the needs of GE businesses.

"If the technologies do not yet exist to make better products and services, we invent them," says Randy. "Design is what we do. We start with an idea of what a customer wants; heed social, economic, and political trends; identify fundamental limits; and design for the future."

On the Job

Randy and his GE colleagues have designed a new controller to make aircraft engines safer and electric generators more efficient. They have also designed tools for better satellites, and created new medical imaging techniques to give doctors more information to treat patients.

Engineering teams are vital to this work. "We pull together the best people we can find from around the world to solve our challenges," says Randy. "Often a product will require input from many different technology areas. I work with electrical engineers, physicists, chemical engineers, biologists, mechanical engineers, computer scientists, and other specialists to design new products. I often manage teams with four to ten people."

The teams use technology to visualize their designs. "It is easier and cheaper to design and build a test model in a computer than to build a physical prototype," Randy says. "Then we use machine learning and artificial intelligence to evaluate and improve the designs."

Randy communicates with his global teams by email, instant messaging, videoconferences, and virtual whiteboards, but he still feels the best communication is face-to-face. Randy spends about a third of his job traveling to meet new people, gather new ideas, and work with project teams.

Inspirations

Randy grew up in Bismarck, North Dakota, where his parents owned a small retail furniture business. From

Randy Rausch: Computer Engineer, General Electric Corporation

time to time, Randy's father brought home new computers and would let him tinker with them. Randy's interest in how things worked led him to discover the power of computer technology. That power whetted his appetite for a degree in engineering and computer science.

Education

Randy earned a Bachelor of Science degree in computer engineering from the University of Notre Dame. He continued with a Master of Science degree from Rensselaer Polytechnic Institute while working for GE as part of the Edison Engineering Development Program.

Advice for Students

Randy thinks the best traits an engineer can develop include the ability to think analytically and problem solve, an open mind to new ideas, and the ability to work in a team environment. A successful career in engineering requires these traits. Randy advises students to look at the total project prior to beginning. "Spend time learning, preparing, or planning before beginning the new design," he says. "Work done up front will save time down the road."

Randy wants students to know that engineering will present a challenging career path. "Persevere," he says. "It is worth it. Ask lots of questions. Seriously, ask questions and most importantly, have fun!"

The GE90-115B is the most powerful engine in the world. Randy Rausch helped develop algorithms in its controller to improve engine safety.

10 Part I: The Engineering Design Process

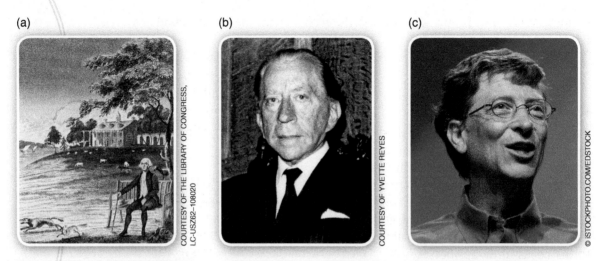

Figure 1-9: (a) George Washington gained power during the Agricultural Age by owning land; (b) J. Paul Getty wielded power during the Industrial Age by using capital; and (c) Bill Gates built a powerful modern company based on the power of knowledge.

The Agricultural, Industrial, and Information Ages have different qualities. For example, power in the Agricultural Age came from *owning land*, power in the Industrial Age came from having *capital*, and power in our Information Age comes from having *knowledge* (see Figure 1-9). Some people think we are now entering a new age, in which power will come from taking an international view of business and industry.

The Industrial Revolution, which began in England around 1750, changed the way people designed and produced products. The Industrial Revolution slowly moved people from a craft-based system to an industrial system. The change to an industrial system required many resources:

- A large supply of workers
- Enough food to feed the workers
- An adequate energy source
- Large amounts of capital (money to invest in the product)
- A transportation system that can move materials to producers and products to markets

Why did the Industrial Revolution start in England? The country had all of these things:

- Large coal deposits for energy
- Excellent fleets to move raw materials and goods
- A government that encouraged private business owners to invest their capital
- A society that was ready to invent and innovate new products

During the early eighteenth century, the American colonies helped England's industrial growth as well. The colonies sent raw materials to England to be turned into products. In addition, the colonists were eager to buy British-made goods.

Around this time, people began to use the French word entrepreneur to describe a group of middle-class merchants. These British entrepreneurs assumed some accountability and inherent risk to purchase raw materials from a variety of sources and then supplied the materials to craftspeople who used them to make

entrepreneur: A person who takes initiative to establish a new enterprise or business and assumes risks in the hope of gaining financial and other successes.

products. When the product was finished, the entrepreneur-merchants found consumers and shipped the goods to those markets.

As production and transportation techniques improved, products became cheaper to make. Lower prices helped the markets grow, increasing the money, or revenue, returned to the business. These revenues, and more importantly the *profits*, were reinvested, helping the industrial revolution expand worldwide. Today we use the term *entrepreneur* to describe someone who takes initiative and risk to create a new business.

The American System

The Great Exhibition (see the boxed article below) showcased new British, European, and American products. The American exhibitors included McCormick, Goodyear, Colt, and Singer and received much attention at the fair. Some critics said the American products were unsophisticated. Others admired the ingenuity of the product design and production. Most importantly, the United States was recognized for innovations in what became known as the American System of production.

The American System that created the McCormick reaper, Goodyear tires, the Colt revolver, the Singer sewing machine, and many other products was driven by innovations that included:

- ▶ Large-scale production for mass markets
- ▶ Products made to precise design standards
- ▶ Use of interchangeable parts
- ▶ Expanded use of machine tools
- ▶ Unskilled labor performing simplified operations

Point of Interest
The Great Exhibition (1851)

The Great Exhibition of 1851 was nicknamed the *Crystal Palace* because it looked like a giant greenhouse. It was 563 meters long by 138 meters wide (1,848 feet by 456 feet) and was large enough to house more than 13,000 exhibits that celebrated English industrial successes.

The great civil engineer, I. K. Brunel, along with other noted designers Robert Stephenson and Scott Russell, led the planning committee. The final design for the exhibition hall was submitted by Charles Fox, a structural engineer, and Joseph Paxton, who designed greenhouses (see Figure 1-10). The Crystal Palace was planned and constructed in just nine months using premade wrought-iron components. It was used for several purposes after the exhibition until it was destroyed by fire in 1936. Today the site of the building and its grounds are known as Crystal Palace Park.

Figure 1-10: The Crystal Palace main exhibition building. Illustration from Tallis' history and description of the Crystal Palace, and the exhibition of the world's industry in 1851.

Your Turn

I. K. Brunel helped plan the Great Exhibition of 1851. What are some of his other engineering achievements?

interchangeable parts: An important development during the Industrial Revolution where parts or components of a product were designed and made to fit in any product of the same type.

The increasing use of **machine tools** and **interchangeable parts** during the Industrial Revolution led to a need for better machine tools and product components. Companies needed better materials and consistent quality. But because there were no national standards, consistency was difficult to achieve.

Think about it: What if you needed to replace your 100 mm inline skate wheels and no standards existed? What if all manufacturers for inline skate wheels used different bearing sizes? These kinds of problems affected every industry. Many professional societies formed committees to establish national and international standards to address these problems.

ENGINEERING SOCIETIES

Engineering societies play an important role in promoting their profession and in setting standards (see Figure 1-11). In 1880, prominent New York mechanical engineer Henry Rossiter Worthington (1817–1880) cofounded the **American Society of Mechanical Engineers (ASME)**. Worthington held patents for steam pumps and worked to create an effective New York water-supply system. Early steam boilers were poorly designed and tended to explode, often resulting in human injuries or death as well as damage to property. It is not surprising then that ASME's first standard, established in 1884, was the *code for the conduct of trials for steam boilers*. Like many engineering societies, ASME helped professionalize engineering in the United States.

Robert Henry Thurston (1839–1903), the first president of ASME, worked to establish college engineering classes. Thurston started the mechanical engineering curriculum at Stevens Institute of Technology in 1871. He also started the first U.S. college mechanical engineering laboratory in higher education for conducting funded research.

OFF-ROAD EXPLORATION

Engineering career information and schools offering academic programs can be found at the Engineering Service Center website at http://engineeringedu.com.

Figure 1-11: **Examples of professional design societies. A complete list of societies, including their purposes and contact information, can be found in the appendix.**

Year Established	Society
1852	American Society of Civil Engineers (ASCE)
1857	American Institute of Architects (AIA)
1880	American Society of Mechanical Engineers (ASME)
1884 (AIEE)	Institute of Electrical and Electronics Engineers (IEEE)
1893	American Society for Engineering Education (ASEE)
1934	National Society of Professional Engineers (NSPE)
1944 (SID)	Industrial Designers Society of America (IDSA)
1950	Society of Women Engineers (SWE)
1957	Human Factors and Ergonomics Society (HFES)
1963	American Institute of Aeronautics and Astronautics (AIAA)
1991	American Institute for Medical and Biological Engineering (AIMBE)

Case Study

Do you know how a sewing machine works?

The Singer sewing machine was one of the earliest American **mass-produced** products (see Figure 1-12). Sewing machines were invented to replace sewing by hand. Originally, sewing machines were operated by a foot treadle mechanism that moved a needle up and down and advanced the cloth under the needle. The first functional sewing machine was invented in France. But it was the American patent issued in 1846 to Elias Howe for *a process that used thread from two different sources* that led to the first practical machine designs (see Figure 1-13). Howe invented a mechanism that creates a stitch—the most important part of a successful sewing machine.

Howe's invention used a needle with an eye and a shuttle underneath the cloth to create what is called a *lockstitch*. Unfortunately for Howe, other machine designers quickly adopted his lockstitch mechanism in their sewing machine designs.

Isaac Singer made his first commercially successful sewing machine in 1851. His machine used a

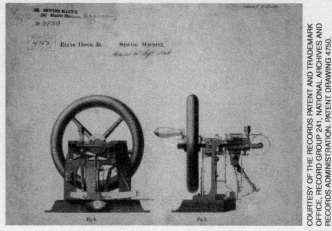

Figure 1-13: *Elias Howe's patent for a lockstitch sewing machine.*

Figure 1-12: *The Singer sewing machine.*

lockstitch device, and added an up-and-down motion mechanism. Elias Howe, believing that Singer's machine violated his patent, sued for patent infringement in 1854 and won more than $2,000,000 as a settlement.

Many successful American inventors and entrepreneurs like Elias Howe have seen their patents violated. Although patent law gives inventors exclusive rights to their inventions, patent law does not prevent someone else from intentionally or unintentionally using the invention. When inventors believe that their inventions are being used in another product, they will likely take legal action regarding the patent infringement.

Mathematics can help you understand financial questions. In 1854, Elias Howe won a $2-million settlement. What is the present-day equivalent of this amount? Is it considered substantial? Here are some suggestions to help you compare:

1. A loaf of bread in the year 1854 cost approximately 7 cents.
2. Go to http://measuringworth.com/. Under Calculators, select "Relative Values – US$" to calculate the current value for the patent infringement award of $2,000,000 granted in 1854. Do not use the $ sign or commas when entering the amount in the calculator. Read *Six Ways to Compute the Relative Value of a U.S. Dollar Amount, 1790–2005* to better understand economic factors that impact relative value.

Point of Interest
The Morrill Act

Abraham Lincoln signed the Morrill Act of 1862, which established the land grant university system. The Morrill Act gave large parcels of land to agriculture and mechanical arts colleges in each state. This explains why today nearly every state university offers undergraduate and graduate degrees (master's and PhD) in engineering, architecture, and other design disciplines.

Figure 1-14: Today's Virginia Tech was founded under the Morrill Act as Virginia Agricultural and Mechanical College, one of the original land grant colleges.

STANDARDIZATION

What do you do if a bolt is lost or broken on your bicycle derailleur or snowboard binding? Make yourself a new one? Go to a machinist to have a new part custom made? No—you go to the hardware store or ski shop to get a standard replacement. Without **standardization**, you could not do this.

Standards affect every aspect of our lives. Without standards, you could not buy a telephone and know that it would work both at home and your office. We could not share computer files with a friend. Fire companies from various areas could not help each other during a major fire because they would not have standard hose connectors. Even time zones are an example of standards. Time zones were originally created by the railroads in the late 1800s to synchronize the scheduling of trains. They are still essential today for modern telecommunications and worldwide business and finance. Standards lower costs to manufacturers and consumers, improve product quality and safety, and help protect the environment.

During the time of the American Civil War, the government recognized the advantages of a standardized railroad track gauge (the distance between the two rails). The United States adopted the British railroad gauge size of 4 feet 8 1/2 inches as the standard in 1886. This gauge was used to build the Transcontinental Railroad between Sacramento, California (Central Pacific), and Omaha, Nebraska (Union Pacific). The railroad was completed on May 10, 1869, with the ceremonial golden spike at Promontory Summit, Utah (see Figure 1-15).

standardization:
A process of establishing a technical consensus agreement that provides a common set of expectations for quality or compatibility of a material or product without creating an unfair competitive advantage in the marketplace.

Figure 1-15: Golden Spike Ceremony, May 10, 1869: The Union Pacific and Central Pacific joined 1,776 miles of rail at Promontory Summit, Utah Territory. This event joined the United States, allowing for western expansion, creating the need for time zones, and setting the stage for economic growth.

Point of Interest
Where do standards come from?

In the United States today, the American National Standards Institute, or ANSI (pronounced *an-see*), develops and oversees standards for products, services, processes, and systems. ANSI also coordinates U.S. standards with international standards so that American products can be used worldwide. A standard is first developed by a governmental agency, consumer group, company, or professional organization and forwarded to ANSI for final adoption (see Figure 1-16). ANSI cooperates with the International Organization for Standardization (ISO) and the International Electrotechnical Commission (IEC) through the U.S. National Committee (USNC).

Standards help ensure that:

▶ The characteristics and performance of products are consistent

▶ Producers and consumers use the same definitions and terms

▶ Products are tested using the same criteria

Figure 1-16: Examples of standards organizations. Can you find any products with these letters? A complete list of standards organizations including their purposes and contact information can be found in the appendix.

ANSI	American National Standards Institute
ASTM	American Society for Testing and Materials
BSI	British Standards Institution
DIN	Deutsches Institut fur Normung
ISO	International Organization for Standardization
NIST	National Institute of Standards and Technology
UL®	Underwriters Laboratories, Inc.

GREATEST ENGINEERING ACHIEVEMENTS OF THE TWENTIETH CENTURY

In 2006, the National Academy of Engineering (NAE) created a list of 20 engineering achievements and posed the following question: *How many of the twentieth century's greatest engineering achievements will you use today?* (see Figure 1-17).

> **OFF-ROAD EXPLORATION**
>
> Information about standards organization information can be found at *http://www.ansi.org*.

By exploring these achievements, mentioned throughout this book, you will learn how engineering design activities shape and change our world each day. As you will see, most achievements of the twentieth century were built on key discoveries and inventions by Edison, Benz, Bell, Marconi, and others during the late nineteenth century.

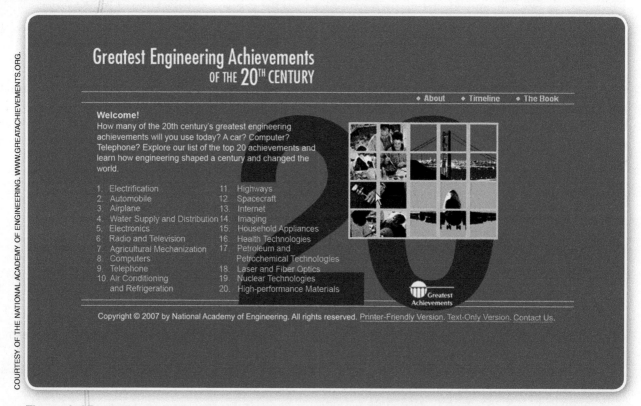

Figure 1-17: The National Academy of Engineering's list of the greatest engineering achievements of the twentieth century.

Electricity in the Twentieth Century

Most people recognize Benjamin Franklin as one of the founding fathers of the United States, but he was also considered a "father of electricity." Franklin's kite experiments in 1752 were recognized by the Royal Society of London for research in electricity. The electricity in our homes and offices, however, is largely due to the work of Thomas Edison, Nikola Tesla, Charles Steinmetz, and entrepreneurs such as George Westinghouse.

> **OFF-ROAD EXPLORATION**
>
> The greatest engineering achievements of the twentieth century can be found at *http://greatachievements.org*.

Between 1880 and 1900, a great debate raged between Edison and Westinghouse over the use of direct current (DC) or alternating current (AC) electricity. In the end, AC prevailed because it was easier to distribute. By the beginning of the twentieth century, engineers and entrepreneurs were rushing to design and develop electricity-generation and -distribution systems. They also began to invent household and commercial products to use this convenient form of energy. Our modern technological society would not exist without electricity.

In electrical devices, alternating current (AC) refers to electrons periodically changing their direction of flow, called a cycle, from forward to backward. Think about the flow of electrons as being similar to the flow of a liquid in a pipe: By turning a valve, the direction of the liquid can be reversed. In contrast, direct current (DC) describes the situation when electrons only flow in a single direction.

The science of electricity tells us that for electrical power lines (conductors), the current (I) is directly proportional to the voltage (V) and is inversely proportional to the resistance (R) of the conductor. This relationship is called Ohm's law and is written as follows:

$$I = V/R$$

Ohm's law is valid for any electrical circuit, from expansive power transmission lines to tiny computer chips.

Until the mid-1930s, electricity was not available to most of the rural United States. Today, almost every person in the United States can switch on a light. The electrification of America's rural areas was the result of a government program created in 1935 by a mechanical engineer named Morris Llewellyn Cooke. President Franklin Delano Roosevelt appointed Cooke to head the Rural Electrification Administration (REA). Under Cooke's program, nearly all farms were wired by the mid-twentieth century.

The Hoover Dam and the electricity that it generated were directly responsible for the growth of both Los Angeles and Las Vegas (see Figure 1-18).

Today, coal, oil, the power of the atom, moving water, wind, and solar energy are all used to generate electricity. With growing concerns for the environmental impact of fossil fuels, and with improving efficiencies, sustainable energy sources represent important new engineering challenges.

Electricity is the workhorse of our modern technological society. To be able to use electricity, public and private companies need to create a power grid. (Private companies are known as investor-owned utilities, or IOUs.) A power grid consists of a source of energy, generators, and distribution lines built to transport the electricity to individual homes and businesses (see Figure 1-19). For the system to work, all components must be constantly monitored, upgraded, and maintained.

Figure 1-18: **The Hoover Dam is a national historic landmark and has been rated by the American Society of Civil Engineers as one of America's Seven Wonders of Modern Civil Engineering.**

OFF-ROAD EXPLORATION

Read George Constable and Bob Somerville (authors), Neil Armstrong (foreword), *A Century of Innovation: Twenty Engineering Achievements That Transformed Our Lives*, Joseph Henry Press, 2003, ISBN 0309089085.

Figure 1-19: A wind farm is a renewable source of electricity that can be connected to a power grid.

But what if the system fails? Everyone knows what it's like to spend an evening sitting in the dark unable to turn on the television or computer. Power failures are often caused by bad weather. A severe storm can cause trees to fall on power lines and disrupt service. Sometimes people are killed when they try to move or cross downed power lines. But most of the time, power failures are simply a reminder of how much we depend on electricity. You can lose computer work, become trapped in an elevator or subway, or find yourself stuck in traffic.

The problem of a major power outage is still a serious concern. The goal of power utilities is to improve reliability to 99.99 percent. Reliability is the percentage of time a customer receives service versus not receiving service over a one-year time period. By using higher quality materials and microprocessor-based monitoring systems, power networks are becoming more reliable.

The production of electricity was important to the economic growth of the twentieth century, leading to new inventions, innovations, and a whole new array of consumer products.

Point of Interest
Blackout!

The largest power blackout in U.S. history occurred on November 9, 1965 (see Figure 1-20). Thirty million people in several northeast states were without electricity for as long as 13 hours. President Lyndon Johnson said of the outage, "Today's failure is a dramatic reminder of the importance of the uninterrupted flow of power to the health, safety, and well-being of our citizens and the defense of our country."

- ☐ Ontario Hydro System
- ■ St. Lawrence–Oswego
- ■ Western New York
- ▓ Eastern New York–New England
- ■ Maine and part of New Hampshire did not lose power during the blackout

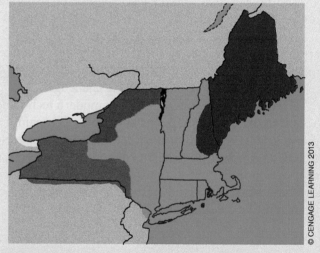

Figure 1-20: Areas of separation at 5:17 p.m., November 9, 1965.

Your Turn

How many hours in a year can power be interrupted and still be considered reliable at the 99.99 percent level?

ENGINEERING CAREERS

The greatest engineering achievements of the twentieth century demonstrate how engineers have worked to improve the quality of life. Who would want to live without electricity or modern transportation or communication systems, or who is not living a healthier life because of clean water or health technologies? In looking at the engineering achievements, it may not be clear which engineering profession contributed to the advancement. Today, over 1.6 million engineers work in manufacturing, construction, communication, or engineering-related jobs in architecture or the government. According to the most recent edition of the U.S. Bureau of Labor Statistics, the top engineering specialties are civil (278,400), mechanical (238,700), industrial (214,800), electrical (157,800)/electronic (143,700), and computer (74,700). Many other engineering specialties exist, including aerospace, chemical, environmental, nuclear, biomedical, and agricultural.

What do engineers do? We have already introduced the idea that engineers and others use design to modify the natural world to meet human wants and needs. If you want to "make something," then any area of engineering may be a good career for you. If you want to "make a difference," then environmental or health and safety engineering may be a good career for you. If you want to "play all day," then engineering new toys or new sports equipment may be a good career for you. Career case studies are presented in many of the following chapters, and a special section on engineering careers has been included in Chapter 3 (Development of the Team).

> **OFF-ROAD EXPLORATION**
>
> Find your engineering dream job! More information about interesting engineering careers can be found at *http://spectrum.ieee.org/at-work/tech-careers/find-your-engineering-dream-job*.

SCIENCE AND TECHNOLOGY

Science and *technology* are terms that are often used together in the same sentence but are not the same thing. Scientists certainly use technology and technologists certainly use science, but there is a big difference between the two. Science seeks answers to questions about the natural universe. Science describes the natural world by asking *why*? *Why do things in nature happen the way they do*?

For example, Sir Isaac Newton asked why planets move the way they do, and why objects and projectiles fall to the ground. His answers described the operating principles of gravity. More recently biologists have successfully described the specific processes involved in cellular reproduction, which led to the ability to clone mammals. Science is very important for understanding the natural world.

Technology, on the other hand, asks *how*? Technologists are less concerned with understanding why things happen in our world than with how to solve problems.

Some technologists have broader goals. They strive to improve a technology that can be used in many ways. For example, electrical engineers are always trying to reduce the amount of power their circuits consume. Such improvements would make a significant difference in many products. Certainly, engineers working on a rugged computer would benefit from circuits that use less power.

Sometimes, the application of a scientific principle comes before the understanding of the principle. Long before anyone knew about the principles of biology, people were cultivating plants and breeding animals. Before organic chemistry, people figured out how to make wine, and discovered herbal remedies for illnesses. The magnifying glass was invented in 1267, although no one at the time could explain why it worked. The telescope was invented 60 years before Newton began studying the physics of optical devices. Designers sometimes find solutions before they understand completely why those solutions work.

> **science:**
> A descriptive discipline that makes observations through experimental investigation to explain physical and natural phenomena. Science seeks to answer the question "Why?"
>
> **technology:**
> The human process of applying knowledge through innovation to satisfy our needs and wants by extending our capabilities and modifying our natural environment.

Point of Interest
Technologists ask: How do you toughen up a computer?

Computers like to be clean, cool, and carefully handled. Most people live where there is dirt, heat, and where accidents happen. Most people in the world don't have much money. To help everyone in the world gain access to the Internet, technologists working with Microsoft cofounder Paul Allen are currently designing a rugged and very low-cost computer that could be used almost anywhere, especially in poor countries.

The rugged computer must:

▶ Stand up to rough handling
▶ Keep out dirt
▶ Use cheap satellite communications
▶ Generate its own power

Experts in areas of plastics molding, physical/mechanical design, and electrical/computer design will work on the computer. These technologists will use scientific principles related to electromagnetic radiation, satellite communication, electric forces, and material expansion and contraction. They will use this scientific knowledge to solve a specific set of problems and create a specific product.

IMPACTS OF TECHNOLOGY

We have already mentioned that every product has an environmental and social impact that designers must consider. There is no such thing as a technology without a consequence. Sometimes, these consequences are predictable; sometimes, they are a complete surprise. But first, let's talk about the four components of a basic **technological system**:

1. Inputs
2. Processes
3. Outputs
4. Feedback

If we want to heat a room in our home, we might light a fire in the fireplace. This is an example of an open-loop system. As the fire burns, the room gets hotter and there are few ways to control the temperature of the room. In contrast to the fireplace, a home heating system utilizes a closed-loop system heating all rooms in the house and controlling the temperature by using a thermostat (see Figure 1-21).

Inputs

In a technological system, inputs help drive the system. Inputs might include resources such as materials, energy, information, people, tools and machines, and capital.

Processes

A system processes material, energy, or information. In many systems, at least two of these are processed, one of which is energy. For instance, leather, rubber, thread, and electricity are processed to create shoes. Can you think of a technological system in which energy is not processed? In fact, every process requires energy, even if it is the energy of human muscle.

Chapter 1: Technology: The Human-Designed World

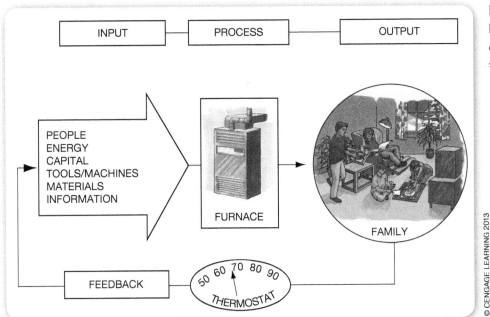

Figure 1-21: **The home heating system is an example of a closed-loop technological system.**

Outputs

The results of the processes are called outputs. The output of a home heating system is a warm house and a comfortable family. But are there other outputs that impact the individual or environment? Just outside the home there is smoke escaping from the chimney vent, and somewhere not too far away, there are clouds of emissions from a power plant's smokestack. There are four types of outputs from all technological systems (see Figure 1-22).

Type I Outputs are both *expected* and *desired*. These are the outputs the system was designed to produce. For example, warm air is the expected and desired output of a home central heating system.

Type II Outputs are *expected* but *undesired*. Some outputs of technological systems are not so obvious. Although the home central heating system provides heat, it can also produce smoke and soot, carbon dioxide, and other unwanted by-products. These outputs are expected but not desired. A chimney is added to the home heating system design to carry the smoke away from the people living in the house. A carbon monoxide detector should be in the home to detect the presence of dangerous carbon monoxide.

Type III Outputs are *unexpected* but *desired*. These are advantages resulting from the use of certain processes or resources unique to that system. A Type III Output is not harmful, but rather a pleasant

Figure 1-22: **Four outputs of technological activity.**

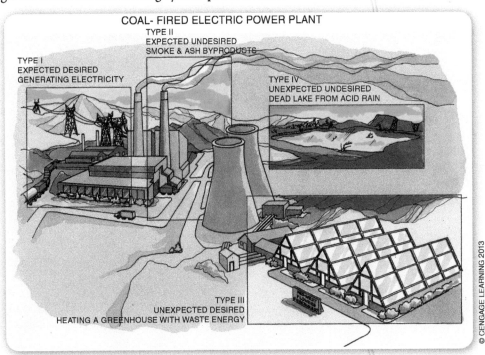

surprise. For example, the waste heat that is produced by electricity-generation plants was considered a problem until someone came up with the idea of selling it to local businesses, such as greenhouse nurseries, for winter heat.

Type IV Outputs give us greatest concern because they are *undesired* and *unexpected*. Engineering or design failures usually fall into this category. Who would have imagined that children could be physically hurt by using a computer? In fact, research is now showing that children are developing musculoskeletal injuries by using an adult-sized keyboard and workstation. Children are reporting discomfort in their wrists, neck, and hand. In this case, the Type IV Output was the failure to predict that children would someday use computers for play and homework, and to account for the human factors of children in their design.

Future design professionals must recognize Type IV Outputs as the most critical concerns for future possible design solutions. For example, burning fossil fuels for electricity was an accepted practice for decades, but has now been blamed for acid rain. Acid rain destroys lakes and defoliates forests. The by-products of combustion are also seen as a major factor in global warming. These undesired and unexpected outputs have created complex environmental and political problems that we are just beginning to address. Although nobody likes acid rain, the problem has given designers a reason to develop new products and systems. For instance, designers have invented *scrubbers* that reduce emissions from power plants. Compare the two maps in Figure 1-23. Did the problem of acid rain increase or decrease between 1994 and 2004?

Figure 1-23: The orange areas in maps (a) and (b) show where higher levels of acid rain occurred in 1994 and 2004, respectively. How did conditions change over ten years?

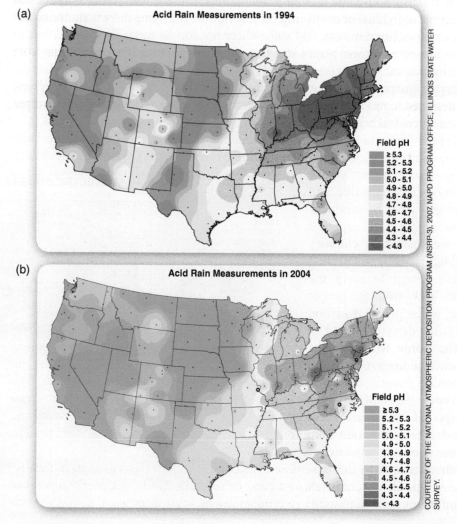

Your Turn

Find an example of a technological system and identify at least one Type I, II, III, and IV Output.

Because all systems produce desired and undesired consequences, trade-offs exist in the design of all new products and systems. With each trade-off, the design professional is recommending that one set of conditions should be selected because it is more desirable than another. There can be serious consequences when that recommendation is wrong. As citizens, we must decide what is an acceptable trade-off. The reality is that each solution can create new problems. When you solve a problem, you also change the original conditions that led to the problem, so you have, in effect, changed the world.

People throughout history have wanted to stop change. The **Luddites** of eighteenth-century England were named after their leader, Ned Ludd. Luddites wanted to halt technological progress by smashing the new weaving machines being introduced into the textile industry of the time. The new machines were taking away jobs that had always been done at home. But smashing a few machines could not prevent change. The new weaving technology helped bring about the Industrial Revolution in Great Britain.

At the other extreme from the Luddites are people who believe today that any problem can be corrected through technological innovation. This *techno-fix* mentality ignores our responsibility to think carefully about the impacts of new technologies. This position also ignores the fact that solving one problem often creates a new problem.

Luddites: An eighteenth-century group from England who wanted to halt technological progress by smashing the new weaving machines being introduced into the textile industry. Today, Luddite is a name given to someone who is against technological progress.

Product Life Cycle

There are five stages in the life of a product (see Figure 1-24):

- ▶ Design phase
- ▶ Manufacturing phase
- ▶ Marketing phase
- ▶ User phase
- ▶ Disposal phase

Until now, we have mostly been discussing the design and manufacturing phases. The marketing phase actually begins during the design phase. During the marketing phase, teams consider how to sell the product and for what price. The user phase begins when consumers buy products, usually directly from the manufacturer or through a retail store. It is important that users fully understand the proper use of the product and that the product functions correctly. From time to time, repairs may need to be made on the product.

The user phase is not the end of the story. No product lasts forever. In the disposal phase, the product is thrown away or recycled. Today, designers must think about how the materials used in a product can be safely disposed. If a product such

sustainable design: Also known as green design, the process of producing products, systems, or environments that are environmentally friendly because they reduce the use of nonrenewable resources and minimize any negative impact on the environment.

as a battery is improperly discarded, the toxic chemicals can cause environmental damage. Sustainable design, also called green or eco-design, happens when designers think about the ecological impacts of the design solution. Examples of sustainable designs include the use of wind to generate electricity, hybrid cars, improved home insulation, and recycling.

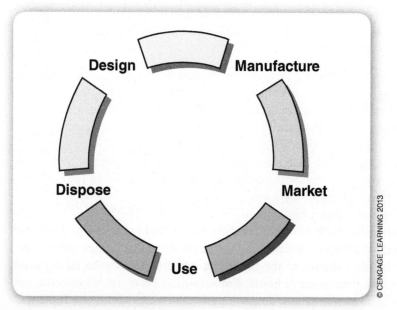

Figure 1-24: Product life cycle.

Point of Interest
Message in a bottle

The plastic bottle was a solution to several problems: Glass bottles broke easily, were expensive to manufacture, and inconvenient to dispose. Petroleum (the primary raw material used to manufacture plastics) was cheap and readily available. In addition, throwing away a bottle seemed so much more convenient than taking empty bottles back to the store. Unfortunately, we now understand that the trade-off of convenience is burying us in garbage (see Figure 1-25). Currently in the United States, only 28 percent of plastic bottles (PET) are recycled and only 40 percent of our citizens have access to recycling facilities. Numerous educational institutions, maybe yours, have banned single-use water bottles within their areas of jurisdiction. Although the plastic bottle industry has fought these laws politically, solutions at all levels are being explored. A recent study in Europe showed that recycling plastic was more energy-efficient than using alternative materials.

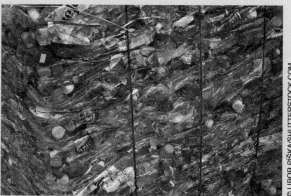

Figure 1-25: Plastic bottles are popular with consumers. These bottles are increasingly contributing to the waste stream.

Individuals, communities, and companies are now looking for better ways to dispose of products. Today, plastic is recycled to make new products such as recycled plastic lumber. Most states have passed laws meant to reduce the amount of garbage sent to landfills. Plans typically include:

1. *Waste prevention:* This means finding ways to reduce the size of the product, to reduce the amount of material used in the product, or to use materials that are more environmentally friendly. New materials are being used to reduce the weight of cars and airplanes and most other consumer products. In packaging, new manufacturing processes use less material to make bottles and cans by reducing wall thickness, and new plastic bags are made to biodegrade.
2. *Recycling:* Most cities have plans to recover materials from the waste stream for use in new products. Of the items recycled, typically 35 percent is paper, 27 percent is plastic, 22 percent is glass, and 6 percent is metal.
3. *Composting:* Many towns and cities collect garden and lawn clippings and fall leaves so they can be composted and reused as fertilizer or mulch.

Our society needs technologically literate consumers who think about the resources needed to produce each product they use; the impact of the product on the individual, society, and the environment; and how the product will be disposed after its useful life is over.

OFF-ROAD EXPLORATION

Information about plastics and recycling can be found at *http://www.plastics.americanchemistry.com/Education-Resources/*.

Point of Interest
Your weight in garbage

As a consumer-oriented society, we produce more than 1,500 pounds of garbage per person per year. That is more than four pounds per person every day! The main components of the consumer waste stream in the United States are paper and packaging.

Your Turn

Make a graph showing the rate of bottled water increasing from 3.3 billion bottles in 1997 to 15 billion bottles in 2002 (assume the growth to be linear). If the charted rate continued, how many bottles would be sold today? If manufacturers produce 31.7 bottles per second for each billion used in a year, how many bottles need to be produced to meet the current demand?

Case Study

On January 28, 1986, the space shuttle *Challenger* disintegrated only 73 seconds after takeoff (see Figure 1-26). All seven crewmembers were killed. Media coverage of the tragedy was intense, and many young schoolchildren watched the shuttle launch on live television.

Citizens of the United States wanted to know how such a dramatic failure could happen at NASA. President Ronald Reagan formed the Rogers Commission to study the disaster. The investigation discovered that a flawed O-ring seal in the shuttle's right solid rocket booster failed to function properly.

The shuttle's engineers created a segmented booster rocket design. The transportation system required a segmented design because no system was big enough to carry a single-piece booster rocket to the launch pad. To join the segments of the rocket together, a rubber-like O-ring was used to create a seal between the segments.

During the second minute of the flight, the O-ring failed and let a plume of flame leak out and burn through the external fuel tank, destroying the *Challenger*. The Rogers Commission found that the O-ring had a technical flaw. After intense study, they also determined that NASA's organization had created a culture in which engineers failed to address the flaw effectively. Investigators also blamed the disaster on unusually cold temperatures, overcompression of the O-rings during assembly, and a lack of proper inspection.

Was there another viable alternative to the O-ring design? How did time and budget influence the decision? Did the design team make it clear to all decision makers that the booster rocket could not be used on a cold day below a certain temperature? These are all questions that have to do with ethics.

Every day, engineers and other technological designers face ethical questions and make decisions. Clearly, a decision to secretly substitute an inferior material in a design is an unethical decision. Unfortunately, when people do not act ethically, results can be disastrous.

Figure 1-26: *The space shuttle* Challenger *disaster is studied frequently as an example of a design failure that raised ethical questions.*

ethics:
A branch of philosophy that considers how to apply concepts of right or wrong, good or evil, and taking responsibility for one's actions.

TECHNOLOGY AND THE EARTH'S RESOURCES

Do you believe that there are sufficient world resources for the next century and beyond, or do you believe that we are rapidly using up the earth's resources? What are some of the resources that have been in the news lately?

As a historian of technology, Melvin Kranzberg noted, "technology is neither good nor bad; nor is it neutral." We all face technological choices and we must all remember to think about the impact our choices will have on individuals, society, and the natural world. Products must be carefully designed, consumers must carefully choose, and government must balance the trade-offs and risks of future technological progress.

Point of Interest

In 1980, as a way of settling this issue, Paul Ehrlich (author of *Population Bomb*) and Julian Simon (author of *The Ultimate Resource*) made a bet on whether humans could use up their natural resources. Ehrlich subscribed to the Malthusian philosophy that population growth would lead to a shortage of resources while Simon subscribed to the Cornucopian philosophy that population growth was good and humans would solve any problems associated with resource shortages. What would be the engineering perspective on this debate? How do you feel about this question?

We can learn a lot from both worldviews. Today, new products are being designed that use less material, thus conserving valuable natural resources. There is a Native American saying: *We do not inherit the earth from our parents—we borrow it from our children*. Do you agree that we should have a worldview that focuses on maintaining a quality environment in the future?

ETHICS AND DESIGN

The impacts of technology are not simple and often involve ethical dilemmas. All design professionals must act ethically, as their work affects individuals, society, and the environment. Sometimes even the best efforts at design solutions fail. The space shuttle *Challenger* disaster is studied frequently as an example of a design failure that raises ethical questions.

Most engineering societies and business organizations have published codes of ethics. A code of ethics describes the behavior expected of the society's members and may indicate how the behavior is enforced, or what happens to someone who acts unethically (see NSPE Code of Ethics for Engineers below).

National Society of Professional Engineers (NSPE)

Code of Ethics for Engineers

Preamble: Engineering is an important and learned profession. As members of this profession, engineers are expected to exhibit the highest standards of honesty and integrity. Engineering has a direct and vital impact on the quality of life for all people. Accordingly, the services engineers provide require honesty, impartiality, fairness, and equity, and must be dedicated to the protection of the public health, safety, and welfare. Engineers must perform under a standard of professional behavior that requires adherence to the highest principles of ethical conduct.

FUNDAMENTAL CANONS: ENGINEERS, IN THE FULFILLMENT OF THEIR PROFESSIONAL DUTIES, SHALL:

1. Hold paramount the safety, health, and welfare of the public.
2. Perform services only in areas of their competence.
3. Issue public statements only in an objective and truthful manner.
4. Act for each employer or client as faithful agents or trustees.
5. Avoid deceptive acts.
6. Conduct themselves honorably, responsibly, ethically, and lawfully so as to enhance the honor, reputation, and usefulness of the profession.

SUMMARY

Throughout all of human history, people have used technology to satisfy their wants and needs:

- The Agricultural Age: Most of human history was known as the Agricultural Age. Until the twentieth century, most technological activity was focused on providing food, shelter, clothing, and other basic needs.
- The Industrial Age (1907–1957): More people began to move to cities and work in industries in which nonagricultural products were being produced. In the United States, a new form of production known as the American system used large-scale production for mass markets. This system used standardized products with interchangeable parts, machine tools, and unskilled labor to perform simplified operations to make new products.
- The Information Age (1957 to present): There are more people in the workforce who manage information about products than there are actually producing goods. This time in history is also called the Technological Age. Some economists argue that another major transformation is taking place that will involve a more international perspective for business and industry.

THE DESIGN CONTINUUM

Today, design professionals work to create the many products we use in our daily lives. We can think of these professionals on a design continuum to see how knowledge is used in their work. Some professionals, such as graphic designers, rely heavily on aesthetic principles. Others, such as engineers, rely heavily on mathematic and scientific principles. Architects and industrial designers rely on an understanding of both aesthetic and mathematic/science principles.

PROFESSIONAL ASSOCIATIONS AND STANDARDS

As the role of the design professional grew in importance, formal training at the baccalaureate level became necessary. Professional associations and societies developed as a result. The mechanical engineering profession was one of the first, forming ASME in 1880. Professional societies play an important role in establishing standards for new products, both nationally and internationally.

ENGINEERING ACHIEVEMENTS OF THE TWENTIETH CENTURY

In 2006, the National Academy of Engineers created a list of engineering achievements that have greatly influenced our lives. Today, electrical, mechanical, civil, industrial, and many other engineering design professionals work to create the new solutions to problems facing society in the twenty-first century.

Technology's Impact

Because technological activity can have positive and negative consequences, it is important to consider the impact of design activity on society, the individual, and the environment. Every design activity has the potential to create expected and unexpected, desired and undesired consequences (Type I–IV outcomes). As we produce new products, we must consider how the product will use existing resources. As communities are becoming concerned about minimizing waste, earth-friendly sustainable products are more popular in the marketplace. However, these products inevitably also become waste. Therefore, although technology can solve problems and improve lifestyle, it can also create problems. Engineers and other design professionals follow codes of ethics developed by professional organizations to guide their work and help ensure the best possible outcome from their design activity. Most would agree that their lives are made better by the products available to meet their wants and needs and that extend their capability.

BRING IT HOME

OBSERVATION/ANALYSIS/SYNTHESIS

1. Develop a questionnaire to assess the understanding of some group on how technology affects their lives. Include at least ten questions. Groups could be students, parents, teachers, or others.
2. Develop a list of design professionals in your local community and organize them by placing them on a design continuum.
3. Select one of the design professionals and prepare a summary explaining the formal preparation needed and the nature of the work associated with that profession.
4. Learn more about why the railroads created time zones.
5. Between 1880 and 1900, a great debate raged between Edison and Westinghouse over the use of direct current (DC) or alternating current (AC) electricity. What were some of the issues that eventually led to AC being the standard for generating and distributing electricity to our homes?
6. As a class or a small group, brainstorm a list of potential technological developments that could become reality in the next fifteen years. By consensus, give each development a rating of 1 (low) to 10 (high) to indicate the group's thoughts about the likelihood of it becoming reality within this time period. Make a list of potential impacts of the development on the individual, society, and the environment. Based on the worldviews of the Malthusians and the Cornucopians, prepare a debate on the issue of how world resource consumption will impact the success of technological development.
7. Develop a list of criteria for judging products as *sustainable*. Compile a list of products that meet these standards using information found at http://www.sustainableplastics.org.
8. Develop a strategy for encouraging recycling in the school, home, or community.
9. Evaluate one of the following for potential ethical dilemmas:

   - Transporting crude oil in tankers
   - Computer database credit ratings
   - Genetically altered foods
   - Burning fossil fuels
   - Recycling of plastic
   - Incineration of waste
   - Personal versus mass transportation
   - Use of pesticides and herbicides

EXTRA MILE

ENGINEERING DESIGN ANALYSIS CHALLENGE

Research one of the following technological design failures:

1. Nuclear crisis at the Japanese Fukushima Daiichi complex
2. The Kansas City Hyatt Regency skyway collapse
3. The space shuttle *Challenger* explosion
4. The Ford Pinto gas tank explosion
5. SUV rollover
6. Boston *Big Dig* tunnel collapse
7. Tacoma Narrows Bridge collapse

Determine what factors contributed to the failure. Was there unethical behavior exhibited by the design professional or team, by management, by the organization producing the product, or by the government? If so, what? What mathematical and/or scientific principles were violated? What technological principles were violated? How did the design failure impact individuals, society, or the environment? How can society help to prevent a failure like the one selected from happening in the future? Using PowerPoint software, make your presentation to the class or a group in your school or community.

CHAPTER 2
The Process of Design

Menu

 Before You Begin
Think about these questions as you study the concepts in this chapter:

1. How are design processes similar for the various designer occupations across the design continuum? How are they different?
2. What is a good example of the problem-solving process?
3. What is a good definition of a design process?
4. What are the three major components of a design process?
5. Why is the order of the design steps, or iteration of the design steps, beneficial?
6. What are the major steps, or phases, of a design process?
7. How are creativity and innovation parts of a design process?
8. What types of constraints are common to most design projects?

INTRODUCTION

Design is the process of planned change. Instead of something changing by accident, design demands that we plan change so that we end up with the results we want. In a **design process** the goal is to minimize undesired effects and control risk. Design also refers to the process used to create something new—to solve a problem. In this chapter, we will be investigating the process of design as a unique form of human activity associated with solving technological problems.

Design processes are used across the design continuum. Engineers at one end of the continuum use more math and science tools in their design process. Engineers design a variety of everyday products, from computer games to advanced medical devices.

At the artistic end of the spectrum, graphic designers use design processes to make decisions about color, contrast, sizing, shape, alignment, and perspective. Graphic designers use design processes to achieve a certain visual appeal, and their designs can include websites, billboards, product packaging, and TV commercials.

Many products, like console computer games (see Figure 2-1) and automobiles, need designers from both ends of the design continuum to add both technical and aesthetic expertise.

The design process can be complicated. It is very important to remember, though, that a design process is nothing more than a logical problem-solving technique. It is a technique that has proven to be very effective and powerful. A good understanding of problem-solving techniques is useful in all aspects of life, not just designing products.

The act of **problem solving** is certainly not new for humans. Effective problem solving, a close kin to the "scientific method," was at the heart of the Renaissance (fourteenth through seventeenth centuries) and the Industrial Revolution (eighteenth and nineteenth centuries) (see Figure 2-2). People are still refining the problem-solving process.

design process: A systematic and often iterative problem-solving strategy, with criteria and constraints, used to develop many possible solutions to solve a problem or satisfy human needs and wants and to narrow down the possible solutions.

problem solving: The process of understanding a problem, devising a plan, carrying out the plan, and evaluating the effectiveness of the plan to solve a problem, or meet a need or want.

The Scientific Method includes these steps or iterations of these steps:

1	Characterization *Observations or measurements*
2	Hypotheses *Theoretical explanations*
3	Predictions *Reasoning using logic*
4	Experiments *Tests of all the above*

Figure 2-1: Gaming systems require both highly technical electrical and mechanical engineers as well as graphic designers.

Figure 2-2: The four steps of the scientific method.

POLYA'S FOUR STEPS TO EFFECTIVE PROBLEM SOLVING

One of the most famous writers on problem solving was George Polya, a mathematician dedicated to improving mathematics education. In 1945, he wrote the book *How to Solve It* to summarize his work on problem solving. Polya's four steps to problem solving are:

1. Understand the problem
2. Make a plan
3. Carry out the plan
4. Look back on the plan; how could it have been better?

A design process is a more detailed version of a problem-solving method. The Museum of Science in Boston has developed a five-step process (see Figure 2-3). This five-step design process includes the following steps:

1. Ask
2. Imagine
3. Plan
4. Create
5. Improve

Can you see Polya's four steps of problem solving within this five-step design process?

All design processes contain three important parts: planning, ordering, and repetition (or iteration).

Planning

Simply put, a design process is a **plan**. Experienced designers take the time to form a plan that lists, orders, and prioritizes items. For simple design problems, like deciding what clothing you are going to wear for the day, you will form a plan in your mind and

plan:
(n) A detailed proposal for doing or achieving something; (v) the act of putting together a plan.

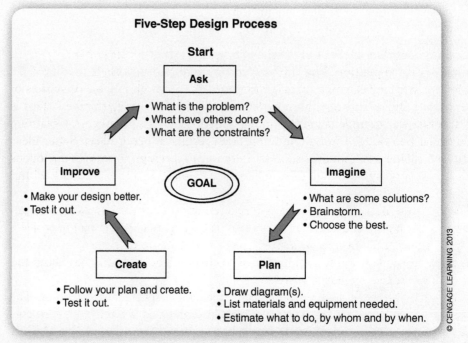

Figure 2-3: *Example of a five-step design process (modeled after Museum of Science in Boston, Massachusetts).*

implement your plan in a matter of seconds. However, as design problems get more complicated, you need to increase your planning to solve the problem efficiently. For example, the problem of constructing and placing a space station into orbit is very complicated and uses sophisticated project-management software that can organize thousands of tasks (see Figure 2-4).

Order

Most design processes place their steps in a certain order. This order is a sequential process. In a sequential process, each step is carried out in order.

Polya's four problem-solving steps are in the same order every time and are therefore sequential. In Polya's process, carrying out a plan comes only after making a plan. Assessing a plan comes only after implementation of the plan.

The five-step design process (see Figure 2-3) also follows an order. The arrows in Figure 2-3 guide you clockwise through the process. However, the order of the steps is intended as a guide only. Designers can use the five-step process in a nonsequential order as well.

Figure 2-4: **The International Space Station.**

A nonsequential process lets designers jump backward and forward between steps to more effectively develop a solution. For example, a designer might think she is ready to select the glue to attach two parts. At some point, she realizes that she lacks enough information to make an informed selection. In a nonsequential design process, she would go back to research additional information about the new materials, and then jump forward to the material-selection stage.

sequential:
Forming or following a logical order or sequence.

Iteration

The third important attribute of a design process is the concept of iteration. Iteration simply means repetition. The fifth step of the five-step process is to "improve." This fifth step invites you to iterate either the whole design process or subsets of the process steps. This repetition of several, or all, of the design process steps is an accurate way to understand the implementation of a design process. Obtaining additional knowledge through iteration often results in better ideas. Better ideas lead to a higher likelihood of success. Iteration in a design process is analogous to a closed-loop system, as defined in Chapter 1. Iteration allows feedback within the process.

iteration:
The act of repeating a set of procedures until a specified condition is met.

When design projects are successful, the iteration of process steps shows convergence. As a project gets closer to the set goals—eventually meeting the project's goals—convergence occurs. When a project is not demonstrating convergence, the team has stalled, and something needs to change. The goals may have been too difficult or the team needs to regroup.

Imagine you are a surgeon on a team that will design a surgical robot. The robot will help doctors perform corrective eye surgery. This elective surgery corrects patients' vision so that eyeglasses do not need to be worn (see Figure 2-5). A very important goal is to have zero major medical failures during a procedure. The procedure cannot injure the patient.

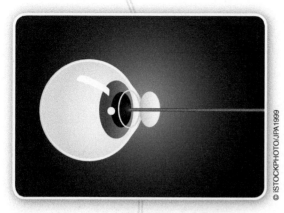

Figure 2-5: During laser eye surgery, a single pulse from an excimer laser reshapes the corneal surface for better vision. A surgical robot could help with the reproducibility and efficiency of this type of surgery.

In the nonhuman testing of the first prototype, the robot performed well, completing almost all of its tasks correctly. However, the computerized control system needed to be reset several times during the one-hour test. Each reset could cause injury to the patient. The robot prototype did not achieve your medical goal.

Over the next several years, the robot design team continued working to fix all major design flaws. Eventually, they reached their goal of zero medical failures. After successive iterations, the surgical robot design team got closer and closer to the goal. By producing many iterations of their design that brought them closer to their goal, this project team showed convergence.

Let's summarize the important aspects about a design process by looking at another design example. Assume you are on a team that will design a seat for a commercial airliner. The three main goals for the seat are: (1) be at least as comfortable as the current seat, (2) have three times longer lifetime, and (3) be 50 percent lighter. The initial motivation for this project was to investigate the use of newer, lightweight, rugged composite materials to lower both fuel and maintenance costs. The design team proceeds, in order, through the "ask," "imagine," and "plan" phases of the five-step design process, but upon documenting their efforts into a plan, they notice some grave mistakes and loop back to the "ask" and "imagine" phases. The team does not continue to the "create" phase because it would be a waste of time, effort, and money to create prototypes of a seat with substantial flaws.

Remember, it is not necessary to follow all of the design process steps in complete, sequential order. The team uses iteration, repeating the first three steps, to obtain a better plan. This flexibility of the design process—the freedom to alter order, iterate steps, and replan—allows a designer, or the whole design team, to be creative and innovative in a controlled and efficient manner. The airplane seat design team chose to be flexible and remained in the first three phases for a surprisingly long amount of time, but also with a surprising result. The team designed an innovative software package that effectively assesses all of the important mechanical properties of new composite materials for seats. It turns out that this software was also very useful for airfoil components like wings, giving the aeronautical engineers in the company a huge advantage in their design work, saving thousands of hours of work and hundreds of thousands of dollars.

OFF-ROAD EXPLORATION

For a review of DARPA's past Grand Challenges, see http://en.wikipedia.org/wiki/DARPA_Grand_Challenge.

What are composite materials and what are their advantages and disadvantages over more standard materials like wood, steel, and aluminum?

Managing a Project

All design projects come with certain challenges, some big and some small. Will the glue be strong enough? Will the microprocessor be fast enough? Will the springs be reliable enough? All challenges come with a certain amount of risk; a challenge that is easy to accomplish has relatively low risk while a challenge that is very difficult to accomplish has higher risk. All design projects have these types of risks. The only time there is no risk in a design project is if the solution is already determined.

Fun Facts

The Defense Advanced Research Projects Agency (DARPA) is the central research organization of the United States Department of Defense. The DARPA Grand Challenge is a prize competition for **autonomous** (driverless and completely self-controlled) vehicles. The mission of DARPA is to encourage research that can help improve U.S. national security. DARPA has sponsored several competitions, all in the area of autonomous vehicles. DARPA is likely to conduct Grand Challenge competitions in other technical areas in the future.

The 2005 Grand Challenge began at 6:40 A.M. on October 8, 2005. The Stanford Racing Team entry, "Stanley," a robotic Volkswagen Touareg, won the race and the team received a $2-million prize, at that time the largest prize money in robotics history. Stanley completed the 212.4-km (132-mile) off-road race with a winning time of 6 hours 53 minutes and 58 seconds. Four other autonomous vehicles also successfully completed the race.

Figure 2-6: "Stanley," autonomous vehicle.

Your Turn

Are there any current DARPA-funded challenges? Describe past DARPA challenges and what their grand prizes were. Do you have any ideas for a Grand Challenge?

Most design problems have many possible solutions. Each solution presents its own challenges and risk level. Throughout the design process, managing **project risk** will drive the team's most important decisions. Managing risk incorrectly can cause design failure, while managing risk correctly usually leads to design success. Design teams should try to remove the biggest risks first, and save the less important risks for later.

Let us go back to the airplane seat example. The team focused on understanding the properties of the new composite materials. They did not wander into the aesthetic or human factors design aspects of the seat early in the process. They focused on the problems associated with composite materials because they knew that the aesthetic and human factors of the design could be effectively solved only after they fully understood the key attributes of the unique composite materials they were considering using. The design team managed risk well by staying focused on only the key aspects first.

It is common for designers to work on what they are most comfortable with first, instead of doing what removes the most risk first. It is simply human nature. Designers should always be asking the question, "What are those key items that are really keeping this project from being completed successfully?" Once you identify the critical roadblocks, you can prioritize them, set your plan, and stay focused.

In summary, you should understand that designing is not always a sequential process but it certainly can be sequential. When you design, you do not always think and behave in separate, sequential steps. The chaotic nature of creativity often requires designers to move between questioning, thinking, acting, and evaluating while reducing risk as effectively as possible.

DESIGN PROCESS

The simplicity of the five-step process makes it a useful model for a design process. However, a more comprehensive model of a design process requires more steps. The 12-step design process discussed in this section lays out all of the possible steps for a project. Just as with the five-step process, the concepts of planning, order, and iteration are still important for the 12-step process.

Figure 2-7 shows the recommended order for the 12 steps. As the arrowed lines show, designers sometimes loop to other steps, resulting in iteration. Following are overviews of each of the 12 process steps.

STEP 1 Defining the Problem

The beginning of the design process is the identification and definition of a problem in need of a solution. What is the problem? On the surface, this appears to be a simple question. However, many projects are conducted without a clear definition and understanding of the problem, and are therefore doomed from the start. An effective definition of a problem requires careful observation and objective points of view. If a client, teacher, manager, or other non-team member identifies the problem, pay particular attention to its nature and the requirements of the solution. Try to understand them as completely as possible.

To help us understand what Step 1 is all about, let us imagine that you will compete in the FIRST robot competition (see Point of Interest). The 2006 FIRST competition involved the scoring of goals by launching a ball at several targets (see Figure 2-8). In this example, the client (FIRST) has already identified the problem and has set several limitations. It is up to the team to determine how they are going to solve the problem.

The first minute of each match in 2006 was an optional portion of the contest. During this time, the robot must operate autonomously—that is, without direction

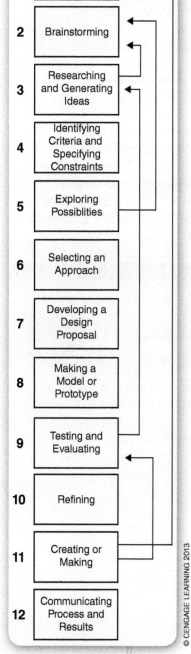

Figure 2-7: Model of a 12-step design process.

Figure 2-8: FIRST® Robotics 2011 national competition.

Point of Interest
FIRST ® competition

Dean Kamen founded the For Inspiration and Recognition of Science and Technology (*FIRST*) organization in 1989 to inspire students in engineering and technology fields. Kamen is best known for inventing the Segway, an electric, self-balancing, two-wheeled human transporter (see Figure 2-9). The organization also runs *FIRST* Place, a research facility at *FIRST* Headquarters in Manchester, New Hampshire, where it holds educational programs and day camps for students and teachers.

Figure 2-10: *FIRST*® robot in competition.

Figure 2-9: Dean Kamen, inventor, riding on a Segway motorized transport.

The *FIRST* organization has founded four levels of competitions that span from high school to elementary school: the *FIRST*® Robotics Competition (see Figure 2-10), the *FIRST*® Tech Challenge, the *FIRST*® LEGO® League, and the Junior *FIRST*®LEGO® League.

FIRST Robotics Competition

The FIRST Robotics Competition (FRC) gives high school students real-world experience working with professional engineers to develop a robot. The first annual FRC was held in 1992. As of 2011, the FRC level of competitions (high school level) consisted of greater than 2,000 high school teams and more than 51,000 students from Brazil, Canada, Israel, Mexico, the Netherlands, the United States, the United Kingdom, and other countries, who competed in the competition. The challenges in the competition change each year, and the teams cannot reuse components from earlier designs. The robots weigh approximately 120 pounds. Details of the game are released at the beginning of January, and the teams have six weeks to construct a competitive robot.

FIRST Tech Challenge

The FIRST Tech Challenge (FTC) was created in 2005 to provide more accessible and affordable options for schools. The FTC robots are approximately one-third scale of their FRC counterparts and require an intermediate skill level. As of 2011, the FRC level of competitions consisted of 1,500 high school teams and 15,000 students.

by a team member. Teams can score important points during this period. After the first minute, the joystick controls are enabled and the team members can "drive" their robot.

Will your team choose to design an autonomous robot? If your team answers yes, then your team will have accepted a more difficult problem. To make a wise choice, your team will need to calculate the potential scoring advantage. Your team might also need to consider if it has, or can get, the necessary expertise to create an autonomous design.

Your Turn

What are the details of the current FIRST Robotics and FIRST Tech Challenge?

Figure 2-11: Exoskeleton mechanical designs combined with advanced electronic systems are being designed for use by the physically disabled. In this image, Jason Gieser and Tamara Mena test the eLEGS system developed by Berkeley Bionics. Both Jason and Tamara are completely paralyzed from the upper chest down. The eLEGS device powers them to a standing position, enabling them to walk again for the first time since their accidents.

In another example, an engineering team may address the unique needs of the physically challenged (see Figures 2-11 and 2-12). Some people, through injury, illness, or a birth defect, cannot adequately function in a world built for people who are more mobile. A physically challenged individual may often require some type of assistance in getting around, seeing, hearing, or in simply picking up common items like a drinking cup or a hairbrush. In what manner can we help a physically challenged person? Notice that this question is asked in a manner that does not preassign a method or technology for addressing their needs. Framing a problem in this way is often very helpful, because it does not prejudice the team into thinking about a problem in a predetermined manner. For example, some type of wheeled device may come to mind, but there are also other possibilities. Are there useful electrical or optical technologies? Are there brace-like designs that would be more useful for certain mobility problems? There are a number of creative possibilities. Taking time to find the root, or very basic problem at hand, is extremely important and sometimes not very easy to do.

Innovation can start at the problem-defining phase. Often, just recognizing a problem is the key to innovation. Let's look at a few quick examples of where problem recognition was an instrumental part of the innovative design process. At some point, someone saw that neither scissors nor files were optimal for trimming fingernails. Hence, we now most often use nail clippers. The inventor of nail clippers recognized a problem; the solution resulted in a substantial change in a routine task for hundreds of millions of people. This is certainly a case where recognizing a problem was an instrumental first step.

Another person recognized the clutter of paperwork in his music sheets; by using a "low-tack" adhesive developed by a colleague, this idea resulted in the invention of "sticky notes." Sticky notes are used all across the world, and consist of a stack of paper held together with a "re-adherable" strip of adhesive (see Figure 2-13). Simply recognizing, and clearly specifying, a problem can be a most critical step.

 STEP 2 *Brainstorming*

Once the problem is well-defined, the design team collaborates. This is a process called brainstorming (see Figure 2-14). In brainstorming, each team member contributes his or her unique ideas to generate solutions to the problem. It is important for each team member to remain nonjudgmental and to not overanalyze ideas presented. An open mind, encouraging attitude, and innovative (out-of-the-box) thinking are also important to the process.

In a brainstorming session, often several team members are assigned responsibilities. It is helpful to have one or more participants responsible for recording all important aspects of the meeting, particularly everyone's ideas and input. It is also helpful to designate a meeting leader. The responsibilities of the leader may include setting up the time and place of the meeting, ensuring attendance, and constructing an agenda for the team to follow. An agenda creates order in the meeting and begins the brainstorming process by validating each topic to be discussed.

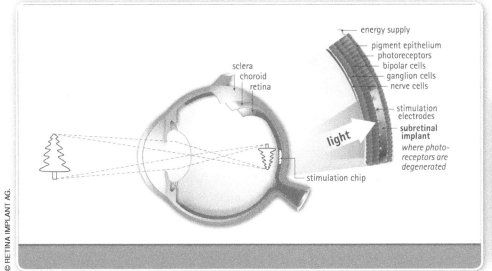

Figure 2-12a: Eye implant. Current research in bio-optics and bioelectronics could one day restore a degree of vision to people afflicted by common forms of blindness. Several engineering approaches are being explored. This approach uses a tiny artificial retinal device being developed by the Retinal Implant Project, a joint research effort between the Massachusetts Eye and Ear Infirmary (Boston), the Massachusetts Institute of Technology (MIT; Cambridge, Massachusetts), and Harvard Medical School (Boston). In this design, an external laser, along with a tiny camera, is mounted onto a pair of eyeglasses that capture visual images. The images are converted to digital signals and transmitted by laser to the implanted chip. The artificial retina chip then transmits the electrical impulses to the brain.

Figure 2-12b: 2 mm ASR device lying on a penny.

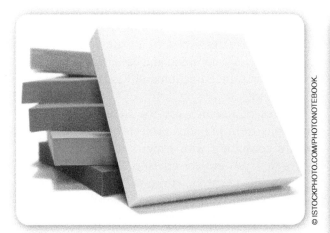

Figure 2-13: Sticky notes are hard copies of paper with "re-adherable" or "multiple-use" glue used to place important comments on documents. The sticky concept was so popular that it is now integrated as an option on computers.

Figure 2-14: Brainstorming is more effective with a number of people.

Fun Facts
Rules of the road for brainstorming

▶ No criticism
▶ Work for quantity

▶ Welcome "piling on" of ideas
▶ Allow free-for-all

The brainstorming team leader runs the meeting, ensuring that it stays on time and that it is open and productive without using excessive control. A simple method for helping control a meeting is to engage the "marker rule"—only the person with the marker can speak. All others must be silent and listen. The meeting leader ensures equal access to the marker.

 STEP 3 *Researching and Generating Ideas*

Investigation and research help people to discover prior knowledge that will be helpful for implementing solutions to particular design problems. Often the trick is finding previously used concepts and applying or modifying them to the new situation. Studying existing devices and solutions, and exploring existing knowledge also help designers develop new ideas for new solutions. Reverse engineering also helps generate ideas (see Chapter 6, Reverse Engineering). This is another phase where innovative, out-of-the-box thinking can occur. Earlier in the chapter, we discussed the simple daily design process of deciding what clothes to wear. Do you perform research to help in your decisions? Yes, one often assesses the weather by consulting the Internet, TV news, or simply looking outside. Living in this Information Age eases access to a wide variety of facts and makes information gathering a common and easier task to perform (see Chapter 7, Investigation and Research for Design Development).

OFF-ROAD EXPLORATION

The following websites have a wealth of information about FIRST robotics competitions, as well as hybrid and all-electric cars, which use substantial automation and robotics technologies:
http://www.usfirst.org/
http://en.wikipedia.org/wiki/FIRST
http://www.howstuffworks.com/hybrid-car.htm
http://www.hybridcars.com/cars.html
http://www.teslamotors.com/

Sources of research are the library, Internet, or documents from previous projects (see Figure 2-15). Directly consulting with experts in technologies or devices that you are considering using can also lead to invaluable research. However, it is always good practice to ensure that you have reliable information by acquiring multiple references from high-quality sources. A surprisingly overlooked source of information is the intended customer for the product. It should be a high priority to interact with actual customers to better understand their needs. It may also be useful to separate the customer's expressed needs, or features, into four categories: (1) Must Have, (2) Strongly Desired, (3) Marginally Desired (not absolutely necessary), and (4) Not Desired (not wanted at all). This structure better quantifies the possible features or

Figure 2-15: **The Internet may be a good place to begin investigating a problem.**

Figure 2-16: **An example of a feature matrix.**

Perceived Need	Feature
Must Have	▶ Same interior space as existing model ▶ 36 mpg (highway), 30 (city) ▶ Front-wheel drive ▶ 10-year warranty
Strongly Desire	▶ 4-wheel drive (as option) ▶ 45 mpg (highway), 35 city ▶ GPS-capable
Marginally Desire	▶ Integrated TV (as option) ▶ Heated seats
Not Desired	▶ Diesel fuel (Currently, the new hybrid technology only works with 89-octane gas) ▶ MP3-capable

© CENGAGE LEARNING 2013

functions because a four-level system is a considerable improvement over a feature simply being either "good" or "bad." Figure 2-16 shows an example of this four-level categorization system for a new hybrid sport utility vehicle (SUV). For this problem, the marketing group of an automotive company has interviewed important customer groups on desired features for the new hybrid SUV. The four levels more accurately describe the needs of customers. For example, features that are "Must Haves" or are "Strongly Desired" probably require inclusion into the design. However, features that are "Marginally Desired" may be eliminated when other constraints are considered. The last category, "Not Desired," captures features that can be eliminated immediately (see Chapter 4, Generating and Developing Ideas).

In researching the problem, it also helps to have an idea of the types of technologies, processes, or materials that may be required. For example, could the solution to the problem require: (1) unique human factors design aspects, (2) special structural or mechanical design considerations, and/or (3) optical or electronic capabilities? Later chapters in this book provide more detailed information on many of these types of capabilities.

 STEP 4 *Identifying Criteria and Specifying Constraints*

This phase requires you to detail what it is you intend to do (**criteria** or requirements) and what the imposed limitations (**constraints**) are. Up to this phase, much of the work on your problem was somewhat theoretical in nature, like asking open-ended questions to help clarify the problem, or researching possible technologies to be used in a solution. However, the "Identifying Criteria and Specifying Constraints" phase requires making decisions, sometimes difficult decisions. For example, will the team attempt to design and build a robot that is capable of autonomous operation? For a new SUV, what is the final set of features to be included? What stays and what goes? If the problem focuses on a mobility device for the physically challenged, will the device be for an adult or a child? Will it be self-powered or motorized? Will it need to be compact, and perhaps fold up? The team must compile a clear list of the criteria, or requirements, for the solution. The criteria set at this stage will determine the level of challenge for the project. Be sure to include vital criteria and to eliminate unimportant criteria—there is no need to spend valuable time and resources addressing unessential criteria (see Figure 2-17).

criteria:
Principles or standards by which something may be judged or decided.

constraint:
(1) A limit to a design process. Constraints may be such things as funding, space, materials, and human capabilities. (2) A limitation or restriction.

Figure 2-17: Microwave cell towers are used to send and receive vast amounts of information, including voice for telephone conversations and Internet data. Shown here are three actual tower designs. These three cell towers function in the same way, but the criteria for design for each one is substantially different.

Documenting and setting the constraints also occurs in this phase (see Chapter 8, Technical Drawing). As opposed to criteria that primarily address the specifications of the product or design itself, a constraint is more of a limitation attributed to the overall project. It is common to have many constraints in a project (even very simple projects). Some examples of constraints are the schedule and budget for the project. Also, the team might be restricted to use only power tools or only hand tools. Or, the team might be restricted to use only nails as fasteners, as opposed to screws or glue.

Let us revisit the FIRST Robotics design challenge. The FIRST Robot Competition is an example of a problem where a client already identified the problem and set several constraints. It is up to the team members to determine how they are going to solve the problem, and perhaps add more of their own constraints. Should there be a limit of the number of team members? How many hours will parents and teachers allow the team to work? What is the budget for the team? The team will also need to understand all other constraints set by the FIRST organization, such as size, weight, or materials. Most likely, many more questions will also arise. Later in this chapter, we discuss the various types of constraints that projects can have. Feel free to use this section on common constraints as a guide to help ensure you have included all of the important constraints.

By the time you have identified criteria and constraints, you will have acquired a lot of information. You will study the information to make some key, and often difficult, decisions. These key decisions lead to clear definitions of both the criteria and the constraints. The design team should record these key decisions so they are able to refer to those decisions at any time.

It is time to write all this information down in a clear and concise manner. The **design brief** summarizes the criteria and constraints for the problem (see Figure 2-18). A design brief is meant to encourage open thinking about all parts of a problem. A design brief does not list, or even suggest, any methods or solutions. Review your design brief to ensure it accurately conveys the work of the design team up to this point.

design brief:
A written plan that identifies a problem to be solved and its criteria and constraints. A design brief encourages thinking of all aspects of the problem before attempting a solution.

Figure 2-18: A design brief for a computer desk that helps with organization of computer cables and common desk materials.

Design Brief	
Client:	Grans Chain, Inc.
Designer:	Ian Mohney
Problem Statement:	Computer desk areas tend to be very cluttered. A variety of desk items (blank paper, printed papers, paperclips, tape, etc.) as well as a plethora of computer cables and attachments are all typically very cluttered and unorganized, causing major loss of work efficiency.
Design Statement:	Design, model, and test a prototype modular commercial computer desk system that will substantially reduce desk and computer cable clutter, leaving ample computer connectivity (USB, Firewire, etc.).
Constraints:	(i) 3-mo. design and prototype, (ii) $175 Max. Mfg. Cost (Qty of 100,000)—base unit, (iii) $250 Max. Mfg. Cost (Qty of 10,000)—fully optioned, (iv) 8 sq. ft. min. surface working area (std. 30" height), (vi) Max. shelving height of 7", (vii) less than 130 lbs—fully optioned.
Deliverables:	(i) Sketches showing breadth of ideas, (ii) Design journal, (iii) Technical drawings, (iv) Mfg. cost estimates, (v) Development schedule and cost estimates.

At the end of the "Identifying Criteria and Specifying Constraints" phase, you probably know enough about your problem to put together a good time frame for all of the major tasks required. It is a good idea to construct an estimated time frame of your plan at the conclusion of this phase. A high-level time frame helps keep the whole team well informed and focused on key tasks and their intended completion dates. Figure 2-19 shows a time line constructed for a "Mega Mousetrap" car design. A spreadsheet software program was used to construct this time line.

Your Turn

Write a draft design brief for a new type of smartphone that you think would be very popular with young adults.

Figure 2-19: A time line for a "Mega Mousetrap" car project.

	Project: Mega Mousetrap Car								
	Wk 1	Wk 2	Wk 3	Wk 4	Wk 5	Wk 6	Wk 7	Wk 8	Wk 9
Research Materials	■								
Define attach. processes	■	■							
Design Brief		■							
Detailed Design			■	■					
Design Proposal				■	■				
Make Design						■			
Test Design							■		
Redesign (if necessary)							■	■	
Final Report									■

  STEP 5 *Exploring Possibilities*

With a good definition of the problem, its constraints and criteria (design brief), and the additional knowledge acquired through brainstorming and research, you are ready to start "sinking your teeth" into the problem by exploring all the possible solutions. This phase requires a lot of attention to detail. How can you best meet all of the criteria while staying within your constraints? In this phase, you should consider all of the ideas from your brainstorming sessions and your research. What basic technologies, materials, designs, or fabrication processes will you use? This phase often calls for substantial compromise. Taken individually, many criteria may be easily achieved. However, when the criteria are taken collectively, conflicts often start to arise. How much information needs to be gathered in the explore phase? The answer to this question is simple: Gather enough information to justify your decisions on which design path to follow. The explore step is completed when you are ready to move to the next phase, "Selecting an Approach."

In the explore phase, you will need to work out your alternative solutions. This means you must be able to see the further choices and problems that these possible solutions will generate. This is important because it is better to pay attention to these details now than to be thoroughly surprised later. As opposed to a high-level view, a more detailed development of the various solutions or design paths is required to fully understand the important consequences.

Developing a number of different solutions to a problem may mean considering the size, shape, placement, or choice of materials at this early stage. Here is where the application of sketching and drawing skills will allow you to better explore your ideas. You may even need to complete a more detailed analysis of a mechanical, electrical, or aesthetic design to see if a solution works. This may involve computer modeling (see Figure 2-20) or discussing your problem with a knowledgeable vendor, both being efficient and accurate methods for exploring.

You should use an engineering notebook to compile all of your valuable information, including sketches, descriptions of ideas, calculations, and computer-modeling results. An engineering notebook is a very important tool. Your notebook is the source for collecting

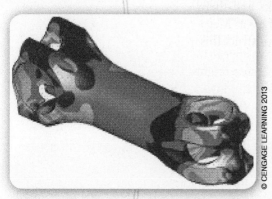

Figure 2-20: **The results of CAD used to explore the effects of stress on a bicycle component (handlebar stem).**

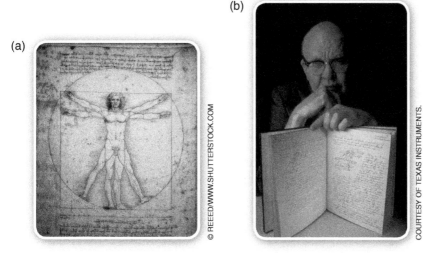

Figure 2-21: Examples of engineering notebook entries. (a) Leonardo da Vinci's famous entries. (b) Jack Kilby's entries on the invention of the integrated circuit.

important information, making it valuable to the designer and the whole project team. An engineering notebook also serves as clear evidence of exact dates of innovative, and potentially patentable, ideas (see Figure 2-21). Engineering notebooks, used frequently in industry, represent legally recognized "hard-copy" evidence of innovation, which can be a deciding factor for both granting a patent and successfully defending a patent. More information about design project documentation is presented in Chapter 8, Technical Drawing.

The exploration phase involves looking into your ideas in detail, requiring intense, focused time for analysis. In this phase, it is common to wander onto the less critical portions of the design that are easier to solve. So, keep asking yourself, "What criteria, or sets of criteria, are particularly troublesome and vital?" Staying focused on these critical areas is key to successfully managing the project's challenges.

Let's revisit our hybrid SUV design team. The SUV design team members have their design brief completed and have finished documenting several good ideas from their brainstorming sessions and research. Using previous knowledge and experience, the team members believe they have clear paths to accommodate the four "Must Have" requirements in Figure 2-16. However, in this phase, they need to take the opportunity to explore their possible design options to confirm this. They need to explore options for these criteria now because in the next phase they commit to specific design paths. For example, even for the criteria that appear to be readily achievable, the design team may wish to ask themselves if there might be certain assembly methods or designs that can result in a substantially lower manufacturing cost or reduction in fabrication time.

The SUV design team is experiencing a very different situation for two of the "Strongly Desired" criteria. The team does not know how to simultaneously meet the criteria of (1) a gas mileage of 45 mpg (highway) and (2) four-wheel drive. From their research, they have several viable paths to achieve 45 mpg with front-wheel drive but the added weight and friction of additional four-wheel drive components limits the gas mileage to, at best, 39 mpg. Having two or more criteria or specifications that are in seemingly direct conflict with each other is a very common occurrence in a design process.

How will the design team solve this conflict? How can they achieve 45 mpg and four-wheel drive at the same time? Will lowering the weight help, and by how much? Can they lower the friction of the extra components required for four-wheel drive? They need to explore these questions vigorously and look for viable solutions. The team decides that they need to loop back and do more research and brainstorming on this problem.

Because of additional brainstorming, a design team member representing the engine design group came up with the idea of using a diesel engine. Up to this point, the design team had limited their ideas to the use of a more conventional gasoline engine. The use of diesel fuel was an out-of-the-box idea. After the brainstorming session, the engine expert explored this diesel fuel option in more detail. After completing a few calculations with a software package designed on another project, the engineer realized that, with four-wheel drive, they could achieve not only 45 mpg, but 50 mpg! This path of using a diesel fuel engine might be viable after all.

But wait, can the engine work with diesel fuel? The engineer, being a combustion expert, thought that the engine could be converted into a diesel engine with some modifications. In talking to the director of the hybrid engine research group, located in another country, the engineer discovered that a diesel version of the engine was actually already being worked on for a special government project. The government project was developing a hybrid diesel engine for use in military transport vehicles. Furthermore, the research group already had working prototypes of the hybrid diesel engine!

Innovative exploration can be very fruitful. The hybrid SUV design team was back on track. Tenacious exploration often yields solutions to the most difficult problems.

> Assume you drive a car an average of 15,000 miles per year and are in the market for a high gas mileage car. The following two cars meet all of your other requirements: (1) Car A gets 32 mpg, runs on standard 89-octane gas, and costs $19,000, and (2) Car B runs on diesel fuel, gets 47 mpg, and costs $27,000. Which car do you buy to save the most money over the approximately eight-year life span of the car? Use the current fuel prices for your geographical region, and perhaps even try to estimate what the fuel prices might be over the next eight years. Is there a price difference between diesel and standard gas fuels?

>>> STEP 6 *Selecting an Approach*

After the explore phase, and perhaps with some iteration through other earlier steps, all of the options to the design team should be clear and well defined. The team is now ready to "down-select" from the available options. By down-select, we simply mean that the team needs to perform an assessment and then choose a design path, or possibly a few design paths if the team has enough resources to do so. The design ideas need to be pared down though a selection or assessment process. A design team rarely has the resources to try all of the design ideas, so the team needs to choose only those ideas that they think will be the most successful or useful. This is often the focal point of risk management—making good decisions here will be a key contributor to project success or failure.

There may be several design options that all appear to be workable, and all may solve the problem, but the team must choose. There are two strategies for helping choose among the design options. You may want to try both strategies. The first strategy is to simply list both the good and bad attributes of the competing ideas and compare them.

assessment:
An evaluation technique that requires analyzing benefits and risks, understanding the trade-offs, and then determining the best action to take to ensure that the desired positive outcomes outweigh any negative outcomes.

The second strategy is more quantitative and involves the construction of a decision matrix. On one axis of the matrix, you list the criteria, or specifications, generated for the design brief. On the other axis, you list the alternative solutions. The internal portion of the matrix is then used to fill in how well each of the possible solutions meets each of the criteria or specifications.

The team needs to determine a scoring system. There are several useful systems. For example, perhaps each entry is simply filled in with a "yes" or "no" (or "good" or "bad"). These are examples of two-level grading systems. You could also use three levels. For example, insert one of the following phrases into the decision matrix: (1) exceeds criteria, (2) meets criteria, or (3) fails criteria. Or simply use + (positive), 0 (neutral), and − (negative) ratings. You could even use a multilevel grading system like that used in school. You could also choose quantitative rankings with values ranging from 1 to 5.

It is also very common to quantify the reality that some requirements are more important than others are. Weighting the criteria or requirements accomplishes this. In Figure 2-22, a decision matrix is shown where each requirement is weighted by its relative importance.

The team documents a summary of the reasons for choosing each particular solution. All of the strategies suggested here should satisfy the team's documentation requirements. Keeping a record of the decision-making process is an important part of an engineer's notebook and may be useful later in the design process.

model:
A detailed visual, mathematical, two-dimensional, or three-dimensional representation of an object or design. A model is typically smaller than the eventual, intended design. A model is often used to test ideas, make changes to a design, and to learn more about what would happen with a similar real object.

> Given the data entered in Figure 2-22, which design path would have been chosen if the criteria were not weighted?

 STEP 7 *Developing a Design Proposal*

The design team has now completed enough work on the possible solutions that they have, with confidence, selected an approach to solve the problem. The team needs to complete all preparations to make a model or prototype. In this phase, the goal of the team is to complete a set of suitable documents for creating a prototype or model of the design. It is now time to "dot all the i's and cross all the t's." Before construction of a prototype can begin, a clear documented description of the design is required. The documentation, called a design proposal, can consist of computer-aided design (CAD) drawings, text descriptions, or hand drawings.

prototype:
A full-scale working model of a design intended to have complete, or almost complete, form, fit, and function of the intended design. A prototype is used to test a design concept by making actual observations and making any necessary adjustments.

Figure 2-22: **Example of a decision matrix that uses weighted requirements.**

Project: Mega Mousetrap Car	Design Matrix					
	Designs					
Criteria/Specification	Weight	1 Trap	2 Trap	1 Trap & CD Wheel	Duel Drive	Special Axle
Cost	1	5	4	4	3	2
Time to build	2	5	2	3	2	2
Time to test	1	5	5	5	4	3
Time to redesign	2	5	3	4	3	2
Peak speed (5 ft/sec.)	2	2	5	2	2	4
Distance (30 ft. min.)	4	2	4	3	5	4
Totals		42	45	39	41	37
					5 = best, 1 = worst	

design proposal:
A written document, or set of documents, that clearly specifies how to fabricate a model, prototype, or final design. The design proposal should include documents that specify all (1) materials, (2) dimensions, and (3) processes used in the construction.

However, the documentation must be clean, legible, and easy to understand. The final drawings will need to specify all (1) materials, (2) dimensions, and (3) processes to be used in the construction of the prototype or model.

In previous stages, the team analyzed many ideas to help determine the design path. However, even more questions may need to be answered before fabricating a prototype. For example, the team will now need to analyze and complete the previous inconsequential decisions. Perhaps the team discovered in the earlier explore phase that the best material for an arm on the FIRST robot would be aluminum due to its lightweight and relatively high strength. However, there are many types of aluminum. Which type of aluminum is best? Perhaps the design requires an aluminum that is more easily weldable. Perhaps the decision is simply a cost or availability issue; what is the least expensive type of aluminum that is immediately available ("in stock")? The type of aluminum is determined in this phase because it needs to be defined on the final drawing. For example, a machinist constructing the part will not make it unless the material is clearly specified.

⟫⟫⟫ STEP 8 *Making a Model or Prototype*

With the design proposal completed, the team is now ready to construct a model or prototype of the solution (see Figure 2-23). All the designs completed to this point need to be tested. Does the solution, or major portions of the solution, work? To answer this question, either a model or prototype, or both, needs to be constructed. This is the construction phase. The team collects materials in this phase, and the physical work of fabrication begins and ends. A model or "mock-up" is, in many ways, less advanced than a prototype because the purpose of a model is to test only a minority of functions or features of the final design. A prototype is much closer to the form, fit, and function of the final design (see Figure 2-24).

Let us return to the FIRST robotics design team. The team decided to make a wooden model of the main frame of the robot, the "box" that holds all of the motors, computers, and power supplies. The frame will eventually be made of metal, but making

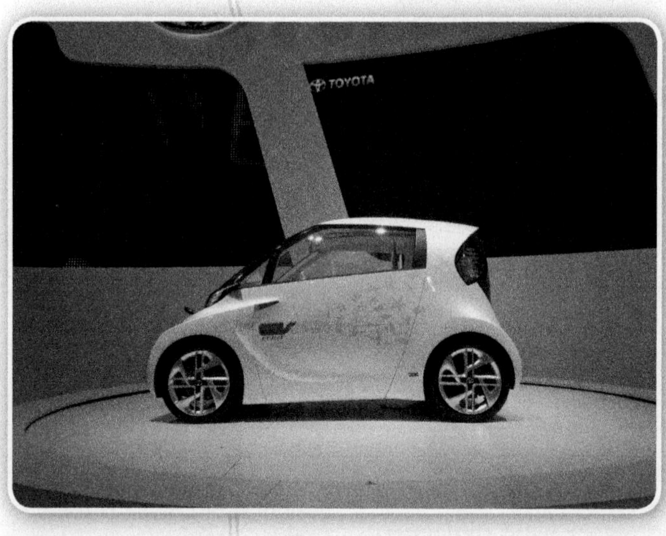

Figure 2-23: Toyota shows the FT-EVII electric city car prototype at the 32nd Bangkok International Motor Show.

Figure 2-24: Three-dimensional (3D) prototyping machine and sample model.

a wooden model of the frame is an easy and quick way for the team to determine how easily the components will fit together and how difficult it will be to work on certain components after assembly. The wooden frame model will help determine physical limitations, but it is certainly not intended to be a preliminary working version of a robot (that is, a prototype).

 STEP 9 *Testing and Evaluating*

In this phase, the team uses the model or prototype to evaluate how well the design meets the criteria set earlier in the project (see Chapter 9, Testing and Evaluating). How will the team test the model or prototype? Under what conditions will members test the model or prototype? What information is important for the team to collect during the test and evaluation phase? How much time will they allow for the testing? If the team has completed the preceding design process steps thoroughly, then the test and evaluation phase should provide extremely valuable information for the team.

What were those "mission-critical" aspects of the design that dominated your time and efforts in previous design steps? The test and evaluation phase should give crucial information about these aspects of the design. In industry, the testing and evaluation phase for some projects is so important that a completely separate design team designs and implements a detailed testing procedure: a "Test Plan." Can you think of a market where the testing phase is particularly important?

The medical field is one such market. The medical industry uses an extraordinary amount of resources for testing and evaluating. Even after years of internal testing within a pharmaceutical company, the U.S. Food and Drug Administration (FDA) requires many more years of external trials before approving a drug for use. Clearly, this high level of testing is necessary to achieve a superior level of safety and reliability. Another field where the testing phase is incredibly important, and complex, is in software or products that include large amounts of software, such as video games, cell phones, smartphones, MP3 players, spreadsheet software, and word processing software.

What are some of the most important considerations for the test and evaluation phase? A few recommendations follow:

▶ Make a list of those attributes that are important to test.

▶ Design a set of experiments that address this list. In this set of experiments, consider testing in two types of conditions: (1) under controlled conditions and (2) in a working environment.

▶ Gather and record your test data. Analyze your data and compare it to the criteria and specification for the design.

▶ Conclude by writing a complete summary of your testing. The summary should identify those major areas of concern that may be the focus of any redesign work.

When the testing procedures are being defined, should the design team test only if the requirements set forth are either met or not met (that is, pass or fail)? Whether a model or a prototype was constructed, the team should strongly consider testing more than simply a pass or fail criteria. It is certainly important to determine if a specification is satisfied. However, it may be even more useful if the team had knowledge of how much a design exceeded, or failed, a specification. If the design performs substantially better than the requirement, the team may

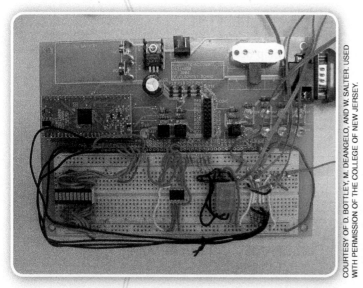

Figure 2-25: A functional electronic C Stamp prototype board for a weather station. This mock-up was prepared by technology students at The College of New Jersey.

Figure 2-26: Testing a toy in the real world.

choose to increase the criteria in future projects. Or perhaps some of this extra margin offers latitude if compromising is necessary in future steps. If the design fails a specification, it is very important to know by how much it fails. Is a major redesign necessary, or can the criteria or specification be achieved with a few minor adjustments?

Some designs require electronic systems to be developed. Electronic systems require special tools for design as well as for testing and evaluating. Figure 2-25 shows a prototype of a microprocessor-controlled weather station. The station measures wind speed, wind direction, and temperature.

Let's review how the FIRST robot design team is doing. They have completed various models, including the wooden frame. After testing and evaluating several models of important subsystems under controlled conditions, the team constructed a fully working unit—a prototype. In addition to testing the prototype under controlled conditions, the team intends to take the robot "through its paces" under fully competitive conditions.

To imitate competitive conditions, the team has invited a nearby competing team to bring their robot so they can simulate a realistic match competition. Testing under real-world conditions should prove very valuable. As an example, if one of the scoring options is to pick up and store balls from the floor, you will need to judge whether your robot does so consistently and reliably. Suppose your robot routinely picks up eight balls out of ten when it is by itself on the field, but when the competitor's robot is also on the field, continually bumping into your robot, your robot only successfully picks up four balls out of every ten. Is 40 percent efficiency enough? Are there adjustments to the design that can increase this percentage?

Design or product testing may require direct involvement with the final customer (see Figure 2-26). In the case of an aid for a physically challenged person, testing may involve giving a device to several individuals to use for a time. The design team may choose to observe these people using the device or possibly videotape customer usage to study the results in more detail. The design team may interview customers at the conclusion of the real-world tests, providing valuable feedback that may lead to important improvements or modifications.

>>> STEP 10 *Refining the Design*

Inevitably, problems arise with all design projects. It is rare that projects are successful on the first try. Therefore, some redesign or "tweaking" typically is necessary. There are two possibilities in this phase. If the results of the test and evaluation phase indicate that all of the criteria or specifications have been satisfied, then the team proceeds to Step 11, "Creating or Making" the design. Otherwise, the team works on redesign.

Chapter 2: The Process of Design 51

Your Turn

What are some examples of designs/products that would require (1) only readily available tools versus (2) very special materials and tools?

In industry, redesign is often included in the design schedule. This phase in the 12-step process directly invites iteration. The team may need to start the design process, not the design, over again. After testing and evaluating, what were those items requiring further attention?

One of the outputs from the testing and evaluation phase will be a list of remaining problems. This is essentially repeating the "Defining the Problem" step—the first of the 12 steps. Usually, the first round of design iteration substantially lessens the difficulty of solving new or remaining issues. From here, the team determines if they need to brainstorm, research, or possibly even to reassess the accuracy of the criteria and constraints.

 STEP 11 *Creating or Making*

The design process is almost complete. At this phase, the design or product is mature enough for final fabrication (see Chapter 10, Manufacturing). The fabrication requires the necessary tools and materials to make the design. Some designs will require only readily available tools and materials (see Figure 2-27) while other designs will require very special tools. The Case Study "Anytime™ Chair" (see Case Study in Chapter 10, Manufacturing) describes how Richard Rose, industrial designer, used a design process to design and produce a commercially available chair.

For a commercial application, this phase translates to being ready for production. There are two general categories of production: (1) mass production and (2) custom production (see Chapter 10).

Mass production is the (typically) rapid fabrication of multiple copies of a design or product. Custom manufacturing is the fabrication of a design or product in much smaller quantities. Examples of mass manufacturing include cell phones, a new microprocessor chip, a video game, or a popular novel where quantities may reach into the tens of millions. Examples of custom manufacturing are the fabrication of only one FIRST robot or the fabrication of a few experimental prosthetic devices.

The decision to mass-produce a product is best made early in the design process and is typically a very important constraint. For example, a constraint on the design brief may read, "The hybrid sport utility vehicle design must be 100 percent compatible with current mass-manufacturing processes and equipment used for the year—2010 standard—gas sport utility vehicle." For mass-manufactured products, the design of the product and the manufacturing systems

Figure 2-27: **Fabricating a part using a drill press.**

used to make it are linked so tightly that it is virtually impossible to mass-produce a product that was not designed for mass production from the start.

It is important in this phase to fabricate the product exactly as the successful prototypes. The test results depend on the detailed choices made for materials, dimensions, and processes. Generally, these components should not be changed. If they do need to be changed, it is the responsibility of the design team, often the manufacturing design team, to validate the changes and confirm that they will not negatively affect the product.

⟫⟫ STEP 12 *Communicating Process and Results*

When evaluation of the solution is complete and the team is sure that the product is performing as required, it is time to communicate the details of the design in a **final design document**. The communication should include one or all of the following: a slide presentation, technical reports, detailed design drawings, or sketches. The final design documentation should communicate clearly and completely (1) what the design is, and (2) how well the design works. Final documentation should include all the necessary charts, graphs, calculations, CAD drawings, modeling, and simulations that, taken as a whole, represent the final design. In industry, the final design documentation can be very extensive, consisting of hundreds of pages of information that comprise a comprehensive description of the final product.

In addition to the final design documentation, communication of the design can include information about marketing, distribution, and sale of the design, which is now "the product." Any patents related to the design are also important communications and should be pursued as appropriate (see Figure 2-28).

A variety of Case Studies of careers that span the design continuum are presented throughout the book: Graphic Design (Chapter 17), Industrial Design (Chapters 6 and 9), and Laser Design (Chapter 2).

final design document: A complete set of design documentation prepared at the final stages of the design process. The final design documentation communicates clearly and completely what the design is and how well the design solves the problem. Final documentation should include all the necessary charts, graphs, calculations, CAD drawings, modeling, simulations, and text descriptions that, taken as a whole, represent the final design.

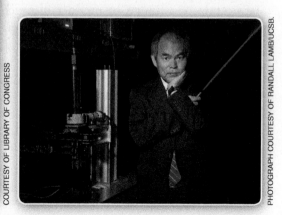

Figure 2-28: Portion of a patent by S. Nakamura of Nichia Chemicals Industries, Ltd. Nakamura used materials based on GaN compounds to create the first blue semiconductor lasers, which are now commercially available. Blue lasers were the technological innovation needed to achieve higher-capacity optical disk storage, up to 25 GB, on a single-layer optical disk. [Search on "Blu-ray Disk Association" for more information.]

Case Study

Ken Dzurko, PhD, Laser Designer

Ken has been involved in laser design professionally for over 15 years. He has worked for several different companies and has professionally consulted for several others, advising them on ways to increase their value by expanding their laser design into new technology and market areas. However, much of his career focused on high-power semiconductor lasers used in a variety of applications, including medical, printing, telecommunications, and marking/machining (see Figure 2-29). Some of these lasers have even been flown successfully on satellites, which have extremely demanding engineering requirements.

Like many engineers, Ken's career started with a strong interest in mathematics and science. Ken earned a bachelor of science in engineering physics, and went on to complete master's and PhD degrees in electrical engineering with a specialization in lasers and electro-optics. Ken specializes in semiconductor lasers—that is, lasers that are made out of semiconductor materials. Computer chips are also made out of semiconductor materials. Semiconductor lasers can be made extremely small (the size of a few grains of salt), with very high performance. They are also inexpensive to make because they can be easily mass-produced.

Semiconductor lasers are at the heart of fiber optics. The Internet would not have been possible without them. Semiconductor lasers are also the most efficient light source known to humankind, with efficiencies routinely above 65 percent and as high as 80 percent. For comparison, common incandescent and fluorescent household lighting has efficiencies of tenths of a percent to a few percent at best.

Using similar semiconductor technology, light-emitting diodes (LEDs) have quickly taken over lighting in cars and traffic lights and are starting to be used more frequently in commercial lighting applications.

To design high-power lasers successfully, Ken needs to understand a lot of the math and physics of semiconductors and optics. In semiconductor lasers where electric current is injected, both light and heat are generated. Therefore, Ken also needs to be adept at working with software packages that calculate the distribution of light, current, and heat within a laser cavity. Furthermore, Ken also had to learn about the physical design of the laser packaging because he must use special packaging materials and processes that efficiently take heat away from a laser. Otherwise, the laser might not live long.

Ken starts many of his laser projects by brainstorming and doing research about previous work in the field. He also finds it very useful to put together a written plan, including a time line chart, of his work as soon as possible. Such a plan includes a clear definition of the problem, the goals (criteria or requirements), constraints, a statement of how to test devices, and an outline for a final report. Such a written plan helps him, and his colleagues, stay focused on the important aspects of the problem. Good time-management skills are very valuable in many fields; laser engineering is no exception.

Engineering notebooks are also a large part of Ken's professional life. He uses an engineering notebook extensively, supplemented by spreadsheet software outputs and PowerPoint files to record progress and ideas. These notebook entries were also useful in patenting several novel laser design concepts.

Ken really enjoys designing and testing lasers. The combination of interesting applications and science, along with the opportunity of working closely on multidisciplinary teams, makes going to work every day exciting. As with many engineers, Ken finds it very rewarding when his designs transition from the lab to a real product used by customers.

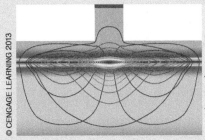

Figure 2-29: *Various calculated electromagnetic modes for a semiconductor laser. Light is emitted from the bright red region in the center.*

CREATIVITY AND INNOVATION IN THE DESIGN PROCESS

As mentioned earlier, a design process is a detailed plan. As outlined in this chapter, the design process contains 12 steps. This is a lot to do. With so much detail involved, can a team achieve all the design steps successfully and be innovative? The answer is yes. Innovative design processes are successfully completed every day—simply look at the multitude of new products that are developed, marketed, and sold month after month and year after year (see Figure 2-30).

Substantial time and resources are dedicated to creativity and innovation when using a designing process as described in this chapter.

Three of the twelve steps almost beg for innovation: Brainstorming, Researching and Generating Ideas, and Exploring Possibilities. Additionally, Step 10 (Refining the Design) calls for the team to repeat much of the process, calling for even more innovation. The last step also invites innovation (communicating) by effectively increasing the size of the "team" to include people outside the immediate design team.

A design process provides focus, which is vital for innovation. Focus and "mindshare" on the problem at hand is very important. Although five of the twelve steps directly call for creativity and innovation, the remaining seven steps are all geared toward keeping the

> **OFF-ROAD EXPLORATION**
>
> The following websites give more information on recently designed airplanes and the use of composite materials:
>
> http://en.wikipedia.org/wiki/Boeing_787_Dreamliner
>
> http://www.aviation-history.com/theory/composite.htm
>
> http://videos.howstuffworks.com/nasa/2153-composite-materials-make-shuttle-stronger-video.htm

(a)

(b)

(c)

Figure 2-30: *The development of the bicycle demonstrates dramatic innovation over the years.*

Figure 2-31: Outdoor sports equipment offers many design challenges. Unlike devices designed for indoor use only, outdoor sports equipment must endure harsh environmental conditions.

design team focused on the problem. If the team members spend effective time thinking about the problem, both creative and innovative thinking can occur, hopefully, resulting in useful innovation. Much innovation has occurred recently with sports equipment, often motivated by the availability of new and unique materials (see Figure 2-31).

> What new materials used in the past 20 to 30 years benefited the design and functionality of sports equipment? What novel materials have substantially impacted the following industries and why: (1) automotive, (2) medical or pharmaceutical, (3) housing, and (4) electronics?

Let's revisit Step 1, "Defining the Problem." In this step, the design team is asked to focus only on the problem definition, resulting in agreement on what the problem actually is. Let's imagine you are in a design group in a company that designs outdoor sporting equipment. After talking about several administrative items in a monthly group meeting, the conversation wanders to a discussion of a possible new type of product: a weather station that can measure temperature, humidity, atmospheric pressure, and possibly other weather-related data.

After the meeting, the mechanical engineer in the group, figuring that the novelty of the product would be in how to attach the weather station to a wall or porch railing, spent hours designing interesting methods of attachment.

Meanwhile, the electrical engineer saw this as an opportunity to use a new low-cost microprocessor able to communicate wirelessly to any home computer.

The electrical engineer was so excited that she called the vendor to get some free samples of a new microprocessor.

> **OFF-ROAD EXPLORATION**
>
> Light-emitting diodes (LEDs) and lasers are very important display and lighting applications. The following sites give good overviews of these optical technologies:
> http://en.wikipedia.org/wiki/Light-emitting_diode
> http://www.howstuffworks.com/led.htm
> http://en.wikipedia.org/wiki/Laser
> http://www.scienceclarified.com/scitech/Lasers/Lasers-in-Science-and-Industry.html

The human factors expert in the group certainly thought that the weather station needed to have a very bright display, making it exceptionally easy to read. Therefore, the human factors engineer, with help from an electrical designer, worked out methods of powering higher brightness displays.

All of these group members were being creative and trying to innovate. Perhaps some of the work these design group members performed will be very useful in the future. However, they had not yet defined the problem, because the discussion in the meeting was simply a discussion about possible product ideas, not a formal session with the intent of defining a problem, or product.

As it turns out, the manager failed to mention that what he really meant to say was that the marketing group was thinking about integrating weather-measuring capability into the company's existing products—items like backpacks, tents, compasses, bike computers, and even some clothing. Ouch!

The design process went awry. Individual members of the design group went steaming ahead, each diving into their own particular areas of interest without focusing on a common, well-defined goal (problem definition). Without focus, creativity can occur but it may not effectively lead to innovation. Many, if not all, of the items that the various engineers independently considered may have been creative, but were relatively useless for the intended new product types.

The manager made sure never to make such open-ended statements again. A design process can definitely be helpful in providing focus and supporting creativity and innovation. However, it has to be properly followed!

DESIGN LIMITATIONS

Clearly defining limitations accurately describes, and effectively solves, problems. Limitations of a design, or design project, can also be referred to as criteria, constraints, specifications, requirements, or sometimes simply as limitations. For most people, the terms *specification* and *requirement* refer to a very specific quantitative measure of a specific attribute or criteria. For example, the specification or requirement for the maximum electrical power consumption for a new console video game may be 25.6 watts.

Similarly, many people understand a constraint is an attribute that is more of an overall limitation—an attribute more concerned with the overall design environment than with the particular part designed. For example, having a project budget of $1,575 or only being given a half bottle of glue would typically be referred to as a constraint.

What a team actually calls a limitation is less important than identifying all of the relevant limitations for a project. If design team members are not acutely aware of their limitations, then they may be in for a big surprise. Let's revisit the FIRST robotics team. The robot is completely done, and from their real-world testing and subsequent improvements, the robot appears to have a superb chance at placing high in the regional competition. However, the team will not be able to compete. The team did not list the competition entrance fee or any travel or shipping expenses as a cost constraint for the design project. This team did not appropriately understand, or manage, the constraints for the project and is, unfortunately, facing the consequences.

Here is a list a design team can consult to ensure that they account for all of the important limitations for their particular project.

Possible Limitations:
1. Resources
2. Human resources
3. Materials and equipment
4. Time
5. Economic factors (all costs, such as materials, labor, fees, etc.)
6. Physical factors
7. Aesthetics
8. Marketability
9. Reliability
10. Manufacturability
11. Safety (human, animal, and environmental in general)
12. Ethics

SUMMARY

Design is the process of planned change. The process of designing is a plan that has clear roots as a problem-solving process. A design process is a detailed and logical problem-solving technique that has proven over time to be a very effective and powerful method for designers. A wide range of design projects successfully use design processes. Engineers and graphic artists, at opposite ends of the design continuum, each use design processes, but they use different tools. Engineers use more mathematical and scientific-based tools while graphic artists use more aesthetics-based tools.

A 12-step design process is comprised of essentially all of the possible steps encountered in a design process. A design process has three important aspects: (1) A design process is a plan with a well-defined problem and limitations; (2) the steps of a design process have a logical order; and (3) design process typically utilizes iteration.

When properly implemented, a design process does not stifle creativity and innovation but rather encourages it. A design process provides focus on the problem at hand, encouraging creativity and innovation all along the way.

BRING IT HOME

OBSERVATION/ANALYSIS/SYNTHESIS

1. List the steps in the engineering design process, and briefly describe what occurs in each step.
2. Overlay the steps of the five-step design process (see Figure 2-3) onto the four-step problem-solving process. Is each of the problem-solving steps utilized by the five-step process?
3. Repeat #2 above but using the 12-step process.
4. Who invented nail clippers and when? Who invented "sticky note pads" and when?
5. Devise a list of criteria needed to test and measure the success of the following:
    a. A toy for a 5-year-old child
    b. A roadside emergency light
    c. A road map
    d. A backpack for college students
    e. A word-processing system
    f. A gaming application for a smartphone
6. Write a plausible design brief for each of the following scenarios:
    a. I need more storage for books and trophies in my bedroom.
    b. When I ride my bicycle to the beach, it is difficult and unsafe to carry my surfboard.
    c. My grandfather is having difficulty walking but cannot travel with a standard cane.
7. Write a problem statement that is associated with problems that occur frequently in each of the following venues: home, school, community, church, vacations, and sports.

BRING IT HOME

(continued)

8. Write a design brief for each of the following problem statements. Use only your existing knowledge about the subject:
   a. Because Uncle Walt has use of only one hand, it is difficult for him to open pop-top cans.
   b. My father can never find his keys.
   c. Books that are required to be stored in the basement are damp and full of mildew.
   d. I have a large box full of old photographs and slides, and I cannot easily find images that I am looking for.
   e. I have a large amount of important schoolwork and photographs stored on an old computer but it often takes a long time to find what I am looking for.
9. Investigate the problem statements that follow and write a design brief that includes a potential client or target consumer and logical and appropriate list of limitations (constraints, criteria, specifications/requirements):
   a. The electric power used to run very high-brightness lights that illuminate large highway signs are often "on-grid." Investigate "off-grid" methods to illuminate large highway signs.
   b. Individuals with arthritis in their fingers often have difficulty grasping small objects or grasping objects tightly.
   c. Barking dogs keep me awake at night.
10. Can you think of five product types, or markets, where the testing phase requires a large amount of time and effort?

EXTRA MILE

ENGINEERING DESIGN ANALYSIS CHALLENGE

1. Rank the following problem statements in order of difficulty, with the most difficult first. Give specific reasons for your ranking. List your assumptions.
   - Many states desire to save money by taking the lights that power large highway signs "off-grid," and are looking for alternative methods.
   - Many people do not have any idea of the quantity of waste we discard or where it ends up and therefore are not motivated to conserve or recycle.
   - Individuals with arthritis in their fingers often have difficulty grasping small objects or grasping objects tightly.
   - Barking dogs keep me awake at night.
   - It takes too long for a professional photographer to level her tripod.
2. Think of three problems leading to exciting design projects that could be completed within the constraints of your class. Write design briefs for the three problems.

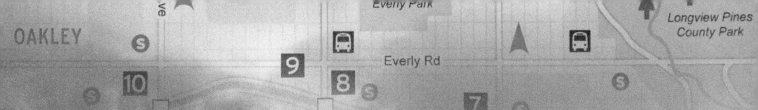

CHAPTER 3
Development of the Team

Menu

 Before You Begin
Think about these questions as you study the concepts in this chapter:

1. How did industry organize work prior to the 20th century?
2. What engineering professionals are on design teams?
3. How does one participate as a design team member?
4. How can successful teams really work while valuing individual differences?
5. What is the role of the team leader?
6. What tools are used to develop positive communication within a group or team setting?

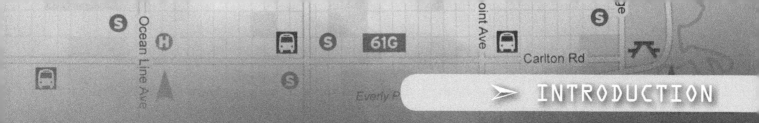

INTRODUCTION

The engineering design process captures the spirit, intelligence, and vision of the human being. When people work together toward a problem-solving goal, they can create innovative solutions. A special energy called synergy emerges when people communicate and listen to each other to explore different ways to achieve the goal together. Good teamwork and communication skills create synergy, and synergy improves creativity. All of these qualities—teamwork, communication, synergy, and creativity—are needed to create and produce any new design.

UTILIZATION OF TEAMS IN INDUSTRY TODAY

Modern industry favors the use of **teams** in almost every aspect of design and production. This is in stark contrast to how industry was organized in prior centuries. During the early Industrial Revolution, Adam Smith (18th-century economics pioneer) was doing research on ways to improve production efficiency. Smith observed that a group of specialized workers, each trained to perform only a single task in the manufacturing process, could substantially increase the level of productivity than when each individual performed all the tasks needed to complete a single unit such as in craft production. This way of dividing a job into a well-defined set of separate tasks has been a model for production and engineering for decades. It is called *division of labor* and it still seems to make sense, at least on paper (see Figure 3-1).

team: A group with a common purpose that achieves a specific goal using each individual's skills and mutual cooperation to produce the end product.

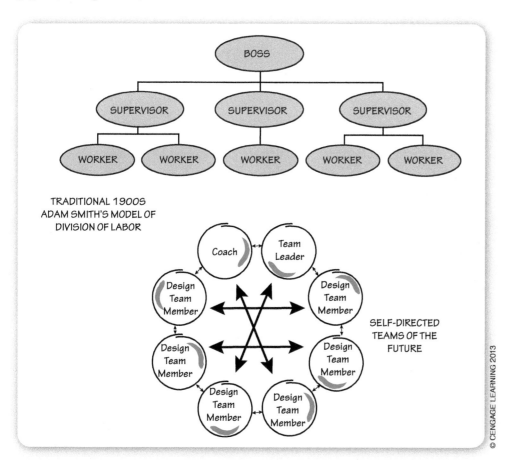

Figure 3-1: Historical view of working styles.

However, the shift in engineering is to have everyone on the design team have a much broader role.

Today, new products are most often produced by multidisciplinary design teams, that contain people with different areas of engineering expertise as well as expertise in business, finance, marketing, and production. For example, a design team for a new smartphone would very likely contain experts in mechanical or physical design, electrical design, computer systems design, optical design (displays and optical interconnects), human factors design, industrial/manufacturing systems design as well as product engineering, customer service, and business representation.

Engineering Professionals

In Chapter 1, we identified engineers as a very large category of design professionals. We also described how engineering societies began to form in the late 19th century, including the introduction of undergraduate and graduate engineering curriculums in most major universities. Engineering societies also play an important role in standardization, which is critically important to economic development and product safety. You learned about the "Greatest Engineering Achievements of the 20th Century," the role of engineering in protecting the environment, and the importance of a "code of engineering ethics" for ensuring a successful design process.

In a recent article in the *IEEE Spectrum* titled "Engineer Your Destiny," high school students were asked to consider if engineering was in their future (see Figure 3-2).

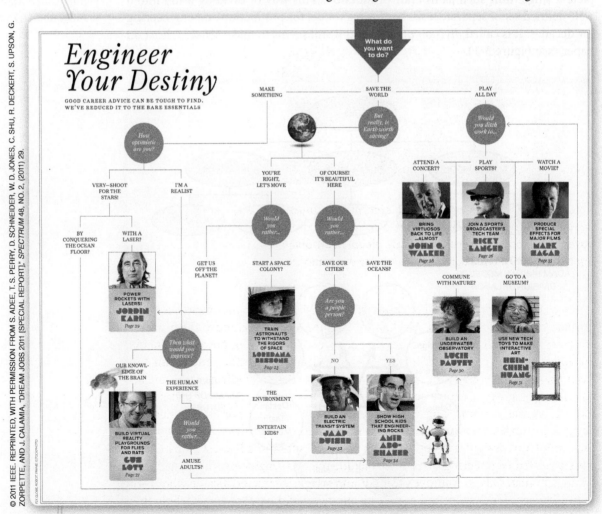

Figure 3-2: Engineering Your Destiny: Essentials of Career Advice.

How would you answer these questions?

1. Would You Like to Make Something? Engineers are about designing and making things. For good reasons, electrification was the first Greatest Engineering Achievement of the 20th Century. Taken as a group, electrical, electronic, and computer hardware engineers are the largest category of engineering professionals. The electrification of the United States in the 1920s led to important developments in lighting and laborsaving devices for work, the home, and leisure. Lighting improves our world by making it safer and by making it more enjoyable. Do you ever enjoy a nighttime baseball game? Think about all the electrical/electronic and computer hardware and software design seen by fans visiting a major league nighttime baseball game.

Electrical engineers design, develop, test, and supervise the manufacture of equipment associated primarily with the technology of electricity and electronic devices. This includes power generation and transmission, machine control, and communications (see Figure 3-3). Examples of products that involve electrical engineers would include electric motors, lighting, computers (microprocessors in general), global positioning systems (GPS), smartphones, and radar systems. Most products that you use daily have at least one component designed by an electrical engineer.

Figure 3-3: Electrical engineers design, develop, and test products such as electric motors, lighting, computers (microprocessors in general), global positioning systems (GPS), smartphones, and radar systems.

Mechanical engineers design, develop, test, and supervise the manufacture of primarily mechanical devices. Mechanical engineers comprise one of the broadest engineering disciplines. They design the mechanical portions of power-producing machines such as internal combustion engines, electric generators, and steam and gas turbines. They are also intimately involved in the design of refrigeration and air conditioning systems, industrial tools, material handling equipment, and people movers such as elevators and escalators. They work in transportation industries designing components for automobiles, trains, buses, and trucks, and in agricultural industries designing components to plant, maintain, and harvest crops. Other closely related engineering disciplines include material engineers who design new composite materials to improve strength and weight characteristics or who work in the aerospace industries and as marine engineers and naval architects designing airplanes, spacecraft, and ships (see Figure 3-4).

Figure 3-4: Mechanical engineers work in a broad array of industries, including aerospace.

Figure 3-5: Chemical engineers might work in the pharmaceutical industry.

Industrial engineers work on design teams to help determine the best and most effective methods for producing products. They are primarily concerned with establishing strong business practices and introducing technologies to help people work productively. They design manufacturing systems using mathematical modeling, financial planning, and cost analysis to ensure high-quality and economically competitive products.

Civil engineers design and manage the construction of roads, bridges, tunnels, commercial buildings, airports, dams, water supplies, and sewage systems. Civil engineering is the largest of the engineering disciplines. Civil engineers must know governmental regulations and potential environmental concerns, including earthquakes and severe weather patterns such as hurricanes and tornadoes. Most civil engineers hold professional licenses that give them the authority to approve designs.

Chemical engineers apply principles of chemistry to solve problems in the design of equipment and processes for large-scale chemical manufacturing. Chemical engineers work in the petrochemical and pharmaceutical industries as well as technologies associated with energy, food, clothing, and paper production (see Figure 3-5). This discipline covers issues associated with chemical processes like oxidation, polymerization, coating, etching, and even nanomaterial design. Environmental, worker, and consumer safety are of special concern for chemical engineers.

Biomedical engineers design devices and procedures to improve human health by combining knowledge from biology, medicine, and several engineering disciplines. They work to develop artificial organs, prostheses, instrumentation, and other aspects of health delivery. Biomedical engineers can work to design advanced imaging systems, automated delivery of medicines such as insulin, or devices to control improperly functioning human systems. Related engineering disciplines are *health and safety engineering*, in which engineers work to ensure safety in the workplace, and environmental engineering, in which engineers consider the role of the impact of technological development on humans and wildlife. *Environmental engineers* work to solve water and air pollution, the management of hazardous waste, acid rain, global warming, automobile emissions, and ozone depletion. Consider problems associated with making something that you could solve by studying in each branch of engineering. How would these solutions lead to new products, systems, or environments?

2. Would You Like to Save the World? Engineers design solutions to help people improve the environment. As an engineer, you could help design a space colony, save a city, or protect the oceans. Think about problems that need to be solved, such as improving urban transportation systems or reducing water pollution.

Consider problems associated with helping people or the environment that you could solve by studying in each branch of engineering. How would these solutions help to save the world?

3. Would You Like to Play All Day? One doesn't normally associate engineering with play, but everything associated with music, sports, and entertainment, to give just a few examples, involves finding solutions to problems. Consider how tennis rackets and golf clubs have changed over the last decade. How have movies changed? Or how has the way music is delivered changed? Consider problems associated with play that you could solve by studying in each branch of engineering. How would these solutions help to make play more enjoyable?

Engineering Education. Engineers enter one of the disciplines by first earning a baccalaureate (bachelor's) degree in one of the engineering disciplines or specialties. Most degree programs are based on civil, mechanical, electrical/electronic, chemical, or biomedical engineering principles. Other specializations such as aerospace or environmental engineering require a strong background in one of the basic engineering disciplines. These programs place a strong demand on understanding mathematics and science because they require applying principles from these and other disciplines for solving new problems. Engineering education involves gaining understanding and finding solutions to practical problems by using computers and having hands-on experience with laboratory equipment. Students in engineering solve design problems in many classes and typically are required to complete a major design project during their senior year.

Graduates can go directly to work or enroll in graduate studies. Typically, graduate studies are needed for those considering working in areas of engineering research or who want to teach at a college or university. Engineering programs are accredited by the Accreditation Board for Engineering and Technology (ABET). Students interested in considering engineering education should have a solid background in mathematics, including calculus, and college-level sciences. Nearly all fields of engineering are projected to have significant growth over the next 5 to 10 years, with the growth rate highest for biomedical and environmental engineering. The median salary for engineering professionals found in the Occupational Outlook Handbook (2011), ranged from a low of $68,730 to a high of $108,020 annually. Engineers are often considered for management positions and, for that reason, a large number of engineers complete a graduate degree in business. The career profiles in this chapter and placed throughout the book will help students better understand what real engineers do.

There is also a national interest in teaching and learning appropriate engineering concepts in elementary, middle, and high schools. All engineering societies and the National Academy of Engineering have expressed the need to increase the level of technological literacy across all levels of society, and one method for doing this is including some engineering education before college. Therefore, so-called "pre-engineering" courses are often offered in middle schools or high schools. Pre-engineering courses give students experiences in design, allowing unique opportunities for thinking and demonstrating the importance of the application of mathematics and science to successful design solutions.

Involving Everyone in the Design Process

Working as a design team member creates a workplace that values and gives recognition to every person involved with a design. A design becomes a final product only after input from all design team members, as well as individuals involved in production and sales/distribution (see Figure 3-3). The competitive demands on a company in today's market economy require maximum productivity from each person. With the team format, each person has the opportunity to contribute to the success of the project, and to gain personal satisfaction in a job well done. Input in the decision-making process and in providing solutions to problems results in more personal responsibility and greater satisfaction.

A design project may involve multiple teams, so the task of each team member needs to be determined early in the project. Each team on a design project may have a different job to do but all members will share the same need for information. This is a must for **collaboration** with each other.

> **collaboration:**
> The process of people working together to achieve a common goal.

Figure 3-6: Input from many sources creates the final product: (a) engineer, (b) design team, (c) manufacturing, (d) sales, and (e) delivery.

Case Study

Teams Make Sense for Two Printed Circuit Board Designers

In this Case Study, John Reynolds and Reina De la Fuente talk about their team experience as engineers for PCB Designer in Fort Collins, Colorado:

One of the traits people look for in potential new hires is the ability to be a team player. The ones that go into an interview boasting "I did this and did that and the company couldn't have done it without me" are the ones to be eliminated first. Why is this? This is because nobody can get to where they are by themselves.

The environment in which there is the most success is one in which there is support, open communication, and the ability to collaborate with all coworkers. We all are responsible for our assigned duties. Many times, in order for teamwork to be truly effective, we must go beyond those duties to gain true satisfaction in our performance. Each individual must be excited about the project, know what he or she can contribute to the process, and be able to get along with other team members. Also, it is important to listen to others, and not need to have your own way.

Many companies offer team-building classes and workshops. These help to make people more aware of the dynamics of success in the workplace. We all feel more confident in an environment where we are respected by and in sync with our coworkers and management.

It takes effort on everyone's part to meet a deadline, develop a new idea, and troubleshoot a crisis. When everyone contributes, employees find more satisfaction in the day-to-day job. They strive to improve production (quality and quantity), and that happens when individuals take pride in what they have to offer and know how to communicate effectively.

Cross-training is an excellent way to use people's skills, while making them aware of how their work affects another. This makes everyone accountable for the work.

Teamwork can be broken down into many types of teams. Our company has two types of teams. One type is a *work team* and is specific to a project: Each member is responsible for a particular assignment within the project. The conclusion of the project is dependent on each individual. The other type of team in our company is a *task team*, which involves indirect teamwork from a wide variety of sources in the company for ongoing projects. Resources from other areas may need to be brought in to accomplish the goals. Charting team progress and setbacks can be demonstrated in diagrams, data sheets, and flowcharts. Round-robin discussions of everyone's input are critical to the development of the team.

Virtual Teams and their Place in Engineering Design

In the past 10 to 20 years, both technology and growing economic globalization has resulted in a new type of team, the "**virtual team**." A virtual team is a geographically dispersed group of task-driven members, with specific and complementary skills, working toward a common purpose using a broad array of communications tools and applications to support the team. The use of virtual teams has become commonplace in companies and has transformed the way many organizations are managing programs and projects. Although there are challenges associated with the use of virtual teams, the benefits typically overcome these challenges.

The typical virtual team in industry averages around 12 to 25 people worldwide. Recent studies estimate that using virtualizing project teams results in overall costs typically dropping 15 to 30 percent within the first year of implementation. Through the use of virtual teams, projects can be designed and delivered in a more effective manner, with regard to both time and cost. Additionally, the use of virtual teams means organizations are no longer limited by physical constraints, such as geography, varying time zones, and access to skilled labor. Because of this approach, companies can more easily form joint ventures, develop alliances, and

> **virtual teams:**
> A group of individuals who work across time, space, and organizational boundaries and are brought together by information and telecommunication technologies to accomplish one or more organizational tasks; members interact primarily through electronic communications. Members of a virtual team may be within the same building or across continents.

Figure 3-7: Virtual teaming relies on continuous communication, often from distant parts of the globe.

integrate more seamlessly with suppliers and vendors. Virtual teaming enables organizations to join unique skills and resources from anywhere on the globe to optimize strategies and development for more efficiently responding to complex and competitive business requirements.

It is important to realize that, even with the obvious differences, virtual teams do share common elements with traditional work teams, such as the need for well-communicated goals and objectives, strong collaboration, prioritization of work, and accountability for specific tasks and activities.

The growth of virtual teams is driven by a number of competitive forces, including:

- Globalization of business and consumer markets
- Enterprise-wide initiatives versus geographically focused projects
- Ever-evolving technical capabilities/remote communication tools (video conferencing, cloud computing networks, IM, and so on; see Figure 3-7)
- The need for reduction in project costs (business travel, physical office requirements, and other expenses)
- Pressure for improved time to market for delivery of products and services (virtual teams have the ability to work continuously across time zones)

cloud computing network:
A process where tasks are solved by using a wide variety of technologies, including computers, networks, servers, and the Internet.

Benefits and Challenges of Virtual Teams

As identified, there are significant benefits to the use of virtual teams, but there are also challenges that must be acknowledged and addressed. The obvious big problem is that team members on virtual teams cannot easily come together in person, which is the preferable method to meet. The challenge with virtual teams is to use communications techniques and technologies to overcome this issue. Note that with any virtual team, trust, project vision, and communication are also critical elements to building highly successful projects. With that understanding, the following are some of the major benefits and challenges that exist when working as part of a virtual team.

Benefits. Some of the most significant benefits of virtual teams include:

- Easier access to expertise. One of the most valuable benefits of working remotely is the ability to hire the best person for a specific role (as opposed to limited hiring availability in a defined region), which results in greater innovation and creativity.
- Allowing people to work together without the limitations of time, physical space, and fixed resources.
- Cross-functional coordination and knowledge sharing among design team members.
- Project work can be performed 24/7, which produces greater efficiencies through reduced costs, faster delivery times, and higher-quality products.
- An ability to build stronger business partnerships and alliances.

- Increased flexibility and agility to compete in the global market.
- Increased collaborative learning and leadership experiences that occur when working as part of a virtual team.

virtual: A term that applies to environments that can simulate physical presence in places in the real world, as well as in imaginary worlds. Most virtual environments are primarily visual experiences, displayed either on a computer screen or other forms of telecommunication technology.

Challenges. Key challenges almost all virtual teams today face include:

- Cultural differences. Teams require an added emphasis on respecting differences and sensitivities to cultural nuances. Politics and cultural work styles may need to be addressed.
- Team integration and socialization. The specific experience and abilities of team members are not always known by others; personal conflicts may be difficult to observe remotely.
- Nonverbal communication. Keep in mind that a large portion of human communication is nonverbal (facial expressions or other bodily expressions) and that virtual team members often do not see each other, so virtual teams need to take extra steps to ensure good communication. When working in virtual environments, you may want to "overcommunicate" in an effort to make sure all team members are receiving the appropriate information.
- Decision making. Team decisions made and communicated are critical to virtual team success but become more difficult with larger and more diverse teams.
- Accountability. It's more difficult to hold team members accountable, as it's much easier for someone to "drop the ball" virtually than to do so in person.
- Language barriers. On a global team, English is the most commonly used language, but may be spoken and/or understood at different paces, so extra time should be allowed. Also, different dialects may impact interpretation.
- Misuse of terminology. This is also very common, and miscommunications are likely, such as confusion about expectations, accountability for specific tasks, time lines, and so on.
- Time zone challenges. Typically, people are spread across multiple time zones (see Figure 3-8), which means the need for flexibility in scheduling meetings (expect adjustments to maintain work/life balance).
- Infrastructure. Computing bandwidth, speed, and technical capacity may differ significantly between geographic regions (that is, voice/video capabilities), thus impacting the efficiency of meetings.

Best Practices: Building a Virtual Team for Success. Building a successful virtual design/project team means incorporating some specific elements into the management of the team. These include first and foremost:

- Building trust by setting proper expectations and clearly communicating what's in it for each team member. Secondly, it's critical to recognize individual differences (on virtual teams, differences are less apparent) and preferred styles of communicating to foster more open and transparent discussions.

Figure 3-8: **Negotiating time zone distances can complicate the logistics of virtual teaming.**

▶ Defining, sharing, and reiterating the team "vision." Start with a specific and measurable vision, and then communicate it often. Virtual team project managers need to have a crystal-clear communications strategy and plan for the team. That vision should be used to run effective meetings, identify team strengths, and conduct regular weekly meetings. Regular meetings are vital to the success of virtual teams, including the practice of:
  a. Making the agenda plan explicit ahead of meeting
  b. Documenting key decisions and call-to-action items
  c. Sending follow-up minutes/summary to the entire project team

▶ Finally, patience and persistence are also crucial elements, exhibited by repeatedly demonstrating and communicating benefits both externally and to the team. The bottom line is that building a high performance virtual team takes time; plan for it and expect to nurture it.

Virtual teams are a reality in today's working world, as they provide cost reductions, improve resource efficiencies, optimize productivity, and deliver quality products and services faster and at a lower cost to markets and customers.

Development of the Team

The formation and structure of a team is based on the team goal, how well the members will work together, and a commitment to a job or product. The team is composed of people who are experts in their own fields and who are capable of adjusting to the changing demands and needs of the project.

A group of people becomes a team when those people have a shared vision. A vision is the goal the team wants to accomplish together. This vision motivates the team to work together through good and bad times to meet this goal.

In industry, a team may have a coach. A coach typically is only needed when the team is experiencing some difficulties, somewhat like the role a coach may play during a championship soccer game. If the team is having difficulties getting shots on goal, the coach can help them see ways to make the needed transition to offense (see Figure 3-9). However, as with a sports team, a design team that is functioning effectively will not need a coach. It is the vision that brings a team to a high level of performance and inspires commitment. An example of a team vision might be the desire of the team to design the thinnest cell phone on the market.

To create a team with vision takes hard work, and all team members must accept the challenge of the vision. The team must have a plan of action to obtain the vision. Team members will have to ask themselves, "Is my vision for success the same as the team's?"

synergy: Results when the unit or team becomes stronger than the sum of the individual members.

Figure 3-9: Successful teams use contributions from each team member.

Team Success

The success of a team depends on synergy. Synergy happens when the team becomes stronger because everyone is working together as a team (see Figure 3-10). The team is then stronger than any one individual on the team. A team can be successful if the members participate and work together to do the job.

Each team member must trust the other members and support any of the team members in time of need. This requires that the team members recognize when help is needed as well as know when to ask for help.

Team success is also dependent on effective leadership, whether this leadership is in the form of self-management or as an appointed team leader. Some teams are led by themselves. This is called a self-directed team. A ***self-directed team*** works on the idea that everyone has an assigned job to do to achieve the team goal. To achieve the goal, different team members may take the role of leader, depending on their area of expertise.

Team failure usually occurs when team members do not understand the team goal, or when a team member is not a team player. If all the members do not understand how a team should function or accept the team goals, failure is certain.

Figure 3-10: When a team builds interdependency, it becomes stronger than the sum of its individuals. This is synergy.

Team success occurs when all team members understand the benefits of the team process. This realization helps individuals to let go of their personal goals and join the team with ideas that support the team goal.

Once a group of individuals becomes a team with a vision, each individual feels the benefits of being on a team. Here are just some of the benefits a team can experience:

- ▶ Synergy is achieved.
- ▶ Individual team members achieve self-satisfaction in knowing their contributions help reach the goal.
- ▶ A team with a vision is a highly motivated team.
- ▶ A common team vision builds better teamwork.

A team is a group of individuals that bring together all different levels of expertise, sensitivity, and a willingness to participate in the team process. The spirit of success for a team will lie in the conversion of individual values and beliefs into team values. This is accomplished through a process called *norming*. The objective established in the norming process must be measurable and realistic for the team to succeed at their mission.

Group Norms

Norms are often well known and established behavior. For example, not running in the school hallway is a well-established norm. A team follows very specific steps to establish group norms.

Norms are very important because they govern the behavior of each individual and ultimately the group. Often norms are already in existence in a group without the group realizing it. For example, the most vocal person may be automatically viewed as the team leader. Sometimes, it becomes acceptable for one person to show up late to a group meeting because that person is always late. If these unspoken norms are not acceptable, it is important that they be discussed or they will likely lessen the effectiveness of team communication. The team, as a team, must agree to change unacceptable norms.

> **norms:** Principles of right action. They are binding upon the members of a group and serve to guide, control, or regulate proper and acceptable behavior.

Your Turn
Establish Your Design Team

Using the following steps for working together with your teammates and your teacher to define your mission, team values, group norms, and member strengths/weaknesses.

Each team is different and the norms it establishes will reflect this uniqueness. For a team to be successful, there are some general norms to consider. For example, get input from all team members, even the quiet ones. Silence is a strong enemy that can undermine a team effort. It is important to establish the method for team agreement—for example, consensus versus majority rule. This can be a norm.

Some other good norms to consider are scheduling of team meetings and recording written agendas and minutes for each meeting. Norms can also address important concerns about confidentiality issues or how the team will handle conflict.

Successful teams do not leave norms to chance. Successful teams have clear, written norms that establish the behavior for each team member. This written document is sometimes called a **charter**.

charter:
Document granting rights.

 STEP 1 *Identify the Mission of the Team*

This is the beginning stage when the team first gets together. The team is identifying the following:

- What is the role of the team?
- What does the team have to do?
- How will they accomplish the task?
- Why are they doing it?
- What resources are available?

A written mission statement clarifies the task to be done. Building motivation starts with a vision. A team's success starts with a shared vision of the mission to be accomplished.

 STEP 2 *Establish Team Values*

The next step is to identify team values. All team members should establish a list of what is important to them as individuals and what they can contribute to the success of the team. **Values** are basic truths that will not likely change over time (see Figure 3-11).

values:
Guiding principle or ideal.

The list of values should be short and concise, and in alignment with the project's values. Team members must come to a **consensus** on the team values. Consensus is an agreement of all the members to act for the benefit of the team, even if an individual team member does not believe that is important to him or her.

consensus:
Agreement in opinion; collective opinion.

Fun Facts
Top of the mountain

A team with a clear mission will accomplish the task. A group hike to the top of a mountain looks and feels very different from a rescue team mission to locate stranded hikers on top of a mountain.

Figure 3-11: *An example of team values.*

 STEP 3 *Establish Group Norms*

The group norms are the guidelines for how team members treat one another. These norms are rules that every team member agrees to follow because they are created by the team. Establish a list of group norms using the following process:

- ▶ Create a list of norms by brainstorming. No norm is a "stupid" norm.
- ▶ Look at each norm and talk about its impact on the team and team goals.
- ▶ Identify the key norms on which everyone can come to consensus. Keep this list relatively short. Establish consequences if norms are broken. People want things to be fair. In fact, people need things to be fair.

- ▶ Take time in this process. Give each team member time to contribute ideas and time to think over the identified norms before final adoption.
- ▶ If a team member has a dissenting vote that is in contrast to the rest of the team, find out what change would be necessary for this person to give total commitment to the process. Adjust the norms to work if necessary.

 STEP 4 *Identify Strengths and Weaknesses of Team Members*

The team is about to embark on the path toward the development of a product or idea. Roles and responsibilities are being assigned and distributed. Now the team needs to identify their individual strengths and weaknesses. This is what makes the team work together as one unit. A team must work together to balance the weaknesses of the group and to identify and utilize the strengths. It is like being on a survival camping trip and knowing what expertise the group can count on and what the limitations are. If only one person on the camping trip can read a map to get the group back home, then that person is given the map-reading responsibility.

Each team member should list his or her talents, skills, and limitations. This process helps identify the need for cross-training, how to assign job duties, who needs more training, and most importantly, it makes the team aware of their total strengths and weaknesses.

Every team member's strengths become a support system for every team member's weaknesses. Again, the synergy of "the whole is stronger than the sum of its parts" comes into play.

Some key areas where differences will arise and may impact how the team can perform will be in communication styles, individual organization and work habits, the level of knowledge and training an individual has on the project, and how each person respects and responds to other team members. To convert differences into strengths, first identify the differences and then discuss how the team can use those differences to create strengths that will help to achieve the team goal. Use the following Table 3-1 as a guide.

Table 3-1: **Example of a team's strengths and weaknesses**

Team	Strengths	Weaknesses
Team Member A	Experience with various aspects of cars CAD: solid modeling	Design analysis (e.g., Finite Element Analysis [FEA]) Knowledge of tolerances Poor handwriting
Team Member B	Off-road racing and mechanical experience Creativity	No FEA experience Report writing/formatting
Team Member C	Manufacturing skills and experience Strong interest in the project	Time management Too persuasive
Team Member D	Analytical and solid modeling Math skills Work experience	Delegating work Physical interpretation of analysis and calculations
Team Member E	Organization and leadership Attention to detail	Inability to compromise Computer knowledge Short attention span

Fun Facts

Decisions are often made slower in a team setting but they result in a higher level of commitment to the decision.

The Cycles of Team Maturity

Now that a team has come together, everything looks good for success. The reality is: The team will have growing pains and will develop at different rates depending on the team's experience in working together as a team.

Most Teams Go through Five Stages of Development

1. Orientation (forming). The members are new to the team. They are probably anxious and excited, but they are unclear about what is expected of them and the task they are to accomplish. This is a period of tentative interactions and polite conversation, as team members undergo orientation and acquire and exchange information.
2. Dissatisfaction (storming). Now the challenges of forming a cohesive team become real. Differences in personalities, working and learning styles, cultural backgrounds, and available resources (time to meet, access to and agreement on the meeting place, access to transportation, and so on) begin to surface. Disagreement, even conflict, may break out in meetings. Meetings may be characterized by criticism, interruptions, poor attendance, or even hostility.
3. Resolution (norming). The dissatisfaction starts to go away when the team members create group norms, either spoken or unspoken, to guide the process, resolve conflicts, and focus on common goals. The norms are rules of procedure and help to establish comfortable roles and relationships among team members.
4. Production (performing). This is the stage of team development the team has been working toward. The team is working cooperatively with few disruptions. People are excited and take pride in their accomplishments, and team activities are fun. There is high orientation toward the task, good productivity, and lots of optimism.
5. Termination (adjourning). When the task is completed, the team prepares to disband. This is the time for joint reflection on how well the team accomplished its task, and on how well the team functioned. And don't forget celebration.

This whole process is often referred to as "norming and storming." A team usually becomes fully productive after a storming phase or two. Team changes can be smooth if the team members look for the positive, share what they have learned from their experience, and celebrate their victories. Another important dynamic in a team might be team member dismissal. If a team member repeatedly shows an unwillingness to effectively participate in the team's vision, the team leader or the team as a whole may choose to dismiss a member from the team. This will change the dynamics of the team. It would be appropriate to revisit team norms, strengths, and weaknesses at this time.

TEAM LEADERSHIP AND TEAM CONTROL

Team leadership and team control can be a sticky discussion. It is so easy in the beginning to turn leadership over to those who already have the power or who naturally talk more. However, as the team matures, the control of leadership moves from the leader to each team member. This is not a smooth process, but everyone gradually assumes responsibilities for their performance (see Figure 3-12).

When a group of students first comes together in a project team, the leader has not been picked yet. In business, management typically chooses the team leader. For your team, the teacher or mentor may choose the team leader or the teacher may ask the team to elect the team leader using a secret ballot. The following are some questions to consider when picking a team leader:

▶ Does the person have good listening skills?

▶ Can the person explain things clearly?

▶ Is this person excited about the project?

▶ Do the other team members respect this person's abilities to lead?

▶ Will this person be able to understand and express all of the project requirements?

As previously noted, sometimes a coach will be called in when a team is having some difficulties. For your team, the coach may be your teacher or a mentor from an outside institution or agency. Gradually, the coach will assume fewer and fewer responsibilities as the team becomes stronger and more secure, and the team and the team leader become well established (see Table 3-2).

Figure 3-12: **Transition of leadership in teams.**

Table 3-2: **From coach to team leader**

Coach	Team Leader
Lead the team from dependence to interdependence on the whole team	Maintain a working role in the team
Manage conflict	Manage conflict
Build synergy and team trust	Maintain synergy and trust
Promote risk taking and help team members identify their personal strengths	Monitor the team purpose, goals, and objectives
Clear the path for the team to function	Facilitate the team meetings
Listen and mediate	Encourage risk taking and individual initiatives for the benefit of the team

Careers in the Designed World

A Stand-Up Job

Anne Fabrello-Streufert built her path to structural bridge engineering with a Lego set. As a child, Anne built anything she could think of. When she played outside, she built bridges and dams on the creeks around her house.

Anne loved to make things stand up. When a high school teacher told her that some people did this for a living, Anne decided to pursue an education in engineering. "I went from building tree forts to building bridges and wharfs, and I get paid to do this!" says Anne. "I get to have fun every day designing and engineering a new structure. Basically, I make things stand up."

Anne Fabrello-Streufert works as a structural bridge engineer for KPFF Consulting Engineers in Seattle, Washington.

On the Job

Anne works for KPFF Consulting Engineers in Seattle, Washington. She is a structural engineer and project manager for the varied projects that this company designs and builds. Anne works on projects that include bridges, retaining walls, tunnels, buildings, docks, and many other types of structures. Anne says, "One of the biggest challenges I face is also one of my favorite parts of my job. This is when my design is actually built by a contractor. The challenge is getting the contractor, who builds the structure, to understand what the design and vision is from the design engineers. When it all comes together and the structure is standing, that is best part of all."

Inspirations

Anne recently managed a project to replace a bridge for the public works department in a small town in western Washington State. The challenge was to replace this bridge with a new bridge that met current standards. The firm had to design the new bridge to sit on an existing curve and fit into the property the old bridge was built on. To meet this challenge, Anne worked with a broad team of experts.

First, they surveyed the old bridge to determine its footprint (outline on the ground). Then a geotechnical engineer tested the soil so that an appropriate foundation could be designed. A water resource engineer made sure the design would let the creek maintain its natural flow path. A transportation engineer laid out the roadway and sidewalks. Civil engineers designed a way for water to drain off the bridge's road surface during rainstorms. Electrical engineers provided lighting and signage on the bridge. Finally, environmental engineers made sure the plants, fish, and wildlife will remain safe while this bridge is in use. As the structural engineer and project manager, Anne combined all the team engineers' results into a final design.

Upon completion by the contractors, the bridge had to meet final inspection standards. "It takes a big team of very skilled people to build a bridge and it is just plain fun to see it all come together," says Anne. "It is my job to make sure that everyone works together and as a team."

Education

Anne received her bachelor of science and master of science degrees in civil engineering from Washington State University. In the summers, she worked in an architectural firm where she made cardboard models and drawings of existing structures.

Advice for Students

"Do what you love!" Anne advises. "Some will say engineering is hard." However, for Anne, the fun of seeing her structures stand up makes the effort more than worthwhile. "Almost anything worth studying takes hard work, but in the long run you will be doing what you have fun doing."

INFORMATION SHARING AS IT RELATES TO TEAMS

For any group or organization to move forward with an idea or dream for an actual product, a plan for action must be in place. This strategy includes mutually agreed-upon goals, standards for performance, and methods and expectations for communication. Communication, if not planned, can be random and haphazard. Poor or inadequate communication usually results in a crisis. This type of communication breeds dissatisfaction, gossip and rumors, and a general demoralization of the group as a whole.

Thus, communication is the tool that keeps all parts of the design process on track. When the goal or vision is not shared or evident, communication breaks down and falls apart. Communication only works when everyone's ideas are heard and respected.

Making Team Communications Work

The level of effective team communication strongly influences how well the design process progresses. Team members must communicate with individuals outside the group as well as with each other. Let us examine the inner dynamics of teams for effective communications.

For teams to communicate effectively, their overall purpose must be clear. This involves understanding the expectations for the team and the team's goals. One way to have effective communication is for the team to have norms that set the standards for team communication. We have already discussed the development of team norms and their importance. Now we want to emphasize the norms that encourage open communication among the team members. Some examples of communication norms are:

1. Each team member shares an opinion whether considered right, wrong, or indifferent.
2. No question is a dumb question.
3. All vital communications will be in writing.
4. The team schedules weekly or biweekly meetings to consider any existing tension in the group.
5. Always verify total team understanding.
6. Celebrate and give positive feedback for good ideas.
7. Reach important decisions through consensus.

The way to verify team understanding is to ask questions. Encourage team members to ask questions at anytime. Another tool for team understanding is progress reports. This tool lets all members know where they are in achieving the goal.

Designers sometimes use scenarios for communication. Scenarios are stories about how something could work. Scenarios are a great way to point out the "what ifs and whys" of a process or design outcome. A design group can use a scenario to identify potential problems a part or assembly might experience under varying conditions. For

Your Turn
Write a Norm for Your Team on Communication

Using the list of examples for communication norms, select one that works best for your team.

example, a NASA space design group may create a scenario that simulates all the possible conditions a space station hatch may have to endure such as vibration on take-off, heat and cold variances, repairs in a weightless environment, and so on. Scenarios help flush out potential problems and are best used in problem-solving activities.

Reaching Consensus

In team communication, you should listen to everyone's opinions and allow all members to have a voice. The expression of opinions allows everyone to have ownership in the process. Agreement among the entire group is unlikely. The process recommended for a group to keep moving forward on an issue is called *reaching consensus*. A consensus decision is one that the entire team can support although they personally may not agree with it. This type of decision is based on a common goal that has already been established and upon which each team member agreed.

You can reach consensus by following these 10 steps:

1. Define the purpose of the decision to be made by the team.
2. Use data and information to make intelligent decisions.
3. Come prepared with your own feelings on the issue at hand.
4. Share your thoughts about the issue.
5. Listen to all team members' viewpoints.
6. Clarify any unclear issues with questions.
7. Value differing opinions. These opinions will help you look at the issue from a different point of view.
8. Confront ideas, not people.
9. Work for a quality decision that the whole team can support.
10. Fully support the team discussion and individually take responsibility to see the project through.

A consensus decision is best for promoting teamwork and team focus. If consensus absolutely will not work, resort to voting and majority rule. The bottom line is to preserve team communication and morale.

Using Feedback for Furthering Open Communications

Teams, by nature, bring varying degrees of skills, abilities, and talents to the group. For a team to function as a unit, everyone must be open to receiving feedback as well as willing to give feedback to one another. Personal growth and team growth occur when positive, constructive feedback takes place. In a team, feedback is a two-way process that includes your point of view and the points of view of others (see Figure 3-13).

Figure 3-13: **Example of team feedback.**

It is important to remember that feedback can be positive or negative. How you give and receive feedback is critical to the trust and emotional health of the team. So feedback must be constructive and positive. The manner in which feedback is provided determines whether an individual will use it to improve or to rebel against the group.

The most important factor in giving and receiving feedback is to keep the focus on the job at hand and not on personalities (see Table 3-3). A team is formed to do a job; the members do not necessarily have to be friends.

When receiving feedback, keep in mind the idea of mutual respect. Other people may see things from a different point of view and deserve the time to share their feelings. Courteous giving and receiving of feedback can result in some fruitful ideas that could benefit the overall project.

Table 3-3:

Giving Feedback
1. Focus feedback to team results.
2. Base your perceptions on specific incidents and facts. Avoid opinions.
3. To avoid defensive positioning, involve the other person in the conversation.
4. Develop a plan of action that is jointly agreed upon.
5. Summarize the communication and the plan of action.
6. Let other people know you appreciate their participation and openness.
Receiving Feedback
1. Listen carefully to the major areas being addressed.
2. View the feedback as an opportunity for growth.
3. Ask for clarification by asking questions and paraphrasing key points.
4. Prove your point of view of the situation while remaining objective. Avoid making excuses.
5. Offer your ideas about how to improve or change the situation.
6. Schedule a follow-up meeting to give a progress report.
7. Thank the person for wanting to help you by giving you feedback.

Team Communications Should Be Organized

Whether your team communication is written or verbal, email or posted, the following is a good format to follow:

What—Explain the goal
Who—Who is responsible?
Why—The reason the task is being done
How—Suggested actions that may help accomplish the goal
Where—Location of the job to be done
When—Project timetable
What—Consequences, rewards, and penalties

Some people call this the "3-1-3" method (www-h-www).

Pressures rise and communication can fall apart when a group is working under a deadline. Table 3-4 offers some tips for positive communication when the behavior is negative.

Table 3-4: **Tips for positive communication when confronting negative or inappropriate behavior**

Always be honest.
Focus on behavior, not on the person.
Keep it team-oriented—the problem is "our" problem because we operate as a team.
Keep the confrontation positive.
Focus confrontation as missed opportunities for the team rather than criticism of individual performance.
Validate others' feelings of frustration and anger.
Remain calm.
Don't get angry in return.
Don't talk loudly.
Build a bridge of commonality to stand on: "We are in this together and I understand."
Be clear on resolutions and individuals' responsibilities.

Listening Skills Complete the Communication Loop

When we are working together and trying to communicate, the biggest frustration comes when we know the other person is not listening. This type of communication is not effective; it's like "talking to a wall." Remember when you played "telephone" with a circle of people and the last message was entirely different than the original? We tend to take listening for granted because we equate it with hearing.

Hearing and listening are not the same. Listening involves a continuous level of understanding of what is being said. You can use techniques to effectively listen. One technique is referred to as active listening. **Active listening** is defined as the ability to hear a message clearly, using your eyes *and* ears, and completely concentrating.

Effective Techniques for Active Listening

- ▶ Send listening signals.
  - Make eye contact.
  - Nod and acknowledge that you have heard.

- ▶ Minimize distractions.
  - Focus on the speaker.
  - Avoid fidgeting or pencil tapping.
  - Avoid yawning.

- ▶ Listen to the entire message.
  - Avoid interrupting.
  - Avoid giving your opinion until the end.

- ▶ Reinforce the communication with verbal cues.
  - "I would like to know more…"
  - "Elaborate on that last point…"

- ▶ Confirm the message.
  - "I hear you saying that this is what took place…"
  - "If I understand you correctly, the following is…"

- ▶ Maintain confidentiality.
  - Do not discuss or repeat what team members said in a meeting.
  - Do not discuss design ideas with non–team members.

One of the biggest obstacles to communication is when we pass judgment on what we hear before we find out what the speaker means. We must recognize that the speaker's insights are real, even if they are not the same as ours.

The effects of poor listening skills have been the cause of lost productivity, wasted time, and critical mistakes in the engineering, construction, and manufacturing process. If you focus more effort on listening instead of arguing or talking, total communication will improve. Active listening is one of the tools for building mutual respect among team members.

Today, a broad range of communication technologies are available to use to our advantage on design teams. There are many tools to choose from for group communication. Each situation will determine which type of communication tool is best to use. Figure 3-14 shows several good examples of tools to use.

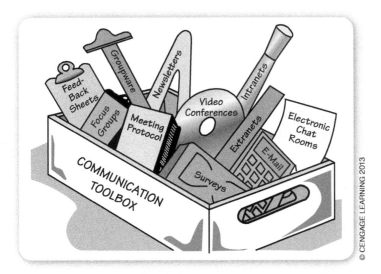

Figure 3-14: **Example of the communication toolbox.**

A well-designed and thought-out communication strategy will become the oil that greases the wheels of innovation: It will help with problem solving and will allow realization of a vision for your design team.

> Technology includes the use of technological devices to enhance human communication. Today, teams have at their disposal many sophisticated technological communication tools. Examples include email, text messaging, Internet and extranet, teleconferencing for virtual teaming, electronic chat rooms and blogs, whiteboards to lay out designs, and handheld computers that allow instant communication, no matter where a designer is.

SUMMARY

Teams and teamwork can work positively to create synergy. Successful teams become an entity that is greater than the sum of its parts.

Therefore, when we put teams in the context of engineering design, each small team becomes part of the larger system to produce a totally integrated process for engineering a new product. Teaming is occurring everywhere, from design engineering companies to factories to cyberspace teams that communicate through electronic chat rooms and email. This major power shift in engineering design has changed how products are developed and how they reach the marketplace.

As the world of engineering continues to develop into an environment that is driven by continuously evolving market desires and needs, the ability to be creative and innovative is crucial to everyone. Engineering is making the skills involved in creativity a high priority. Businesses now design questionnaires and interviews to determine if potential employees possess the characteristics necessary to promote creativity and innovation.

Creativity is a required component of staying ahead in today's design world. To achieve this, individuals must practice the individual habit of thinking creatively and must learn to operate within the team/group environment. In our ever-changing, information-based society, utilizing our most valuable resource—people and their collective creativity, knowledge, and ability to work together—will create the designs that will rise to the top.

Creativity is a life skill that must be continually cultivated. Remember, being creative is *looking* at the same thing as everybody else but *seeing* something different.

When it comes to communication, here are some tips to help you keep up with all the information that will impact you and your design project. Take responsibility for your part of the process and attend all informational meetings, virtual or real. Take the time to locate all potential resources for information, and communicate to your fellow teammates what you know and can do. Listen to all of the ideas and suggestions from your team.

BRING IT HOME

OBSERVATION/ANALYSIS/SYNTHESIS

1. Name the most common system U.S. manufacturing companies used to organize work processes in the 19th century. Describe how this system differs from most industrial organizations today.
2. Name three industries in which a mechanical engineer might find employment.
3. Describe how an industrial engineer contributes to product design.
4. Name three professionals you might find on an engineering design team, and describe their contributions to the product.
5. Describe in your own words the role of a team leader.
6. Write a complete paragraph to describe the term "storming and norming."
7. List five tools teams can use to develop positive communication.

EXTRA MILE

ENGINEERING DESIGN ANALYSIS CHALLENGE

With your teacher's guidance, form into a design team of four to six people. Determine what your design project is. Based on the design project, determine and establish the following:

- Write a team performance goal for the design of the product. For example: The new cell phone headset design will be made to a specific size to fit into a motorcycle helmet. The helmet must be produced in time to meet the fall release of new helmet designs for Harley Davidson.
- Make a list of team norms and values (keep the list to five for each).
- Write a complete list of team members. Include any outside sources for materials, manufacturing, suppliers of fittings, and so on.
- Develop a job schedule to be completed to meet the team performance goal.
- Develop a skills strength and weakness chart to include each team member.
- Establish job assignments based on the team's strengths and weaknesses.
- Identify sources for research for the product design.
- Write three team performance statements that the team would use to determine the success of the team's performance goal. For example: The team will complete the design phase of the new product within two weeks.
- Using PowerPoint software, make a team presentation to the class or to a group in your school or community.

CHAPTER 4
Generating and Developing Ideas

Menu

Before You Begin
Think about these questions as you study the concepts in this chapter:

1. What are some useful strategies to help think creatively about a problem?

2. Ideas need development to make them practical and useful. What can you do to help refine design ideas and make them workable?

3. Does the appearance of a product have an impact on the success of a product in the marketplace?

4. What significance does material choice have on what a product will look like?

5. What are the design elements and principles used in the aesthetic design of a product?

6. How do you go about choosing the best design?

INTRODUCTION

Scientist and Nobel laureate Linus Pauling said, "The best way to have a good idea is to have lots of ideas." Nothing could be truer. With many ideas, you improve the chances that one of them is a good one. There is little chance that only one idea is the best idea.

Ideas are incredibly important to business. New products come from ideas. Improved techniques come from ideas. More effective management comes from ideas. Ideas are the cornerstone of innovation and successful business.

CREATIVE THINKING

Much of the process of designing involves creativity, so individual designers and all members of the design team will need to apply many of the techniques presented in this chapter. But, without some training and practice, creative thinking does not come easily to most people. This is because most of our education experience and decision-making situations involve convergent or deductive thinking, a type of thinking that involves a defined logistic order. We are often taught to look for the "right answers" to our problems, and we are taught also that not being right means failure. In creative thinking, there are no right or wrong answers—only ideas. Later, when we have a lot of ideas, we may use logic to sift through them. Edward de Bono, a recognized expert in the field of thinking and creativity, coined the terms *vertical thinking* and *lateral thinking* to describe these concepts.

In vertical thinking, each idea rests on another idea in logical form, like a house resting on a foundation. Edward de Bono describes it as *high probability thinking*, which means we use this kind of logical thinking to function day to day. Instead of analyzing each and every action we take, vertical thinking allows us to make assumptions based on past experience, such as when we see a doorknob, we reach out and turn it without analysis. Life would be nearly impossible without vertical thinking.

Lateral thinking, however, is *low probability thinking*; that is, it follows quite unconventional paths. Lateral thinking allows new ideas, while vertical thinking follows previous paths. Learning how to think laterally, or "out of the box," is an important step to creativity and design. As you look at the steps in the design process, can you tell which steps emphasize logical or vertical thinking, and which steps emphasize creative or lateral thinking?

> **creativity:**
> The ability to make or bring a new concept or idea into existence; marked by the ability or power to create.

> **lateral thinking:**
> Thinking that follows unconventional paths, sometimes called *low probability* thinking.

> **OFF-ROAD EXPLORATION**
> For more information on creative thinking, see *Cracking Creativity: The Secrets of Creative Genius* by Michael Michalko (ISBN-13: 978-1580083119) or try *Lateral Thinking: Creativity Step by Step* by Edward de Bono (ISBN-13: 978-0060903251).

GENERATING DESIGN IDEAS

Generating alternative solutions is a key stage in designing for individuals and for design teams. In fact, some clients require designers to submit alternatives before actual project development begins. For example, a graphic designer will often submit several design ideas for approval because the client expects to have an opinion of the final design before it goes to the printer. Similarly, landscape architects will often submit several design ideas for approval by the client before committing to actual work (see Figure 4-1).

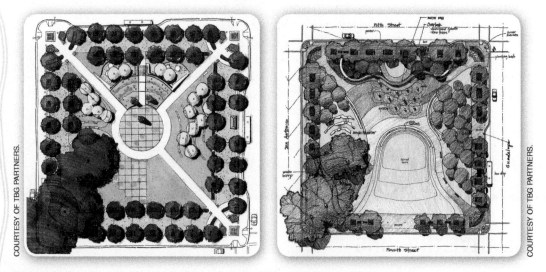

Figure 4-1: **Two landscape designs for Republic Square in Austin, Texas.**

Lateral Thinking

You can develop your creative thinking potential by employing lateral thinking techniques pioneered by de Bono. One of these techniques involves the identification of the *dominant idea* in the situation in which you are attempting to find a creative solution. An example might be useful. Personal desktop computers were historically architected to consist of three major components: (1) a computer housing (that held the actual microprocessor-based hardware), (2) a monitor, and (3) keyboard and mouse components. The "dominant idea" in the marketplace was to keep this type of format, or architecture, and to provide improvements for newer computers by improving one or more of these three major components. However, the computer company, Apple Inc., has consistently challenged the dominant idea of "separate components" by designing all-in-one computers that combine the microprocessor-based hardware with the monitor. An all-in-one format, such as the very successful Apple "iMac," provides customers a higher ease of use and a product with less "cable clutter."

The dominant idea has been compared to a hole dug in the ground. People may enlarge the hole or they may dig it deeper, but it is easier to stay with that hole than to begin digging a new hole elsewhere. "Breaking new ground" is a common phrase that simply means to dig a new hole. Identifying and understanding the dominant idea allows you to step away from it to pursue other paths of reasoning.

A good example of a dominant idea in the area of science can be seen in the work of Sir Isaac Newton. For centuries after Newton, his ideas about the mechanics of the universe served both science and technology very well, and still do. However, there were small inconsistencies in his theories, but these inconsistencies were largely dismissed as a lack of understanding or error in observation. However, as time passed, experimental measurement accuracy got better and these inconsistencies could no longer be ignored. It took someone who both knew the work of Newton and could step away from the "accepted way of thinking" to form a new theory that fit the evidence. Einstein's willingness to question one of the most fundamental assumptions of the human mind—*time*—led to one of the biggest revolutionary leaps in our understanding of the universe, resulting in the current theories of general and special relativity, which deal with curved space-time.

After you recognize the dominant idea, you can find different ways of looking at the problem. Leonardo da Vinci (1452–1519) said, "All our knowledge has its origin in our perceptions." The point da Vinci is making is that perceptions of reality, as opposed to reality itself, can dominate your way of thinking. You must try to find ways to go beyond your perceptions, however. You must uncover other viewpoints that can help you see aspects of the problem that logic does not easily permit. Your point of view can have a profound effect on your ability to see both the obvious and the obscure. Some of the following idea-creation techniques can help you develop other ways of looking at a problem.

Analogies

An analogy is a similarity between two unlike things. In problem solving and design work, analogies can play a helpful role in generating ideas. It is said that Johannes Gutenberg's invention of the printing press in 1455 was, in part, a result of the analogy he saw in the coin punch and wine press of the day (see Figure 4-2). He was able to visualize a machine that combined the principles of both devices to print words on paper using individual letters.

In your design work, can you see similarities to other problems? Do the solutions to these problems hold useful ideas for your problem? If you are working on a problem involving structural design, can you find analogies in the structure of plants or the human skeleton, or in other areas that are totally unrelated to your design problem? Seeing analogies may provide you with design possibilities.

Figure 4-2: Gutenberg's printing press has been called a combination of the coin punch and the wine press of the day.

Brainstorming

Brainstorming is built on the belief that creative ideas differ from conventional wisdom. In a brainstorming session, two or more people get together to exchange ideas and use their ideas to stimulate more ideas. They try to get away from conventional wisdom by letting their imaginations run wild and by reaching for outrageous solutions. This helps break down the logic of convention (vertical thinking) and provides an atmosphere that encourages creativity. In brainstorming, participants must be open and adopt a friendly environment in which new ideas will be welcome (see Figure 4-3).

brainstorming: A group technique for solving problems, generating ideas, and stimulating creative thinking by unrestrained spontaneous participation in discussion.

Figure 4-3: A brainstorming session involves a number of people.

Figure 4-4: **Organization of a brainstorming session.**

Organizing a Brainstorming Session
Brainstorming can be an effective strategy for generating a lot of ideas for solving a problem. Here are some important rules for brainstorming. 1. Work in a group of at least three or four people. 2. One person must take notes; recording of emerging ideas is critical for allowing revisiting of earlier inspirations. 3. Define the problem well, and make sure that each person understands it. 4. Set relatively short time limits on each problem (30-60 min.). 5. Be spontaneous, be outrageous, be imaginative. 6. Listen carefully to other people's ideas, and build on them. 7. Do not criticize, evaluate, or even elaborate. [Important!] 8. Go for quantity to ensure quality. 9. Evaluate only after your idea creation ("ideation") session has ended.

Nothing stops the flow of creativity more effectively than criticism, so participants in a brainstorming session are not permitted to criticize. There is time for evaluation of the ideas generated after the brainstorming session. The following are the basic tenets of brainstorming:

▶ The more ideas that are generated, the better the odds that a quality idea will emerge.

▶ Group ideation is more valuable than individual ideation because group members stimulate one another; brainstorming is usually ineffective with one individual.

▶ A nonthreatening environment is necessary for free expression of ideas; therefore, criticism is deferred.

▶ A limited time frame enhances the creative environment.

Often, when students attempt brainstorming, they meet with little success. This is usually because of two factors. First, participants are inexperienced with the process and do not take it seriously. Second, participants are inhibited and are afraid of looking foolish. To be successful, you will need to practice the technique. Figure 4-4 describes how a brainstorming session is organized. In addition to these rules, some of the following special brainstorming techniques can help your team break through conventions to inspire more creative thinking.

Mindmapping. This is a method that combines free word association and brainstorming. The idea is to generate as many ideas with a central theme or topic as possible. Using the theme or topic as a starting point, begin to write down thoughts that occur to you. Do this as quickly as possible; do not analyze anything. You will have words/ideas related to your central theme (main branches) and then words/ideas related to these branches (see Figure 4-5).

Breaking the Rules. One of the simplest ways to get new ideas is to reverse or twist previous ideas. This method of "breaking the rules" allows for innovative solutions to common problems. For example, tradition dictated that designers place elevators on the inside of buildings. However, a creative architect considered putting the elevator on the outside of a building. This **innovation** created a more beautiful view, provided a striking architectural feature, and saved interior building space—all at the same time.

mindmapping:
A method that combines free word association and brainstorming. The idea of mindmapping is to generate as many ideas as possible by beginning with a central theme or topic.

innovation:
An improvement of an existing technological product, system, or method of doing something.

Chapter 4: Generating and Developing Ideas   91

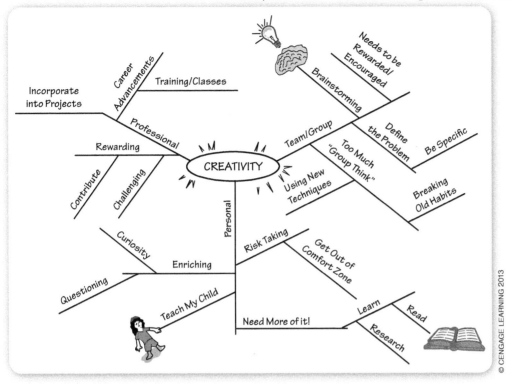

Figure 4-5: Example of the mind mapping model of creativity.

The Nominal Group Technique. The nominal group technique (NGT) is a structured decision-making process designed to involve all group members, encourage multiple ideas, ensure thorough consideration of ideas, and generate an optimal group decision.

 STEP 1 *Silent Generation of Ideas in Writing*

The first step in an NGT meeting is to have group members write out ideas silently and independently. The following are benefits of this step:

1. It allows adequate time for thinking and reflection on the idea.
2. It avoids interruptions.
3. It avoids competition between team members and allows all ideas to be heard.
4. It avoids choosing among ideas prematurely.

 STEP 2 *Round-Robin Recording of Ideas*

The second step of NGT is to record the ideas of group members (preferably on a flip chart). Round-robin recording means going around the group and asking for *one idea* from *one member* at a time. The recorder writes each idea on the flip chart and then proceeds to ask for one idea from the next group member in turn. The benefits of round-robin recording are:

1. It encourages equal participation in the presentation of ideas.
2. It can promote depersonalization—ideas can be separated from the person who wrote it.
3. It introduces a larger number of ideas and alternatives.
4. It increases inclusion of conflicting ideas.

5. It encourages "hitchhiking." (An idea listed on the flip chart by one member can stimulate another member to think of an idea he or she had not written on the worksheet during the silent period. Group members are free to add the new idea to their own worksheets and report it when their turns arrive.)

Research evidence shows that sharing all ideas and equalizing participation increases group creativity and generates more alternatives. The structured format of engaging each member in turn to elicit ideas establishes an effective behavior pattern.

 STEP 3 *Serial Discussion for Clarification*

The third step of NGT is to discuss each idea in turn, for the purpose of explaining the idea. Serial discussion means taking each idea listed on the flip chart in order, and allowing a short period of time for explanation or elaboration. The recorder points to Item 1, reads it aloud, and asks if there are any questions or requests for clarification. The recorder allows time for clarification by the group member who proposed the idea, and then moves on to Item 2, Item 3, and so on. The main purpose of this step is to clarify, *not* to win arguments. The benefits of this step are:

1. It avoids focusing unduly on one or only a select few ideas.
2. It helps to eliminate misunderstandings.
3. It records differences of opinion without arguments.

 STEP 4 *Serial Discussion for Evaluation*

The fourth step of the NGT is to evaluate the ideas. The recorder points to Item 1 and asks for observations, opinions, statements of agreement or disagreement, pros and cons, and so on. The recorder allows for discussion and moves on to the next items in sequence. The purpose of this step, again, is not to win arguments, but to articulate the advantages and disadvantages of ideas. The benefits of this step are that it provides opportunities to:

1. Discuss the rationales for ideas.
2. Discuss the pros and cons of a particular idea.
3. Air disagreements and differences of opinion.
4. Consider the feasibility of a given idea.

 STEP 5 *Vote on Idea Importance*

Each group member ranks in order of priority the ideas on the flip chart according to his or her perception of its importance to the issue or question under consideration (assigning #1 to the idea of highest importance). Lowest collective averages identify the most important ideas for the group.

An average NGT meeting generates over 12 items per group during its idea-generation phase (compared to 3 or 4 in a traditional "interacting group" meeting). Through serial discussion, group members come to understand the ideas, the rationales for those ideas, and the arguments for and against each one. By *voting*, however, the group aggregates the judgments of individuals to collectively determine the relative importance of individual ideas and, thus, arrives at a group decision.

Synectics

Synectics is a technique used for uncovering perspectives. The goal is to make the "familiar strange and the strange familiar." What this means is that synectics is useful for viewing a problem in a new way. In one technique, the designer role-plays the part of the product or device used to solve the problem, asking *"Who affects me and whom do I affect?"* After establishing a list, the designer role-plays each of the affected people or objects involved and tries to see the other person's point of view or object's role. This strategy works better with a design team rather than just one person.

For example, a basketball shoe is familiar to most people, and those who wear them regularly do not give them much thought. If you were involved in athletic-shoe design, you might imagine yourself as the shoe, being worn by a player during a game. You could feel your layers of foam and sole materials compressing together as the wearer prepares to leap up for a jump shot, or the repeated compression and expansion as you pound your way to the other end of the court. Although this may sound a bit bizarre, it is a clever and effective strategy for getting away from vertical thinking and into generating new ideas.

Sketching and Doodling

Written and spoken language has many rules, rules that can be limiting to designers. Designers have found it more effective to use a communication technique that is more intuitive and spontaneous. Drawing is the language of all designers. Quick, freehand drawing, called sketching, is a fast and efficient way to get ideas out of your head as fast as possible, before they disappear.

The development of your ability to sketch will help you in your creative efforts of problem solving and design. Chapter 5 introduces you to techniques for sketching, as well as a number of practice exercises to help you develop your skills. Sketching helps you imagine and refine devices and systems that could solve a problem (see Figure 4-6). Sketching forces you to develop your ideas in terms of the relationships between components and parts of a system. Just as sketching an object forces you to look more closely at that object, sketching something from your imagination forces you to address details.

For example, if you were trying to develop solutions to a problem involving door-opening devices for people in wheelchairs, you might sketch the possible ways in which doors can open: slide, rotate, pivot, fold, and so forth. Your sketches then might include simple systems for opening doors, such as motors, levers, pneumatic cylinders, and others. As you continue to sketch, a little more detail emerges until you begin to see if any of these ideas have merit.

Figure 4-6: **A sketch for a solar home, and its realization.**

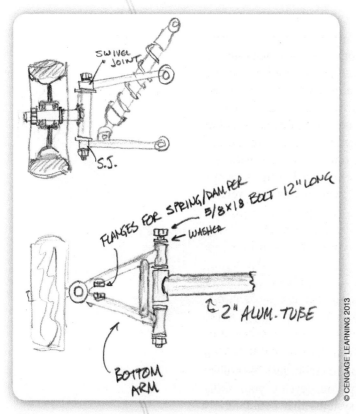

Figure 4-7: An example of an effective technical sketch.

Figure 4-7 shows a basic sketch of a front suspension system for a go-cart. The sketch shows some details about how a number of parts work together. The designer is just beginning to address details about materials. It is a good example of an effective technical sketch.

The use of instruments for formal drafting practices will inhibit the creative flow of your ideas. The structure imposed by these instruments and practices are just like the rules of the spoken and written word, coming between your ideas and the paper, so most designers consider the best way to sketch is to use just paper and pencil. However, the sophistication of design software is beginning to make sketching on the computer possible, if still a bit awkward.

Incubation Period

The mind is always working, even when we are not consciously thinking about a problem. It is therefore valuable to provide a period of time during a project for ideas to "incubate" or work through in our minds. It is not uncommon for someone to struggle with a problem then, frustrated that an answer has not been found, go to sleep that night, only to wake the following morning with a solution. This also holds true for students who study for a test. Studying the night before and getting a good night's sleep is more effective than studying all night without sleep. During sleep, our mind sifts through information and categorizes it.

Although it is unrealistic to expect that your design solutions will suddenly become clear before breakfast, it is usually beneficial to get away from the problem for a day or two and let it incubate in your mind. If you become stuck in some phase of the design process, go do something else. Often, when you sit down to work on the problem again, you will find new insights and ideas that you had previously missed. It is important, however, to record ideas when they come to you. These "flashes of insight" can be easily lost if you do not put them on paper immediately.

Your Turn

A ratio is the relationship between two quantities, such as the two adjacent sides of a rectangle. Sometimes these are expressed as aspect ratio, in a format of x:y, where the x is the width and y is the height. There are important examples of standard rectangle ratios in the world around us. For example, older televisions had a screen aspect ratio of 4:3 (1.33:1). High-definition televisions use a screen aspect ratio of 16:9 (about 1.78:1).

A golden rectangle is a rectangle whose lengths of sides have a ratio of 1:1.618 (1.618 is approximately 0.5(1+s), where s is the square root of 2). Using the Internet, look up (1) what is special about a golden rectangle (how a golden rectangle is defined), and (2) how to construct a Golden Rectangle using basic geometrical concepts. (Search for "constructing a golden rectangle.")

DEVELOPMENT WORK

Ideas need development at two important stages. The first stage is when one has a number of generated ideas, each having some level of good potential for the project. In this first stage, each of these multiple ideas will need to be developed to the point of either elimination or inclusion into your project design. The second stage is when a final solution path has been selected. More development at the "final" stage is needed simply because all of the details of the exact design have to be worked out. The amount of development effort will vary with the complexity of the solution: Simple ideas may sometimes need little development, while complex ideas almost always need much development. The goal of development during early stages of the design process is to get the idea into a shape that will tell you if it is a workable solution to the problem. Do not stop when you achieve the first workable solution; develop your other solutions in a similar manner.

After developing a number of workable ideas, you will choose which one to implement. At this stage, the idea you have chosen undergoes further development. The goal at this stage is to make the idea producible. You may need to make detailed decisions about size, shape, materials, fasteners, finish, or other considerations. In the case of a nonmaterial-oriented project, such as one involving programming, electronics, or media, decisions may involve flowcharting, logical sequencing, storyboarding, or other considerations.

If your solution involves the construction of a physical product, designers use technical drawings to enable communication among other professionals and for the actual making of the product or system. Chapter 8 includes more detailed information about technical drawings.

> **technical drawings:** Drawings needed to actually produce the component or system.

Construction Kits

A very useful strategy for developing early ideas can be found in the use of the various construction kits available commercially. If your product involves mechanical systems, Lego, K'NEX, Fischertechnik, Meccano, and others provide easy and quick ways to test your ideas.

Some construction kits allow complex mechanisms to be modeled. The construction kit approach allows you to easily fabricate sophisticated structural and mechanical designs that otherwise would be too difficult and time-consuming to build from scratch. You may need to design and build a gear reduction mechanism that will increase the torque of a small electric motor to perform a lifting task. Modeling from standardized parts and materials to build a mechanical system and framework structure to hold it would take much skill and time. But this task is easily done with most construction kits, once you are familiar with the parts and joining and fastening techniques of the particular kit. Prototypes made with construction kits also enable important engineering design analysis. For example, building a particular gearbox can be used to mathematically and experimentally verify the mechanical advantage required for a design.

Perhaps the greatest strength of the construction kit is the ability to work out ideas in three dimensions (3-D). Drawing and sketching are used for the development and refinement of technical ideas in two dimensions (2-D), on paper. But what appears to work in a sketch may not work in actuality because of the inexperience of the designer or lack of specific detail in the design.

96   Part I: The Engineering Design Process

Figure 4-8: VEX construction kit used to create a robotic vehicle.

Figure 4-9: A "torque amplifier" for an analog computer constructed from Meccano.

Visualizing potential solutions in three dimensions is possible with construction kits. Using kits allows you to work out problems and ideas and to get immediate feedback on the viability of the design (see Figures 4-8 and 4-9). This is a crucial step toward the development of any device that includes motion of parts, and is a valuable tool for projects intending to be constructed of standard industrial materials. This is the power of kit modeling.

Computer-Aided Design

The use of the computer has revolutionized design. The entire design enterprise has been turned upside down with the introduction of the desktop computer and the development of software that is not only for design, but also analysis, testing, and information research. Computers allow teams of product developers, including industrial designers and engineers, suppliers, and subcontractors, to all take part in the product-development process. Computers also make the distance between key team members less important, allowing design team members almost any place on the planet to take part.

Figure 4-10: Engine assembly created in CAD.

Like any tool, the capabilities of the computer and software limit what designers can do. It is said that when architects began to move to computer-aided design (CAD), an experienced eye could tell that particular buildings had been designed using CAD because the software limited the things that the architect could do with shape and style. Of course, newer CAD software is much more capable and offers designers a lot more latitude in their designs. However, it is not only the sophistication of the software that may limit design options. The capability of the person using the software is also a limiting factor. The more you know and the more skilled you are with the design tools, the more options you have in your own designs (see Figure 4-10).

In idea development, the use of parametric solid modeling has made the development of virtual 3-D models possible. With programs such as Inventor, Pro/ENGINEER, SolidWorks, and others, the creation of individual parts and whole products made up of individual parts is possible. Unlike drawings, parts can be rotated and checked for fit with other parts. Clearances between parts can be measured and adjusted if necessary. Alternative shapes and sizes can be built into a part, sometimes called configurations, and designers can look at and analyze many "what if" scenarios almost instantly to optimize the design solution.

Many solid, or 3-D, modeling packages allow the development of mechanical and structural systems and will help you analyze them. In a very simple example, the design of a model rocket can be analyzed for the placement of the center of gravity (CG), an important factor in the stability of the rocket (see Figure 4-11). The CG is the point around which the rocket will rotate—the balance point, like a seesaw. A second factor is the center of aerodynamic pressure, or CP. If the CG is too far forward, or too far rearward of the CP, the rocket will tend to topple over in flight.

Of course, there are now specialized CAD modeling programs that have been created by model rocket enthusiasts that are dedicated to model rocket design. These programs, such as RockSim and SpaceCAD, allow hobbyists to predict stability and flight characteristics of their model rocket designs (see Figure 4-12).

CAD took over as the primary design tool in many professions and allowed designers to create photorealistic images that help communicate design ideas as well as sell ideas to management and customers. Industrial designers and other designers also use programs such as Adobe Photoshop to create striking images (see Figure 4-13).

Figure 4-11: CAD drawing of model rocket.

Appearance Development

The term *aesthetics* is often thought to be related to art, but aesthetics is an integral part of almost every aspect of our lives. Aesthetic is the word used to describe something that seems pleasing to us. Whether we are aware of it or not, aesthetics is considered in the decisions we make about what we like and do not like, and what we buy and do not buy. In the design work we do, aesthetics is an important consideration.

More information about product appearance is covered in Chapter 17, Design Styles. Today, all companies are concerned about the visual appearance of their products and, as such, typically involve an industrial designer on the design team. Aesthetics and the visual design principles associated with product development, including marketing and individual or team concerns for making more effective presentations, are covered in Chapter 18, Graphics and Presentation. As noted in Chapter 1, when the Industrial Revolution progressed and factories began to produce consumer goods, the competition for market share increased. Companies that produced better-looking (and hopefully better-quality) products sold more. But it was still difficult to create some 3-D forms in quantity. Manufacturers only had copper, brass, iron, and a handful of other metals that could be heated and poured into a mold for economically creating complex surface forms in quantity. Consequently, these products were limited in design, and were often quite heavy.

An important factor that led to the design-style changes of the last century or so was the introduction of new materials into the marketplace. When plastics as molding compounds were introduced in 1927, it became possible to produce consumer products in much more varied shapes than ever before. Like metal,

aesthetics:

Having to do with appearance; a branch of philosophy that deals with human response to visual stimuli leading to judgments about the things we see. These judgments usually result in expressions of like or dislike and are related to cultural, economic, political, and moral values.

design principles:

The rules of thumb that describe how a designer might put together the various design elements to create a finished product.

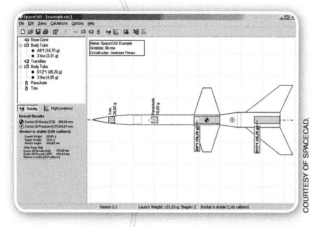

Figure 4-12: SpaceCAD design showing CG and other characteristics.

Figure 4-13: Industrial design rendering of bicycle concept using Photoshop®.

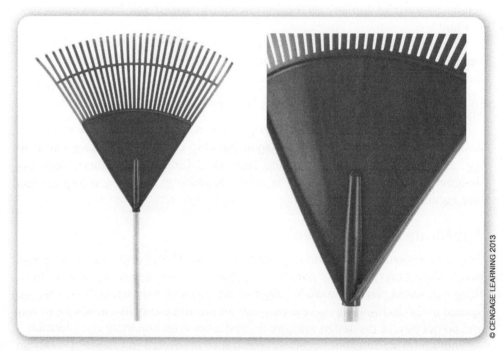

Figure 4-14: Material choices have an impact on product design and manufacture. Shown here, a lawn rake made from plastic.

plastics can be cast in molds, but unlike metal, most plastics are light in weight. The development of cellulose acetate, polyvinyl chloride, low-density polyethylene, polystyrene, and others changed product design and what was aesthetically possible and practical to produce.

Material choices drive how a product is produced as well as how it looks. Figures 4-14 and 4-15 show two lawn rakes, one made from plastic and one from steel. Although they may look similar on first inspection, they are constructed in very different ways. The plastic rake is made in one piece by injection molding, and it can be made in almost any shape or color. The steel rake is assembled from 24 separate tines that must be coated (painted or powder-coated) to prevent rust, and reinforcing members used near the middle and at the top where the handle is connected. Setting aside issues such as durability and cost, the plastic rake gives the designer a much wider range of aesthetic possibilities.

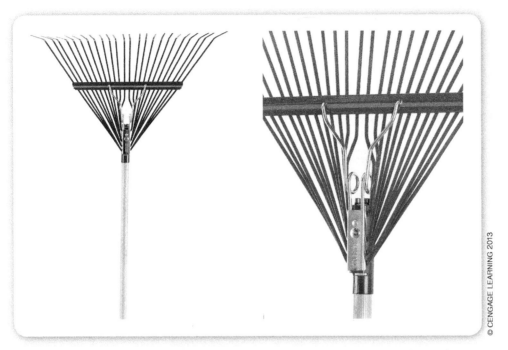

Figure 4-15: A lawn rake made from steel, although intended to perform the same function as the plastic rake, is of a completely different design.

The rake is just a simple example of the idea that material choice can drive aesthetics. There are many other important reasons for material choice, including strength, durability, cost, manufacturing processes, environmental impacts, and others.

Visual Design Elements

You may want to include a number of design considerations in your development work. These elements often serve both aesthetic and functional purposes, as in the folded surface of a metal or plastic case. The folds provide a visual element of line and at the same time impose rigidity to an otherwise flexible material. Here are some design principles you might consider.

Line. A common way of decorating surfaces is to use line. Notice the linear ridges or indentations applied to computers, radios, calculators, and even the stripes on your sweat socks. The high-power cycle light in Figure 4-16 is a good example of the use of line as a design element.

Line can break up a dull or uninteresting surface. It can give the impression of motion or speed. Line was used extensively in Art Deco design in the early part of the twentieth century.

Color. Possibly the most accessible graphic design element is color. We can react to color strongly on emotional levels, and it directly affects our impressions. Physically, color has several dimensions, including wavelength, which determines the name or "hue" (red, green, yellow, and so on), value, and intensity, all of which can be manipulated by a designer. Refer to Chapter 5 for examples of how color can be effectively used.

Figure 4-16: High-power LED cycle light by CatEye has a line design element.

Science of color: Radio signals, light, and heat are terms we use for different frequencies of electromagnetic radiation. When we see different colors, we are actually seeing different frequencies of electromagnetic radiation in the narrow band of frequencies called the visual light spectrum. As particular frequencies stimulate the retina of the eye, we see particular colors. If all visible light frequencies are present, we see white. If almost no visible light frequencies are present, we see black. The letters on a printed page appear black because the characteristics of black ink are to absorb almost all light frequencies and reflect almost no visible light radiation at all.

Technology of color: Creating pigments and dyes involves complex technical processes that draw on knowledge from organic and inorganic chemistry and other science disciplines. These compounds are used to create paints, tinting dyes for vinyl and other plastics, coatings for metal, and a wide range of inks for the printing industry. To ensure consistency of one batch of dye to another, computer-controlled machines have been developed that measure the frequencies of reflected and transmitted light of the dye mixture to control the amount of various pigments that are added during production.

Engineering color: Teams of engineers design the machines and systems that control the production of dyes and pigments. Knowledge of physics and chemistry, as well as a strong background in machine design and electronics, is necessary to successfully produce these devices. It would be rare for a single engineer to possess all these qualities, so a successful design team makeup includes a number of contributing specialists working on the development of the product or system.

Mathematics of color: It is impossible to talk about color in a precise way without mathematics. Even using a computer drawing program, color is described by the percentage of red, green, and blue, or by hue, saturation, and value. More precise color definitions involve the wavelength or frequency of the color. Red is described as a frequency of 384 to 482 terahertz (THz) or a wavelength of 622 to 780 nanometers (nm).

Form/Shape. When a line drawn on a sheet of paper closes back on itself, you get a shape. Organic shapes are those that seem random, meandering, or "natural." Rectilinear shapes are those that have straight sides meeting at corners and look human-made. Shapes can have hard or soft edges, be light or dark, opaque or transparent, colored or textured. The three-dimensional counterpart of shape is called *form*. All physical objects have a form, and either they or we have to move in space and/or time to appreciate it. The same characteristics apply to form as to shape, but much of the functioning of products depends on form. The overall size, weight, materials, and purpose constrain the form of a car, aircraft, building, or sculpture.

Form is something that can go through fads. If you look at automobiles from the 1940s, you see similar forms: flowing, rounded forms, such as those in Figure 4-17. In the 1980s, the boxy form was in vogue (see Figure 4-18). Today, a number of car companies have introduced retro styles that go back to the 1930s and '40s.

Figure 4-17: 1940 Ford Sedan with rounded form.

Figure 4-18: 1987 Nissan Maxima with a more "boxy" form.

Figure 4-19: The rear of the Lotus Europa is an example of symmetrical form.

Figure 4-20: This home design offsets various elements to achieve asymmetrical form.

The form can also be symmetrical or asymmetrical. The rear of the Lotus Europa is symmetrical (see Figure 4-19), while the asymmetrical home shown in Figure 4-20 uses various **design elements**, such as the roofline, doors, and window placement, to achieve this form or shape.

Space. The use of three-dimensional space is also an important design element. For example, is your classroom or bedroom cluttered, having objects haphazardly distributed through its space? Or is the space designed and organized to be uncluttered and therefore a more effective use of space? The effective use of space is important in many designs. However, in architectural design, the effective use of space is of vital importance. For example, what if the space used for the hallways of your school was one-half or one-third as wide? If the hallway space were made smaller in this way, it would certainly severely restrict the movement of students and teachers, disabling the function of the school.

> **design elements:**
> The factors that the graphic designer has to work with and can manipulate, including line, shape, form, value, color, and texture.

Part I: The Engineering Design Process

Your Turn
Analyze a Simple Mechanism through Sketching

Most doors within a home have a mechanism consisting of a round knob or lever and a latch that protrudes into the jamb. Rotating the knob (or pushing down on the lever) causes the latch to retract or extend.

1. Sketch how you think this mechanical action might be accomplished. Include as much detail as you can to describe the mechanism.
2. If you can, carefully disassemble a door mechanism and sketch in detail how it actually works.

Figure 4-21: *Textured surface on a set of tires provides friction to grip the road.*

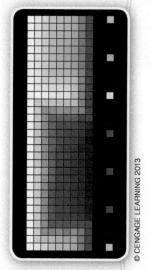

Figure 4-22: *Color and value scale.*

Texture. For functional and aesthetic purposes, texture is extremely useful. As an alternative to value and color, which make surfaces interesting, texture is applied to objects that might be touched or at least might take advantage of light and shadow. Rough textures on tires, tool handles, surfboards, and skateboards provide needed friction for grip (see Figure 4-21).

Value. In graphic terms, *value* relates to the relative blackness of a color. For example, the value of the color red might vary from a very light ("white") red to a very dark or blackish red. It can be more generally applied to the extremes, and the steps between them, of anything. Value always indicates things that are relative; value is seen in contrast to something else. Visually, a light, or high-value, feature makes a dark, or low-value, feature more visible. Figure 4-22 shows values for various colors within the extremes of white and black.

CHOOSING THE BEST SOLUTION

Once you have generated a number of workable ideas, you will need to evaluate them to find the optimum solution. **Optimal** (which has the same meaning as optimum) means the best solution that can be achieved, considering all the requirements and constraints on the design, such as functionality, cost, durability, reliability, safety, and other important and sometimes conflicting factors. In a complex device or system, this decision is often more difficult because of all the factors involved. Previously in this chapter, we discussed the importance of creative thinking when developing design ideas. However, choosing and developing the best solution frequently involves more **analytical thinking** as well.

Criteria Selection

The idea of "best" or optimal can mean many things. It is important to base this choice on realistic and well-defined criteria. Some solutions, although better in concept, may not be realistically tackled in the time permitted for a school project; others may require equipment or expertise that is not available to you; still others may require expensive materials or materials that are difficult to obtain. Although school-based projects may have more constraints, businesses must also consider many of the same important restrictions on their designs.

A good starting point is to review the original specifications set early in the design process. In engineering, design criteria will most likely be very specific and provide extensive guidelines for the direction of the project.

SUMMARY

Generating and developing ideas are central to the design process. Those who are creative imaginative, and good at developing their ideas are valuable assets to any company. Both beginners and professionals can use strategies to generate ideas and think laterally, as well as vertically.

The need to generate as many ideas as possible cannot be overemphasized. It must be understood that the greater the number of ideas generated, the greater the probability of finding the best possible solution to the problem. In technology and engineering, designers are challenged to develop a variety of thinking skills. De Bono's model for vertical and lateral thinking was used to show how new solutions can be found by using traditional and nontraditional thinking skills. By using techniques such as the dominant idea, analogies, brainstorming, synectics, and sketching and doodling, additional ideas may be found, and then analyzed. Most ideas need an incubation period to fully develop.

After a designer decides that sufficient ideas have been generated, some developmental work on each promising idea must begin to see if the idea still has merit when looked at more closely. The purpose of developmental work is to determine if an idea has a good chance of working. Usually, the structure, function, and aesthetic appearance for each idea are developed. The visual design elements include line, color, form/shape, space, texture, and value. In developing ideas, a designer uses a variety of materials and techniques, such as computer design programs, sketching and drawing, and commercial construction kits. The human user and the environment should be considered at this point, as these concerns often are the deciding factors for a successful design.

At this point, the designer will select an optimum solution based on a variety of factors, including cost, engineering, appearance, ergonomics, manufacturability, and ethical and environmental impacts, among others. The selection process should be based on established criteria that are agreed to early in the product development process. If the process of generating and developing ideas has been done well, then there will be confidence that the idea selected will be a successful solution to the identified problem.

BRING IT HOME--

OBSERVATION/ANALYSIS/SYNTHESIS

1. Make a list of at least 10 examples of vertical thinking in your everyday life.
2. Make a list of the steps in the 12-step design process and label which steps emphasize vertical thinking and which steps emphasize lateral thinking. Write a short reason for why you believe this is the case for each step.
3. Organize a three-minute brainstorming session with at least four other people on one of the following topics: (a) ways to make use of discarded plastic soft drink bottles; or (b) ideas for fund-raisers for a local Technology Student Association (TSA) chapter or other school organization. Submit your list of ideas to your teacher.
4. Research on the Internet and find at least five different examples of the use of *line* as a design element in a commercial product. Create a page in a word processing program and use the design element *alignment* to place these images. Develop a caption for each that briefly describes how line is used.
5. Research on the Internet and find at least five different examples of products that have symmetrical balance in the layout of controls and displays. Include these photos as page two of the project in question number 4.
6. Determine the aspect ratio for (a) the output of a digital camera when the photo is printed full size; (b) letter and tabloid-sized paper; and (c) the computer screens in a school computer lab.
7. Analyze a consumer product in terms of its aesthetics by taking several photographs of it and printing the photos about 4- by 6-inch on a letter-size page. Use the surrounding white space to annotate the use of design elements in the product, such as line, shape, form, value, color, and texture.

EXTRA MILE

ENGINEERING DESIGN ANALYSIS CHALLENGE

- Construction kits are useful for trying out ideas for mechanisms and simple machines to see if the ideas should be further developed. In the following challenges, first create a detailed sketch of how the mechanism will work before building the model. When completed, take a digital photo of the model if possible. Write a short, one-paragraph description of how well the planning sketch represents the actual model and what changes (if any) you made to the model to make it function as required. Include the sketch, photo, and description when you hand in your assignment. Using a construction kit (Lego, Fischertechnik, K'NEX, and so on):

1. Design and build a mechanism that can take rotary motion as an input and produce oscillating motion (back-and-forth motion in an arc) as an output. The input should be a crank that is continuously rotated by hand.
2. Design and build a mechanism that can take rotary motion as an input and produce a reciprocating motion (linear back-and-forth motion) as an output. The input should be a crank that is continuously rotated by hand. (Hint: research "rack-and-pinion" devices.)

CHAPTER 5
Drawing to Develop Design Ideas

Menu

 Before You Begin
Think about these questions as you study the concepts in this chapter:

1. Why is freehand sketching an important skill for designers, scientists, or engineers?

2. Why do engineers and other designers need to document their work?

3. What different abilities do the right and left hemispheres of the brain offer to artists, designers, or engineers?

4. What are the six "visual design elements" that can be integrated into an effective design?

5. How can you use color to better communicate design ideas?

6. What drawing techniques can be used to record and present ideas in engineering notebooks and design portfolios?

7. How are engineering notebooks and design portfolios developed?

INTRODUCTION

In Chapter 2, you were introduced to the *design process*, the central activity of all technological and engineering design. In Chapter 5, we look at some of the communication techniques designers use, especially at the beginning of the process, where ideas are just forming in the mind and need much further development.

Design is derived from the Latin word *designare*, "to mark out," which became the French word, *dessiner*, "to draw." Therefore, design is historically associated with drawing.

Drawing, or making graphic images by some means, is an integral part of being a professional designer or engineer (see Figure 5-1). Drawing can also be very useful to "everyday technologists"—homeowners, parents, or businesspeople solving practical problems.

Figure 5-1: A young artist showcases a digital sketch of a concept car during an international motor show in Kuala Lumpur, Malaysia.

THE MANY USES OF DRAWING

Drawing plays several roles in the design process (see Figure 5-2). These roles fall into three major categories: exploration, idea development, and documentation. For the purposes of this chapter, drawing includes **sketching** (rapid, **freehand** drawings) as well as more exacting depictions such as those required for technical work, possibly using instruments like compasses and straightedges. Sketching is shorthand for artists. It is used to get information down quickly, as in visual brainstorming. Detail drawing, for example, tends to be more careful and time-consuming.

sketching: Creating a rough drawing representing the main features of an object or scene, often as a preliminary study.

freehand: Done manually without the aid of instruments such as rulers.

Exploring the Visible World

Exploration is how we gather and make sense of information. The act of drawing requires us to look very hard at a subject, to really focus our attention on the details. Up to 90 percent of the information humans take in is visual, and yet rarely do we take a hard look at the world. Looking with the intention of drawing requires intense concentration. Drawing is an excellent investigative tool; it can help us understand how things work and how parts relate to one another.

Developing Ideas

We develop solutions to design problems by generating and manipulating ideas. Our first ideas for solving problems are usually rather rough. Sketching allows us to give form to ideas and to think more clearly about how to build on those ideas. Creating successive drawings allows us to capture each improvement. Drawing is one of the handiest tools at our disposal for putting down an idea and then changing it until we have something we like.

Exploration and idea development are the building blocks of creativity. The freehand sketching and drawing explained in this chapter provide a direct bridge from what's in our minds to the world of physical objects and products. It is the immediacy of hand drawing that allows us to capture a creative insight or a moment's vision for future use.

107

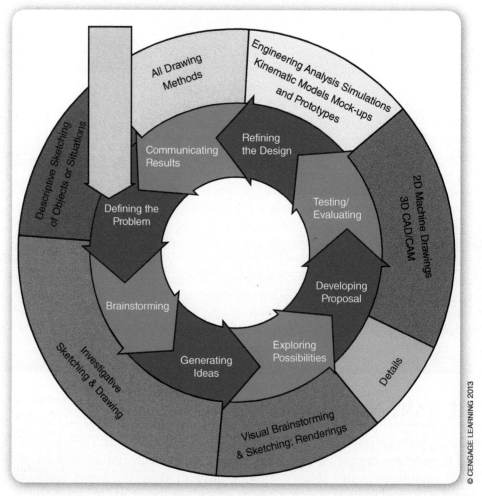

Figure 5-2: How drawing and modeling support the design process.

perspective:
A form of pictorial drawing in which vanishing points are used to provide the depth and distortion that is seen with the human eye; perspective drawings can be drawn using one, two, and three vanishing points.

Once an idea is on its way to being a product, we use technology to extend our drawing skills. Not long ago, draftspersons labored over drawing boards, hand-inking production drawings and blueprints, measuring precise dimensions, and adding notations. In Chapter 10, we will see that today those drawings are largely done by computer-aided design or computer-aided drafting (CAD) programs that allow designers to work in perspective using color, lighting effects, and realistic surface renderings.

Documenting the Process

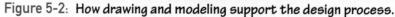

Documentation is the activity of collecting the evidence of the thinking and problem-solving process that has gone into a designed innovation. Figure 5-2 shows how drawing supports the design process. The outer circle suggests different drawing and modeling techniques that allow designers to see and develop their ideas. When you do documentation in a pre-engineering class, you are developing basic design skills and providing your teacher with a window into your mind.

Keeping a record of ideas and development work is an integral part of designing (see Figure 5-3). The term designer is used for a wide range of occupations. Engineers do technical designs for the products and systems we use every day, from our cell phones to the water we drink, the games we play, and the vehicles we use to get to school, work, or vacation. Architects and civil engineers are designers of structures; graphic artists works with elements on a page, such as typeface, page

documentation:
(1) The documents required for something, or that give evidence or proof of something; (2) drawings or printed information that contains instructions for assembling, installing, operating, and servicing.

layout, and photos to design print materials; industrial designers work with physical products; and fashion designers work with clothes and accessories. These and many other occupations involve design, and a record of work is a necessary part of the job.

In the world of business and industry, the **engineer's notebook** is a careful record of ideas, calculations, thoughts, and plans for a particular project (see Figure 5-4). It is not unusual for an engineer to be pulled from the middle of a project, only to pick it up again six months or several years later. The documentation within the engineer's notebook saves countless hours of rediscovery when the engineer returns to a task.

In a similar manner, the records the engineer keeps may well be the legal basis for awarding a **patent** when several similar patent applications are submitted. Questions such as "Which applicant developed a critical idea or design first?" can be answered with documentation evidence. Contrary to popular belief, ideas do not generally blossom forth fully formed; they evolve and grow. The documentation of the development of an idea over time is generally a collection of original sketches, notes, calculations, and other evidence.

Engineers must preserve the evidence of that evolution of thought, not just for legal purposes, but perhaps more importantly, so that they can "backtrack" when the inevitable "false trail" is taken. Being sure about where you have been makes future progress possible, because we all are forgetful and designing is a very involved process. Good documentation helps designers keep track of important facts and information, ideas, and the details of design solutions.

Design Portfolio. People in creative fields and students preparing for work in a creative field should keep records of their work. Although the engineer's notebook serves an important technical and legal role, a **portfolio** consisting of a collection of materials that document the thinking and physical work of an individual may also be created. Architects show a portfolio to prospective clients so that the clients can evaluate their previous projects. If a prospective client sees evidence of quality work and style compatible with the project's requirements, then the deal may be "closed." Throughout a project, an architect will use drawings or photographs

Figure 5-3: **An architect shares concept drawings with a client.**

engineer's notebook:
Also referred to as an engineer's logbook, a design notebook, or designer's notebook. Used as (1) a record of design ideas generated in the course of an engineer's employment that others may not claim as their own, and (2) an archival record of new ideas and engineering research achievements which can provide proof of an idea for patenting purposes.

portfolio:
A set of pieces of creative work intended to demonstrate a person's ability or to document the development of an idea over time.

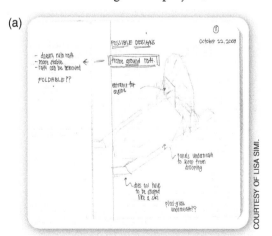
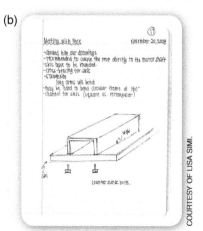

Figure 5-4: **Examples of drawings from a bound engineering notebook, a) using sketching to show a possible solution, b) using a sketch to communicate an idea for a meeting with a machinest.**

Figure 5-5: This artist's online portfolio allows visitors to navigate through the portfolio's pages.

to communicate the progress of the job to the client. The portfolio plays an important role in the designer-client relationship.

Like architects, other design professionals know the importance of portfolios (see Figure 5-5). Designers can show their portfolio to prospective clients interested in commissioning work, or can use it to apply for grants and exhibits. For many years, designers have photographed their work and developed slide portfolios. More and more designers are now documenting their works digitally and posting their portfolios on websites.

Engineer's Notebook. Engineers or other designers may develop a kind of portfolio as well, describing projects and original work through photographs, sketches, illustrations, and publications. These are used for job promotion and job search, and may even be used to obtain funding for a project or to open one's own business.

The patent system in the United States rewards the first person who invents a new product. An engineer's note book or journal helps you prove that you were first (see Figure 5-6). An engineer's

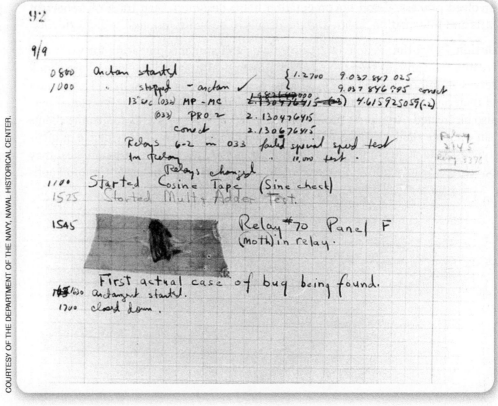

Figure 5-6: Page describing the first computer "bug" from an inventor's log of Rear Admiral Grace Hopper, inventor of COBOL computer language.

notebook is officially a bound notebook on which each day's activities are recorded, with the date and any drawings that show the evolution of the problem solution. Often major benchmarks are signed and witnessed to prove that the claims for the design and the dates are authentic. An engineer may not be trying to protect new methods for patenting purposes, but may keep a log to record decisions made and benchmarks in the development of a project to replicate the results and evaluate the process and product.

Student Design Portfolio. In education, the portfolio is taking on new importance. Many critics of education are saying that paper and pencil tests are of only limited value when it comes to assessing what students have learned. Because tests only assess a student's knowledge at one point in time, they provide little evidence of what someone actually knows; we have all learned something for a test and then promptly forgotten it when the test was over. Within the portfolio, however, is evidence of what an individual has actually done (applied) over a period of time. This evidence does not simply appear the last week of school; it is accumulated over weeks, months, or even years.

The portfolio is an especially powerful tool in technology and pre-engineering courses (see Figure 5-7). Because much of design and development work is graphic in nature and because the portfolio lends itself to graphic evidence, it becomes a central part of the design process on two levels:

1. Each problem tackled is tracked and documented with a portfolio; and
2. The individual portfolio of each project is incorporated into a cumulative portfolio, which can act as evidence for application to a college or professional training program, for a job, or an award, honor, or competition.

The documentation of your design work is an integral part of both the study and understanding of design. By means of the portfolio, you will gather evidence of your ideas and your creative and developmental work.

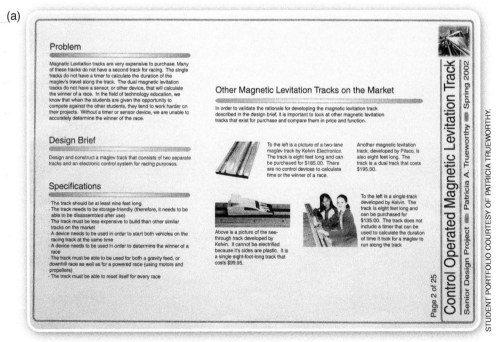

Figure 5-7: *Pages of a student design portfolio.*

112 Part I: The Engineering Design Process

Figure 5-7: (Continued)

(b)

(c)

(d)

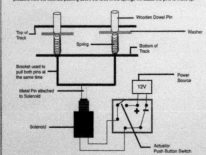
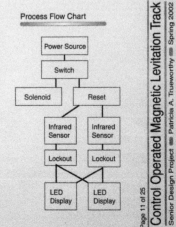

Figure 5-7: (Continued)

(e)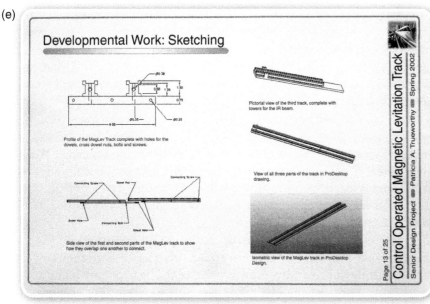

(f)

(g)

COMMUNICATING THROUGH DRAWING

Drawing provides a language that allows individuals to describe the visible world and to visualize, refine, resolve, and communicate ideas. It is a skill that can be learned. Although some people have a natural talent for sketching and drawing, everyone can pick up the techniques and develop the basic skills.

You do not have to be an artist to draw and sketch. Although a few people find it very easy to learn, most of us must practice the techniques and "tricks" to be effective visual communicators. Quick, preliminary drawings are called sketches. Typically, they are used to capture the essence of an object or scene for later reference, or to roughly describe an idea that will need further development. Getting impressions and ideas down on paper as quickly and effectively as possible is a major goal of sketching.

Before we discuss the techniques of sketching, it is important that you become aware that most people must overcome barriers before they can learn to sketch and draw successfully. One barrier is the fear of not being able to sketch and draw, and looking foolish. Although it is rare that someone will admit they have this fear, just ask someone to sketch an idea or even a map. The first words out of that person's mouth are usually excuses about why their drawings will look awful. Overcoming this barrier takes some work, but the results can be very satisfying. After all, a picture has been said to be "worth a thousand words."

Whole-Brain Drawing

Nearly everyone has the manual dexterity to draw, but by adolescence, many people do not "remember" how to see. Drawing requires that we shift from thinking in a verbal and sequential way to a more intuitive approach. Scientists believe that verbal thinking happens on the left side of the brain, and visual thinking happens on the right side. Everyone uses this visual thinking quite naturally as a young child, but because most of our school experience by seventh or eighth grade has focused on reading and math, we forget how to see details and relationships that we need for drawing.

Exercises and activities can help people purposely activate the right side of the brain—the side that artists use when drawing, painting, and sculpting. One of the main strategies is to trick the overdeveloped left brain into inaction by challenging it to do things it's not good at. The ultimate goal is to be able to use your whole brain, with both halves working together, each doing what it does best.

Warm-Up Exercises

The following exercises can be done on plain white drawing paper or even copier/printer paper. Use a medium-dark pencil, like a 2B (that is, #2), or B or HB, very sharp.

Have you ever seen pictures that can be interpreted as two different things, depending on how you look at them? One example is the "faces and vase" puzzle (see Figure 5-8). At any given moment, you can see either the faces or the vase, but not both. Psychologists use this example to illustrate that the brain can flip back and forth between the two images, but it can only make sense of one set of information at a time.

Figure 5-8: *Faces and vase puzzle.*

Exercise 1: Drawing Mirror Images. Let's analyze the "faces and vase." Whether you're seeing two faces or a vase, all the same shapes and lines are present. Notice that the contour (outline) of the profile is also the outline of the vase. The object shape (positive shape) in one drawing becomes the shape of the background space (negative space) in the other.

It just so happens that drawing a puzzle like this can help us mobilize our right brain. Let us draw our own version of the "faces and vase."

Step 1. Hold a piece of 8 ½ by 11 inch paper vertically. At the top of the paper, a few inches from the right edge, begin drawing a face of a witch in profile. This is not any specific witch, just a vision you have in your mind.

Step 2. Start at the forehead and use a continuous line to draw the profile large enough so that you end with the neck going off the bottom of the page, just below the chin (see Figure 5-9). Note: The entire profile must stay within the right half of the page. You can talk yourself through this drawing by saying, "First I draw the forehead, then the eyebrow ridge, then I go down the nose and tuck under; next I make the upper lip and some ugly teeth, then the lower lip and down around the chin, in to the neck and off the page." This should not be difficult to do. It is a very "left brain" activity, because you can name all the parts and use the symbols you have in your mind for "witch's chin, nose, forehead, lips," and so on.

Step 3. Now, think through the right-brain part. Along the left side of the paper, draw a mirror image of the face you just drew. It should look as though identical witches are staring at each other. This time you cannot just call on your mental symbols. You have to take an existing outline and flip it right to left (see Figure 5-10). It is no good talking to yourself the same way you did in the right profile drawing. Producing a mirror image is all about seeing relationships and visually measuring distances.

When you have finished, take a look at your picture. Which witch was easier to draw? Did they come out looking pretty much the same? Were you aware of having to use some different strategies to draw the second profile than you needed for the first? (No fair folding the paper and tracing. That is an example of your sneaky left brain trying to take over.)

By the way, take a look at the shape between the twin witches, it should look like a sort of vase or urn. (To use this drawing as an example of an optical puzzle—the standard faces and vases conundrum—you may want to modify some of the contours, which will, of course, change how your witches look.)

Exercise 2: Turn It Upside Down. The faces and vase exercise gives you a little glimpse of trying to call upon your right brain for some drawing help. If you found it easy, you are probably already in touch with your right brain. If you found it rather difficult, then you probably have a very well-developed left brain, and it wants to stay in charge.

Step 1. Next we will look at a line drawing of a woman's face (see Figure 5-11). On a piece of paper the same size as the drawing, try copying the picture (see Figure 5-12). Give yourself about fifteen minutes to complete the drawing. In case you find this tedious, do not worry, that is not surprising. When you are finished, set the drawing aside.

right brain:
The right hemisphere of the cerebellum where it is believed simultaneous, holistic, spatial, and relational information processing are favored.

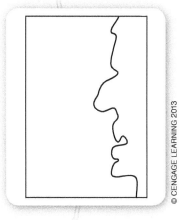

Figure 5-9: Witch's profile.

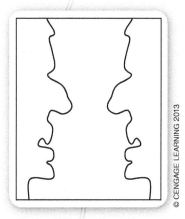

Figure 5-10: Staring witches.

left brain:
The left hemisphere of the cerebellum, where it is believed linear, verbal, analytical, and logical information processing are favored.

Figure 5-11: **Face right side up.**
REPRINTED WITH PERMISSION FROM *DRAWING: A CONTEMPORARY APPROACH*, SIXTH EDITION, BY CLAUDIA BETTI AND TEEL SALE. COPYRIGHT © 2008 WADSWORTH.

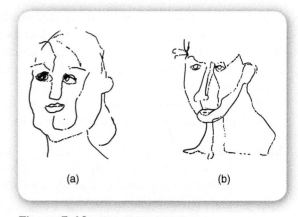

Figure 5-12: **Student drawings.**
REPRINTED WITH PERMISSION FROM *DRAWING: A CONTEMPORARY APPROACH*, SIXTH EDITION, BY BETTI AND SALE. COPYRIGHT © 2008 WADSWORTH.

Step 2. Now turn the picture you were copying upside down (see Figure 5-13). Once again use a piece of paper the same size as the drawing, but now copy the upside-down drawing (see Figure 5-14). Your drawing will be upside-down too. Allow yourself about 15 minutes for this drawing. (You may want to set a timer or ask someone to let you know when the time is up, because the right brain is not very interested in time and is not likely to be keeping track.)

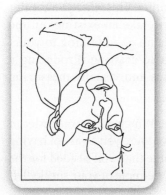

Figure 5-13: **Face upside down.**
REPRINTED WITH PERMISSION FROM *DRAWING: A CONTEMPORARY APPROACH*, SIXTH EDITION, BY CLAUDIA BETTI AND TEEL SALE. COPYRIGHT © 2008 WADSWORTH.

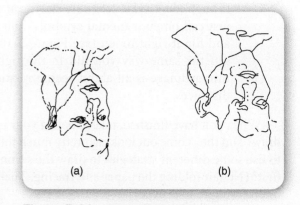

Figure 5-14: **Student drawings.**
REPRINTED WITH PERMISSION FROM *DRAWING: A CONTEMPORARY APPROACH*, SIXTH EDITION, BY BETTI AND SALE. COPYRIGHT © 2008 WADSWORTH.

Once you get started on this activity, you will find that *what you are drawing* (eye, contour of the cheek, ear) becomes irrelevant. It is all a matter of lines and spaces and their relationships—higher, lower, distance apart. This is the kind of silent activity that the right brain really likes.

Step 3. When you have finished this drawing, turn it right side up. Now compare it to the copy you made from the right-side-up picture. Which one looks better? In most cases, the second drawing, done upside down, will be far more pleasing. It will look much less self-conscious, more relaxed. Just look at the kind of line used in each drawing. The upside-down drawing probably uses a much smoother line. The proportions may be off a bit, but seeing size relationships improves with practice, once you have let the right brain know that you welcome its help.

If you try a few more upside-down drawings, you will get used to seeing the world in terms of visual relationship. You will be able to call on that ability whenever you need it. In fact, you already use your right brain for many of your more accomplished activities—like playing a sport, dancing, riding a bike, or driving a car. Anywhere you have to judge distances and spatial relationships quickly and continuously, your right brain is quietly taking the lead. And if you are good at those activities, you probably find them exhilarating and relaxing at the same time. That is how artists feel when they are drawing. If we learn to draw using the right brain, we develop a very useful communication skill that is also relaxing and stimulating. It is a win-win situation.

Exercise 3: Blind Contour Drawing. If you are up to the challenge, try a third activity that takes you further into the world of the right brain. Some artists describe this kind of activity as a combination of seeing, feeling, and drawing. It is an exercise to help you learn to follow the shape of an object; it is contour drawing.

 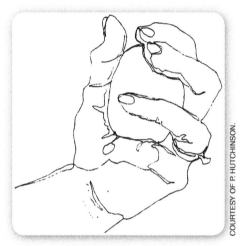

Figure 5-15: **Hand and egg (photo) and continuous line drawing.**

Step 1. If you are right-handed, place an 8 1/2 by 11 inch sheet of paper on a table to your right. Tape it in place so you don't have to worry about it moving around. Sit sideways at the table so that your right hand can draw on the paper, but face ahead or slightly to the left so that you are not easily able to see your paper. Hold a small object (a bottle, egg, spoon) in your left hand (see Figure 5-15). This is what you will be looking at as you draw. If you are left handed, you now get more brain exercise. Reverse the directions in Step 1 so that the paper is on your left and the object in your right hand.

Step 2. Position your pencil at a point toward the bottom of your paper. Focus your eyes at a point on your wrist just below your hand.

Step 3. With your eyes, travel along the contour (outline) of your hand, following the line as it goes around each finger, perhaps moving in a bit at a fold where the finger bends, back out around the outline of the finger, along the parts of the object you can see, and so on until you get to the other side of the wrist.

Step 4. With your pencil, follow the same path your eye is taking around the hand. Do this without looking at the paper. This is very difficult at first, because it is hard to judge distances without feedback from your eyes. Force yourself to try, and if you feel yourself going off the page, try to retrace your pencil line until you get back on the paper, and then continue. This will probably

Figure 5-16: Student contour drawing.

Figure 5-17: Advanced student contour drawing.

Figure 5-18: Seeing and drawing negative shapes around the scissors.

make you want to draw smaller, which is fine. You should finish the drawing at the same time you finish the visual journey around your hand.

Step 5. Take a look. It is very unlikely that you will have produced a well-proportioned portrait of your hand, but you probably have a nicely flowing line around shapes that "feel" like the fingers and object you were drawing (see Figure 5-16).

Contour drawing is not something you do for immediate gratification. However, if you are willing to try a few more times, you will be surprised how much better you become at:

▶ Seeing details,

▶ Judging distances, and

▶ Controlling your drawing hand and the marks you make on your paper.

Try drawing your foot (barefoot, and then in a sneaker), a paper bag, and a sandwich on a plate. Another good way to practice is to copy outlines from maps. Try your state or the map of the United States or North America. (Maybe you will discover a talent for cartography!)

These first three exercises are primarily meant to get you in touch with your right brain. If you thought you would be drawing like Rembrandt at this point, your expectations were much too high. You should, however, be looking at things and noticing details you may have previously missed. You should also have seen that you *can* occasionally draw lines that are confident and flowing, if only your left brain is not trying to criticize (see Figure 5-17).

Exercise 4: Positive and Negative Shapes. Most of the time, we think of the world as full of objects, surrounded by nothing—empty space. For purposes of drawing, the space that surrounds objects needs to be visible to us. It is a very useful tool for drawing objects effectively and in proportion. One way to make negative shapes more visible is to put a frame around the objects you want to draw, because now the shapes of space have an outer edge. Drawing negative shapes is a right brain activity, because they are often hard to describe in words.

Step 1. Fold a piece of light cardboard in half and cut a square or rectangle, about 8 by 8 inches or 6 by 8 inches out of the middle. Place this frame on the glass table of an overhead projector and project the square or rectangle of light onto a light-colored piece of paper taped to the wall. Position the projector so that the projected shape of the frame is contained on the paper. Trace around the frame shape on the paper.

Step 2. Open a pair of scissors and place them on the projector table, within the frame. Move them around until they overlap the edge a bit on at least two sides of the frame. (If they are big enough, they may run off all four sides, which is fine.)

Step 3. Look at the shadow image projected. You see the shape of the scissors, but surrounding it are a number of shapes of light, the negative shapes, or shapes of negative space. Now trace around the negative shapes, and then turn off your projector. Look at the shapes on your paper. Lightly number all the negative shapes and cut them out, and then discard the shape of the scissors. Now, on a sheet of contrasting colored paper, arrange the negative shapes and use a glue stick to glue them down. If you can re-create the way they were arranged on the original drawing, you will have produced the image of the scissors (the positive shape) in the background color (see Figure 5-18).

Using negative shapes to create a positive image usually produces quite a striking pattern. If you were to alter any of the negative shapes, it would change the positive image. This exercise makes you aware of both kinds of shapes. To apply what you have learned to your own drawing, make yourself a small viewing window by cutting out a little frame with a 1 by 1 inch or 1 1/2 by 1 inch opening. Move this back and forth in front of your eye to frame things you want to draw—an object or a scene. The frame helps you eliminate a lot of the confusing information surrounding your subject. If you can quickly sketch the positive and negative shapes you see in the frame, you can capture your image very effectively.

After some practice, you will be aware of the negative shapes around things without having to use a frame. If you are drawing a subject and having trouble with the proportions, you can always switch to right brain mode and concentrate on the negative shapes. If one of them looks wrong, fixing it may make your positive shape look better too.

DRAWING BASICS

When you first set out to draw an object, you probably pick up a pencil or pen and begin to draw a line. And yet, nothing in the real world is surrounded by a line. An outline is a useful way to block out the space an object occupies, but to really see the world, you also need to be aware of shading, color, texture, types of shapes and forms, and how objects relate to other things around them in space.

These aspects of the physical world become the "visual tools" of drawing, and later the elements of design:

▶ Line
▶ Shape and form
▶ Value (shading)
▶ Color
▶ Texture
▶ Space

element of design:
A basic visual component or building block of designed objects (for example, line, shape and form, value, color, texture, or space).

The more time you spend looking and drawing, the more useful these elements will seem.

Line

A line defined mathematically has only direction and length, but no width. In sketching, of course, the width of a line is not only real but also important to the appearance of the sketch. To achieve a good line, an HB pencil is fine for most sketching, because it can be used to make faint lines or bold lines, according to how hard you press it against the paper (see Figure 5-19). It will become blunt quickly, so it is important to maintain a sharp point. Experiment with different qualities of line: straight, curved, sharp, fuzzy, uniformly thick, or varied.

Figure 5-19: **Different kinds of lines.**

Shape/Form

Lines are used to enclose a space. The **shape** of an object is the **two-dimensional** space enclosed within the lines. Shapes can be natural or **organic** (things in nature, like clouds, leaves, and rocks), geometric (squares, triangles, circles, or combinations of these mathematically generated shapes), or some combination of the two. Shapes made with only straight lines are called **rectilinear shapes** (see Figure 5-20).

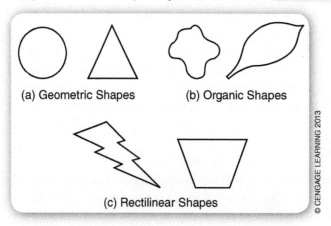

Figure 5-20: **Types of shapes.**

When a shape is given a third dimension, it becomes a **form**. In sketching, this can be done by using lines, shading, and/or color. The geometric forms of the cube, cylinder, cone, and sphere are developed from the square, circle, and triangle (see Figure 5-21). They are the basis for many of the manufactured objects seen every day (see Figure 5-22).

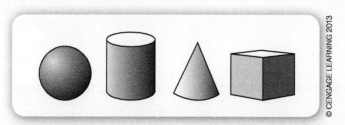

Figure 5-21: **Basic forms: sphere, cylinder, cone, cube.**

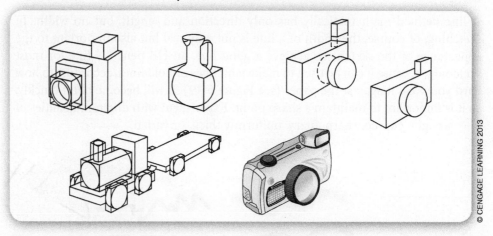

Figure 5-22: **Each of these objects is made from a combination of geometric forms. Artists block in the shapes, then remove the overlaps and sketch lines, and finally add shading, color, and background to make the objects look real.**

Many other geometric forms are also common, such as the tetrahedron and various **prismatic** shapes (see Figure 5-23). Look for more examples of these forms in the designed world.

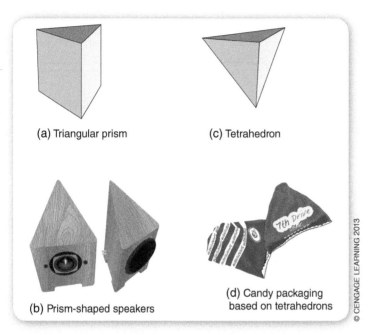

Figure 5-23: (a) Triangular prism. (b) Prism-shaped speakers. (c) Tetrahedron. (d) Candy packaging based on tetrahedrons.

In addition to geometric forms, natural forms are also used as a basis for creating products and structures, such as the buildings pictured in Figure 5-24.

Figure 5-24: (a) The Xanadu House in Florida represents a possible direction for housing in the future. Demolished now, it was made from a combination of organic forms. (b) The Guggenheim Museum in Bilbao, Spain, is made from crumpled and twisted geometric forms.

Your Turn
Value Scale

Step 1. Draw a vertical rectangle, 2 inches wide × 7 inches long. Mark off seven bars, 1 inch high and 2 inches wide. Leave the top bar the white of the page.

Step 2. Shade the bottom bar as black as you can with your pencil, making it as smooth as possible, so you cannot see your pencil strokes.

Step 3. Shade in the five bars between white and black in equal steps from very light to very dark gray. Try to make each bar just one shade, very even and smooth. A light pressure on the pencil will give you a light tone; a heavier pressure can give you a very dark tone. These principles apply to both regular and colored pencils. As you move the pencil back and forth, you can blend in the tone, going back over places you have missed. Getting a smooth, even tone takes practice and patience (see Figure 5-25). It is better to use a lighter pressure and go over an area several times than to use a heavier pressure and end up with a streaked, uneven effect. Caution: It is best to start at the top of the scale (lightest gray) and work your way down the scale. Be conservative—it is always easier to darken than to lighten.

Step 4. Step back and see if your eye can travel up and down the seven-step scale smoothly—no step should be greater than any other. Go back and adjust as needed, until you are satisfied.

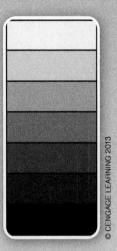

Figure 5-25:
Value scale.

Value

Although the curved line at the base of the cone and the three visible sides of the cube in Figure 5-21 are the first clues that these are **three-dimensional** forms, it is the shading that makes the forms convincingly realistic.

The shading is the result of light falling on the surfaces of the objects. The greater the range of shades, or **values**, on your drawing, the more three-dimensional the forms will look. When you draw, think of the white of the paper as the lightest value available to you. The blackest black your pencil can make is your darkest value. Between these are many shades of gray. Shading successfully requires two things: controlling your pencil and seeing the widest possible range of values.

Light Source and Shading. When we look at an object, we see a range of tone, according to how much light strikes a particular surface. Because light most often comes from one main source, such as the sun or a lamp in the room, the surfaces of an object toward the light source reflect the most light. Surfaces at an angle to, or away from, the light source do not reflect as much light, so they appear darker. Even when there are several sources of light, one usually dominates the object we are viewing because of the intensity of that light or its proximity to the object we are looking at.

The cube in Figure 5-26a is drawn by outlining the edges where two surfaces meet. This cube looks rather dull and lifeless. In Figure 5-26b, however, the

Chapter 5: Drawing to Develop Design Ideas   123

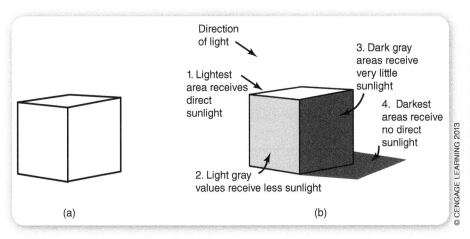

Figure 5-26: The cube shown in part (a) receives depth and greater realism from addition of shading in part (b).

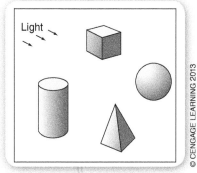

Figure 5-27: The direction of the light coming from the source and the distance from the source to the object affect how light or dark a surface will appear.

same cube has been shaded with pencil to represent the amount of light reflecting on the three surfaces we can see. A cast shadow has also been added to define the surface the cube is sitting on. This second cube has depth and appears more realistic. This is because it more closely resembles the way in which we see actual objects.

For sketching purposes, it is handy to think of the light source as coming over your left shoulder (see Figure 5-27). This will help you visualize which surfaces are lighter and which are darker.

Shade your forms along their long axis, holding the pencil at a low angle so that most of the length of the pencil lead contacts the paper (see Figure 5-28). This will give you a granular type of shading. If you choose, the tone can be made smoother by smudging with a tissue along the same axis.

Highlights. Highlights and reflections are visible when light strikes bright and shiny objects. In drawing, the highlights and reflections are often exaggerated to give the object more impact or a look of realism. A shiny, flat, horizontal surface has vertical highlights (see Figure 5-29a). These are easy to add by using pencil shading, markers, or other techniques. Curved or rounded shapes, such as spheres or cylinders, tend to distort the light hitting them (see Figure 5-29b). A cylinder tends to stretch the highlight along its length, and even the inside of the cylinder, being curved, reflects light. A sphere will reflect light in a circular pattern. Often people use a window on a sphere as a highlight. Notice that the square window bulges in all directions as the sphere's surface turns the window into a kind of circle (see Figure 5-29c).

Figure 5-28: Using a pencil to get a blended shade on a cylinder.

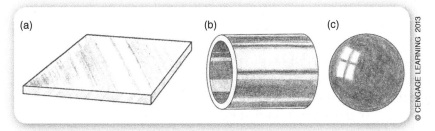

Figure 5-29: (a) Flat, reflective surface; (b) highlights on a shiny cylinder; and (c) a spot reflection on a sphere.

Color

Color is an important part of life and affects us on a number of levels. It has found its way into our everyday language to help us express our emotions. "Feeling blue," "green with envy," and "seeing red" are only a few of many examples of how color has become part of our language. Color has profound physical and psychological effects on our moods and feelings of well-being. Certain "cool" colors are used in hot environments and "warm" colors in cold environments to help people cope with the temperature extremes. Calming colors are used in hospitals, schools, and even in prisons, to help minimize agitation and disruption. Color is a part of tradition, religion, and all aspects of our everyday lives.

We see color because of the wavelength of the light that reflects from an object to our eyes. Photoreceptors in the retina of our eyes, called rods and cones, are sensitive to the value and intensity of colors. Generally speaking, the rods distinguish the level of lightness and darkness of colors, while the cones perceive their brightness or dullness. Where the level of light is dim, as at twilight, we see little color. When light is bright, colors also appear intense.

Hue, Chroma, and Value. Hue refers to a specific color's name—for example, red or mauve or viridian. Hue refers to the actual wavelength of the light on the spectrum. Reds have the longest wavelength (lowest frequency 700 nm). The other end of the spectrum contains the blue hues, which have short wavelengths (highest frequency 400 nm) as shown in Figure 5-30a.

Value sometimes refers to the lightness and darkness of a hue. A light value of a hue is called a tint, while a dark value is referred to as a shade. Value describes the amount of light the color actually reflects. The value of a color can be altered by adding either white or black to the hue. Tints and shades are shown in Figure 5-30b.

Brightness or intensity of a hue is referred to as chroma. A pure hue or color is the most intense, but the color may be neutralized, or dulled, by mixing it with a color on the opposite side of the color wheel, as shown in Figure 5-30c.

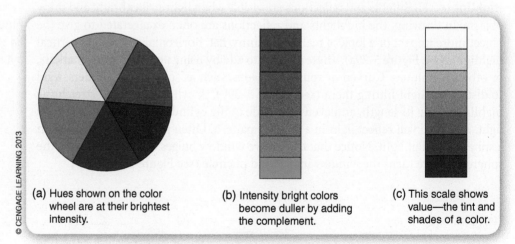

(a) Hues shown on the color wheel are at their brightest intensity.

(b) Intensity bright colors become duller by adding the complement.

(c) This scale shows value—the tint and shades of a color.

Figure 5-30: (a) Hues shown on the color wheel are at their brightest intensity. (b) Intensity bright colors become duller by adding the complement. (c) This scale shows value—the tint and shades of a color.

Primary Colors. When painters, printers, and video engineers talk about primary colors, they are discussing the basic colors from which all other colors are made. But because each of these individuals works with color media and light differently, the primary colors they refer to are different. Color theory is presented in more detail in Chapter 18 (Graphics and Presentation). Color can be added to drawings using pencils and paints that are made with pigmented clays, resins, and other substances that absorb light (see Figure 5-31).

Secondary and Tertiary Colors. When two primary colors are mixed, a secondary color results. Mixing red and blue yields violet; red and yellow, orange; and yellow and blue, green. When a primary color and a secondary color are mixed, the result is called a tertiary color (a third color). All the possibilities so far add up to twelve colors, and these are often illustrated in what is called a color wheel.

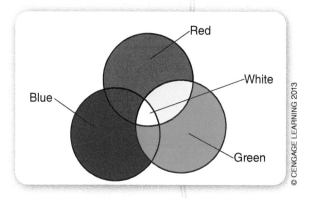

Figure 5-31: Mixing colors by adding pigments together subtracts light, making each new color darker.

The color wheel (see Figure 5-32) is a handy tool for choosing colors. The color wheel contains the three primary colors, the three secondary colors, and the six tertiary colors.

Most artists use several different hues of the primaries if they want to mix a full range of secondary and tertiary colors. It is difficult to mix both purple and orange from the same red, so artists often have a "warm" red with which to mix oranges and a "cool" red with which to mix purples. *Warm* means that the red leans toward yellow, while *cool* suggests that the red already has a purple (or bluish) cast. Crimson, scarlet, and carmine are all names for different hues of red (see Figure 5-33).

Colors close to one another on the color wheel, such as blue, blue-green, and green, are called analogous colors (see Figure 5-34). They are related because they share components. Interior and fashion designers often use analogous colors to create harmonious color schemes.

Figure 5-32: Color wheel.

Figure 5-33: Pure, warm, and cool reds.

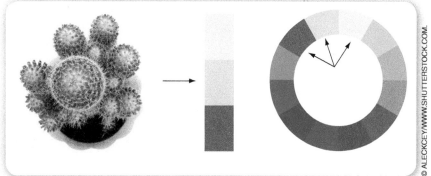

Figure 5-34: Analogous color scheme in nature.

Colors from opposite sides of the wheel are complementary colors; because they contrast with each other, they can be used to create emphasis in a color scheme. It is important to understand how this contrast works if you want to create visual effects in your design work. For instance, the use of green shapes on a red background causes a surface to vibrate, visually. This may work well for the wallpaper of a nightclub, but because red and green are close to the same value, green letters on a red package will be hard to read from a distance. Contrast of value is more important than contrast of color for distance vision, so if you use dark green letters against a white background, you will be able to read it from some distance. If you mix a bit of red into the white, creating just a slight pinkish tinge, the green will appear brighter by contrast to the complementary color. Even the subtle use of color can be very important as you illustrate your design ideas (see Figure 5-35).

Figure 5-35: Green lettering against a red background and a slightly reddish-white background.

Color Science and Technology

Renaissance artists theorized about color from the fifteenth century onward. Isaac Newton attempted to describe the nature of color from a scientific standpoint in his 1704 treatise, *Opticks*. Our understanding of color continues to expand, thanks to computer artists, optical scientists, and industrial chemists.

From an optical perspective, we know that our eyes are sensitive to light that lies in a very small region of the electromagnetic spectrum labeled "visible light." This visible light corresponds to a wavelength range of 400 to 700 nanometers (nm) and a color range of violet through red. The human eye is not capable of "seeing" radiation with wavelengths outside the visible spectrum. The visible colors from shortest to longest wavelength are violet, blue, green, yellow, orange, and red. Ultraviolet radiation has a shorter wavelength than the visible violet light. Infrared radiation has a longer wavelength than visible red light. The white light is a mixture of the colors of the visible spectrum. Black is a total absence of light.

Our ability to see in color is a marvelous adaptive and survival tool. Technology can help us to extend our physical ability. Thanks to infrared film and goggles, for example, we can see images beyond the visible spectrum.

Texture

The texture of a material is an important feature. In products, it is useful for both functional and aesthetic purposes. It can provide a nonslip "grip" on the handle of a tool, a floor surface, or a skateboard. It can also add an interesting visual element to a product. Texture can be the result of the nature of the material itself, or it can be a result of a production process, such as molding (see Figure 5-36).

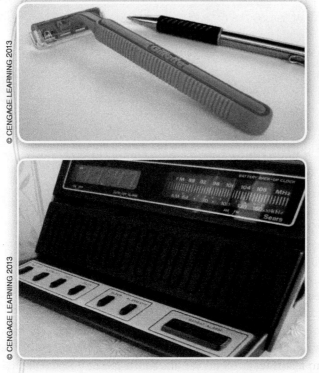

Figure 5-36: Many objects use molded lines and other features to achieve texture. Some textures are functional, others purely decorative.

Sketching the texture of a material is not difficult. Architects use a number of graphic standards that represent materials like concrete, earth, and foam insulation (see Figure 5-37). These and other techniques can add a great deal of impact to a drawing and can give viewers a better idea of what the object you are sketching actually looks like.

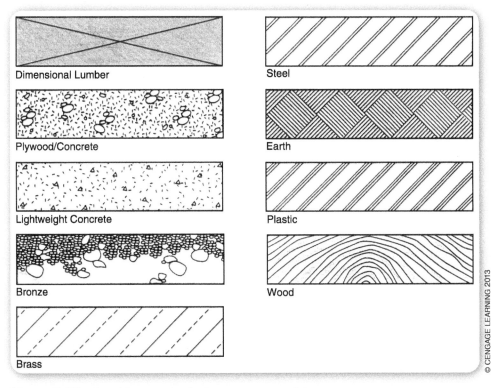

Figure 5-37: Graphic standards for some materials.

One of the easiest materials to sketch is wood. With a little practice, you can even sketch different kinds of wood, such as pine, mahogany, and oak. The different grain characteristics of these woods make it easy. Other textures are also easy to achieve (see Figure 5-38).

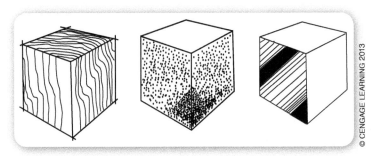

Figure 5-38: Simple sketching techniques to achieve the appearance of texture.

Artists have a number of tricks for simulating texture (see Figure 5-39). One is to make rubbings over an actual texture, such as sandpaper or embossed metal. Some media, such as pastels (artists' chalks) or colored pencils are particulate and can be used on rough-surfaced paper to create interesting textures. Shapes can then be cut from this paper and attached to a surface to **simulate** fabric, concrete, or stone. These simulated textures, as well as real textures, can be placed on a scanner and printed, and then incorporated into a drawing.

Figure 5-39: **Examples of simulated textures.**

Space

All of the objects we see exist in space. We may isolate them to work out the details, but when we explore a scene visually, or when we try to show someone how a product we have designed will look in use, we need to understand how to show space. At least five spatial cues help make sense of a scene:

- ▶ High and low position,
- ▶ Large and small relationships,
- ▶ Overlapping,
- ▶ Lines converging as they move away from us, and
- ▶ Atmospheric haze that makes close things sharper than faraway things.

These cues allow us to navigate in the world without bumping into things. They also help us make effective drawings.

SKETCHING AND DRAWING TECHNIQUES

When someone needs to represent an object in space, one of the following perspective drawings techniques can be used. **Perspective drawing** techniques are used to represent three-dimensional objects in a two-dimensional space. All perspective drawing techniques use a **vanishing point** to define where lines of an object converge in the distance. **Isometric drawings** represent a three-dimensional object in two-dimensions but do not use a vanishing point. For that reason, this drawing technique does not show objects as we actually see them but the technique can be used to show objects quickly. In addition to these perspective techniques, objects can be represented using **technical drawing** techniques where the object is represented with a high level of precision and accuracy and containing all the information needed to produce the object. Technical drawing will be explained in greater detail in Chapter 8.

Many tools are available for freehand sketching and drawing. Pencils, pens, and markers are among the handiest and most versatile for freehand sketching. When more precision is desired, straight edges or rules, t-squares and triangles can be used with a drawing board or drawing machine. Grid papers are also available or drawing software can be used. Recommended sketching pencils range from soft (6B) to medium (2B) to hard (2H). Markers come in very fine to very broad points: Some are water-based, others are spirit-based (see Figure 5-40). The texture of the paper you use affects the control you achieve in your drawing. If you need to capture some small details or make notes, use a smoother paper. On a very hard paper surface, softer pencils will smear.

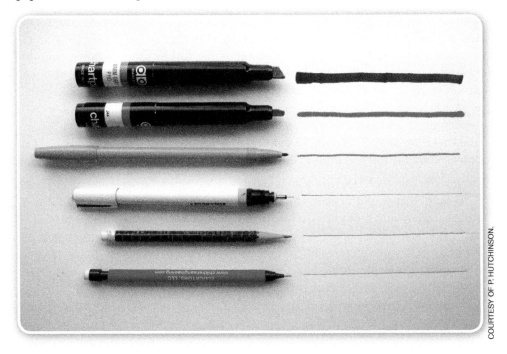

Figure 5-40: **Different drawing implements provide different line qualities.**

Perspective Drawing

In the world of three-dimensional space, we take into account that things seem smaller the farther away they are. A car in the distance appears very small; as it moves closer, it appears much bigger, even though it is the same car. We often do not take notice of this phenomenon, because we take it for granted.

Figure 5-41: Follow the roof, windows, ends of column, and fence posts to find the vanishing point.

The same principle applies to a long building (see Figure 5-41). Standing near one end, we see the full height of the near wall. The other end of the building, in the distance, seems smaller, and the roof connects the two by slanting down from the near wall to the far wall. At the same time, the ground at the bottom of the near wall almost seems to rise to meet the bottom of the far wall.

A photo of the building, taken from where we are standing, shows what is happening. If we draw a line along the roofline and one along the ground line, we find that they meet. This point of convergence is called the **vanishing point**. Lines drawn along the top and bottom edges of the windows converge at the same point.

Linear Perspective. During the Renaissance, scientists/artists/engineers developed rules that would explain the perspective that their right brains understood quite naturally, and the system of **linear perspective** was invented. Using terms like *horizon line, vanishing point, key edge,* and *front face,* masters could teach apprentices how to lay out pictures of complicated scenes. The system of linear perspective can be cumbersome, but it is a good way to represent space realistically. In the 1800s, when Nicephore Niepce produced the first permanent photographic images, the camera's "vision" confirmed the rules of perspective developed centuries earlier.

Point Perspective. Drawings that use only one vanishing point are called **one-point perspective** drawings. The vanishing point is located on a **horizon line**, which is a horizontal line representing your eye level. Placement of the horizon line is important for how the object will appear in the drawing. Figure 5-42 illustrates how the placement of the horizon line affects the appearance of the object.

In one-point perspective, the vanishing point is used to depict one or two faces of an object as well as the **front face**. The front of the object is drawn "head-on." The vanishing point should not be placed too far to the side of the object or the object will look distorted.

linear perspective:
In drawing, an approximate representation, on a flat surface, of an image as it is perceived by the eye. Typically, objects are drawn smaller as their distance from the observer increases, and items are somewhat distorted when viewed at an angle.

one-point perspective:
A method of realistic drawing in which the part of an object closest to the viewer is a planar face, and all the lines describing sides perpendicular to that face can be extended back to converge at one point, the vanishing point.

Chapter 5: Drawing to Develop Design Ideas   131

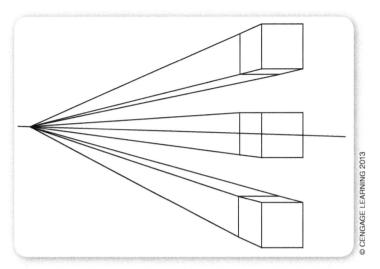

Figure 5-42: *The placement of the horizon line affects the view of an object.*

Exercise 5: Cube in One-Point Perspective

Step 1. Use a ruler to make sure your lines are straight. You can also draw on graph paper to keep corners square. Draw a square, a line, and a point in the positions shown in Figure 5-43. The square is the front face, the closest side of the cube you will draw. The line, called the horizon line, stands for your eye level. The vanishing point is the point where parallel lines converge. It tells you where you are standing, because it is projected back from your eyes onto the horizon line.

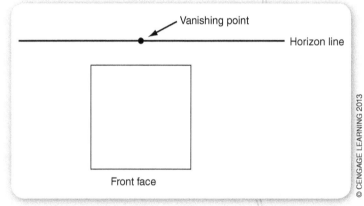

Figure 5-43: Step One

Step 2. Draw a dashed line from each corner of the front face to the vanishing point. These are imaginary lines. Next, draw a dashed line parallel to the top edge of the square about one-third of the way back to the horizon line. Place the dashed line between the diagonals going back to the vanishing point. Note that it is shorter than the top edge of the square. This makes it appear to be at some distance back from the square. If you draw a dashed line parallel to the sides of the square dropping down from the points where the horizontal line meets the diagonals, you create the top corners of the back face of the cube. Where those lines meet the diagonals going back to the vanishing point from the bottom of the front face, draw a dashed line parallel to the bottom of the front square. For the time being, all the lines other than the edges of the front square and the horizon line should be dashed. What you have drawn here is a transparent cube (see Figure 5-44).

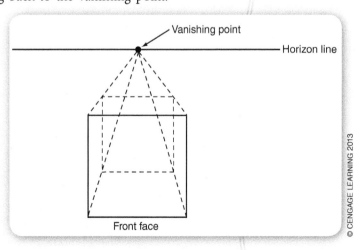

Figure 5-44: Step Two

Step 3. The position of the horizon line and vanishing point tell us that you are directly in front of the cube and slightly above it. What you see of the cube is the front and top faces. Outline those in solid black lines (see Figure 5-45). This way you are showing what is seen and what is not seen but is understood.

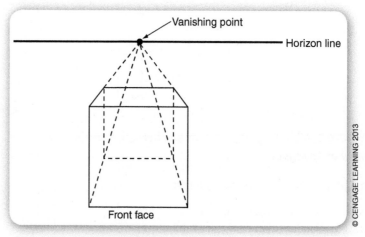

Figure 5-45: **Step Three**

Interior Views. One-point perspective is also useful for drawing the interior of a room (or an entire outdoor scene). When depicting a scene from a standing position, the horizon line is placed just slightly above the center, and the vanishing point is usually centered on this line (see Figures 5-46 and 5-47).

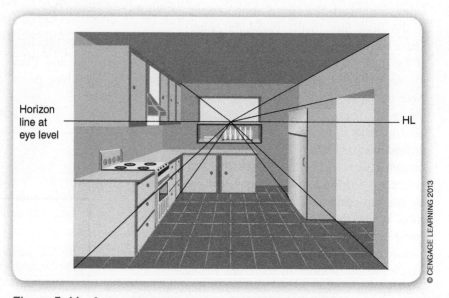

Figure 5-46: *One-point perspective of a kitchen interior.*

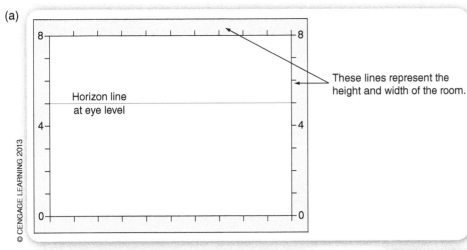

(a) Always start a one-point perspective with a front face. In this case, the front face is the page itself. Next draw a horizon line slightly above the center. If the drawing is of an interior of a room, locate the horizon line at standing eye level. In the sketch (a), the walls have been scaled in feet to determine an 8-foot ceiling at an eye level of 5 feet 6 inches.

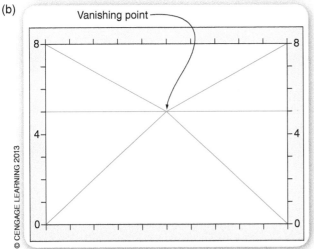

(b) Locate a vanishing point on the horizon line in about the middle of the drawing. Connect the near corners of the room to the vanishing point. A room can also be drawn with the vanishing point off-center. As long as the vanishing point is within the walls of the room, the one-point perspective will work.

(c) A smaller rectangle with the same proportions as the original outline of the room is drawn to represent the back wall of the room. Its corners are points on the converging diagonal lines. You will need to estimate the length of the room and use this to determine the size of the box.

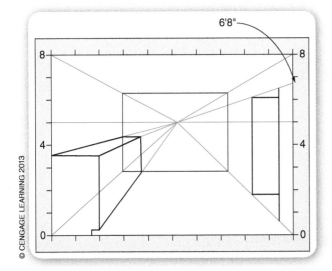

(d) Draw cabinets, doors, windows, and other features of the room using the vanishing point. Each of these objects is a small one-point perspective drawing. Start with a front face—whatever side of the object is facing you is the front face. In this drawing, the cabinet height and depth are scaled on the outline and connected to the vanishing point. These lines represent horizontal lines in the room.

Figure 5-47: Interior of a room in one-point perspective.

Figure 5-48 shows an outdoor scene viewed from a one-point perspective. If you follow the lines of the sides of the buildings back into space, they will converge in a single point.

Figure 5-48: Outdoor one-point perspective.

two-point perspective:
Two-point perspective is a realistic way of drawing objects in three dimensions using a horizon line, a key edge, and two vanishing points.

Two-Point Perspective. Two-point perspective is similar to one-point perspective, but is used for the more common situation in which you are not right in front of the object you want to draw. In a two-point perspective, you are at an angle to the object; so the closest part of the object will be an edge rather than a face. This closest edge is called the key edge. When viewing an object at an angle, it is necessary to provide two vanishing points. These determine the outlines of the two or three sides you see (see Figure 5-49).

Placement of the horizon line depends on the eye level you are trying to represent in your drawing. If you wish to present a "bird's-eye view," the horizon line should be high on the page, well above the space in which you will draw your object (see Figure 5-50). To view the underside of the object being drawn, place the horizon line low on the page, below your drawing space.

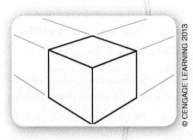

Figure 5-49: A box in two-point perspective.

The placement of the two vanishing points will have an effect on the appearance of the object you are drawing. In many cases, the vanishing points will need to be off the paper on which you are drawing to give the object a realistic appearance. Placement of the vanishing points too close together distorts the appearance of the object. Figure 5-51 illustrates how placement of vanishing points can give you different effects. A good rule to follow is that the bottom angle of the box you are drawing should be no less than 90 degrees unless you are trying to achieve a very dramatic effect.

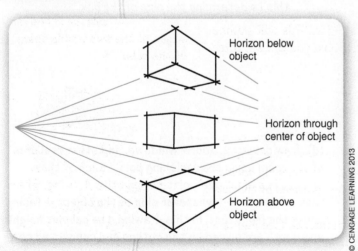

Figure 5-50: The horizon line is your eye level. A low horizon line means that you are below the object and can see the bottom. A high horizon line gives you a bird's-eye view.

Chapter 5: Drawing to Develop Design Ideas   135

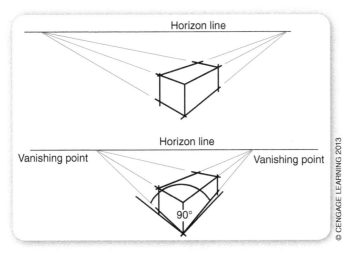

Figure 5-51: Close and far-apart vanishing points.

Exercise 6: Horizon Line and Vanishing Points

Step 1. To draw a box in two-point perspective, start with a vertical line (the key edge) rather than a square face; add a horizon line and two vanishing points (see Figure 5-52). The key edge is the closest corner of the box you want to draw. The horizon line is once again your eye level, and the vanishing points are situated on the horizon line to the left and right of the key edge. If you place them fairly close together, you will get a result that is quite dramatic (in fact, sometimes uncomfortably distorted) and suggests that you are close to the box. If you place them quite far apart, you will get a much more comfortable view, as if you are at some distance from the box.

Step 2. Draw a light line from the top and bottom of the key edge to each vanishing point (see Figure 5-53).

Step 3. Now, draw the vertical lines that represent the far edges of the two side walls on either side of the key edge (see Figure 5-54). Outline each side of the box. The side that is closer to a vanishing point will be shorter than the other side.

Step 4. Draw a light line back to the opposite vanishing point from the back corners of the two visible sides (see Figure 5-55). This will outline the top of the box. Because you are above the box, you can see three faces—two sides plus the top.

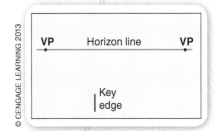

Figure 5-52: Draw a horizon line about one-quarter way down the page. Place a vanishing point near each end of the line. Draw a short vertical line, the key edge, at the bottom of the page, as shown, just left of center.

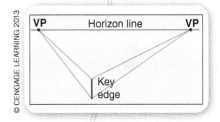

Figure 5-53: Connect the ends of the key edge to both vanishing points.

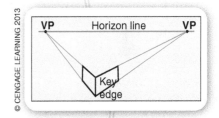

Figure 5-54: Sketch in the far edges of the two visible sides of the cube.

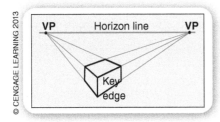

Figure 5-55: Lightly connect the far corners of the cube to the opposite vanishing points. Darken the outlines of the cube.

The building exterior (see Figure 5-56) and the room interior (see Figure 5-57) are both viewed from a two-point perspective. Following the edges of the walls, ceiling, roofs, door, and window frames on either side of the key edges back to two points where they converge allows you to appreciate the two-point perspective view. A line drawn through the two points is the horizon line (your eye level).

Figure 5-56: Exterior two-point perspective.

Figure 5-57: **Interior two-point perspective.**

Linear perspective has many rules and is time-consuming to learn and use. It is a left-brain drawing technique, making it ideal for translation into computer code. Many of the drawing programs available for architects, industrial designers, and engineers can generate perspective drawings very rapidly, allowing clients to see realistic views. Typically, changes are then made on isometric or trimetric views and translated into perspectives by the software.

Perspective and Mathematics

Perspective as a realistic way to describe what we see has its roots in the Renaissance. Before this time, medieval artists typically sized objects and characters according to their importance, not relative distances.

Around 1300, the artist Giotto di Bondone used an algebraic method to determine the placement of distant lines in his paintings. His method was generally accepted until the twentieth century. In the early 1400s, Filippo Brunelleschi demonstrated the geometrical method of perspective by painting the outlines of various Florentine buildings onto a mirror. Extending the building's outlines beyond the edges of the mirror, he noticed that all of the lines converged on the horizon line.

Soon, nearly all of the artists in Florence applied geometrical perspective to their paintings. All horizontal lines converged to a vanishing point, and the rate at which the lines receded into the distance was graphically determined. Not only was perspective a way of showing depth, it was also a new method of composing a painting into a single, unified scene.

Today, CAD software and some computer games (especially games using 3-D polygons) use linear algebra (in particular matrix multiplication) to create a sense of perspective. The scene is a set of points, and these points are projected to a plane (computer screen) in front of the viewpoint (the viewer's eye). The mathematical problem of perspective is simply finding the coordinates on the plane that correspond to the points in the scene. By the theories of linear algebra, a matrix multiplication directly computes the desired coordinates, thus bypassing any descriptive geometry theorems used in perspective drawing.

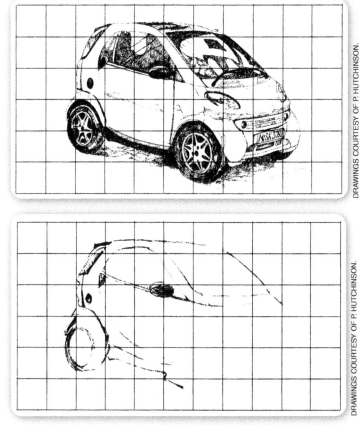

Figure 5-58: *Using a grid to copy a drawing.*

Understanding Perspective Visually. The right brain of the artist seems to be able to bypass the verbal rules of perspective, capturing spatial relations at a glance without plotting all the points necessary in a CAD program. That is because the right brain is holistic and looks at the overall visual relationships rather than trying to explain to itself what is happening to forms as they extend into the distance.

Two very simple visual aids that artists have used for centuries are the transparent grid (see Figure 5-58) and the paper frame (see Figure 5-59). Placing a grid over a scene allows the scene to be broken down into smaller parts. The grid helps organize what we see, so that we can concentrate on one manageable dose of information at a time.

A paper or cardboard framing viewer held between the eye and the subject helps filter out surrounding information that competes for our attention.

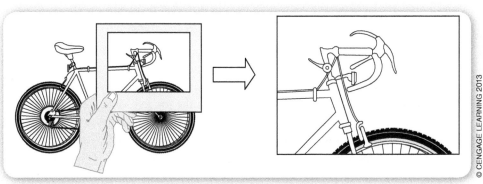

Figure 5-59: *Using a framing viewer to eliminate confusion and narrow focus.*

Exercise 7: Using a Grid to Capture the Whole Scene

Step 1. On a piece of clear acetate, make a two-inch grid using a fine-tipped permanent marker. Draw a grid the same size on drawing paper. Make a space (where you can draw) right in front of a window through which you can see buildings (some closer than others) or a street scene. Place your drawing paper in a comfortable place. You may want to prop your pad or drawing board on a book so that it's at an angle, because you will want to be able to avoid moving your head too much on this drawing. Now, tape the transparent grid to the window and look through it at the scene you want to draw (see Figure 5-60).

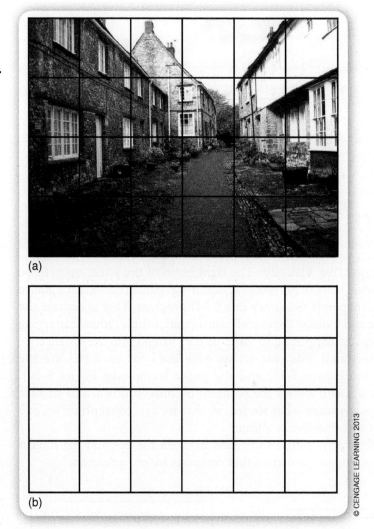

Figure 5-60: Reproduce the scene viewed through the transparent grid in the squares of the gridded paper.

Step 2. Draw what you see through the grid, square by square. In other words, transfer the lines and shapes you are seeing in the top left square of your window grid to the top left square of the paper grid, and so on. (Notice that when you broke the scene into small pieces with the grid, you moved from left to right brain mode, because you could not really name the things you were drawing.)

Point of Interest
Sketching as investigation

The best way to prepare to draw your ideas—things that exist only in your mind—is to practice drawing the world around you.

Although most of us use our sight effectively to avoid tripping over furniture and running into doors, we seldom look at details unless something catches our eye. A closer look at everyday objects often reveals features we have missed. These features may be proportion, material, a design element such as line or texture, or the way in which one part fastens to another. Art students spend hours drawing things in classes and then often keep a notebook with them to practice sketching the things they see outside of class.

Isometric Drawing. Using isometric grid paper, you can sketch all the objects we have previously drawn, but without the horizon line and vanishing points. In isometric view, objects are seen at an angle, so that the greatest number of sides (three) can be seen at once. All parallel lines are drawn in parallel—so no vanishing points are needed. Diagonal lines representing horizontal edges are drawn 30 degrees from a horizontal base line. Measuring on these drawings is simple, because lengths do not diminish in the distance.

Isometric drawings do not show objects the way we actually see them, but designers use this technique to explain an idea quickly. As with perspective drawing, all of the basic shapes can be drawn in isometric view, and more complex forms can be made from these forms.

Isometric construction can be used for quick sketching and comes in handy, particularly for visual brainstorming, which is typically done from ideas and not physical objects in front of you. Using Exercise 8, practice visual brainstorming. Because isometric drawing is a drafting convention and a standard feature of mechanical design software, much more attention will be given to this approach in Chapter 10.

Skills for the Future

Some futurists project that virtually all of the logical and sequential abilities of the left brain will eventually be taken over by the computer. In his 2005 book, *A Whole New Mind*, author Daniel Pink notes that many of these qualities make up more than half of the work done in fields like engineering, law, medicine, and accounting. The right brain abilities—visualizing, empathizing, thinking holistically—are less likely to be replicated by computers, and these will be the skills most valuable for future professionals and entrepreneurs. So, honing these skills now could give you an advantage in tomorrow's job market.

Exercise 8: Isometric Drawing

Step 1. Draw the front vertical edge of the cube (see Figure 5-61).

Step 2. The sides of the box are drawn at 30 degrees to the horizontal to the required length (see Figure 5-62).

Note: All lengths are drawn as actual lengths in standard isometric.

Step 3. Draw in the back verticals (see Figure 5-63).

Step 4. Draw in the top view with all lines drawn 30 degrees to the horizontal (see Figure 5-64).

Step 5. Now try drawing the remote in isometric view (shown in Figure 5-65).

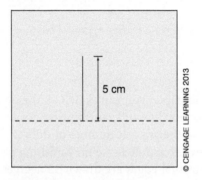

Figure 5-61: Step 1.

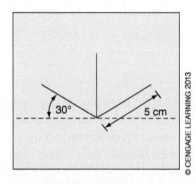

Figure 5-62: Step 2.

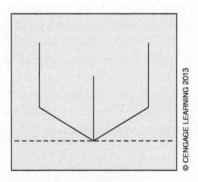

Figure 5-63: Step 3.

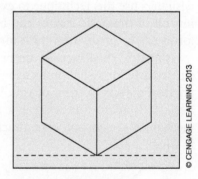

Figure 5-64: Step 4.

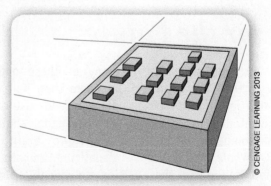

Figure 5-65: Two-point perspective remote.

Types of Drawings

- **Orthographic projections** show three different sides of an object as though seen head-on (see Figure 5-66).
- **Cutaways** use a jagged edge to show what is under a part of the facade of an object (see Figure 5-67).
- **Exploded drawings** show how the parts of an object fit together (see Figure 5-68).
- **Sectional views** show an internal slice of the object, much like a CAT scan image (see Figure 5-69).

Because they help isolate an aspect of the subject you are working with, it is helpful to use these approaches from time to time in your sketching.

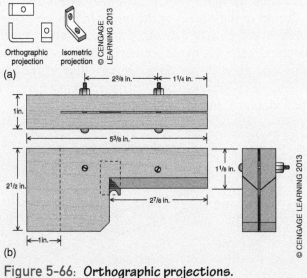

Figure 5-66: Orthographic projections.

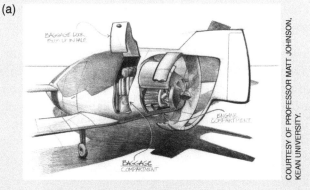

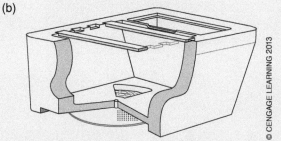

Figure 5-67: Cutaway drawings.

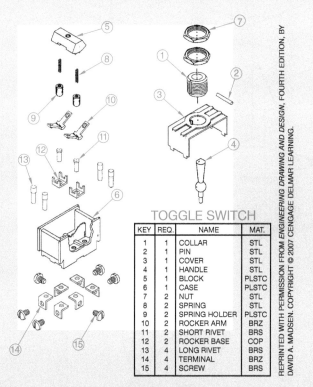

Figure 5-68: Exploded drawing.

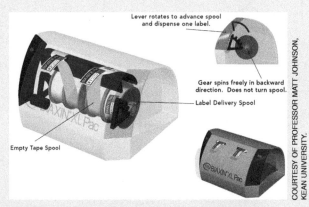

Figure 5-69: Sectional drawing.

OTHER DRAWING CONVENTIONS

Besides perspective and technical drawing techniques, many other drawing conventions are used by designers to record or convey ideas.

Crating

Crating is the process of visualizing the object you want to draw inside a box or a crate. First draw a cube in one- or two-point perspective. The cube can be used to help you develop the other basic shapes: the cylinder, cone, and sphere (see Figure 5-70). Diagonal lines are used to find the center of the crate side. In the cylinder, the center of each end is connected and becomes the center axis of the cylinder. The tip of the cone is at the center of the top plane of the crate, and the circumference of the cone bottom touches the center of each side of the bottom of the crate. The sphere touches one point at the center of each side of the crate. More complex objects can be constructed from a basic form or combination of forms, as in Figure 5-71.

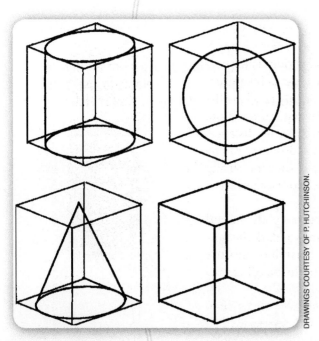

Figure 5-70: Basic forms drawn in relation to crates in isometric view.

(a)

(b)

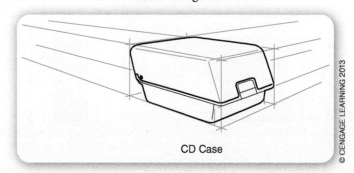

Figure 5-71: A box or crate is used to help sketch an object.

Natural (or organic) shapes are more difficult to think of as drawn in perspective, because they do not have straight edges. Imagine this novelty tape dispenser as a combination of basic shapes. Then think of (visualize) it within a snug box or crate. Following the edges of the box back to a vanishing point makes the perspective more obvious (see Figure 5-72).

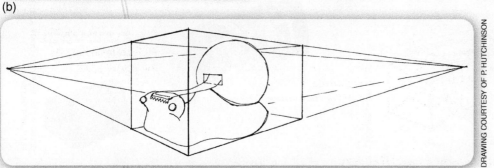

Figure 5-72: A novelty tape dispenser (a) drawn as basic shapes, and (b) shown in two-point perspective.

Sighting for Proportion

Drawing the parts of a person or object in correct proportion can be difficult at first. **Sighting** helps to determine relative points in a drawing. Artists know they need tools to take the **visual measurements** they need, and they find those tools all around them.

Try drawing a cordless phone like the one in Figure 5-73. Look at the relative sizes of the different features. The light-colored upper oblong shape is about the same length as the darker, lower keypad area. The screen is about three buttons wide. The whole length of the phone is about five antenna lengths. Close up, you can actually measure these relationships and then use the details as units of measurement. By checking the sizes of the components against one another, you can keep your drawing in proportion.

The same approach can be applied to a building seen from some distance. Doors and windows can be used as the units of measurement, and distances can be measured using your pencil. This only works, however, if you keep your measuring unit a constant distance from your eyes. You may even have to close one eye and squint at your subject.

Hold your pencil, blunt end up, at arm's length in front of you and place the end of the pencil touching the top of the door frame. Use your thumb to mark the bottom of the door on the pencil (see Figure 5-74). You have created a unit of measure to help you find the proportions of your building. The facade of the building may be 1 1/2 door lengths high and 2 door lengths wide, and so on.

When drawing people, artists often talk about a figure being seven heads high, with arms that are three heads long. They are using what mathematicians call **nonstandard units** of measure, which will help you find proportions consistently on your drawing subject.

The drawing techniques that follow can be used throughout your portfolio, as needed. There is no one right place for a given technique—it must simply fit the purpose for which it is being used. Quick sketches, for example, should be done freehand, without rulers and straightedges; linear perspective is not necessary in visual brainstorming.

Outlining

Outlining in a drawing can help express spatial qualities of a form (see Figure 5-75). The use of dark lines surrounding an object can make it stand out from its background. Outlining the closest planes of an object can make the form look more three-dimensional. Using both shading and outlining on a sketch can be confusing, because shading describes how light falls on a form in space, and outlining tends to reduce the drawing back to a flat symbol or cartoon.

Figure 5-73: The keys and screen on this cordless phone can be used as units of measurement when measuring visually to draw the object.

Figure 5-74: Sighting using a pencil as a measuring tool.

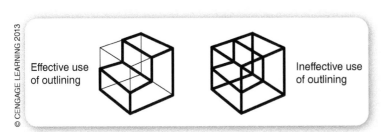

Figure 5-75: Outlining to help express special quality.

Figure 5-76: Compare the three drawings. Which makes the object easiest to see and understand?

Adding a Background

A contrasting background always helps isolate and focus attention on an object. Use bold, rapid strokes with a black or dark-colored marker to add a backdrop. It is not enough to just outline the object, because the outline will then compete with the object itself. Extend out from the outline so that the dark shape becomes a background. The outer edge of the background should not draw attention to itself or, once again, it will compete with the object you want to "spotlight" (see Figure 5-76).

Colored Pencil Techniques

The use of colored pencils is a relatively easy way to make sketches and drawings more interesting, effective, and informative. They may be used to make a colored background and to outline, highlight, or shade. Colored pencils lend themselves to shading and soft transitions, and they are good for representing matte surfaces (finely textured but nonreflective).

Begin with a sharp point, and hold the pencil as you would for shading. The idea is to create a smooth blend of color by going over an area a number of times in several directions (see Figure 5-77). Light pressure must be used to obtain an even color application.

Figure 5-78: Shading a sphere with colored pencil to give a 3-D appearance.

Figure 5-77: Using a colored pencil to make a smooth transition in tone.

Colored pencils are ideal for shading to give an object a 3-D appearance (see Figure 5-78). Notice how the color on the sphere is dark and sharp at the edges and softer as it moves toward the internal highlights. Carefully blend the color to get a crisp, neat drawing.

Color Marker Techniques

Color marker rendering is an effective and dramatic technique that looks more difficult than it actually is. Although it takes some practice, there are "tricks" that you can use to make your drawings look as professional as Figures 5-79 and 5-80. Simple techniques are described in Figure 5-81.

Figure 5-79: Shading with markers.

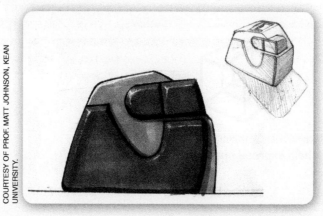

Figure 5-80: Concept drawing using markers.

Chapter 5: Drawing to Develop Design Ideas **145**

(a) Sketch a cube in perspective or isometric. Use light pencil lines. Put two additional pieces of paper under your original drawing to prevent "bleeding" through to the work surface underneath.

(b) Use two pieces of paper as a mask. Place the mask over your sketch so that only the top surface of the cube is visible. Use a medium-value marker, and work from one side of the cube's top surface to the other. Begin each stroke on the mask, and use bold, rapid, vertical strokes that end past the object lines of the cube. You are working out from the "V." Don't go back over areas you have missed.

(c) Rotate the page to give yourself a comfortable position in which to use the marker. Adjust the mask to leave a thin space between the colored top surface and the left surface you will render. This space will be the corner of the cube between these surfaces. Work out from the "V," and use the same technique, but wait about a minute and give this surface a second coat of marker. This will make this side darker than the top surface.

(d) Again, rotate the page, and reposition the mask to expose only the right side. Be sure to position the mask pieces so that a thin, uncolored space is left between cube surfaces. Again, work out from the "V." Give this side three coats of marker, waiting a minute or so between coats. These three coats will make this surface the darkest side.

Figure 5-81: Using color markers to add appeal to drawings.

(e) To complete the rendered cube, use a black wide-tipped marker and smooth, bold strokes along the six outside object lines. This technique will make the object stand out from the page and clean up the outside edges.

(f) Overlapping the marker strokes gives the object a sharp, designer-rendered appearance. You can use this technique to outline pencil-drawn objects, too.

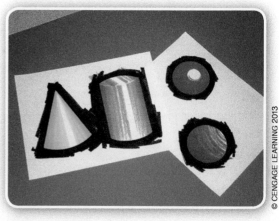

(g) Another excellent technique for marker rendering is to cut out the cube and use stick glue to paste it on another sheet of paper. A background can be applied first with another contrasting color marker.

(h) A cylinder, cone, and sphere are shown rendered here. Notice how the marker lines are drawn along the axis of the cylinder and how the value goes from dark along the upper edge to light, to dark along the bottom edge. A cutout placed on the end surface to mask that area is shown. The marker lines on the cones begin at the top and radiate out as they reach the bottom.

Figure 5-81: (Continued)

The two main purposes for hand drawing are (1) to help you describe what you can see, and then (2) to use that experience to describe ideas that exist only in your mind. The concepts and techniques that have been discussed in this chapter should provide you with some tools for drawing effectively. The problem identification, initial research, and idea-generating steps of the design process use these skills heavily. As a designer's ideas gel, the questions to be answered become more technical, and technical drawing and modeling approaches are used. Technical development, testing, and evaluation using CAD/CAM and engineering simulation software, as well as graphing, spreadsheets, and other computer programs are explained in other chapters.

All of these tools contribute to the documentation of the design process, and you will want to use as many of them as are available when creating your own design portfolio.

USING DRAWINGS IN THE DESIGN PROCESS

The drawing techniques we have discussed so far are designed to help you draw what you see. This is particularly useful for studying the world around you and understanding how things work, their components, and how they fit together. This kind of drawing serves the information-gathering aspect of the design process and allows you to create visions of what a developed solution might look like in use.

Preliminary Sketches

Freehand drawing allows you to develop and present your ideas. The first ideas you put down on paper are usually in the form of preliminary sketches (see Figure 5-82). These are done quickly so that you do not lose your train of thought or flow of ideas. This process should not be encumbered with straightedges, rulers, compasses, or other drawing instruments. These instruments only slow down the process that, by its very nature, is free and creative.

modeling: Creating a visual, mathematical, or three-dimensional representation in detail of an object or design, often smaller than the original. Modeling is often used to test ideas, make changes to a design, and to learn more about what would happen to a similar, real object.

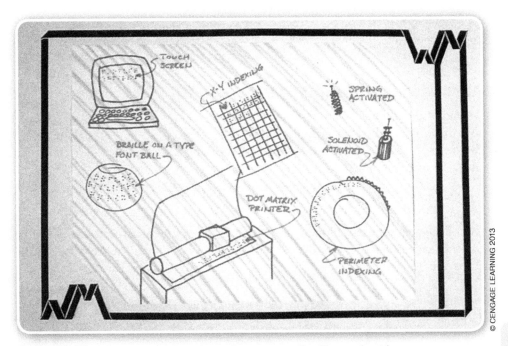

Figure 5-82: Preliminary sketches of computer Braille reader ideas.

At the beginning of the design process, it is helpful to be able to pull ideas out of our minds and put them on paper as quickly as possible. Then we can develop and communicate those ideas. There are ways to simplify that process by using symbols and techniques that communicate without all the details, much as a stick figure can symbolize a person and a sketch map can provide directions to a destination.

Visual brainstorming, like verbal brainstorming, is a method for stimulating and expressing the most possible ideas in the least possible time. Try this in the Your Turn exercise. Begin by drawing an object that you would like to improve upon, for example, a skateboard (see Figure 5-83).

visual brainstorming: A method of ideation in which drawing (in contrast to verbalizing) is used to generate large numbers of ideas. First, an existing object is drawn, then variations on that object are drawn, and then variations on one of those ideas are drawn, and so forth.

Your Turn
Visual Brainstorming

Step 1: Sketch the skateboard in Figure 5-83 as it now exists.

Step 2: Do five variations on this sketch. If a photocopier is available, you can make five photocopies of the original sketch and then draw over the parts where you would make simple modifications.

Step 3: Choose the variation that seems the most interesting and make five variations on that version.

Step 4: Choose the best of these and develop the idea in more detail.

This kind of activity should be done as spontaneously as possible, so you do not get bogged down in details.

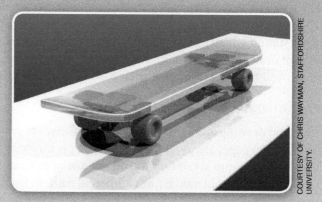

Figure 5-83: Visual brainstorming led to the design of this skateboard innovation.

Annotated Sketches

The addition of notes about materials, fasteners, and other features (**annotations**) to a drawing is called an **annotated sketch** (see Figure 5-84). These remarks are reminders to yourself about what you were thinking at the time you developed the idea. Ideas are often lost if they are not recorded immediately.

Developmental Sketches and Drawings

The next step in designing is the further development of the ideas in the preliminary sketches. These **developmental sketches** contain more detail as the ideas are refined (see Figure 5-85). At the completion of the developmental

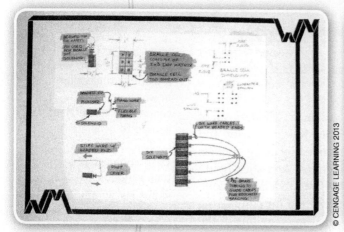

Figure 5-84: Annotated sketch of computer Braille reader components.

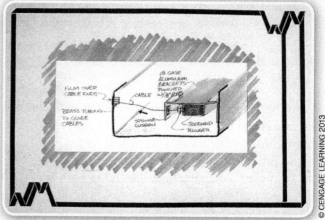

Figure 5-85: A developmental drawing showing the solenoid control in the final Braille reader system.

sketches, the solutions should be workable, with features such as mechanical linkage or electronic circuitry worked out. Annotations are often a part of developmental sketches. Developmental sketches are done both freehand and using CAD programs. Software approaches to product development will be detailed in Chapter 8.

Production Drawings

In the final stage, production drawings are developed that contain the information needed to actually make thesolution (see Figure 5-86). These are often drawn to scale, so that size, proportion, and location of features such as holes can be finalized. Working drawings will be discussed in a later chapter.

(a)

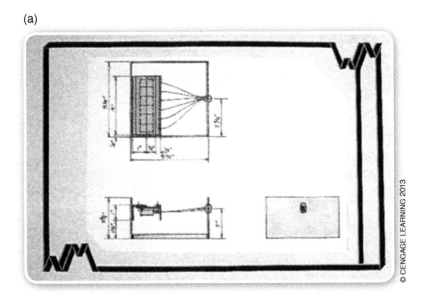

(b)

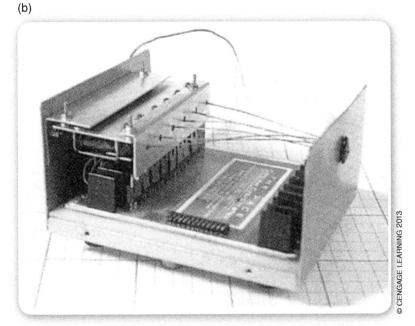

Figure 5-86: A production/working drawing showing (a) final placement of components and (b) the model of the Braille reader system.

Point of Interest
How to document your design process

Design Step	Documentation Technique
Defining the Problem	Description of a technological situation explaining the problem. Write, draw, annotate. Use computer-generated text or neat hand lettering.
Brainstorming	Include a collection of the ideas considered including ideas from different members of the design team.
Researching and Generating Ideas	Notes, sketches, letters, interview tapes, bibliography, and photos. Use photocopies of catalog pages and other reference materials. Include diagrams that explain mechanical or electronic principles to be applied.
Identifying Criteria and Constraints	Actual brief with specifications, conditions, and requirements. Use computer-generated text or neat hand lettering.
Exploring Possibilities	Notes, preliminary sketches, and development sketches with annotations, 2-D models. Use pencil, color pencil, marker, technical pen, and other mediums. This should be a well-documented section.
Selecting an Approach	Notes, matrix comparing requirements to solutions, checklists, and so on. Presentation drawings using color pencil, marker, or other color medium.
Developing a Proposal	Photos of evolutionary models: wood, plastic, cardboard/paper, foam core, metal, ceramic, found objects; computer simulations, kits, clay, VHS, DVD, and so on. Working drawings, such as orthographic projection. Color rendering of final appearance of solution.
Modeling or Prototyping	Photos of various stages of completion; descriptions of necessary adjustments and changes, and so on.
Testing and Evaluating	Data checklists, graphs, charts, photos, slides, VHS, and DVD. Evaluation (solution evaluation description of test results and self-criticism: "press type" and self-evaluation), computer-generated text, neat hand lettering, and so forth. In addition, a title page is always appropriate, and, if the portfolio is lengthy, a table of contents may help the organization. The portfolio sections correspond with the design steps outlined in earlier chapters.
Refining the Design	Include a list of problems to be addressed in the redesign of the project and show evidence that the problems were resolved.
Creating or Making	Show evidence that the production process will result in a product that meets all design specifications or show how the production process has been modified if necessary.

DEVELOPING AN ENGINEER'S NOTEBOOK AND DESIGN PORTFOLIO

As noted earlier in this chapter, an engineer's notebook is typically bound with numbered pages (see Figure 5-87). The purpose of the engineer's notebook is to keep a written record of all the important work completed on a project or a range of projects. Notebook entries could include calculations, graphs, ideas, concerns, and results of testing or research. An engineer's notebook is where patentable ideas are described in full detail. In this format, a patent idea in a notebook is most often signed by a witness that has read and understood the idea. All notebook entries should be dated. More about developing an engineer's notebook will be presented in Chapter 6, Reverse Engineering.

Design portfolios are used to show the design process to clients, and therefore the content and format are less important than the aesthetic presentation. For students learning about design, the portfolio may be used to show the design process and solutions to a teacher or a review board. Both the design portfolio and the engineer's notebook will contain critical information from the design process. Following is a list of sections and appropriate documentation strategies.

PORTFOLIO

Because design portfolios serve a less formal purpose as compared to the engineer's notebook, no special formatting, binding techniques, or content are required. A design portfolio should contain a number of sections, depending on the complexity of the project. Simple design problems will have fewer sections, often combining several from the previous "Point of Interest" list. Major projects will use all or most of the sections, or will combine the sections into another logical format.

Figure 5-87: **Bound Engineering Notebook.**

Title Page, Page Numbering, and Table of Contents

Your portfolio should have an attractive title page that clearly identifies the portfolio's purpose and author. Each page of the portfolio should be numbered. A design portfolio should have a table of contents listing the design steps and where to find them. This will need to be the last step in organizing your portfolio. Portfolios will be most attractive if they follow sound graphic design principles. More information about graphic design principles can be found in Chapter 18 (Graphics and Presentation).

Page Orientation

A horizontal or landscape orientation of the paper is well suited for portfolio work (see Figure 5-88). Choose a paper size that is large enough to allow for a border and a title block and still give you plenty of room to draw. A standard 11 by 17 inch page is large enough without being too cumbersome to carry around, but other sizes are acceptable. You may also need to develop or purchase a cover or case to keep pages from tearing and to protect your work from wind, rain, and other hazards.

Logo

A company that wishes to be easily recognized develops a symbol for use on its products, correspondence, and advertising called a logo. You may develop your own logo for your design work and incorporate it into both your portfolio and your final design solutions. In the portfolio, the logo may appear on each page.

Binding

Pages may be bound on either the left side or the top, using plastic bindings, a modified three-ring binder, or one of a variety of other methods.

152 Part I: The Engineering Design Process

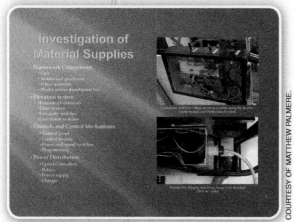
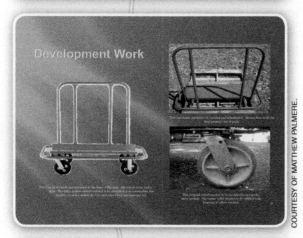
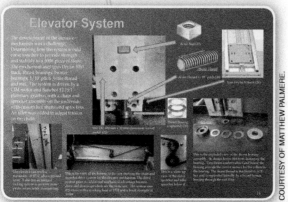

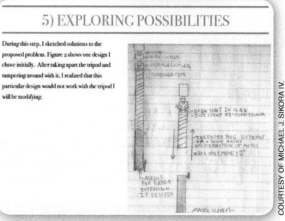
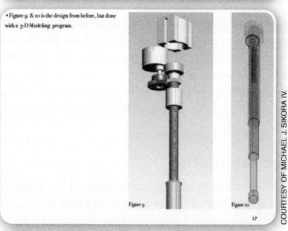
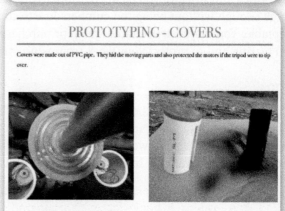

Figure 5-88: Selected pages from senior design portfolios.

Chapter 5: Drawing to Develop Design Ideas

Page Content

It is important to strike a balance between putting too little and too much on a page. A typical page in a portfolio might have a number of drawings (see Figure 5-89). There is no reason, however, that all of these drawings must originally have been drawn on the same piece of paper. You can redraw from original sketches onto one page or cut out a number of different drawings and glue them on a portfolio page. These techniques will allow you to present ideas in an interesting and concise format.

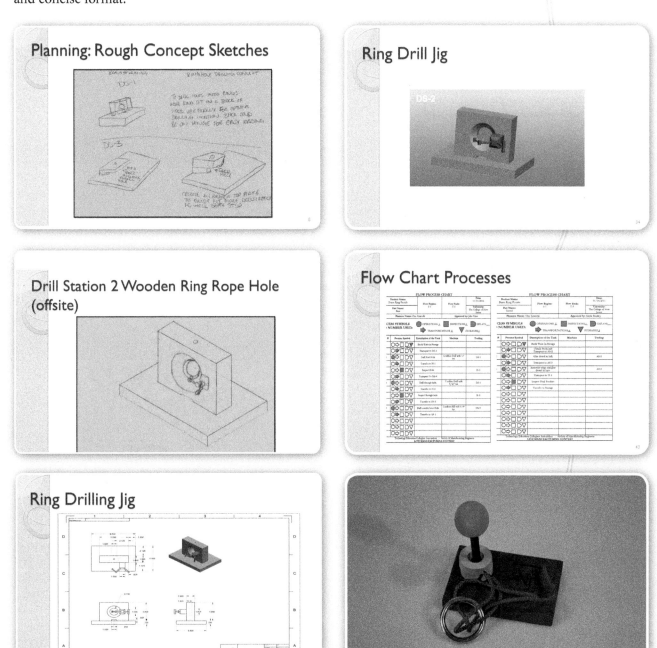

Figure 5-89: *Pages of a design portfolio showing examples of drawing techniques used during the design process. a) sketching, b) perspective, c) orthographic, d) rendering, e) charting, f) final product.*

COURTESY OF JULIE RYAN, CLOE KAWECKI, KEVIN BRADLEY, TANNER WILSON, JOE CARSON, AND DALTON FOWLER, TCNJ TEAM MEMBERS, TECA EASTERN REGIONAL MANUFACTURING DESIGN CHALLENGE

Portfolio Page Layout

Organizing the appearance of the page is an important part of portfolio presentation. Figure 5-90 illustrates a number of strategies you can use to make a page look interesting.

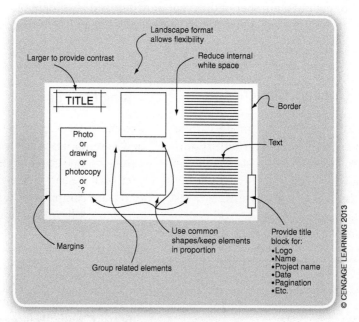

Figure 5-90: A well-organized page is easy to read and understand. It follows rules of alignment.

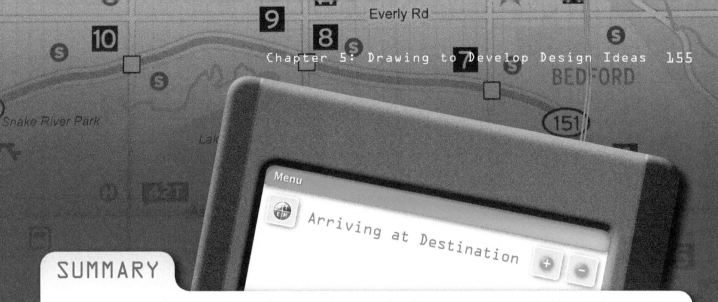

SUMMARY

Drawing is integral to designing and investigating. Ideas, which are the seeds of the scientific hypotheses and technological innovations that define our modern world, cannot grow unless they are exposed to the "light of day." Investigating the wings of birds through sketches allowed the Wright brothers to formulate their theories of flight. Each of Leonardo da Vinci's inventions was developed through a series of drawings. Modern innovators like James Dyson and Jonathan Ive develop their inventions through drawings.

Today, the computer has provided a powerful graphic tool. CAD and solid modeling programs tempt us to sidestep the discipline of drawing. But software cannot replace the human element in designing. When we gain confidence with freehand sketching, we acquire a valuable tool for examining and communicating our thoughts. That confidence comes from learning to use our whole brain to interpret what we see. Like writing, keyboarding, texting, or using a computer mouse, sketching is also a matter of motor skill. All of these develop through practice.

Awareness of the elements of drawing: line, form, shape, value, color, texture, and space, as well as techniques like linear perspective, shading, color mixing, gridding, crating, and exploded and cutaway drawings make us more fluent scientists, designers, and inventors. We can use these abilities as students to document our work; as professionals communicating with collaborators and clients; and as homeowners, family members, and citizens in the technological world.

BRING IT HOME

OBSERVATION/ANALYSIS/SYNTHESIS

1. Using an HB pencil, create a one-point perspective sketch of one of the following: VHS cassette tape, audio cassette tape, "Walkman"-type personal stereo, iPod, BlackBerry personal organizer.
2. Using colored pencils, make a cube stand out by shading around it.
3. Using colored pencils, give a cube a 3-D appearance by shading with three values of one color.
4. Using a scale and calipers, sketch a full-size or half-scale orthographic projection drawing with a front, top, and side view of one of the following: iron, hair dryer, electric drill, personal stereo, cordless telephone.
5. Research and reproduce a number of symbols used for a special-purpose map, such as a geological survey or aeronautic sectional.
6. Use visual brainstorming techniques to explore five improvements on the design of a tent for camping. Choose your favorite of these and develop five variations on that solution.
7. Scan and print or photocopy five copies of your best tent design and show five different color schemes that might appeal to five different groups of users (for example, families, teens, retirees, Girl Scouts, or military).
8. Design and produce a master portfolio page, including an original logo, company name, and a block for project name, date, and other necessary information.
9. Use the techniques in this chapter to fully document a design problem.

CHAPTER 6
Reverse Engineering

Menu

Before You Begin
Think about these questions as you study the concepts in this chapter:

1. How does reverse engineering lead to new understanding about a product's function?
2. How can reverse engineering decrease product waste?
3. How do patents protect a company's products?
4. What does a reverse engineering process look like?
5. Why reverse engineer a product?
6. How is reverse engineering done? How is information gathered through product disassembly?
7. What information is uncovered by functional, structural, materials, and manufacturing analysis?

Chapter 6 is contributed by Donna Matteson, Associate Professor, State University of New York at Oswego, and Master Teacher for Project Lead the Way.

INTRODUCTION

Have you ever wondered how a product continues to stay popular in the marketplace? Have you noticed that someone else soon makes the latest clothing style? Successful graphic designs, car styles, cell phones, and in fact, all new products reflect some feature of a similar product. If you look at products associated with the engineering achievements mentioned in Chapter 1, you can trace a historical line of technological innovations. Consider Alexander Graham Bell's invention of the telephone, which evolved through innovation to become the modern cell phone.

All great engineering achievements involve some form of new design work. Invention refers to design work that results in something unique or novel. Often an invention is based on some earlier development. For example, electricity had to be harnessed before Bell's telephone could be invented (see Figure 6-1). Other inventions such as the automobile, airplane, radio, television, and computer required major breakthroughs in science or technology.

To legally protect an invention from theft or misuse, individuals and organizations apply for a patent. Chapter 7 (Investigation and Research for Design Development) presents more information on patents.

A period of product innovation usually follows inventions. *Product innovation* is a process of improving or modifying an existing product. This important design process is evident throughout history and is a cornerstone of our modern industrial economy. Companies must constantly improve their products to remain competitive in the marketplace. They must also design new products to diversify their product line.

As members of the design community, we sometimes want to learn how devices work. The reasons for this vary. Curiosity is possibly a reason, or perhaps the product is malfunctioning and we want to try to fix it. Maybe we are trying to design a new product and we want to learn from other products designed for similar purposes. Gathering information about existing products can help us design a product that will successfully compete in the marketplace.

Designers apply reverse engineering when they disassemble a product to learn more about how it works and how it was made. Reverse engineering allows the development of everyday items such as cell phones. For example, other companies soon replicate a new, thin cell phone, made possible by newly designed internal parts and the arrangement of those parts. How do you think the competition learned how to make a similar ultrathin cell phone?

Figure 6-1: *An early telephone.*

Many companies apply a reverse engineering process to closely examine and analyze a product's external and internal features. The knowledge gained from this process often leads to new or improved product design. In 1997, Philippe Kahn was waiting in a maternity ward for his daughter to be born when he had the idea to send a picture of her to friends and family through a wireless connection. Kahn made a close examination of his existing cell phone and digital camera to determine if he could somehow connect the two. While he sent for parts from a local Radio Shack store, he wrote a program on his laptop to connect the two devices. After several hours of wiring and programming the new device, Kahn was able to send a photo of his new daughter—wirelessly over a cellular connection—to family and friends around the world. Today, the widespread use of camera cell phones allows firsthand documentation of major events, including weather disasters and crime (see Figure 6-2).

Figure 6-2: *Cell phones can be used to conveniently capture pictures and record video.*

> **reverse engineering:**
> A strategy used to find answers to questions about an existing product that are later used in the design of another product.

REVERSE ENGINEERING LEADS TO A NEW UNDERSTANDING ABOUT PRODUCTS

What attracts us to a new product? Product function, ease of use, dependability, appearance, and cost are all concerns of design teams when designing new products. Well-designed products are valued by consumers, and are profitable to the manufacturers who make them.

Research is an important step in the design process. When graphic designers want to design a new greeting card, they may go to a card store to look at existing cards. A civil engineer assigned to design a new bridge may request drawings from a previously built bridge. An electrical engineer designing a new cellular device circuit may disassemble or benchmark the circuit of an existing device. When a mechanical engineer prepares to design a new mechanical device, she will most likely use reverse engineering to discover how a similar product functions or was produced. They are all on a mission to gather a better understanding of successful devices, structures, or products.

Understanding how a product functions through reverse engineering is one reason to research a competitor's product. However, to be competitive in the marketplace, companies will research the demographics, perspectives, and preferences of potential consumers. Chapter 7 (Investigation and Research) covers techniques used to study consumers. Companies must also consider current trends in product style and package design. Attractive, well-packaged products are more likely to be big sellers. Chapter 18 (Graphics and Presentations) will discuss attributes of effective graphic design and package design.

The process of reverse engineering is usually applied to consumer products, and involves taking the object apart to discover the internal components and their operation. Unlike a traditional engineering design process that begins with a problem and works toward a solution, reverse engineering begins with an end product, system, or process and works toward understanding, documentation, and possibly a new idea. Reverse engineering is used to analyze a product's function, mechanical features, structural integrity, and materials, and to understand the manufacturing processes used.

This analysis can lead to design changes that strengthen or enhance the product's features or performance.

OFF-ROAD EXPLORATION

Did you know that Motorola passed on the offer of Kahn's camera phone technology? How do you think the Motorola company was impacted by that decision? Can you recall some major events documented through camera phone technology? Visit the following website for answers to these questions and additional information about the first camera phone and its inventor:
http://www.usatoday.com/money/industries/technology/maney/2007-01-23-kahn-cellphone-camera_x.htm

Technology: A basic principle followed by a design team is to begin by investigating similar products.

Reverse Engineering Decreases Product Waste

Manufacturers in the die and mold industries have traditionally used reverse engineering to replace worn or broken parts on old equipment. This gives new life to a product or machine that is no longer in production. Without reverse engineering to develop replacement parts, many machines would be scrapped.

Reverse engineering is also used to improve manufacturing efficiency. When drawings for old products are no longer available, or when the existing drawings are inaccurate, companies disassemble products to create valuable documentation and electronic data. Designers use manual precision measurement tools, coordinate measuring machines (CMM), and 3-D laser scanners to collect geometry for input into computer-aided design (CAD) programs. CAD programs are then used to generate solid models and new production drawings.

Case Study

The Ohio Art Company and Etch A Sketch

In this case study, you will learn how the Ohio Art Company used the techniques of continuous improvement to keep the Etch A Sketch popular for the last fifty years. As the chapter continues, you will have several opportunities to apply the reverse engineering process to examine the popular Etch A Sketch toy.

In 1960, the Ohio Art Company first introduced the Etch A Sketch to the marketplace (see Figure 6-3). Arthur Granjean invented the original toy in the late 1950s. The story goes that one day Arthur, an electrician in Paris, was installing a switch plate. After peeling off the clear label affixed to the plate, he used the label to make a note and noticed that the image came through on the other side. This was the inspirational moment for Granjean's L'Ecran Magique—the magic screen.

Figure 6-3: An early Etch A Sketch.

THE ETCH A SKETCH® PRODUCT NAME AND CONFIGURATION OF THE ETCH A SKETCH® PRODUCT ARE REGISTERED TRADEMARKS OWNED BY THE OHIO ART COMPANY.

Following development of his mechanical drawing screen, Arthur took his toy to the International Toy Fair in Nuremberg, Germany. The high price tag caused many toy companies to pass on the novel toy, but the Ohio Art Company invested in the invention and spent a year of research and development (R&D) to address mechanical and component fit concerns (U.S. patent #3,760,505). The company renamed the toy Etch A Sketch, and introduced it to consumers in 1960.

When the advertisement for the new toy appeared on television, the demand was so high that the company continued to manufacture the Etch A Sketch until noon on Christmas Eve in order for customers in California to receive their toy by Christmas.

Ask a parent or grandparent if they played with an Etch A Sketch as a child. Although the toy has been on the market for over fifty years, the internal workings have changed very little (see Figure 6-4).

Figure 6-4: A current Etch A Sketch that might be seen on store shelves today.

THE ETCH A SKETCH® PRODUCT NAME AND CONFIGURATION OF THE ETCH A SKETCH® PRODUCT ARE REGISTERED TRADEMARKS OWNED BY THE OHIO ART COMPANY.

The Ohio Art Company, as is true for most companies, uses **continuous improvement** methods to maintain and enhance consumer interest. Today, in addition to the Classic Etch A Sketch, Ohio Art produces the Pocket Etch A Sketch and the Travel Etch A Sketch. The company continues to evolve and expand the product line with products such as the Etch A Sketch Freestyle that has a controller that swivels 360 degrees shown in Figure 6-5.

Figure 6-5: Etch A Sketch Freestyle features a single swivel controller to easily create smooth arcs and loops and intricate designs.

THE ETCH A SKETCH® PRODUCT NAME AND CONFIGURATION OF THE ETCH A SKETCH® PRODUCT ARE REGISTERED TRADEMARKS OWNED BY THE OHIO ART COMPANY.

The Classic Etch A Sketch is still available on store shelves. However, to remain successful in the marketplace, the product has been redesigned to make use of new materials and manufacturing processes. By applying reverse engineering principles, you can discover these changes and perform research to determine why they occurred. As we research the Etch A Sketch, you may be surprised to learn that some parts are assembled by hand. You will discover which parts and why later in this chapter.

continuous improvement: A strategy used by industry to innovate or modernize existing products.

Your Turn

In this chapter, we will use the Etch A Sketch to illustrate the reverse engineering process. Do you know how the Etch A Sketch works, or what materials and processes designers used to make the product?

Most items can be reverse engineered through disassembly and detailed research, testing, and documentation. The reverse engineering process will give you the opportunity to enhance your research and documentation skills while expanding your understanding of the product. The process may also include the creation of virtual part models. Understanding the sequence of steps in the reverse engineering process is much like understanding the steps in the design process. Once you learn them, you can easily apply them to other situations.

finite element analysis (FEA):
A computerized numerical analysis technique used to solve mechanical engineering problems, such as stress and thermal distributions in parts.

Solid models developed with CAD may undergo finite element analysis (FEA) such as stress analysis and mass properties calculations. Some measurement systems acquire dimensions that are accurate up to 0.005 mm (0.0002 inch). CAD data can then be converted to computer numerical control (CNC) code and sent electronically to prototyping machines to create detailed molds for the remanufacture of parts. These precise, integrated production practices, generally referred to as computer-aided manufacturing (CAM), greatly reduce the amount of product waste.

REVERSE ENGINEERING AND PATENTS

Reverse engineering is a process industry uses to analyze competitive products. Design teams often reverse engineer existing products to develop new understanding, which may be applied to new or similar products. However, what protects a company from having its product copied as a result of reverse engineering? Patent law and professional codes of ethics were designed to help protect inventions; however, there are no guarantees that designs will not be copied. If a company holds a patent, which they believe a competitor has violated, it is their responsibility to sue the competitor for patent infringement. Eli Whitney, credited with inventing the cotton gin, did not protect his patented invention from being copied (see Figure 6-6). Although the invention was very successful, farmers who were unhappy about paying user fees began to produce their own machines.

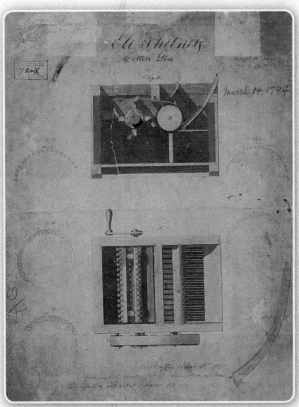

OFF-ROAD EXPLORATION

Visit the following website to view an animated patent drawing showing how the cotton gin works:
http://www.eliwhitney.org/new/sites/default/files/minisites/cotton/patent.html

Figure 6-6: Eli Whitney received a patent for his cotton gin on March 14, 1794 (later numbered as X72 and validated in 1807).

Because others easily duplicated the design of the cotton gin (see Figure 6-7), and because Whitney failed to defend his patent in court, his company went out of business in 1797. Companies strive to maintain market advantage through continuous product improvement.

Design professionals who belong to professional societies must adhere to a code of ethics. The code deters professionals from violating the rules that are associated with **intellectual property**. In addition to professional codes, businesses have written policies that promote ethical employee behavior. Consider the Code of Ethics of the National Society of Professional Engineers, presented in Chapter 1. The code does not cite uncovering understanding of a product as a violation. Rather, the relevant general standards include these statements: "3) Issue public statements only in an objective and truthful manner, 5) Avoid deceptive acts, and 6) Conduct themselves honorably, responsibly, ethically and lawfully . . ." Clearly, violating a patent raises legal issues and can force a company to take expensive legal actions to defend patent infringement. An ethical violation could lead to a person's censure by a professional society, or dismissal by his or her employer.

Figure 6-7: A cotton gin based on Eli Whitney's design.

Engineering: Engineering societies play an important role in the profession by setting high standards of professional ethics.

Your Turn

Can you think of an example of how a professional code of ethics for engineers might impact national security? Visit the website http://en.wikipedia.org/wiki/Tupolev_Tu-4 for an example of how reverse engineering was used in WWII on the Boeing B-29 Superfortress, a four-engine bomber aircraft (see Figure 6-8).

Figure 6-8: The Tupolev Tu-4 Soviet bomber, a reverse engineered copy of the U.S. Boeing B-29

REVERSE ENGINEERING: THE BIG PICTURE

Reverse engineering consists of a series of logical steps as shown in Figure 6-9. The steps are not necessarily followed in a linear fashion. Most designers and engineers usually move back and forth between steps as illustrated by the yellow dashed lines in Figure 6-9.

Reasons for Reverse Engineering

The reverse engineering process starts when a design team or management team identifies a need or concern that could be addressed through reverse engineering inquiry. Typically, designers use reverse engineering for one or more of the following reasons:

▶ To research similar products in an effort to discover possible ways to make a more competitive product

▶ To test a product or design to determine what is failing or causing a failure

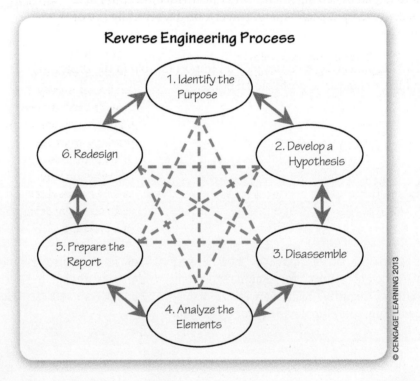

Figure 6-9: Reverse engineering process flowchart.

Your Turn

While preparing the report (Step 5), you may discover that you need to gather additional information about the tensile strength of a component part. Notice the green or dashed yellow lines connecting all of the steps of the reverse engineering process. By using either the green or the yellow lines, you are free to move to the next logical step to accommodate the situation. To which step would you move to gather additional information? Let us try another example. During the disassembly process (Step 3), you suddenly determine the cause of product failure, and the solution for redesign is apparent. To which step would you proceed?

- As part of a company's continuous product improvement policy
- To provide documentation for product components when original drawings are no longer available or accurate
- To educate design professionals on the topics of function, structure, manufacturing, and aesthetics
- For equipment repair, through the design of replacement parts, for products no longer in production
- To develop CAD and CNC electronic data for computer-enhanced manufacturing processes

A company's continuous improvement strategy may dictate that the purpose of reverse engineering is to look for ways to improve function or cut production costs (see Figure 6-10). If this is the case, reverse engineering may lead to changes in a product's size, shape, material, manufacturing, assembly, packaging, or shipping. It is important for everyone to have a clear understanding of the purpose. Knowing the purpose of the task will help everyone to stay focused. However, it is possible to rethink or redirect the purpose, especially during the product analysis phase.

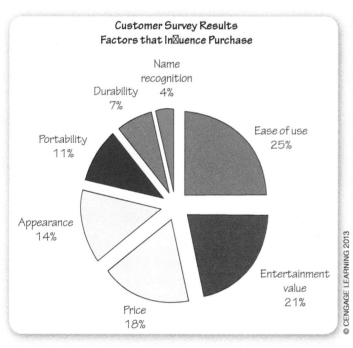

Figure 6-10:
Pie chart showing customer survey results from a company's continuous product improvement plan. This information may indicate a need for reverse engineering of a product.

engineer's notebook:
Also referred to as an *engineer's logbook*, a *design notebook*, or *designer's notebook*. Used as (1) a record of design ideas generated in the course of an engineer's employment so that others may not claim the idea as their own, and (2) an archival record of new ideas, product modifications, and engineering research achievements.

REVERSE ENGINEERING: STEP BY STEP

STEP 1 *Identify the Purpose*

When you are ready to reverse engineer a product, begin by recording your purpose in your engineer's notebook. What do you want to learn about the product? Think about questions to ask, areas of research, people to contact, and tests to be completed. It is important that you keep accurate and detailed documentation throughout the entire reverse engineering process. Your engineer's notebook will provide evidence of your process, thoughts, and findings. Whenever possible, you should add supporting documentation, such as annotated sketches. Ultimately, your notebook will support your findings and may serve as evidence to support legal proceedings or a patent application.

Case Study

The Etch A Sketch

The Ohio Art Company must continually search for ways to keep its product profitable without sacrificing quality. Large discount stores often establish the price point, or retail price of a product. This is determined by considering the price of competitive products, what customers are accustomed to paying, and the psychological effect of the pricing. With the development of our global economy, these price points have made it difficult for companies to continue manufacturing products in the United States. Due to less expensive labor and material costs in other countries, many companies have been forced to move manufacturing or assembly to countries such as China.

The Ohio Art Company continued to manufacture the Etch A Sketch in the United States until 2001. Although financially costly, the family-owned company chose to be loyal to its employees, many of whom had been with the company for thirty years or more. Finally, when the majority of its production workers reached retirement age, the company moved its manufacturing first to Thailand, and then to China.

Your Turn

The following list shows some of the past and current Etch A Sketch product concerns addressed by the Ohio Art Company. Which of these concerns could you address with the reverse engineering process?

1. Expand the consumer base to include younger children
2. Look for collaborative ventures and/or licensure opportunities
3. Increase product safety
4. Improve knob control to improve drawing quality
5. Improve case seal
6. Maintain price points

hypothesis:
(1) An assumption based on limited evidence as a starting point for further investigation; and/or
(2) a proposed explanation for an observation. A hypothesis is an educated guess, which forms a basis for investigation or analysis.

STEP 2 Develop a Hypothesis

Because reverse engineering is most appropriately used to determine how something functions and how the components work together to achieve that function, the next step in the process is to describe your hypothesis about the product function. A hypothesis is a statement that suggests a possible, unproven answer to a question. In our Etch A Sketch example, our unknown relates to the mechanism that controls the drawing stylus, the part that draws the lines on the screen. Is the stylus driven by gears, a series of levers, or some other device? It may help to refer to a book on mechanical systems. Chapter 12 (Designing Mechanical Systems) shows many illustrations to assist in developing a hypothesis.

> **Science:** Scientists often develop hypotheses (predictions or unproven ideas) that require thorough research and evaluation. Technologists and engineers have adapted that process to engineering design: They conduct laboratory experiments, testing, and observations to confirm or disprove their hypotheses.

How do you think the Etch A Sketch works? Do you think the stylus (drawing point) is controlled through a pulley mechanism, or is it gear-driven? Using a small amount of testing and observation, you will discover that turning the knobs controls the vertical and horizontal movements of the stylus, which scrapes off a substance from the reverse side of the screen. So, how do the knobs actually control the stylus? If you are

Your Turn

Record your hypothesis in your engineer's notebook. Use sketches with arrows and **annotations** to describe your thoughts fully. You may want to ask someone to read your entry to see if he or she understands your hypothesis. Remember, your engineer's notebook serves as a historical record of your reverse engineering process. Clear, concise, and complete documentation is necessary, because in the future you may not always be available to explain or decipher your notes and sketches. For example, after you move to another company or retire, someone else may have to interpret your engineering notebook for continuous product improvement.

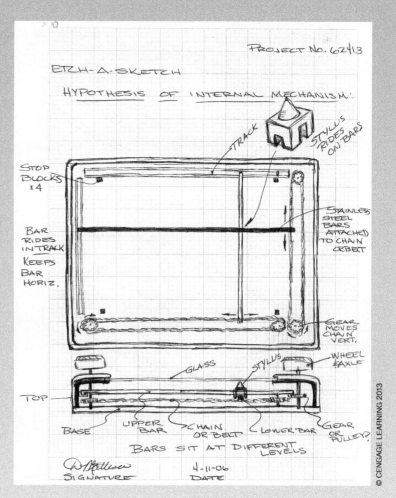

Figure 6-11: A sample sketch entry from an engineer's notebook to explain a hypothesis.

very patient, you can scrape off a small portion of the screen by drawing a set of lines very close together. This will allow you a glimpse inside. Can you guess the material used to coat the glass when the toy is shaken? Write down these and other questions you may have and think about possible answers before moving on to the disassembly stage.

 STEP 3 *Disassembly*

Once you have documented the purpose of inquiry and hypothesis, you may carefully disassemble the product to uncover the internal components and mechanisms. This process is often called **teardown**. When someone in an industrial setting uses tools or enters a material laboratory, he or she must wear safety gear. Safety is taken very seriously in business, industry, and education. General safety rules include wearing safety glasses, ear protection, and proper protective clothing. It is also common practice to remove all jewelry. If dust or other environmental contaminates are present, respiratory protection is required.

Common disassembly tools (both metric and U.S. customary) include:

▶ Screwdriver set

▶ Allen wrench set

teardown: The process of product disassembly to expose individual parts for analysis and enhanced understanding.

166 Part I: The Engineering Design Process

Safety Note

Whenever you disassemble a product, wear all necessary protective gear, and follow all safety instructions the teacher provides.

- Needle nose and regular pliers
- Spring compressor and tweezers
- Ratchets, sockets, and adjustable wrenches
- Drill and index
- Utility knife and cutting mat
- Wire strippers, tin snips, and side cutters
- Small handsaw, coping saw, or miter box and backsaw
- Micro pry bar and nail puller
- Strip heater or heat gun

The disassembly process (product teardown) is a very delicate process; you should do it slowly and with great care. Have you ever disassembled a broken item with the hope of repairing it, only to have parts go flying across the room as soon as the cover was removed? The person in charge of product disassembly must not lose any of the internal parts. Small plastic bags or envelopes and masking tape may help to organize parts. It is very important to take careful notes on the internal parts found within the product. You can use a data organizer, shown in Figure 6-12, to record information during disassembly.

Figure 6-12: A data organizer form.

Data Organizer

Name: _____
Date: _____
Product: _____

Testing Properties
- Chemical
- Electrical
- Magnetic
- Mechanical
- Optical
- Thermal

Manufacturing Processes
Combining	Conditioning
Joining	Machining
Molding/ forming/ casting	Pulverizing/ separating
Shaping/ reshaping	Shearing

Part Name	Material & Features	Dimensions	Weight	Mfg. Process	Noticeable wear	Connecting or mated parts

Case Study

Disassembling the Etch A Sketch

Figure 6-13 shows an annotated sketch of the Etch A Sketch toy made before disassembling it. The notes include speculations or facts about the external components.

When we carefully disassembled the Etch A Sketch, we removed the knobs and top cover and discovered a plastic sheet over a glass plate and another cover. We documented our findings by numbering each part with masking tape and recording the information onto the data organizer sheet. Through research, we discovered the cover was joined using sonic welding, which bonds the plastic parts together at a molecular level, making it very difficult to disassemble. An Internet search of sonic welding uncovered an interesting fact related to our case study of the Etch A Sketch. In the 1960s, the Ohio Art Company was the first company to test sonic welding of plastic on the Etch A Sketch product. Sonic welding is performed by using a movable "horn," which is operated at ultrasonic frequencies to produce high-frequency sound waves. When applied to the surface of a plastic part, the pressure and heat welds the two parts together.

One of the critical concerns of manufacturing an Etch A Sketch is creating an airtight product so that the contents will not seep out. Sonic welding was a solution selected by the Ohio Art Company to create a tight and durable seal of the plastic case (see Figure 6-14).

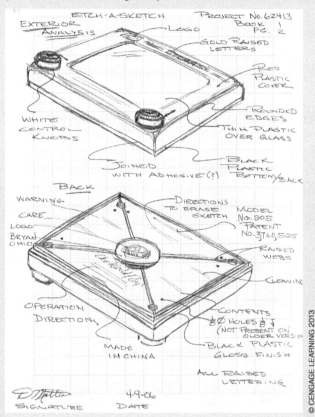

Figure 6-13: Annotated sketch of the Etch A Sketch prior to disassembly.

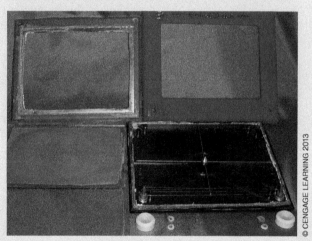

Figure 6-14: A partial disassembly of the Etch A Sketch.

Think back to a time when you disassembled an item for repair. How successful were you in putting the pieces back together? Did you remember where everything went? Keeping this in mind, you can see how important it is to carefully record information in an engineer's notebook. You should keep records on all part locations, interaction with other parts, and the purpose or function of each part. You can label parts by using masking tape to name or number the parts, which are cross-referenced on the organizational chart and engineering notebook sketches and entries.

Designers often use digital photographs to record information at several stages of disassembly and place them in their engineer's notebooks. They must keep in mind the purpose of the inquiry as they perform the disassembly process. If the purpose is to reduce

OFF-ROAD EXPLORATION

Research the Internet to discover the difference between adhesion and cohesion. Record the definitions in your engineer's notebook. Continue to search the Internet to learn more about sonic welding and its applications: Would you classify sonic welding as adhesion or cohesion process? What are some of the advantages sonic welding offers over glues or fasteners?

What other bonding or joining techniques could be used to create an airtight seal?

What other products might require airtight containers?

the return rate due to product failure, a list of consumers' comments should be reviewed prior to disassembly. The person disassembling the product will look for ill-fitting components, parts with excessive wear, or other indicators of product failure. Thoughtful, deliberate analysis and documentation must occur during all stages of disassembly.

>>> **STEP 4** *Analyze the Elements*

analysis:
A detailed examination of the elements or structure of something.

Analysis is the most crucial part of the reverse engineering process. During the analysis step, engineers attempt to answer all of the questions originally posed. There are four main categories of product analysis:

- Functional
- Structural
- Material
- Manufacturing

Analysis requires detailed research of each category. Research results are recorded in the engineer's notebook along with sketches and digital photos to provide clarity for detailed information. During analysis, some products may require partial reassembly to observe the interaction of functional components. For example, in the Etch A Sketch, the wire, rods, and pulley components, which control the movement of the stylus, remained assembled to trace the mechanical function from the control knob to the brass stylus.

Functional Analysis

Simply put, how does the product work? Usually functional analysis involves answering questions about a mechanical system or how an electronic circuit works. As you will learn later in Chapter 12 (Designing Mechanical Systems), nearly all mechanical systems contain at least one of the following elements:

- Lever or crank
- Wheel or gears
- Cam
- Screw
- Object that transmits tension or compression, such as a pulley, belt, chain, spring, or hydraulic fluid line
- Object that transmits intermittent motion, such as the ratchet

Look for these elements when analyzing a product's mechanical function. During analysis, an engineer will take measurements or perform tests on a product's components. Many tools have been developed to aid in measuring parts. Accurate measurements are important, and care must be taken to use each measuring tool correctly.

Micrometers and calipers are widely used to take precise measurements of material thickness, internal and external diameters of shafts, and the depths of slots and holes (see Figures 6-15 and 6-16).

Modern measurement instruments include the dial caliper shown in Figure 6-16 and digital caliper shown in Figure 6-17. Both are easy to use to obtain dimensions.

Precise measurement and testing with quality instruments increases the reliability of the data collected during the reverse engineering process. Figures 6-18 and 6-19 show various instruments used to take measurements. Measurements provide a better understanding of part interactions and clearances. If you are reverse engineering to design a replacement part or improve an existing part, accurate measurements are vital to a successful outcome. Some tools used to obtain measurements include:

- Micrometers
- Calipers
- Radius gauge
- Feeler gauge
- Protractor or digital angle gauge
- Digital Scales
- 3-D scanner

Structural Analysis

All structures, regardless of their purpose, must support internal and external loads, and must hold parts in place. One of the first steps in structural analysis is to determine the purpose of each part and how it interacts with the other parts.

Structural analysis addresses all or some of these properties (interactions). The way a product is constructed must provide:

- Support for internal parts
- Housing
- Container
- Protection
- Transportability

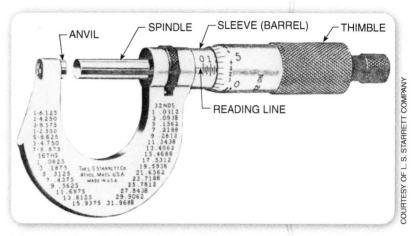

Figure 6-15: *Micrometers are used to take external, internal, and depth measurements.*

Figure 6-16: *Dial caliper.*

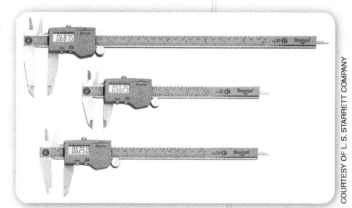

Figure 6-17: *Digital caliper.*

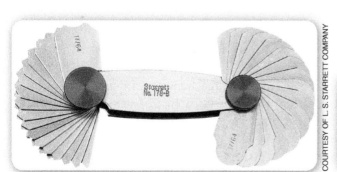

Figure 6-18: *Radius gauges are used to measure rounds and fillets.*

Figure 6-19: *Feeler gauges are used to measure gap width and clearance between two parts.*

Case Study

A Functional Analysis of the Etch A Sketch

Look closely at Figure 6-20, which shows the complex wire, pulley, and rod system as the internal mechanism of the Etch A Sketch. This mechanical system converts the rotary motion of turning the control knobs into the linear motion of the stylus (drawing tool). The vertical and horizontal movement of the rods is controlled through wire tension. The rods slide through holes in the brass stylus—as the rods move, so does the stylus. By turning both knobs at the same time, a variety of diagonal lines and arcs can be drawn.

What would happen if the holes in the stylus were too small for the rods to pass through smoothly? How would the performance be affected if the opposite situation occurred where the holes were much larger than the rod? How does part material impact the amount of wear that could occur over time? What would happen if the wires of the pulley system were too tight or too loose? Will the device continue to function if one of the wires breaks?

During the mechanical analysis of our case study, many drawing entries were added to the engineer's notebook. Colored pencils were used to clarify and visually separate the two independent pulley systems contained inside the Etch A Sketch, as shown in Figure 6-21.

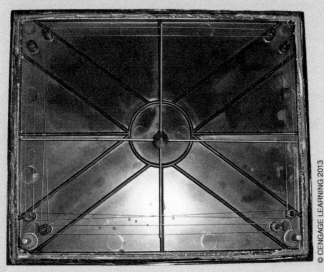

Figure 6-20: *The detailed string and rod mechanism inside the Etch A Sketch.*

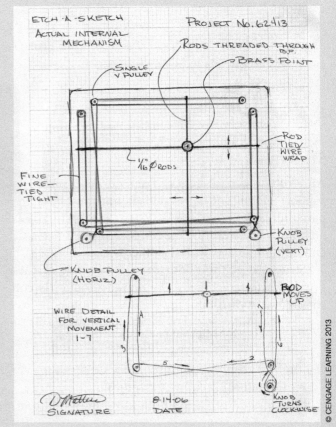

Figure 6-21: *An engineer's notebook sketch entry of the internal mechanism.*

Structural analysis involves principles of mechanics and material properties. Finite element analysis (FEA) software can be used to learn more about the structural qualities of the product being reverse engineered. You will learn more about structural analysis in Chapter 11 (Designing Structural Systems).

For the reverse engineering process, we will research and identify material strength of the structural component and decide if those properties match the structural purpose. Often computer programs with finite element analysis provide precise structural and stress analysis information.

Case Study

A Structural Analysis of the Etch A Sketch

The main structure for our Etch A Sketch toy is the case. It *houses* the internal component parts, and must *protect* and keep the parts in place for the toy to function properly. The stylus is a moving part *supported* by two rods. These rods must be of a material durable enough to hold the stylus without bending as the stylus glides back and forth. The stylus must be constructed of material that will not be damaged by repetitive movement. If the stylus breaks, the toy is useless. The stylus in the Etch A Sketch is made of brass, identified by its softness, compared to many other metals, and muted gold-like color.

You might wonder why brass was chosen for this part. Although basically soft, brass, an alloy of copper and zinc, is stronger and harder than pure copper. Remember that the stylus must ride back and forth continually on the rods without wearing out. The function of the stylus is to remove the aluminum powder from the glass surface to form a clean sharp line without scratching. What would happen if the stylus was made of a harder metal?

During the structural analysis, solid modeling programs can be applied to produce finite element analysis (FEA) of individual parts (see Figure 6-22).

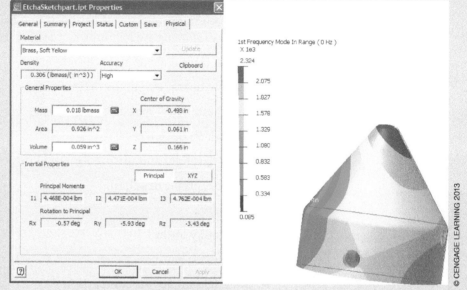

Figure 6-22: *Finite element analysis (FEA) of the brass stylus and the supporting documentation obtained by using the Autodesk Inventor™ solid modeling program.*

Materials Analysis

The choice of material greatly affects a part's performance, and the material's properties must be correctly matched to the part's application. At a most basic level, we can identify a material by its common name. Materials are usually identified by type, such as wood, metal, plastic, ceramic, or composite. Designers need to understand material properties and how these properties contribute to performance and durability (see Figure 6-23). It is also useful to know what kind of manufacturing processes are used to manipulate materials into desired forms.

Analyzing materials requires understanding basic material properties, because materials are also identified by scientific properties. These properties include:

▶ Mechanical
▶ Electrical
▶ Thermal
▶ Chemical
▶ Optical
▶ Acoustical

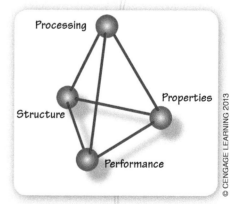

Figure 6-23: **Material science tetrahedron.**

Material properties are measured in various units. For example, Young's modulus is a mechanical property that describes dimensional change in a material as it is subjected to an applied load (pressure or weight). The values are measured in pounds per square inch (lbs/in^2 or psi) or in newtons per square centimeter (N/cm^2). Material property values are determined through standardized tests. The values of these material properties are compared for design purposes. For example, structural steel has a Young's modulus value of approximately 30,000,000 psi. The value for aluminum is approximately 10,000,000 psi. Therefore, an engineer can look at these values and recognize that steel is three times stronger than aluminum. This is one reason why builders construct the skeletal structures of skyscrapers with steel instead of aluminum. More information about material properties is included in Chapter 10 (Manufacturing).

The following items are helpful when testing material properties:

- Digital force scale or balance
- Laser temperature gun or thermometer
- Strip heater, heat gun, or oven
- Stop watch or timer
- Small flashlight and mirror, for visual inspection
- Material properties chart
- Magnets
- Continuity tester
- Durometer hardness gauge
- Material samples for comparison

When engineers analyze a product, they add careful notes in the engineer's notebook about the material characteristics of each part. As engineers examine the parts, they also look for signs of wear or other indicators of potential failure.

Manufacturing Analysis

Manufacturing is a broad term used to describe the application of tools and processes to the transformation of raw materials into finished goods. Designers always consider manufacturing processes during the design process. Professionals who help design the most efficient way to produce a new product are *manufacturing engineers*. Manufacturing engineers are either part of the design team, or act as consultants during the design process.

Many different systems to produce products are available. In Chapter 1, you learned about the *American system*, which became very successful during the late nineteenth century. Other methods include mass, just-in-time, lean, and sustainable manufacturing. Manufacturing systems will be covered in Chapter 10 (Manufacturing).

The reverse engineering process determines how the product was manufactured. Although it may be difficult to determine the actual production system used to make the parts, the manufacturing processes are relatively easy to identify. Materials are manipulated by three different methods—forming, separating, and joining:

1. *Forming* methods use heat and/or pressure to reshape a material into a desired form.
2. *Separation* methods carve a desired form from an existing block of material.
3. *Joining* methods combine two or more objects together.

Case Study

A Material Analysis of the Etch A Sketch

The screen-clinging powder, which puts the "magic" in the Etch A Sketch, has very specific properties. When examining the fine powder, we began our search by matching the material to a known item, or sample. The color and reflective property of the powder was similar to aluminum foil, aluminum cans, aluminum fishing boats, airplanes, and trim on storm windows. The major characteristic noted was a gray color and silvery reflectivity as it was spread across the fingers, as shown in Figure 6-24.

Internet research showed that the properties of this material are consistent with the properties of finely powdered aluminum.

Recently, the Connecticut-based company that produced the aluminum powder for the Etch A Sketch shut down, and a new source had to be located. Before a new powder could be purchased and used, a multitude of tests were performed to determine the material's reliability, performance, and safety. Fortunately, the exhaustive process led to a less expensive, finer particle-sized material, which worked even better than the original material!

Figure 6-24: A close examination of the "magic" powder and the polystyrene beads used to distribute the powder onto the glass screen.

To remain competitive and offset ever-rising material costs, the Ohio Art Company continues to research and test alternative materials for the Etch A Sketch. High-impact polystyrene (HIPS) recently replaced the original plastic cover made of acrylonitrile butadiene styrene (ABS) plastic. High-impact polystyrene is not quite as shiny or resilient to scratches, but it still meets the needs of the application.

Your Turn

Keep your purpose of inquiry in mind as you perform a material analysis. As you reverse engineer the Etch A Sketch, you will observe a thin sheet of plastic covering the screen (see Figure 6-25).

Use www.wikipedia.org and search "plastic," and then select "common plastics and their uses." You are looking for a plastic film listed as polyethylene terephthalate (PET). You will discover that one of the trade names of PET is Mylar. Research the properties of Mylar and record the information in your engineer's notebook. As part of continuous product improvement, the Ohio Art Company added Mylar film to the Etch A Sketch in the 1970s. What do you think is the purpose of the Mylar covering? Record your thoughts and the results of your research in your notebook.

Figure 6-25: In response to changes in safety laws and concern for product safety, a Mylar sheet was placed over the glass Etch A Sketch screen.

Table 6-1: **Common industrial processes.**

Forming Processes	Separation Processes (material removal)	Joining Processes
Casting	Drilling	Welding
Molding	Shearing	Brazing
Forging	Sawing	Soldering
Rolling	Electrical discharge	Sintering
Extruding	Milling	Adhesive bonding
Pressing	Turning	Fastening
Bending	Broaching	Stitching
Piercing	Shaping	Stapling
	Planing	Finishing
	Honing	Electrochemical
	Sanding	Chemical

Other joining methods are used to create extremely strong and lightweight composite materials, such as those found on modern aircraft. See Table 6-1 for a list of common industrial forming, separation, and joining processes.

The seven resources of technology are people, capital, tools, machines, time, information, and energy. Along with these resources, the quantity of parts required, environmental impact, and government regulations all enter into the determination of a suitable manufacturing process. If large numbers of products are to be made, a mass production technique may be most efficient. For smaller production runs, some assembly by hand may be appropriate. Once you have identified the material and manufacturing process, you will need to analyze the product's assembly. During this step, you will closely examine the product and review your notes from the disassembly phase to determine the sequence of assembly. Can the entire product be assembled through automation, or are there delicate parts that may need to be assembled by hand? Which parts will need to be assembled first? Can you think of a way to streamline the assembly process? Record your answers and ideas in your engineer's notebook. These considerations will greatly affect the product's manufacturing cost.

In our case study, the use of brass for the stylus is essential in the process for another reason. The Ohio Art Company tested several other materials, but none functioned as well as brass in removing the aluminum powder from the glass to create a clean, crisp line.

STEP 5 Prepare the Report

In the report stage of the reverse engineering process, you will present your findings. Start out by confirming that you have addressed the purpose of the reverse engineering inquiry by revisiting Step 1. For example, if the purpose of the reverse engineering process was to develop CAD drawings for conversion to computer numerical code (CNC), have you gathered the information necessary to create them? If you reverse engineered a product to determine the durability of the internal mechanisms, what did you discover? Did any parts show wear? Were you able to determine the cause of the wear and how to correct it? If your purpose was product improvement or evolution, how did the information gathered about the product function lead to new understanding? In preparation for your report, remember to revisit your hypothesis. Were your assumptions correct? What information did you discover most helpful in addressing the purpose of inquiry? Once you answer these questions, you are ready to prepare your report for presentation.

Case Study

A Manufacturing Analysis of the Etch A Sketch

Let's explore a few of the manufacturing processes used on the Etch A Sketch. The exterior case was formed by molding. Sonic welding, a joining process, was used to permanently attach the components together.

Earlier in our case study, we deduced that the stylus was made of brass for its durability, weight, noncorrosive properties, and because it is malleable (easy to shape).

Brass can be machined into precise shapes like the cone-shaped stylus needed for the drawing point. The stylus is also drilled to allow the vertical and horizontal rods to pass through. See Figure 6-26.

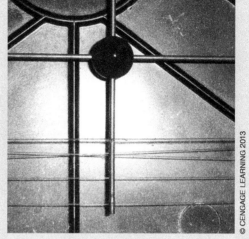

Figure 6-26: *Brass stylus.*

Fun Facts
Assembling an Etch A Sketch

You may be surprised to learn that the crucial operation of stringing and tying the pulley system is completed by hand to ensure the wire filaments are tight and knotted at proper locations. No machine operation can replicate this delicate procedure satisfactorily. The tight stringing of the wires provides a quick stylus reaction when the knobs are turned (see Figure 6-27). All assembly workers completing this important function must undergo detailed training to ensure a quality product.

Figure 6-27: Close-up of the wire stringing.

Communicating the results of your reverse engineering process in a clear and concise way requires careful consideration and planning. Determine which method of delivery will be most effective in presenting the relevant information. Design teams often use a PowerPoint program to support their presentation. Sometimes teams use a poster session to graphically present information on a series of poster panels. Other support documentation might include charts, graphs, photographs, video, or a combination of these. You will learn more about presentational techniques in Chapter 18 (Graphics and Presentation).

Case Study

Following are some examples of documentation you might choose for developing the reverse engineering report and presentation of the Etch A Sketch product (see Figures 6-28, 6-29, and 6-30).

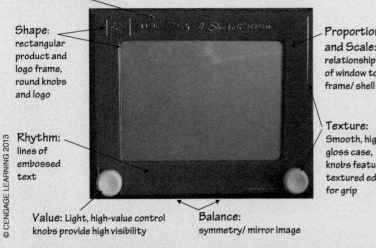

Figure 6-28: Example of a PowerPoint slide to document a product's exterior design analysis.

Figure 6-29: Presentation poster showing product disassembly and analysis of parts.

Case Study

(continued)

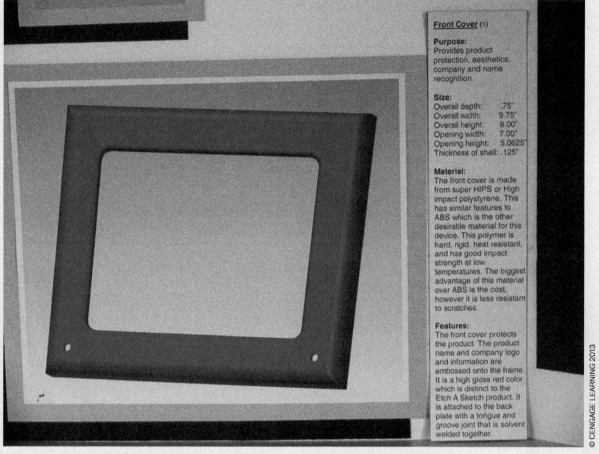

Figure 6-30: Detail of the presentation poster showing typical part documentation.

The Etch A Sketch Today

The reverse engineering of our case study did not uncover any mechanical product weakness. The product functions well and has stood the test of time. However, the product has undergone continuous improvements. For example, designers redesigned the knobs that control the movement of the stylus for better control by child-size hands (see Figure 6-31). Researchers collected data from a test group of young users to determine control knob diameter, gripping surface, amount of force needed to turn the knobs, and dexterity needed to use it.

Can you see another improvement of the newer design shown in Figure 6-31? Children have to think about which knob to turn to achieve the desired lines. If you have ever tried to write your name using an Etch A Sketch, you know how much hand-eye coordination is required. Consider the difficulty of making the drawing in Figure 6-32. A relatively new release from Ohio Art, the Etch A Sketch Freestyle, now features a single control.

(continued)

Case Study
(continued)

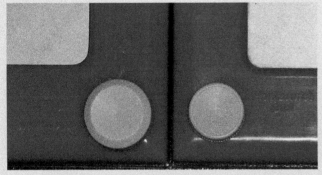

Figure 6-31: The redesigned control knob, shown on the left, features improved size and shape to provide better control for small hands.

THE ETCH A SKETCH® PRODUCT NAME AND CONFIGURATION OF THE ETCH A SKETCH® PRODUCT ARE REGISTERED TRADEMARKS OWNED BY THE OHIO ART COMPANY.

Figure 6-32: Etch A Sketch drawing. The Etch A Sketch product name and configuration of the Etch A Sketch product are registered trademarks owned by the Ohio Art Company.

THE ETCH A SKETCH® PRODUCT NAME AND CONFIGURATION OF THE ETCH A SKETCH® PRODUCT ARE REGISTERED TRADEMARKS OWNED BY THE OHIO ART COMPANY.

A continuing challenge for the Ohio Art Company is to produce the Etch A Sketch at the price point cost and still make a profit. The company continues to research and test packaging options as one way to decrease product cost. Another focus of redesign is to increase sales by adding new Etch A Sketch products that will appeal to younger children. Art Clark, vice president of manufacturing of the Ohio Art Company, tells us that new products include Etch A Sketches featuring SpongeBob SquarePants and Dora the Explorer (see Figure 6-33). The company continues to work on innovations to keep the product in constant demand.

Figure 6-33: SpongeBob SquarePants and Dora the Explorer Etch A Sketches.

THE ETCH A SKETCH® PRODUCT NAME AND CONFIGURATION OF THE ETCH A SKETCH® PRODUCT ARE REGISTERED TRADEMARKS OWNED BY THE OHIO ART COMPANY.

STEP 6 *Product Redesign*

Depending on the initial purpose behind the reverse engineering and the outcome of the process, redesign may occur. If continuous improvement or part failure initiated the process, then the data you have gathered will lead to brainstorming of ideas for redesign. Did the reverse engineering process identify a weak part that will require redesigning? Will you recommend different materials or a new manufacturing process?

At this point in the reverse engineering process, you could go in several directions. Based on your findings, what recommendations would you make? Do not forget to use charts, graphs, or a CAD program to support your research and ideas. Refer to Chapter 2 (The Process of Design) and Chapter 4 (Generating and Developing Ideas) to assist you. The redesign may be as simple as correcting a weakness by changing the size or part material. However, it may be as complex as redesigning into a completely different product.

Many times a quality product inspires or leads to other developments by other companies. You will often see many similar items on store shelves. Smart consumers will examine the features closely to compare options before making a purchase. Sometimes a product leads to the development of accessory items, which are used in conjunction with the original item. One example from this category includes cell phone covers. They are designed to fit a product made by another manufacturer and there are many styles and materials available. Do you think designers used reverse engineering to gather size specifications of the cell phones?

Fun Facts

In the following example, the Etch A Sketch inspired SIGGRAPH (Special Interest Group on GRAPHics and Interactive Techniques) to create a huge Etch A Sketch, which could be controlled by three thousand people during a "groupthink" collaborative experience (see Figure 6-34). If you look at the photo closely, you will see that their drawing is not in perfect alignment with the teapot overlay. Why do you think the audience had trouble drawing the teapot?

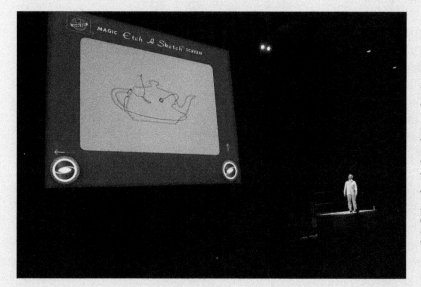

Figure 6-34: The world's largest Etch A Sketch was unveiled at the 33rd SIGGRAPH international computer graphics and interactive techniques conference and exhibition in Boston. Using Cinematrix's patented audience participation technology, the left half of the audience turned the left knob while the right half turned the right knob to create a drawing.

THE ETCH A SKETCH® PRODUCT NAME AND CONFIGURATION OF THE ETCH A SKETCH® PRODUCT ARE REGISTERED TRADEMARKS OWNED BY THE OHIO ART COMPANY.

Careers in the Designed World

ART CLARK: VICE PRESIDENT, THE OHIO ART COMPANY

Just an Ordinary Guy

Art Clark describes himself as "an ordinary guy from South Jersey," but his career has been anything but ordinary. From military service, to hourly work in a fiberglass insulation plant, to managing a toy company's global manufacturing operations, Art has always kept his career moving.

Art is the vice president of manufacturing in the Bryan, Ohio, facility of the Ohio Art Company. Art oversees production operations and the development and manufacture of new toy products. His responsibilities include human resources, plant maintenance, and close work with engineering, quality, and research and development personnel to design and manufacture toy products. "In addition to developing our new toy concepts, inventors bring us toy concepts, usually in rough prototype form, and we either develop the product further or we learn from an idea or concept and think about ways to transform it into something else."

On the Job

Art travels to China and Hong Kong to supervise production operations, product design revisions, material testing, and new technological manufacturing and assembly processes. With product contributors expanding across the globe, Art makes sure that communication between all parties is effective and timely. When a product is in development, Art spends time in daily and weekly meetings to work with personnel from engineering, finance, quality, R&D, sales, and marketing. Each discipline brings something to the table as they deliberate product and package design, tooling, mock-ups, prototypes, costs, schedules, licensing, safety, approvals, and other development considerations.

Inspirations

Even while Art was working at the insulation plant, he was taking steps to improve his career path. While working full time, Art attended night school to obtain his associate's degree. Art's work ethic helped him move up the corporate ladder until he became the plant manager. After twenty-six years of service in the company, a leadership change led to downsizing, and Art lost his job. Although many people would describe downsizing as a career setback, Art saw it as an opportunity. He went back to college to complete a degree in behavioral sciences.

Eventually, Art's new career path led him to the Ohio Art Company. "This was quite different from anything I had done before," Art recalls.

Advice for Students

When asked to describe one thing he brings to the table, Art replies, "I try to bring fresh creative ideas, and challenge the team to pursue alternate avenues and directions as required. Changes are necessary to keep up, as there are constant challenges to meet objectives such as targeted costs, production capacity, and on-time product delivery to our customer base. The major issue encountered during the product development process is related to change."

Art Clark's job and responsibilities have changed tremendously during his career, but the one thing that holds constant is his ability to adapt to change by working successfully with others and thinking creatively. By creating a collaborative environment, Art continues to provide fresh ideas and solutions within the framework of continuous improvement.

Betty Spaghetty, an Ohio Art Company toy.

SUMMARY

To remain competitive in the marketplace, companies frequently review their existing product lines to see how they can improve their products. Companies also look at competing companies' products to develop a better understanding about product function, structure, materials, and manufacturing processes. This new understanding can lead to something completely new, an invention, or product improvements (product innovation).

Reverse engineering is a process by which an individual or a company can learn how a product works and how a product is made, which can lead to a better understanding of new processes, techniques, technology, and emerging markets. Most technology and engineering students will reverse engineer a product at some time during their education. The reverse engineering process will help students gain insight into how the product was created, and to learn about various factors that impact product design. Many times, the investigation and research phase of a design process will include reverse engineering. Reverse engineering can result in the development of CAD/CAM documentation for detailed repair or accurate replication of historical products or parts.

Designers begin the reverse engineering process by determining what needs to be learned from the product in question. Typically, designers use reverse engineering to do one of the following tasks:

- Research similar products in an effort to discover possible ways to make a more competitive product
- Test a product or design to determine what is failing or causing a failure
- Change or add to a company's continuous product improvement policy
- Provide documentation for product components when original drawings are no longer available or accurate
- Educate design professionals on the topics of function, structure, manufacturing, and aesthetics
- For equipment repair, through the design of replacement parts, for products no longer in production
- To develop CAD and CNC electronic data for computer-enhanced manufacturing processes

An engineer will perform a reverse engineering process to gain insight about a product or its components. The process typically starts by identifying the purpose of the inquiry (Step 1). Then the engineer makes some guesses about how the product functions—development of the hypothesis (Step 2). The engineer carefully disassembles the product (Step 3) and begins to analyze the components (Step 4). Designers document the results of the reverse engineering in a report and often communicate them in a formal presentation (Step 5). Finally, depending on the initial purpose and the outcome of the process, they may redesign the product (Step 6). An important fact to remember about the reverse engineering process is that it does not necessarily proceed in a sequential manner. Several times throughout the process, an engineer may need to jump back to an earlier step to gather additional information.

The outcome of reverse engineering is a better understanding of product function, structure, materials, and manufacturing processes.

We used the Etch A Sketch toy as our case study to demonstrate how a real product has kept its competitive edge in the toy market. The case study featured several steps in the reverse engineering process and provided examples of documentation from an engineer's notebook.

BRING IT HOME

OBSERVATION/ANALYSIS/SYNTHESIS

1. Make a list of discarded, broken, or unused items. Choose three items from your list and make a chart to record answers to the following questions: What was the initial purpose of the item when originally purchased? What caused the item to fail? Why is the product no longer in use? What could be done to repair or improve the item?
2. Choose one item from the previous list and use the Internet to help you identify the product's materials. Make a list of some of the material properties that may have been a consideration in the design phase. If you were to reverse engineer this product, what other materials would you research as a possible replacement? Is the material readily available, easily machined, durable, and recyclable?
3. Create a reference sheet of manufacturing processes identified in this chapter to assist you in the reverse engineering of future projects. Include a clear definition, product examples, and the advantages and disadvantages of each process.
4. Take a digital picture of your mailbox or another product approved by your teacher. Use the Internet to trace its history and create sketches that show various stages of the product's evolution, as it underwent continuous improvement.
5. Make a list of mechanical products found around your home. Identify the type of mechanism used to make the product function as intended.
6. Select an inexpensive product to reverse engineer. Document all steps of the reverse engineering process and make a report to your class on your findings.
7. Use your computer-aided design system to create solid models of individual parts of a 3-D brainteaser puzzle or Transformer. Assemble the parts and create an animation to illustrate how the item can be manipulated.
8. Reverse engineer an item, such as a desk accessory, that you could ship disassembled to save packaging. Create your own set of instructions with an exploded view to explain how to assemble the item.

EXTRA MILE

ENGINEERING DESIGN ANALYSIS CHALLENGE

- Carefully disassemble and reverse engineer an inexpensive item with two or more mechanisms. Develop the computer-aided design documentation for each part. Prior to disassembly, record your hypothesis about the mechanism and its function.

- Reverse engineer a broken or malfunctioning product to determine its weakness or cause of failure. Create a solid model and technical drawings for a replacement part.

- Use your computer-aided design program to create a solid model of a part that has failed, and perform a finite element analysis (FEA) on the part. Make recommendations for redesign based on the results of the FEA.

- Reverse engineer and compare two inexpensive competitive products to determine why a customer would choose one over the other. Develop your own data organizer to document your findings.

CHAPTER 7
Investigation and Research for Design Development

Menu

 Before You Begin
Think about these questions as you study the concepts in this chapter:

1. Why is investigation and research necessary?
2. How did H. J. Heinz use research to develop the Top-Down™ ketchup bottle?
3. How do I decide what questions to ask?
4. What types of research might I need to do?
5. How do I conduct the research I need to answer questions about my design problem?

INTRODUCTION

The purpose of investigation and research is to collect data and information that will assist you in the design process. Asking the right questions and consulting the right sources are keys to being a successful designer. Investigation and research helps you move forward. It provides a pathway to follow and it helps you know when to pause and look for more information. Sometimes it takes you in new directions that you never initially considered, which we identified in Chapter 4, Generating and Developing Ideas, as lateral thinking. Jennifer thinks about the problem she has with the lighting in her room. Dave's family travels a lot and enjoys riding bikes, but they always have trouble finding space for their bicycles when they pack for their trips. Jose enjoys sports and would like to improve his soccer shot. Natalie's mother is opening a new store and needs a sign that will hang above the front door. These are all problems or opportunities that raise many new questions (see Figure 7-1).

Each of these students will need to answer many obvious questions, but, more importantly, they need to identify and ask additional questions. Probably the most important part of conducting investigation and research is asking the right questions. Effective design team members know how to ask the right questions and are skilled at using resources to get the right answers. Whether it is a student learning to do design or it is a company making a new product, finding the right information is critical to the success of the project.

Whenever a company like Kellogg's introduces a new cereal or Trek produces a new bike, a great deal of research is already completed (see Figure 7-2). These companies spend as much as 50 percent of the project time initiating the new product design process. They also spend a great deal of money gathering and analyzing information about how the `consumer` will respond to a potential new product.

As you learned in Chapter 3, Development of the Team, the design team must find information to support the design process. Considerable research goes into how best to produce and market the product. The design team involves many people and resources as they move toward a final solution and product.

Early in the design process, the team must learn about market potential. This is done through

Figure 7-1: Asking the right questions and finding the right information.

Figure 7-2: Trek bicycle.

`market research`, which informs the team about industry trends and the `demographics` of the people who will purchase the product. The team will also use the Internet or trade publications to locate information about successful products. Let us consider the development of a new package for Heinz ketchup. Consider what information the H. J. Heinz Company design team needed to be able to bring the Top-Down bottle successfully to the marketplace.

185

Case Study

Designing a New Ketchup Bottle for the H. J. Heinz Company

HISTORY OF KETCHUP

Ketchup originated in Asia as a sauce made from pickled fish—try putting that on your french fries! Evidence shows early recipes for tomato ketchup printed in Virginia in the early 1800s.

By 1875, the food processing industry was trying to earn the public trust for foods packaged in cans and bottles. Henry Heinz, along with his brother John and cousin Frederick, was one of the first to bottle tomato ketchup in the late nineteenth century (see Figure 7-3).

Figure 7-3: The Heinz family began to bottle ketchup in the late nineteenth century.

Background

Ketchup was originally packaged in amber or green glass bottles. In 1869, Heinz was the first company to introduce the clear glass bottle, thus allowing consumers to see the contents inside. An important milestone for the food processing industry was the passage of the first Pure Food and Drug Act in 1906. In addition to the company support for the new act, Heinz became the first company to employ scientific control and laboratory testing to ensure product quality in 1918.

Some restaurants still use glass ketchup bottles, but have you noticed that it is difficult to get the ketchup out of those glass bottles? Polyethylene plastic replaced glass in the 1980s, making it easier to pour the ketchup and eliminating the danger of broken glass if the bottle was dropped. Unfortunately, consumers still complained about the initial "watery" first squirt and pouring remained difficult as the bottle became empty.

Kids and adults love ketchup. Forty-six percent of all people surveyed said that ketchup was their favorite condiment. Heinz produces more than one billion ounces of ketchup per year. With annual sales of over $1 billion, Heinz ketchup sales represent twice the market share of the nearest competitor.

Problem Identification

Bill Johnson, chairman and CEO of H. J. Heinz, explained to the design team that the company must always look for ways to innovate their older brands. Mr. Justin Lambeth, Heinz brand manager, was told to come up with the "next new product for ketchup!"

Investigation and Research

Lambert studied consumer market trends and found that convenience was the number one factor or trend in other product categories. He noted that consumers complained about difficulty getting the ketchup out of the bottom of the bottle and the dried ketchup made a mess around the cap or lid. Research studies reported that nearly 25 percent of the consumers used a knife to get the ketchup out, something that was very common with the older glass bottles. Of great interest was the fact that 15 percent of consumers stored their bottles upside down—a finding that would later lead to the new bottle design.

Anytime a company makes a major change in their product, they take a risk that the consumer will not be happy. New designs can confuse consumers, erode **brand** recognition, and reduce market share. According to Casey Keller, managing director for Ketchup, Condiments, and Sauces at Heinz North America, the company uses primary market research to guide their product development process. In the case of the new ketchup bottle, Heinz knew that kids used more ketchup per capita than any other age group, so they wanted to find ways to make using ketchup a fun experience for them. In a brainstorming session with a focus group of children ages 6 to 12, they found that the children were using the bottle to draw art on their food—the inspiration for the new package design was born!

Case Study

(continued)

The standard Heinz ketchup bottles are made of polyethylene terephthalate (PET) plastic with a 0.25 inch orifice cap. This bottle design delivers a full serving of ketchup in one or two squeezes. Unfortunately, the existing closure system would not work for an inverted bottle design.

The design team studied other product packaging systems and found an upside-down concept used to package shampoo. If the ketchup bottle was inverted, a new cap such as the one used for the shampoo bottle would have to be designed. Further investigation found that self-sealing caps using a silicone valve closure design had been available in the United States for over 10 years. Because the technology already existed, it could be used at a lower cost.

The design team also considered human factors (ergonomics) issues. It would be necessary to decide what population to consider for the bottle shape. Adults have larger hands than children or adolescents. Comfort was a primary concern, and the bottle shape needed to fit smaller hands and feel softer. Grip was also a consideration for the bottle design so that children could hold and control the bottle. Grip and comfort are different concepts but are related to material used (slipperiness and feel) and overall shape and size.

Aesthetics, or the look of the bottle, was also important. Although a new design could be a fun food experience, Heinz also recognized that the consumers wanted to be reassured that it was the same Heinz ketchup they were purchasing. As a result, the designers decided to keep the traditional keystone label easily associated with the Heinz brand.

Refinement of Preliminary Ideas

With the preliminary design concept of the upside-down package approved, the team learned that the new package would need a more stable base. A bigger base would give more stability but would also increase cost—a trade-off the team needed to consider. They decided that the diameter of the cap in relation to the bottle size was a critical factor and that the lid needed to be very flat. A CAD/CAM program was used to determine the center of gravity of the bottle designs using different amounts of ketchup.

Testing and Evaluating

The team tested the silicone valve closure for performance (leaking and sealing) using the standard ketchup thickness. They also evaluated the cost of the new closure. The 0.1-inch opening of the twist-shut closure from Creative Packaging Corporation was finally selected to give the consumers more directional control from the higher-pressure flow and allow children or adults to have fun "drawing" with the ketchup.

During the test marketing, consumers indicated that avoiding the mess around the traditional lid of the upright bottle was a very attractive feature.

The team also found that consumers were willing to pay a premium price for the convenience of the new package design (see Figure 7-4).

Figure 7-4: *Heinz Top-Down™ package solution.*

IMAGE IS OWNED BY H.J. HEINZ COMPANY AND USED WITH PERMISSION.

Post Log

According to Keller, the new Top-Down package introduced in 2000 reflected the collaborative spirit of the design team. Even companies with instant brand recognition like H. J. Heinz must invest and adapt to keep up with new consumer expectation. Colored ketchup products including green, purple, pink, orange, teal, and blue were added to the new product line but have since been discontinued.

In 2005, Heinz formally dedicated a new $100 million, 100,000 square foot "Global Innovation and Quality Center" near Pittsburgh, Pennsylvania. Teams of researchers, engineers, package designers, chefs, food technologists, and nutrition and quality assurance professionals will work to deliver new products, new packages, better nutrition and product taste, and improved consumer value. In addition to direct product improvement, the center will conduct basic research on understanding the value of antioxidant lycopenes on consumer health, while botanists and agronomists work to develop tomato hybrids to improve yield, color, and flavor.

Your Turn

The Heinz team asked questions and gathered information based on the following techniques and questions:

1. Corporate mission/goal, finance: Do we need to reinvigorate the mature brand?
2. Market research: What are the current market trends?
3. Focus group research: What are the problems with the current product package?
4. Market research: Will a new package confuse the consumer and erode brand recognition?
5. Human factors and stability: What shape should be used for a new package?
6. Engineering design: What material should be used for a new package?
7. Engineering design: Will a new cap or closure need to be designed?
8. Graphic design: How should a new label look?
9. Marketing: Can we successfully market a new bottle?

Which of these techniques and questions might you use to complete your next design project?

ASKING QUESTIONS

How many school reports have you written? In writing the typical school report, students usually only need to ask **closed questions**, which generate specific answers. For example, reports titled "What were the causes of the Civil War?" or "How did Thomas Edison invent the light bulb?" require looking up answers to these closed questions in a variety of library sources. The typical report-writing process usually involves taking careful notes on the specific topic and preparing a well-organized and thoughtfully written report.

Research for designing is different. Asking the right questions is at the heart of finding appropriate solutions to problems. An **open-ended question** can have more than one answer and often leads to more questions. For example, "How can I design a bicycle that can be transported in the trunk of a car?" Answering this question requires research on many topics, including the design of bicycles and car trunks. Dave needed to consider what bicycle designs already existed (see Figure 7-5). Do you think he should consider a bicycle that folds to fit into the

closed question:
A question that has only one correct answer.

open-ended question:
A question that has many possible correct answers.

(a)  (b)

Figure 7-5(a) and (b): **Dave found this collapsible bicycle design on the web.**

trunk rather than hanging it on the outside of the trunk? Answers will vary, and each answer will lead to more questions about materials, structures, and design aesthetics. Being able to ask the right questions is crucial to all research.

The first questions asked should be general in nature. Does the design team need to consider changing an existing product or creating a totally new product? Should the team recommend that the company drop the product line and move in another direction? For example, the Heinz design team needed to consider what changes in the ketchup package consumers would like.

> Design teams use market research to find information about potential customers. Research companies use quantitative research techniques to identify samples of the population who will then answer a questionnaire concerning future products. These companies then analyze the data using statistical measures and prepare a report for the design team.

USING MARKET RESEARCH

Market research is a good place to start and usually involves asking questions about consumers. Companies generally need to know about the market potential before considering a new or revised product. Some forms of market research are very general and available to consumers and businesses. For example, if you are considering purchasing a new cellular telephone, you should check out how the newest products on the market are rated by one of the many consumer reports. A quick check of the Web will locate these reports (some free of charge and some for a fee). Reports by an independent organization such as the Consumers Union are usually the most reliable. Their annual buying guide is available at any library. If you are a company considering your next cellular telephone design, reviewing the *Consumer Reports* on your model and competitive models is very helpful.

Sometimes companies and students want to find answers to specific questions. For instance, Heinz wanted to know what consumers did not like about their existing ketchup bottle. To answer this type of question and others, research companies may mail out specific questionnaires to consumers, conduct focus groups, or interview certain consumers. Generally, each individual is asked the same set of questions. The data are then collected and analyzed. The information received from market research is critical to the success of new product designs.

Design teams utilize market research early in the project formation and throughout the design process. Research provides information about industry trends and the demographics of the people who will purchase the product. Common research questions are: Who will use this product? Will the appeal of this new product be its price, appearance, uniqueness, or ease of use? What similar products are already available? Will this new product be worth the time and money that it takes to produce it? Other questions about government regulations, economic trends, and technological advances should also be asked. Information about market potential can be found in both primary and secondary sources.

primary source: Information that is original and has not been summarized or reported by someone other than the person or group responsible for the information.

USING PRIMARY SOURCES

A primary source refers to original information not previously summarized or reported by someone other than the person or group responsible for the information. Primary source information is the most reliable information that designers can use. Examples of information from primary sources are researching patent information or reading a

Figure 7-6: H. J. Heinz Company website.

OFF-ROAD EXPLORATION

For information about the recyclable packaging PlantBottles being developed by Coca Cola and used by the H. J. Heinz Company, go to http://www.thecoca-colacompany.com/citizenship/plantbottle_partnership.html.

research article written by a person or company. Primary sources are sometimes available in databases that can be accessed online. For example, the H. J. Heinz Company website offers profiles, product information, financial data, distributors, and names of company contacts (see Figure 7-6).

Using or creating primary research information requires much more commitment of time and money. The design team can use existing studies or conduct their own telephone surveys, online bulletin boards, informal focus groups, or direct-mail surveys; conduct laboratory studies on materials, processes, or systems; or conduct test marketing to determine the desirability of the potential product. The team can also hire market research companies to question consumers about their purchasing habits. This information allows the business decision makers to make thoughtful decisions about the development of a product. Market research and laboratory studies are essential tools of product development, and the advertising and sales forces can later use the collected information.

Consumer-Based Information

There are many ways to ask questions. You can get a group of people together who seem to fit the demographics for the product you are planning and ask them specific questions. Another way for students to get information from consumers is to post a question on an Internet blog. Companies and students get answers to questions by conducting telephone or face-to-face surveys. Some companies use test marketing where a product is made available in a small area to determine general acceptance before producing it for a broader market. Sales forecasting can determine the expected level of sales given a certain level of demand. Other companies conduct brand name testing to see how consumers feel about a particular brand. Advertising and promotion research helps companies judge the effectiveness of advertising and the ad campaigns on purchasing behavior.

Careers in the Designed World

ANITA BARNICK, DEPUY SPINE, A JOHNSON & JOHNSON COMPANY

Engineering Improvement

It is not surprising that Anita Barnick's engineering career includes a lot of medical research and teamwork with surgeons—both of her parents are physicians. However, her work to improve the quality of life for patients with serious spinal injuries and conditions is extraordinary.

Anita works at DePuy Spine, a Johnson & Johnson Company, as the group manager for research and development in the Minimally Invasive Spine Surgery group. During minimally invasive spine surgery, doctors use special instruments to see inside the patient's body and correct spinal pathologies without making a large incision. Minimally invasive techniques can reduce the patient's pain and scarring, and speed recovery time.

"My department is responsible for design and product development of innovative implants and instrument systems," says Anita. "I manage a group of nine product development engineers at various steps in their career paths." Anita provides direction and coaching to the engineers, and helps identify her company's strategy for developing minimally invasive medical applications.

On the Job

Anita's product development team manages projects from the design conception stage through design verification and ultimately to the launch of a system in the market. Product development engineers work with cross-functional teams that represent design, quality testing, manufacturing, regulatory agents, legal departments, marketing, and sales.

"We work very closely with real practicing surgeons to know what new design is needed in the medical profession," says Anita. "A new design often starts with our Thought Leaders." Thought Leaders represent the top 10 percent of all surgeons whose ideas are revolutionizing the medical profession. "We call them the Pioneers of the Future," says Anita, "because they are able to predict into the future what the medical profession will need."

From there, Anita's team engages Design Surgeons, who will work closely with the engineers to develop the Thought Leaders' new ideas. Design Surgeons "help translate the futuristic design ideas into practical products for use in medicine. They will tell the engineers how they would want a technique to be done," says Anita. That process defines the new design.

Anita's team develops 3-D CAD models for the designs, and then uses rapid metal injection molding to create a prototype. Rapid metal injection molding uses laser hits to cure a metal powder into a molded prototype. The procedure is similar to stereolithography, but uses metal, and can produce a prototype within 24 hours. "This process allows the design engineer to put a new product prototype in front of the Design Surgeons within two to three days instead of the six weeks it used to take," says Anita. "If changes are to be made, it is then a quick turnaround to see the results. This new process allows products to be brought to market in a much timelier fashion than ever before."

Once the team completes the prototypes, the process then involves a third group of surgeons. These surgeons are called the Consulting Surgeons and represent 90 percent of all surgeons in the medical profession. The team asks the Consulting Surgeons if they would use the new design and when it should be released into the field.

Inspirations

Anita has developed a career path that lets her use her scientific and analytic skill to improve the quality of life for patients young and old.

Anita's customers are the surgeons who will implant the devices, and her greatest challenge is to know what they want before her competition does. She is inspired by how technology assists her in the design process. Today, engineers can take new design ideas from surgeons in the field and make them into hands-on prototypes in a short period of time.

Education

Anita earned a bachelor of science in biomedical engineering with a mechanical concentration from Rensselaer Polytechnic Institute. She continued her studies at Drexel University to earn a master of science in biomedical engineering with a materials concentration. Anita continues her education with courses from Johnson & Johnson in management and leadership development.

Advice for Students

"Work in the field that you are passionate about," says Anita. "Build your core skills in technology and be a team player." She urges students to foster their creative spirit, to be innovative, and to "think out of the box."

Point of Interest
Patent law

The Constitution of the United States gave Congress the power to enact laws relating to patents in Article I, Section 8: "Congress shall have power... to promote the progress of science and useful arts, by securing for limited times to authors and inventors the exclusive right to their respective writings and discoveries."

Patent Information

intellectual property: A term used to describe certain types of ideas and information for the purpose of determining the legal right of ownership.

Personal and corporate rights to the ownership of designs and other forms of intellectual property are established in patent, trademark, and copyright law.

The American Inventors Protection Act (AIPA) enacted the first patent law in 1790, with major revisions to the law enacted in 1952 and again in 1999. The law also established the United States Patent and Trademark Office (USPTO), which grants and administers the many rules and regulations associated with patent law.

The USPTO issues a patent for original and unique inventions, designs, or industrial processes. Patents grant the owner property rights generally for a period of 20 years and apply only to the United States, U.S. territories, and U.S. possessions. These rights give the patent holder "the right to exclude others from making, using, offering for sale, or selling" the invention in the United States or "importing" the invention into the United States. The key phrase in patent law is the *right to exclude others*. Unfortunately, the enforcement of that exclusionary right falls on the patentee, not the USPTO.

patent: A form of legal protection granting exclusive rights to the inventor of a unique new product or process.

According to the USPTO, there are three types of patents:

OFF-ROAD EXPLORATION
For more information about patents, go to the USPTO patent website at *http://www.uspto.gov/*.

1. Utility patents may be granted to anyone who invents or discovers any new and useful process, machine, article of manufacture, or composition of matter, or any new and useful improvement thereof.
2. Design patents may be granted to anyone who invents a new, original, and ornamental design for an article of manufacture.
3. Plant patents may be granted to anyone who invents or discovers and asexually reproduces any distinct and new variety of plant (visit http://www.uspto.gov for more information).

Trademark and Copyright Protection

A trademark or service mark is a word, name, symbol, or device that is used to market or sell a product or service. A trademark indicates the source of the goods while a service mark identifies the source of a service. The term *mark* can be used to refer to both trademarks and service marks. Companies use logos and other forms of unique design to create a corporate identity in the marketplace (see Figure 7-7). Companies use the symbol ™ or ® to designate a trademark. Trademark law protects these companies against others who would use similar marks to deceive consumers into buying their products or services.

Figure 7-7: **Registered Coca-Cola trademark.**

Did you ever ask for a Coke in a restaurant and have the waiter or waitress ask, "Is Pepsi OK?" The reason the waitstaff must clarify your request for a specific brand of cola drink is that the Coca-Cola Company trademarked the term *Coke* in 1945, to refer only to the Coca-Cola brand drink.

A copyright is similar to a trademark but relates to protecting original works such as literary, dramatic, musical, artistic, and certain other intellectual works, both published and unpublished. Material that has been issued a copyright is marked with the © symbol. The owner, such as the author of a book, is granted exclusive rights to reproduce the copyrighted work, to prepare derivative works, to reproduce and distribute copies of the work, to perform the copyrighted work publicly, or to display the copyrighted work publicly. Their copyright is registered by the Copyright Office of the Library of Congress.

When people use copyrighted material inappropriately, they are breaking federal law. Plagiarism is a common form of inappropriate use of another person's work. Plagiarism results when someone uses a direct quote or paraphrases the work of another person without giving credit to the author(s). Students who plagiarize can fail a course where the material was used or can be expelled from their college or university. Some students may not be aware of the seriousness of plagiarizing, but need to know that people who plagiarize in the marketplace are commonly fired from their jobs.

Patent Searches and Independent Inventor Resources

Design teams may want to know what patent(s) have been issued for similar products or components of products. Although there are organizations that specialize in patents and patent law searches, students can search the USPTO database on the Web at http:/www.patft.uspto.gov/. Searches for patents issued between 1790 and 1976 generally provide issue date and patent number. Searches of patents after 1976 provide considerably more information, including the inventor's name, title, abstract, date of issue, patent number, descriptions, drawings and images, and other associated patents. The Google patent search engine is also useful for searching over seven million patents at http://www.google.com/patents.

In addition to being able to learn more about existing patents, the Inventors Assistance Center (IAC) provides help for independent inventors to learn more about the patent and trademark process. Material is also available regarding problems and businesses that engage in fraudulent invention development and marketing practices. Information about IAC can be found at http://www.uspto.gov/inventors/iac/index.jsp.

Visiting Stores

A great way to learn about new products is to visit stores or go online. Although information about a product can be quickly and easily found online, looking at or handling the actual product may be important. If you would like to do more than look at or sketch a product in a store, it is important that you get permission from the merchant. Ask before you photograph, operate, measure, or in anyway handle the product. The store owner may not want you to operate the device but may be willing to personally demonstrate the operation of the product for you. Store owners usually are cooperative and helpful if you are courteous and explain why you are interested in learning more about the product (see Figure 7-8).

Figure 7-8: **Getting help from a store owner.**

Laboratory Studies

Existing primary and secondary sources sometimes provide all the necessary information available about your question. Design teams may use laboratory studies when required information to answer an essential question is unavailable (see Figure 7-9). Laboratory studies involve direct observation of a material or process. For example, the shampoo packaging used silicon closures but the ketchup packaging did not. Clearly, the designers had no available data regarding how well the closure would work with ketchup, so the team needed to conduct a laboratory study.

Fortunately for the Heinz design team, the closure worked for ketchup without any modification. Laboratory studies may involve testing the strength of a material or the reliability of a new process. Sometimes laboratory studies use typical consumers to test products. Again, the Heinz team wanted to observe consumers of various ages using an upside-down ketchup bottle. Not only did they learn that the bottle functioned well, but they also learned that the consumers, especially young users, had fun "drawing" with the smaller cap orifice. Because laboratory studies are expensive, they are one of the last forms of research and investigation design teams use.

Figure 7-9: Collecting laboratory data using a compression testing machine. Pictured is the Instron® Model 3342 testing system with Bluehill® 2 software.

USING SECONDARY SOURCES

secondary source:
Information that has been previously published usually by someone else.

A **secondary source** refers to information that has been previously published, usually by someone else. Authors who are reporting on a topic write most newspaper and magazine articles. They hopefully review primary source information and then write an accurate summary about the topic. Unfortunately, the article is their interpretation of the information. The report may be accurate but may not contain critical information for the design process. Secondary sources are good places to start research because they often represent broad summaries of information on your topic. Do not overlook the library; it may be a better place to start than just plugging in a few **keywords** into a search engine.

Your Turn

Write keywords that you think will help you find information on your design problem. Review your keywords with other members of your design team or your teacher and use feedback received to revise your keyword list.

Library Homepages

One of the best ways to get started is to go to the homepage of a local public high school or college library. Keep in mind that poor research techniques usually result in irrelevant and even discouraging information. Done incorrectly, initial research may lead you to believe that there is no information available to answer your design problem when actually you are simply not asking the right questions or looking in the right places (see Figure 7-10).

Figure 7-10: The College of New Jersey library homepage.

By starting at the library homepage, you will find all the search options available to you, arranged in an organized and user-friendly manner. From this site, you can launch your search for books, magazine, newspaper articles, databases, and even the Internet. For example, you might not realize that there are **trade journals** available on your topic or that the library has several good books to check out that give an overview of your topic. Most libraries allow remote access to their catalogs so students can check on the availability of books or reserve or check out a book from a home or school computer. Libraries are good for much more than getting books. Library homepages can simplify and improve the quality of your initial research.

Web Portals

There are also online libraries such as Refdesk.com, Internet Public Library (ipl.org), Virtual Reference Desk (virtualref.com), and Libraryspot.com. These **Web portals** organize databases in categories and are considered portals or doors to research. They are invaluable and will enable you to find information that you never realized was available (see Figure 7-11).

Encyclopedias

It is always a good idea to begin research with an online encyclopedia such as Britannica.com or Wikipedia, as these articles provide an overview and define terms. These are brief articles and usually written in simple language. As you read about your topic, you should be looking for keywords that you can use as you do further research. This is especially important as your next source will probably

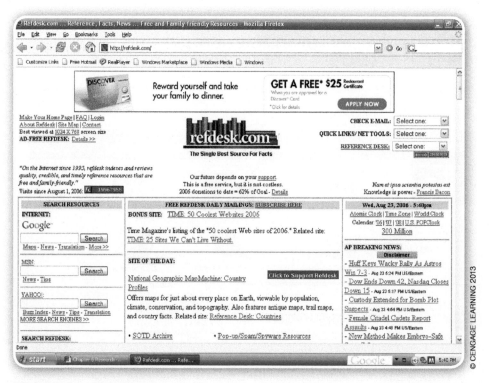

Figure 7-11: Web portal.

Safety Note
Be sure to follow all school and family regulations when using the Internet and do not give out personal information.

be the Internet, which requires the exact spelling of the keywords that you use to search your topic. For example, the alternate spelling of ketchup, c-a-t-s-u-p, yields different results, and a misspelling of the word such as k-e-t-c-u-p takes you to an unrelated site altogether.

Internet

Research in the 21st century requires the use of the Internet and a search engine. Quick general searches provide an overall idea of available information on your topic; however, honing your searching abilities is necessary to gather more useful information. Many search engines are available, such as Google, Yahoo, and MSN. Search engines are very helpful, but carefully sort through your findings and determine the information most relevant for you. Search engine help desks assist in narrowing a search. Search engines all work differently, and many are designed so that a company pays to have its site on the first page of the results of the search, which may not represent the best information for your project. Look through the first few pages of search results and select the sites sponsored by government agencies, educational institutions, or professional societies that most often provide quality and accurate information.

To be useful, gathered information must be relevant to the topic. When you are searching for information, you should start with a few keywords and then expand or narrow the results by changing the keywords you are searching. What do you get if you search on the keyword *ketchup*? Unfortunately, you get over 15,000,000 references! This is too much information, so how would you narrow the search topic?

Google Scholar is an excellent tool for searching for scholarly articles in many disciplines worldwide. Most search engines, including Google Scholar, allow for

Figure 7-12: *Google Scholar: Advanced Scholar Search.*

Boolean logic searching and the inclusion or exclusion of information through the use of operators such as AND, NOT, and OR. When you get too much or too little information in your search, use these Boolean operators or go to the site's advanced search section. There you can narrow the topic by date, exact phrase, or the type of sources you want to include (see Figure 7-12). For example, a search of the exact phrase "Heinz ketchup" results with approximately 215,000 references and takes you to different sites other than just "ketchup." Searching "Heinz" alone would give you quite different information. Good searching can take time, but if you are careful, you will get better results.

Magazines, Trade Journals, and Newspapers

Not all of the best information is found by using a general search engine. Some specialized search engines search databases of magazines and newspapers. For example, the Heinz team needed to know more about new materials being used for plastics packaging. Possible solutions to these questions could have been found in a trade publication such as *Plastics Additives & Compounding* (see Figure 7-13). They would also have been interested in the latest trends in designing package labels. Examples of graphic design trends were available in a graphic design publication such as *I.D.* magazine.

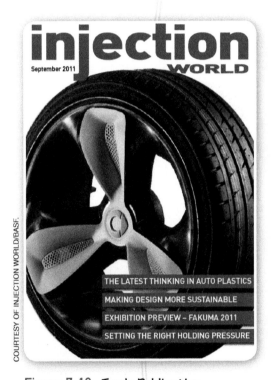

Figure 7-13: **Trade Publication.**

Many high school, college, and public libraries have online resources that lead to magazine indexes. You can search the databases using the same keywords helpful with your Internet search. Many of these magazines are available in full text so you can read it online, upload to your email, or print to save. These resources are paid sites and usually require a library user name and password. To access these sites, you may use a local library site where you are eligible to receive these services free of charge.

Your Turn

Write an appropriate reference in the MLA style for this textbook using Noodletools. Go to http://www.noodletools.com/login.php and follow the easy-to-use instructions.

Point of Interest
Avoiding plagiarism

You must give credit to another person's work, called intellectual property, whether you are using a direct quote or a paraphrase of his or her work. Intellectual property includes written ideas, statistics, drawings or graphs, photographs, art, and music. When in doubt as to what is common knowledge or original property, it is always better to cite the source. It is also essential to use one consistent style of referencing.

Saving Information and Citing Sources

Designers need to document search results throughout the design process and provide references in the final report. One way to keep track of your work is to **bookmark** relevant information and organize it in files. Carefully organized reference material can save a lot of frustration later in the design process. Google Docs allows you to create and share your documents and is especially useful where multiple people, perhaps the whole design team, need to simultaneously access documents. Google Docs is available at https://docs.google.com.

OFF-ROAD EXPLORATION

Learn more about Google Docs at *http://docs.google.com,* and go to "Try Google Docs Now." Be sure to get permission from a parent or guardian before registering or providing any personal information such as your email address.

Each organization, including your school, expects accurate and complete references on all materials used in support of the design process. To provide accurate information, you need to know what **reference style** is required. Reference styles such as the American Psychological Association (APA) or Modern Language Association (MLA) ensure proper citation of referenced material. Reference information will vary by type of resource (book, magazine, or website), but all references require at least the author, title, and date of publication. It is best to record complete references as you find pertinent information. The information is unusable to the design team if the source cannot be cited. NoodleBib Express on the Noodletools website is a free reference that can help you create all the details of the citations for the works cited page of your report.

All the information found in secondary sources will help the design team develop the big picture of the potential consumers and the general focus of the product design. This data collection allows the company decision makers to make good business decisions and allocate resources, including the development of a budget.

Human factors information provides data about the potential user of the product to the design team. The data is part of the scientific study of humans known as anthropometrics. The data is presented to designers as statistical measures, including averages, percentiles, and standard deviations of a given population of people.

Using Human Factors Information

One question that will require some research for nearly every problem situation involves the question concerning the human user. For example, would you like to play golf with golf clubs designed for an 8-year-old as shown in Figure 7-14? We expect to live and work in environments and to use products that "fit" us as human users. The degree to which the things in our environment are comfortable is a measure of good planning and the application of human factors engineering principles. Human factors engineering, or ergonomics, the term more commonly used in Europe, is the design of products, as well as of working and living space, to fit the needs of humans. Said another way, it is the application of the knowledge about people and their environments to the design of products, systems, and environments. In fact, human factors engineering is the synthesis of knowledge from psychology, anthropology, physiology, biology, and engineering. The more we know about the human user, the better we can plan the environments we create.

Figure 7-14: Products must be designed for the intended user.

The Myth of the Average Person

Most people believe that things are designed for the average person. This is not always true. For example, Lance Armstrong has a bicycle that was designed specifically for him, and Sally Ride has a spacesuit that was made specifically for her (see Figure 7-15). Is a standard door, which is 6 feet 8 inches tall, sized for the average person? Obviously not! The door is designed to allow nearly all adults to pass through without bending over. In human factors, this is called a *clearance measurement*.

Human Scale

Because almost all design problems involve the human user in some way, the design team needs to be able to get information about human users. Keywords such as human factors or ergonomics in combination with the product such as a mountain bicycle will render valuable information. More information on human factors is presented in Chapter 15, Human Factors in Design and Engineering.

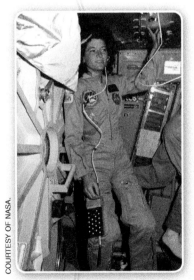

Figure 7-15: Sally Ride's spacesuit was custom designed.

SUMMARY

The purpose of investigation and research is to collect data and information that will assist you in the design process. The keys to being a successful investigator are asking the right questions and consulting the right sources. Investigation and research helps designers move forward. It provides a pathway to follow and it helps you know when to pause and look for more information. Sometimes it takes you in new directions that you never initially considered, which is known as divergent thinking.

We learned that companies spend a great deal of money gathering and analyzing information about how the consumer will respond to a potential new product. Companies also need to find information to support the design process. Finally, considerable research will go into how to best produce and market the product. The design team will involve many people and resources as they moved toward a final solution and product. The Heinz design team had to find information about reinvigorating brand products, market trends, new plastic packaging materials and closures, graphic design trends, and much more.

Companies usually start with market research by asking questions about consumers. Consumers can also use market research to assist in making well-informed new purchase decisions. Primary source information includes consumer surveys, patent searches, or laboratory studies and is the most reliable source of information for designers. Secondary source information helps people to answer questions using articles that report other researchers' works. Secondary sources are good places to start research because they often represent broad summaries of information on your topic. Do not overlook the library; it may be a better place to start than just plugging in a few keywords into a search engine. Students should start by using a library homepage to begin looking for books, magazines, newspaper articles, databases, and even the Internet. Using the correct keywords helps get you to the information you need. Appropriate use of the Internet includes using Web portals, encyclopedias, published materials, and databases. Most design teams will realize that, at some point in the design process, human factors information will be needed about the population of people who will be using the product. Once gathered, design team members must save, record, or summarize the information and prepare correct reference citations for future use. Design teams have the best chance of success when they ask effective questions and gather pertinent information.

BRING IT HOME

OBSERVATION/ANALYSIS/SYNTHESIS

1. Write questions about your design problem that you think need to be answered. Review the questions with other members of your design team or your teacher and use the feedback to revise the questions.
2. Visit your school library and have the librarian review specific research tools available for your research.
3. Check out the homepage of your local public or college library. Check to see what you need to do to use their materials.
4. Modern ketchup was modified in the early years of the twentieth century as a result of concern over the use of sodium benzoate as a preservative. What were the concerns about sodium benzoate and how was the problem solved? Look up the keywords sodium benzoate and Harvey W. Wiley to find the answer.
5. What can you find about patents for bicycle derailleur mechanisms? Try the USPTO patent full text website, http://www.patft.uspto.gov, or the Google patent search site, http://www.google.com/patents. You should record key information about the patent, including images and other related patents.
6. You are the member of your design team assigned to look at solar power, with the goal of finding the most efficient products to be installed at your school. Compile a list of references used to find information about this problem.

EXTRA MILE

ENGINEERING DESIGN ANALYSIS CHALLENGE

- Have students in the lunch room use the Heinz EZ Squirt ketchup bottle. Be sure to get permission to make this study. Make observations and develop a spreadsheet summarizing the information found about students' hand size and ease of use, whether the bottle stayed clean, whether the students enjoyed using the bottle (for example, to draw on the food), and other observations.
- Use human factors data to describe information about students' hand size or gripping capability.
- Use the spreadsheet to create a statistical summary of student hand sizes. For example, what hand size encompasses the middle 50 percent of the students, or the smallest 10 percent of the students?
- Prepare and give a class presentation summarizing what you found and stating how that information could be used to design a new product or give further direction for more research. Include appropriate documentation for all references used.

CHAPTER 8
Technical Drawing

Menu

Before You Begin
Think about these questions as you study the concepts in this chapter:

1. What is a technical drawing?

2. How is measuring used in engineering design?

3. How are technical drawings used by engineers, designers, and consumers?

4. What standards are used to draw and dimension orthographic multiview drawings, and who sets these standards?

5. How do technical drawings fit in with the engineering design process?

6. Why has computer-aided design (CAD) been adopted as the standard method to produce most technical drawings?

7. What are isometric and oblique projection drawings?

8. What are orthographic multiview drawings?

9. How are lines and dimensions used in orthographic drawings?

10. What are sectional and auxiliary views and why are they used?

11. What are the characteristics of solid modeling CAD software modeling programs that make them such a powerful tool?

INTRODUCTION

Design work takes creativity, research, trial and error, and teamwork with other people who have a stake in the solution to be successful. Designers often begin the design of parts and products only to find that their ideas do not work out. So, they start again, sometimes from the very beginning. Remember that the design process was described as an iterative process, meaning it repeats if necessary. Good design must balance many different requirements, and sometimes those requirements conflict. This is what can make the work so interesting and challenging.

Not long ago, *technical drawing* simply meant drawings created by draftspeople. It was a time when people participated directly in the production of parts and the assembly of products. Over the past several decades, with the increased use of computers in the process of design and manufacturing, the definition of technical drawing has broadened to include communication graphics. In this chapter, we will look at both traditional technical drawings and some of these newer strategies for the graphic representation of technical ideas.

Design communication is critical in every stage of design. In Chapter 5, we looked at drawing and sketching as a means of idea development and communication between members of a design team. This type of communication occurs in the early stages of design. In this chapter, we will look at design communication that makes it possible to produce what has been designed. Detail and precision are very important in this kind of graphic communication.

TECHNICAL DRAWINGS

The term *technical drawing* is often used to describe a number of different drawing types. However, it is most often used to refer to *orthographic multiview drawings*. Orthographic drawings are drawings made from looking straight down on an object perpendicular to the plane of the object. Because a drawing must provide correct information about the product, including accurate and precise measurements, it is only possible to get those measurements from a view that is perpendicular to the plane being measured. To understand this principle, hold a piece of 8.5- by 11-inch notebook paper in front of you, with the large face as perpendicular to your eyes as possible. Now begin to rotate the paper about an imaginary line that is parallel to the ground. That is, move the upper portion of the paper away from you and the lower portion toward you. Notice how the paper no longer appears to be the same height. If you rotate the paper 90 degrees, the piece of paper appears to be very thin. However, the paper is still 8.5 by 11 inches. To illustrate this point, no accurate measurements can be taken from the digital camera in Figure 8-1 because the camera is being presented as a pictorial drawing and no view is perpendicular to a plane for measurement. However, the same camera—now shown as a orthographic front view drawing in Figure 8-2—does provide true length information such as the overall length and width of the camera and the diameter and position of the camera lens. Being able to determine when a drawing is presenting true length measurements is an important design skill.

Multiview drawings are often called three-view drawings because simple objects take three views (the front, the top, and one side) to capture all the information necessary to fully describe the object (see Figure 8-3). Take a minute and see what information is presented on this drawing. The title block tells us who did the design, the date, scale, and tolerance. The notes on the drawing provide additional production information and the drawing itself provides all the necessary measurements for the overall shape and size of the part, including location and size of holes and the keyway. In this chapter, you will learn more about what knowledge and skills are needed to make accurate and complete technical drawings. A three-dimensional (3-D) or **isometric view drawing** of the same object is shown in Figure 8-4.

204 Part I: The Engineering Design Process

Figure 8-1: CAD model of a digital camera prototype.

Figure 8-2: Orthographic view of the camera in Figure 8-1.

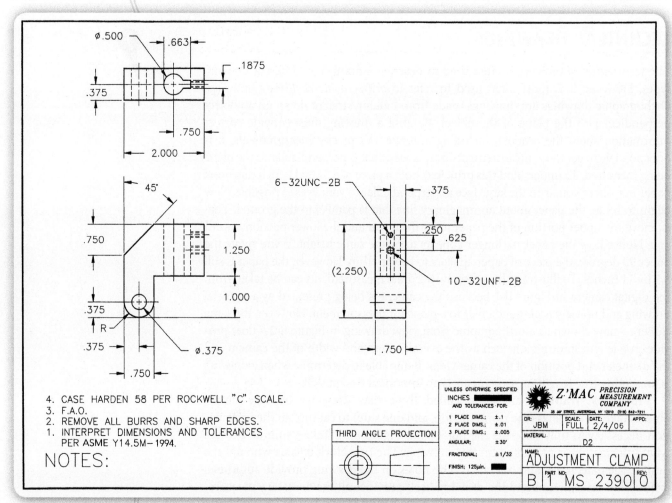

Figure 8-3: Industry drawing with three views used to describe the part.

Figure 8-4: Isometric view of the object in Figure 8-3.

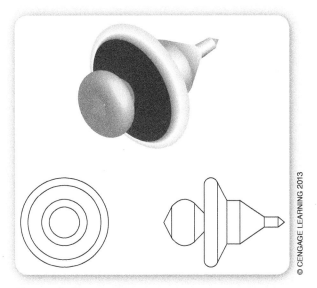

Figure 8-5: Sometimes a two-view drawing is all that is needed to describe an object, such as this shaded object.

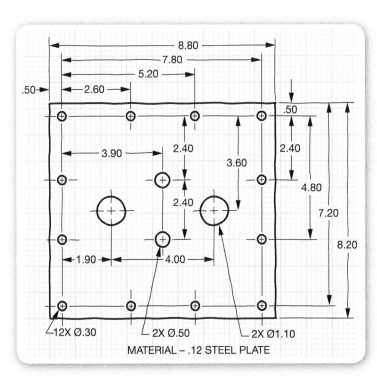

Figure 8-6: Objects made from sheet metal or thicker metal plate can often be described in one view, as shown here.

REPRINTED WITH PERMISSION FROM *INTERPRETING ENGINEERING DRAWINGS*, SEVENTH EDITION, BY CECIL H. JENSEN AND JAY D. HELSEL. COPYRIGHT © 2007, 2002 DELMAR CENGAGE LEARNING.

Many objects, however, require more than three views because there are elements of the design that are not perpendicular to one of the normal planes or there is something in the design, such as some interior structure, that cannot be shown in one of the three views of the normal orthographic drawing. Auxiliary view drawings will be covered later in the chapter. There are other instances in which an object can be fully described in only two views, but these objects are commonly circular, like the kind of model you would make on a lathe (see Figure 8-5). When a part is being made from a piece of sheet stock, such as the sheet metal plate shown in Figure 8-6, a single view may be all that is necessary to communicate the necessary production information.

Technical drawing is a very important part of engineering and product development. Today, most companies develop technical drawings automatically from 3-D **computer-aided design (CAD)**. The technical understanding and skills needed to create the CAD design still take years to acquire, but designers now work to create a final model from which individual drawings are made automatically rather than directing someone to make drawings of components to be used to assemble the final product. With the creation of many technical drawings at the "touch of a button," a draftsperson's job is rapidly disappearing. Small companies may still employ people to generate technical drawings by hand, but these positions are increasingly hard to find.

Orthographic drawings are precise. That is, these drawings provide readers with the exact specifications to make the part. Typical orthographic drawings show the size and shape of parts, the location of features, and the surface finish, and may provide information related to machining or other industrial production processes, such as grinding or welding (see Figure 8-7).

A complete set of technical drawings will provide all of the information that a skilled craftsperson needs to know to make the part. For a large commercial architectural project, it is common to have more than 50 drawings to describe all the work that will need to be completed on the project. When projects are large, planning and scheduling the work is critical to the success of the project. Project management tools like Gantt charts can be used to schedule work and include information about each task, start and end dates, duration, and definition of the critical path for the project.

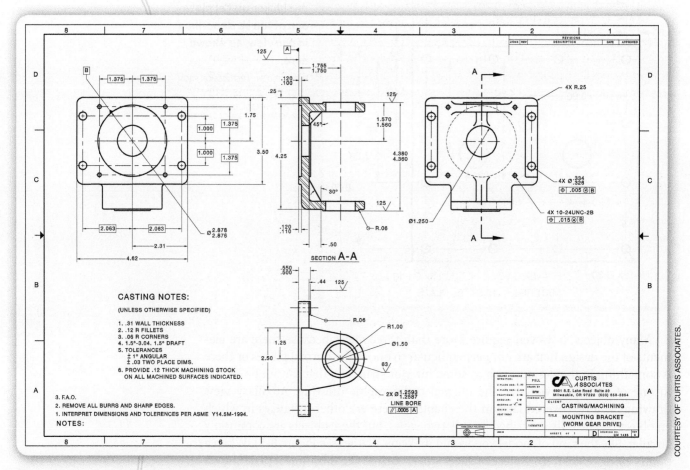

Figure 8-7: Drawing with casting and machining information. A machinist could use this drawing to machine the rough casting to finished dimensions.

Good drawings should answer all design and production questions. Craftspeople should not need to consult a designer when making the part. Increasingly, rapid prototyping machines and computer numerical control (CNC) equipment use files created in 3-D design programs to drive them. Most drawing programs can export design as stereolithography (.stl) files. These files are then processed into control codes that manage the Cartesian coordinates of both the part and the cutting tool, as well as other important parameters like delivery of coolant and feed rates. These control codes are referred to as the "G and M codes." Three-axis CNC milling machines control the three rectilinear coordinates x, y, and z. Three-axis CNC milling machines are used to form relatively simple parts. Four- or five-axis CNC milling machines add the ability to rotate the machined part or the cutting tool, enabling the formation of very complicated parts.

Orthographic drawings are also used heavily in quality assurance (sometimes called quality control) to verify that parts have been produced to the necessary specifications. Quality-assurance workers carefully measure and test components against orthographic drawing specifications. Of course, design-team members still use these drawings to share information.

Technical drawing is a kind of language. Engineers, architects, machinists, and other design and production professionals depend on this language to communicate complex ideas. Technical drawing as a formalized process of communication evolved along with the development of the engineering and machinist professions and has been around since the Industrial Revolution. Prior to this time, individuals still used drawings to record their ideas and to communicate them to others—this is clearly evident in the notebooks left by Leonardo da Vinci (see Figure 8-8). However, the process of technical drawing became formalized in the early decades of the twentieth century. The process became so specialized that it evolved into its own profession called drafting. In 1948, professional draftspeople developed their own society, called the American Design Drafting Association, which is still active today.

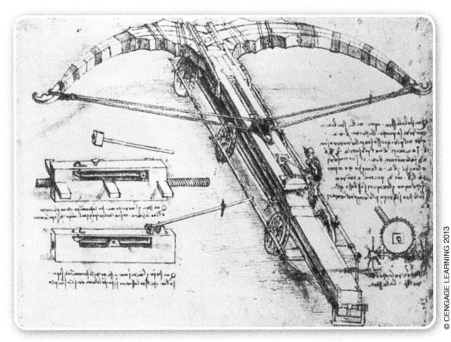

Figure 8-8: Leonardo da Vinci (1452–1519) made extensive use of technical sketches in his many notebooks.

208   Part I: The Engineering Design Process

> **Technical Drawings and Math**
>
> Geometry concepts are common in design and technical drawings. When drawing by hand or through the use of computer-aided design programs, designers need to be able to use mathematics to successfully create technical drawings. Designers commonly use terms such as *parallel, perpendicular, intersecting, tangent, eccentric,* and *concentric* when drawing. Angles and arcs are important parts of many drawings, as is the use of polygon shapes, such as triangles, squares, rectangles, pentagons, and hexagons. Of course, designers must also be able to add, subtract, and multiply or divide numbers in decimal or fractional form to complete the drawing or design tasks at hand.

MEASUREMENT FOR ENGINEERING DESIGN

Having a standard system of measurement and tools for measuring has been critical to all civilizations. Archeologists have found clay vessels used to measure volume for items such as grain and stones for measuring weight. For example, the carat used to measure gem stones today was derived from the carob seed. Many past cultures used human body parts such as arm length or hand width to measure length. The Egyptian cubit was the length of the forearm from the elbow to the tip of the middle finger. The Romans divided the foot into 12 unciae (inches) and established the mile at 5,000 feet. Queen Elizabeth I later changed the mile to 5,280 feet (8 furlongs), which is used today. King Henry I made important contributions by establishing standardized units for measurement. These standardized units were officially established by the British Weights and Measures Act of 1824. The British imperial system (inch for length) is used extensively throughout North America.

The metric system, which needs to be understood by all engineers and other design professionals, answered the need for a universal worldwide measurement system. By the late 18th century, the French Academy of Sciences proposed that a base 10 or decimal system be created. The commission established the meter as the unit of length, which was based on the length from the North Pole to the equator. Other units of SI (International System of Units) included the gram for mass, second for time, ampere for electrical current, Kelvin degrees for temperature, and mole for substance. An act of Congress established the metric system as a preferred system of weights and measurements in 1988. Engineers and architects often need to read high-precision engineering rules, architects' scales, vernier calipers, and all other measuring tools in the imperial and metric systems (see Figure 8-9).

In the imperial system, the inch is often divided into "divisor of two" fractions such as 1/2, 1/4, 1/8, 1/16, or smaller fractions. For example, a 1/2-mark is one-half of one, 1/4 is one-half of one-half, 1/8 is one-half of one quarter, and so on. To more easily distinguish the different measurement lines on a rule, different line lengths are used. For example, on a rule marked in inches, typically all the lines marking the integer inches and 1/2 inches are the longest lines while all the lines that mark 1/4 inches are second longest. Similarly, lines that mark 1/8 inch positions are the third longest, and so on (see Figure 8-10). Determining the correct fraction of the

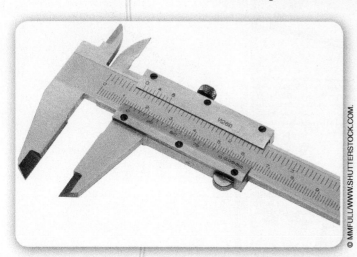

Figure 8-9: A high-precision differential micrometer being used to accurately measure the outer diameter of a metal part being turned on a lathe.

Chapter 8: Technical Drawing   209

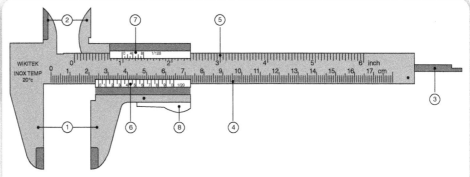

Figure 8-10: Measurement lines on a rule marked in inches.

1. **Outside jaws:** used to measure the width or outside diameter of an object
2. **Inside jaws:** used to measure internal width or diameter of an object
3. **Depth probe:** used to measure depths of an object, hole, or channel
4. **Main scale:** metric scale marked in cm (large divisions) and mm (small divisions)
5. **Main scale:** Imperial scale marked in inches (large divisions) and fractions (small divisions)
6. **Metric Vernier** gives interpolated measurements to 1/10 mm or better
7. **Imperial Vernier** gives interpolated measurements in fractions of an inch
8. **Retainer:** locks the slide or movable part of the caliper so that measurements can be read off the work piece

Figure 8-11: Reading a vernier caliper.

inch is easy because it is always a matter of counting the number of divisions of the inch. Estimates must be made if the object being measured falls between two lines. Any fractional rule limits the accuracy and precision of the measurement. Adding and subtracting fractions can be cumbersome so fractions are rarely used in engineering drawings. Instead, fractions are converted into decimal form before being included into a drawing.

The imperial inch may also be divided into tenths, hundredths, thousandths, or smaller units, resulting in a decimal imperial system that can sometimes be more useful. Engineering rules marked with tenths divisions are commonplace but are limited in measurement accuracy and precision because a tenth of an inch is still a somewhat large dimension. More accuracy can be obtained with a vernier caliper, invented by Pierre Vernier in the early 17th century. Vernier, a French mathematician, developed a caliper capable of measuring to the thousandths of an inch (see Figure 8-11). A thousandths of an inch is typically referred to as a "mil," and is often used in the United States. The vernier caliper can be used to make both internal and external measurements. Vernier calipers are available in both mechanical and digital forms. Achieving accurate and precise measurements is something engineering students must be able to do.

ARCHITECTURAL SCALE

Another important measuring tool is the architectural scale (see Figure 8-12). The architectural scale is usually triangular in cross-section, allowing six scales to be included in one instrument. An arcitectural scale has two main purposes: direct measurement and scaling. Scaled drawings are necessary when the object being made is large, and it would be impossible to draw the object to full size.

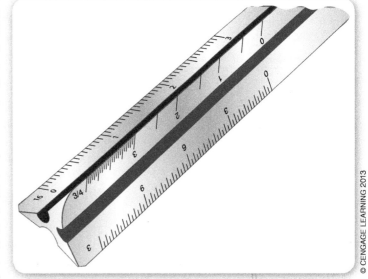

Figure 8-12: Architectural scale.

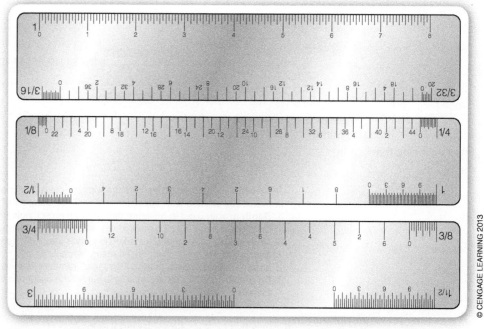

Figure 8-13: Reading an architectural scale.

Because drawings are typically scaled, it is efficient to have markings that can correspond to common scaling factors. Some common scaling factors are 1:10, 1:20, 1:40, and 1:60. One of the six measurement scales includes a standard imperial inch scale. Two other sides are imprinted with inches scaled to smaller units such as 1/8, 1/4, 1/2, 3/8, and 3/4 inch, as well as units that are decimal fractions of an inch. Also included are larger units such as 1.5 and 3 inches, which can be used to manage 1:6, 1:4, and 1:2 scales. Because there are 12 inches to the foot, the scaled sides of the rule have 12 divisions (see Figure 8-13).

MEASUREMENT: THE REAL WORLD OF VARIABILITY

Nothing in the real world can be measured or fabricated with infinite accuracy, so variability must be considered in design. **Variation** is a critical concept for designers to understand. Variability is a measure of the extent to which a dimension or parameter is expected to vary.

All dimensions have variability. Successful design and manufacturing depend on engineers paying close attention to each critical dimension. There can be hundreds of critical dimensions, and their associated variabilities for devices such as MP3 players, TVs, or cell phones. More complex designs like bridges, power plants, and spacecraft, can have thousands or millions of critical dimensions. One bad measurement can be a major problem. Why do engineers or technologists need to consider variability? As the following example shows, engineers consider variability to make sure parts fit and fasten together correctly. The correct use of variability is also vital to ensure that designs are reliable and can be manufactured.

Consider the case of a simple box that needs to fit through a simple hole (see the blue box in Figure 8-14). Perhaps the box represents a piano that needs to fit through a doorway into a music room located on the fifth floor in a building. How wide is the door? How

variation:
Also known as *variability;* a measure of the extent to which a dimension or parameter is expected to vary in magnitude.

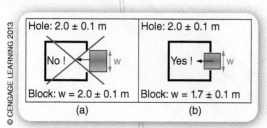

Figure 8-14: A "box-through-a-hole" example illustrates the need to accurately define variability to achieve successful designs. Unless the variances in both the hole and box are correctly accounted for, the box will not fit through the hole, resulting in failure.

Your Turn

Select an item in your engineering classroom to be measured. Have different students measure the object. List their individual measurements and determine the amount of variability in the measurements. Consider repeating the measurements with different measuring devices. How does the variability change with different devices, or with different people or with different processes? List items where the acceptable variability is (1) very small or (2) very large.

wide is the piano? These are very important questions. Certainly, a professional piano mover would measure both the piano and the doorway before committing resources (muscles) to lift and carry the piano up five floors. Clearly, if the piano is too big for the doorway, either the doorway must be enlarged or a smaller piano must be used. How accurately must the measurements be to convince the piano mover to move the piano? If the piano measured 2.0 meters wide and the doorway measured 2.01 meters wide, would you commit to moving the piano up five floors by hand, or would you demand a more clear definition of how much each of these dimensions could vary? Figure 8-14 shows how designers need to account for variation, or variability, to ensure success. As the diagram shows, to ensure 100 percent success, the dimension of the box (the piano) must be such that its largest possible size is still smaller than the smallest possible size for the opening (the doorway). Consider that the piano mover is the only one taking the measurements and is using the same tape measure for both measurements. This ensures the measurements likely have less variability. However, it should be noted that it is not generally the case that the same person is always using the same measurement device because parts for products are often made in many different factories all over the world with sometimes very different processes. This is why so much effort worldwide has gone into establishing clearly defined standards.

The "box-through-a-hole" example could easily apply to more high-tech devices, such as the following: (1) male and female connectors on cables used in MP3 players, cell phones, or computers; (2) doorways on space stations and shuttles; (3) heat exchangers on nuclear power plants; or (4) pistons on car engines.

In a math or science book, a line of length L is simply and precisely a line of length L. Similarly, a rectangle of length L and width W is simply and precisely a rectangle (see Figure 8-15). However, for an engineer, dimensions typically require an accurate estimate of their variability. As shown in Figure 8-15, engineers need to view even the simplest dimension with some variability. For example, engineers view a rectangle as having a length (L $\pm$ ΔL) and a width (W $\pm$ ΔW), with the values ΔL and ΔW representing the maximum variability of each dimension. Therefore, engineers have at least twice the numbers to take into account: the nominal values and their variability. (The Δ symbol is the capital Greek letter delta, or "d," which engineers use to symbolize a difference.) As an example, in Figure 8-15b the length of the rectangle is nominally 9.7 cm but has a variability of 0.3 cm. Therefore, the length could be as large as 10.0 cm or as small as 9.4 cm. The width of the rectangle is known to tenfold higher accuracy, to the hundredths place (3.54 $\pm$ 0.03 cm). It is important to note that when quoting variability, it is routine, and definitely expected, that both the nominal value and the variability be quoted

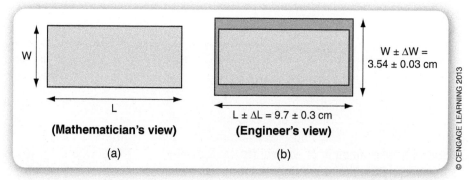

Figure 8-15: *Math texts correctly describe a rectangle as a perfect structure formed with perfect imaginary lines. However, to be successful, designers must go one step further and account for variability in both dimensions.*

to the same accuracy, or the same number of decimal places. Take, for example, a dimension that is quoted as follows: $X = 2.3456 \pm 0.3$ mm. It is very misleading to quote the nominal value to 4 decimal places while the variability is quoted to only one decimal place. Do readers assume that $X = 2.3 \pm 0.3$ mm, or do readers assume that $X = 2.3456 \pm 0.3000$ mm? Readers cannot know which interpretation is correct and must therefore ask the designer for clarification. If the dimension X is only known to an accuracy of 0.3 mm, then the appropriate statement of the dimension would be $X = 2.3 \pm 0.3$ mm. Given that calculators can easily calculate to many decimal places, quoting values with too many digits is unfortunately a common occurrence. One should try not to make this confusing error. An example of calculating variability of the volume of an ancient gold brick is included in Chapter 16, Math and Science Applications.

Accuracy and Precision

A discussion on measurement is not complete without defining two very important attributes of a measurement or calculation: **accuracy** and **precision**. Accuracy is the degree of conformity of a measured, or calculated, quantity to its actual value while precision is the degree to which several measurements, or calculations, show the same, or similar, results. *Repeatability* and *reproducibility* are also terms for precision.

The results of measurements, or calculations, can be (1) accurate but not precise, (2) precise but not accurate, (3) both, or (4) neither. A measurement, or calculation, that is both accurate and precise is referred to as *valid*. Figure 8-16 shows graphical examples of precision and accuracy.

We have discussed that skill with variability is necessary for successful technology development. However, how is the variability itself determined? A detailed answer to this question requires knowledge of statistics, which is covered in more detail later in Chapters 15 and 16. However, to summarize, variability in measurement is determined in one (or any combination) of three ways: (1) guessing, (2) estimating given the limitations of the instrument being used, and (3) performing multiple measurements to acquire a statistically valid variability. A few simple rules are useful in determining variability in measurements.

accuracy:
The degree of conformity of a measured, or calculated, quantity to its actual value.

precision:
The degree to which several measurements, or calculations, show the same, or similar, results. *Repeatability* and *reproducibility* are also terms for precision.

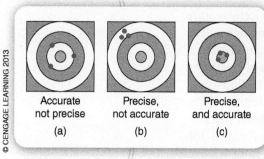

Figure 8-16: *Graphical bull's-eye, examples of measurements that are (a) accurate, (b) precise, and (c) both accurate and precise.*

Simple Rules for Variability of Measurement

1. When using a physical (comparative) device such as a ruler or measuring tape, take a fraction of the smallest viewable marked dimension as the variability. The worst-case variability would be a full smallest dimension, while an aggressive value would be approximately 1/3 to 1/5 of the smallest dimension. For example, if the smallest dimension on a ruler is 1/10 of an inch in your measurements, you might feel that, in your situation, you can only ready to 1/2 of this dimension, resulting in a variability of ± 0.05 in.
2. When using a commercial instrument, use the published accuracy. (However, it is typical to verify the variability by completing a few measurements, including a known standard if possible.)

Defining variability in design drawings is also very important. In design drawings where the intention is to describe how to fabricate something, all dimensions have an associated variability. In design drawings, variability is typically referred to as *tolerance*. Figure 8-17 shows an example of a drawing that includes the allowed tolerance for each dimension. Most machinists will neither accept a drawing nor give a financial price quote unless all dimensions are set with a tolerance. This makes sense because the machinist needs to understand what tools or processes are required to make every aspect of a part correctly. The choice of tool and process are critical for both the machinist and customer because tools and processes vary substantially in both capability and cost. For example, cuts made with a band saw, a CNC milling machine, or wire- or laser-EDM (electrical discharge machining) will give very different precision at vastly different costs.

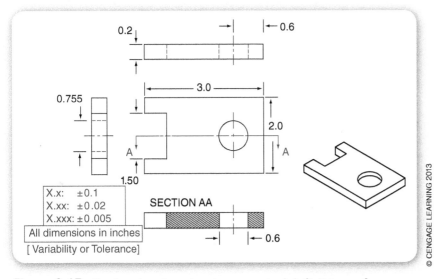

Figure 8-17: Design drawing showing expected definitions of variability for all dimensions. For drawings, "variability" is often referred to as "tolerance."

Who Uses Technical Drawings?

Products begin with ideas. Initial ideas are "fuzzy," and rough sketches often reflect them. Just as da Vinci used paper and pencil to work through his ideas, contemporary designers also use this technique to work through design solutions. Sometimes they are elaborate as in Figure 8-18, and sometimes they are basic as in Figure 8-19; however, this is generally where technical drawings begin.

Figure 8-18: Sketches showing ideas for a new chair design. The designer has drawn multiple possible solutions for both adjustable and nonadjustable designs.

As ideas develop and become more concrete, design decisions are made and technical drawings evolve into a more formalized graphic language. This intermediate step prepares the way for the creation of CAD or orthographic multiview drawings. As products are generally created from a number of parts (sometimes a very large number of parts), the design of each part and the relationship between parts must be worked through. This often involves more precise sketches that describe these parts and their relationships (see Figure 8-20).

A variety of individuals and groups involved in product development use technical drawings. Because they reflect a specific technical language, they are not often used to communicate ideas to the general public. An exception to this is the use of architectural floor plans. However, many people are not able to visualize what a house might look like when shown such drawings (see Figure 8-21).

Those who will make or check a component most often use technical drawings. Orthographic drawings, such as in Figure 8-22, show a fan component that will be stamped from a sheet metal blank. These drawings allow a tool designer to develop the tooling to make the part and a quality-control person to check if the part is within specifications.

Figure 8-19: A design team comprised of industrial designers; mechanical, manufacturing, and materials engineers; and marketing experts use technical drawings to plan aesthetic treatment of product details.

Chapter 8: Technical Drawing 215

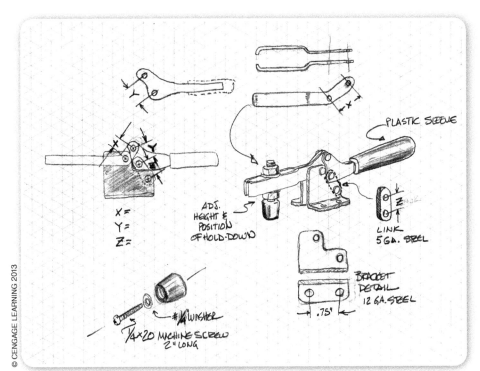

Figure 8-20: As solution ideas become more concrete, more elaborate sketches are often made to describe the object more completely.

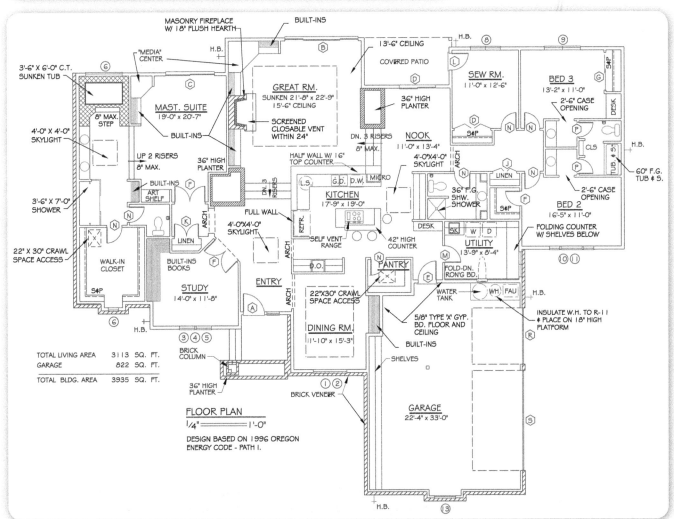

Figure 8-21: An architectural floor plan.
REPRINTED WITH PERMISSION FROM *RESIDENTIAL DESIGN, DRAFTING, AND DETAILING*, BY ALAN JEFFERIS AND JANICE A. JEFFERIS. COPYRIGHT © 2008 DELMAR CENGAGE LEARNING.

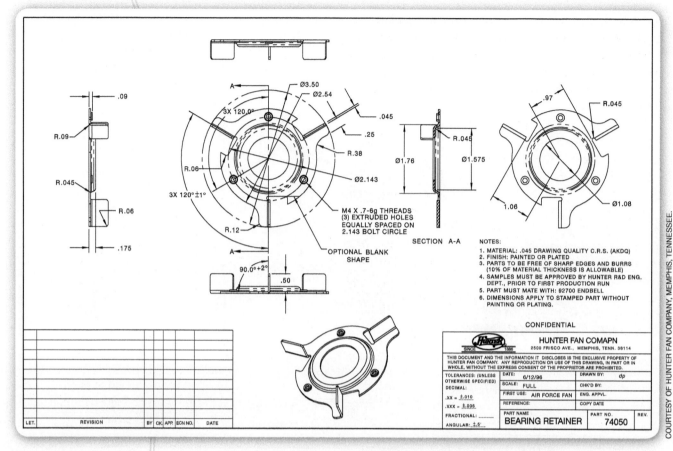

Figure 8-22: Detailed drawings of a fan component provide a complete description of the part.

Technical Drawings and the Consumer

You may have used technical drawings when assembling a new product. The "some assembly required" warning on the package scares many people, because they do not understand the language of technical drawing. To make assembly easier for the public, some manufacturers produce elaborate pictorial images or use photographs in their instructions.

In most cases, along with the parts of the item to be assembled, there is a set of instructions (usually pictorial). Figure 8-23 shows an example of a technical drawing that is included in the assembly instructions for a small table.

In Figure 8-24, an assembly drawing shows the main components for a wood cabinet. In this case, the drawing serves as a guide to the craftsperson when building and assembling the table. Note the annotations throughout the drawing. This pictorial is an example of an exploded assembly drawing.

Technical Drawing Standards

A convention refers to the way in which something is usually done. Drawing conventions have been identified and standardized through professional organizations, such as the American National Standards Institute (ANSI), to give order to the ways in which technical drawings communicate information.

ANSI has established standards for technical drawings. Engineers in the United States, Canada, and several other countries follow these standards. The International Organization for Standardization (ISO) is headquartered in Geneva,

annotations:
Explanatory notes on drawings that provide important additional information to the reader.

Switzerland, and has developed a similar standard for almost 80 of its member nations. ISO standards are based on the **metric system**. Some North American companies use both ANSI and ISO standards to ensure accurate communication with both domestic and foreign business partners.

Figure 8-25 shows paper-size specifications as identified by both ANSI and ISO standards. For example, a standard ANSI B sheet is the one you would choose if you want to use an 11- by 17-inch paper size (also called tabloid size). An ANSI A sheet size describes the standard 8.5- by 11-inch letter size paper that you use in a computer printer. The ISO A4 sheet is the metric equivalent of the ANSI A letter size. You may have noticed the letters "A4" written on a printer paper tray or next to the window on a scanner or copy machine. The United States and Canada use ANSI standards; however, most of the rest of the world uses ISO standards.

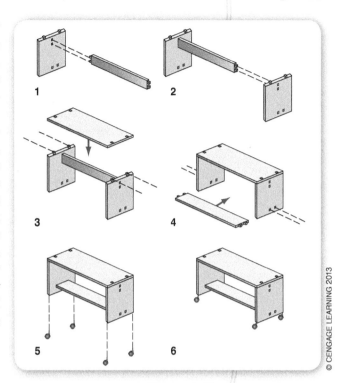

Figure 8-23: Assembly instructions for a common consumer product.

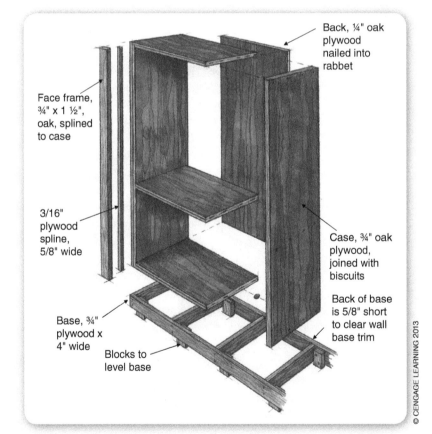

Figure 8-24: An example of a technical drawing that is used to describe the main details of a project. This particular drawing is hand-drawn and is "exploded" to show the construction and assembly details. The drawing also contains annotations that provide additional information.

Paper Sheet Size	Inches (WxH)	Millimeters (WxH)
A0	43.8 x 33.1	1189 x 841
A1	33.1 x 24.4	841 x 595
A2	24.4 x 16.5	595 x 420
A3	16.5 x 11.7	420 x 297
A4	11.7 x 8.3	297 x 219
A5	8.3 x 5.9	219 x 149
ANSI A	11 x 8.5	279.4 x 215.9
ANSI B	17 x 11	431.8 x 279.4
ANSI C	22 x 17	558.8 x 431.8
ANSI D	34 x 22	863.6 x 558.8
ANSI E	44 x 34	1117.6 x 863.6

Figure 8-25: ANSI and ISO standard paper sizes for technical drawings.

There are standards that dictate how and where designers place **dimensions** on technical drawings. This includes how numbers and dimension tolerances are represented. Other standards address the size, shape, and types of information found in drawing **title blocks**. We discuss dimensioning and other standards later in this chapter.

ANSI standards also exist for a wide range of technical drawings that use schematic symbols to communicate information about electrical, mechanical, pneumatic, hydraulic, plumbing, heating, and air-conditioning systems.

Technical Drawings and the Engineering Process

Although sketching is used to help get ideas out of your head so you can evaluate and improve them, the primary purpose of technical drawing is to communicate a solution between various members of a design and production team. Initial concepts are conceived and developed by product designers using both sketches and elaborate renderings and are proposed through the design team (see Figure 8-26). After designers agree on the initial design, they develop these ideas into a detailed design solution, which they then communicate through technical drawings. Many types of engineers and other designers, including mechanical engineers, industrial designers, and manufacturing engineers, create technical drawings. These groups and individuals determine the exact specifications for the shape of the product, the necessary materials for the various parts, and the manufacturing processes for the individual parts.

These technical drawings offer ways for various members of the design team to clearly communicate their piece of the solution. Team members update the drawings as changes occur. They indicate such changes by including a dated **revision** note on the drawing and by issuing an **engineering change notice (ECN)**.

As part of this process, the engineering team will also coordinate with a machine shop or modeling studio, and ask them to produce prototypes of the individual parts. Today, many companies send computer files directly to rapid prototyping or CNC machines to make various parts of these prototypes. This minimizes the need for some technical drawings. However, technical drawings are still needed for checking the **accuracy** of components and for assembly instructions. When it is time to mass-produce the product, the engineering team uses technical drawings to design the production methods used to manufacture the parts and assemble the product.

As components are prototyped, the engineering team consults with both the product designers and the manufacturing engineers to ensure that the parts adhere to the design requirements, and can also be manufactured in an efficient and cost-effective manner. The team uses technical drawings to evaluate the prototype and plan production.

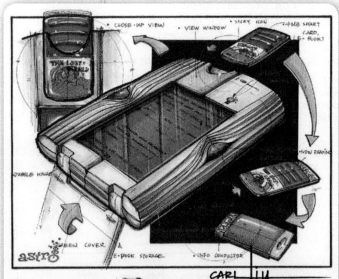
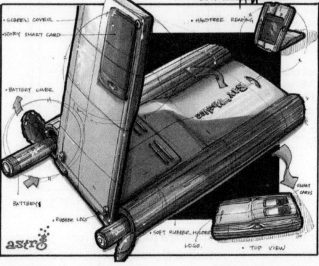

Figure 8-26: **Industrial designer concept drawings of an e-Book.**

DRAFTING AND CAD

Prior to the development of the computer, drawings were made on vellum (a translucent, finely textured paper) using pencil and ink in a process known as drafting. Draftspeople produce their drawings on a drafting table, and use a series of drawing tools that included T squares, triangles, shape templates, lettering guides, compasses, and various other instruments. For most of the twentieth century, college and university engineering programs required courses in drafting.

In the early 1980s, technological advancements introduced the personal computer to the workplace. Advancements in output hardware resulted in monitors capable of displaying graphics. In November 1982, AutoCAD Version 1.0 was demonstrated at the COMDEX trade show in Las Vegas and began shipping the following month. In the months and years that followed, AutoCAD and other similar programs became the standard method for the production of engineering drawings. Major automotive and aircraft manufacturing companies that maintained a large engineering staff and had the resources to invest in costly computer systems were the first to adopt CAD technologies.

After close to a century of development and refinement, the era of manual drafting was replaced by CAD in the course of a decade. The profession of drafting continued, but its tools had changed forever. The use of programs like AutoCAD dramatically reduced the amount of time needed for engineers and draftspeople to produce engineering drawings. These changes allowed one draftsperson to do the work of many. Figure 8-27 shows an example of an engineering drawing produced using AutoCAD.

During the decades since the introduction of AutoCAD, the world has witnessed unprecedented development in computer technology. Rapid improvements in computer clock speeds, display technology, and computer memory technologies have been accompanied by decreases in the costs of computer systems. System

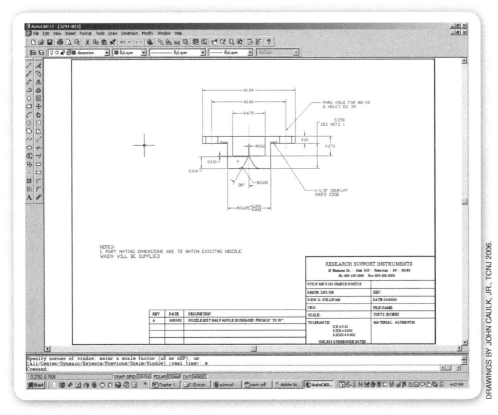

Figure 8-27: AutoCAD drawing. Drawing by John Caulk, Jr., TCNJ 2006.

COURTESY OF JOERG ROTH, UNIVERSITY OF APPLIED SCIENCES, NUREMBERG, GERMANY.

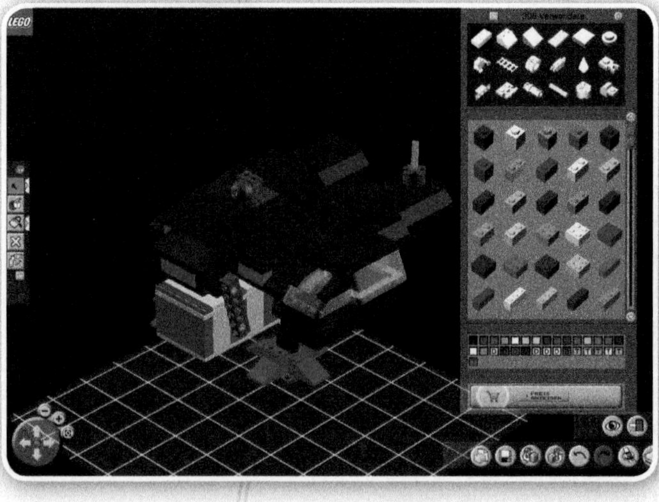

Figure 8-28: **Creating objects in LEGO Digital Designer.**

graphic capabilities, with a large amount of the development credit going to makers of computer games, have become extremely sophisticated, which has contributed to the Internet becoming the medium of choice for sharing information. Major computer companies continue to develop hardware and software systems that maximize the benefits of these advances.

With the development of design software for kids, 3-D solid modeling tools have now entered the realm of play. The LEGO Corporation developed applications that allow users to create 3-D LEGO models by picking and manipulating virtual LEGO bricks (see Figure 8-28). Although this program is not nearly as sophisticated as engineering CAD software, programs like this help introduce this design concept to younger users. There is also a parallel development of computer games that employ sophisticated graphics and require users to navigate through a 3-D virtual world. Game development has been both influenced by and has influenced engineering design software.

The solid modeling process allows for the modeling of individual parts, which can then be brought together to form assembly models. For example, Figure 8-29a shows a solid model of a spur gear that is part of the Vex Robotics Design System. The design team brought this part together with other Vex component solid models to create assembly models of robot designs. Figure 8-29b shows examples of these design assemblies.

7

ISOMETRIC AND OBLIQUE PICTORIAL DRAWINGS

In Chapter 5 you learned that designers use perspective drawings to portray objects or scenes on a two-dimensional surface in a way that gives the illusion of three-dimensional reality. Perspective is very useful for rendering buildings and products when the objective is to present an overall idea, or show what something looks like in a life-like setting. However, because perspective projection distorts

COURTESY OF WILDSTANG.

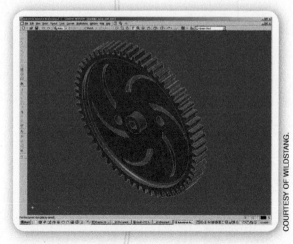

Figure 8-29a: **Vex gear developed in Autodesk Inventor.**

DRAWINGS BY JOHN CAULK, JR., TCNJ 2006.

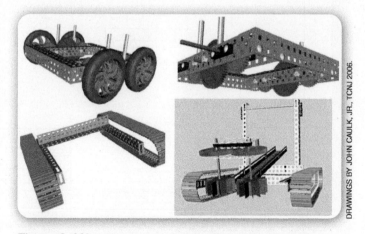

Figure 8-29b: **Examples of Vex robot designs that were developed as part of a student project using the Autodesk Mechanical Desktop solid modeling program.**

an object's size as distance increases from the eye to the object, perspective is not very useful when the objective is to accurately represent distances. Even very small distances are distorted, as in the cube in Figure 8-30. However, there are other projection views that, although they distort distances, do so equally.

Figure 8-30: A cube in two-point perspective.

There is a distinct difference between a drawing and a projection. A projection is an exact representation of a 3-D object projected onto a plane from a specific location. The general term used to describe the different projected views is known as **axonometric projection**. Three common projections are isometric, dimetric, and trimetric (see Figure 8-31). These are known as pictorial drawings because they closely resemble realistic pictures. Although you will not often come across dimetric or trimetric projections in hand drawings, you may encounter them in CAD work.

An **isometric drawing** is created when an object is rotated 45 degrees and tilted toward you at a 35.3-degree angle. This results in a figure where the two bottom edges will occur at 30-degree angles to the horizontal in the case of a cube. Figure 8-32 shows a cube in isometric orientation. Note that the three edges that share the upper close corner in the center of the image are equally spaced at 120-degree angles. This is why the term *isometric* means equal (iso) measure (metric). If the cube were cut at an angle, a nonisometric plane would result with edges that do not occur at 30-degree increments to the horizontal.

In Figure 8-33, the planes represented by the sides labeled 1, 2, and 3 are called *isometric planes*. The lines that define those sides are isometric lines and are measurable because they are true length. Lines that are not parallel to the isometric planes (nonisometric lines) cannot be measured.

Isometric Grid Paper

Sometimes designers can easily create isometric sketches using isometric grid paper, pictured in Figure 8-34. The light lines on the paper represent the three principle axes of the cube, as seen in Figure 8-32. The distances between the intersection points, as measured along the lines, are equal in all three directions. This allows you to establish your own nonstandard units, or transfer measurements directly to the grid. In either case, the grid helps you keep your sketch geometry proportional.

As an example, each unit on the grid in Figure 8-35 is equal to one inch. Therefore, the overall size of the object shown is four inches wide, six inches deep, and three inches high. There is a three-inch wide by one-inch deep notch in the top of the block. You can draw a fairly accurate sketch by dividing the distance between each intersection by half or quarters if needed.

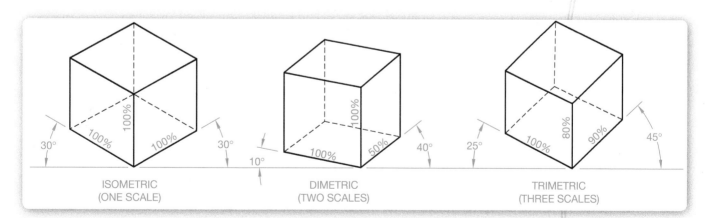

Figure 8-31: **Three types of axonometric projections.**

REPRINTED WITH PERMISSION FROM *ENGINEERING DRAWING AND DESIGN*, FOURTH EDITION, BY DAVID A. MADSEN. COPYRIGHT © 2007 DELMAR CENGAGE LEARNING.

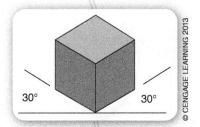

Figure 8-32: A cube in isometric projection.

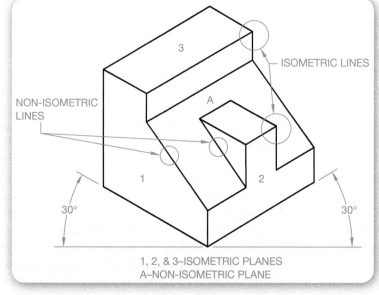

Figure 8-33: Isometric and nonisometric planes.
REPRINTED WITH PERMISSION FROM *ENGINEERING DRAWING AND DESIGN*, FOURTH EDITION, BY DAVID A. MADSEN. COPYRIGHT © 2007 DELMAR CENGAGE LEARNING.

Figure 8-34: Isometric grid paper used to rapidly sketch isometric drawings.

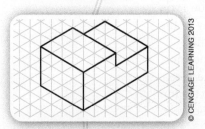

Figure 8-35: Using the isometric grid paper to sketch.

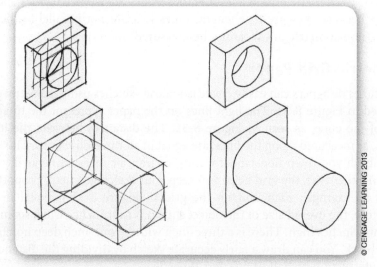

Figure 8-36: Techniques of sketching in isometric projection. Note how a square (distorted by isometric projection) is used to sketch a circle.

If you want to draw a more accurate isometric pictorial drawing, use the isometric grid paper under drawing paper and develop your drawing using a straight edge, a protractor, and a scale (see Figure 8-36). With heavier-weight paper, a light table allows you to see the grid through the drawing paper. "Crate" the object (as discussed in Chapter 5) using a starting point, a horizontal line, an intersecting vertical line, and two 30-degree (from the horizontal) lines. With the scale, measure the overall width, depth, and height to create the crate (see Figure 8-37).

Oblique Views

An **oblique projection drawing** has one face parallel to the viewing plane. The other two object planes are shown at an angle, which is usually 30 degrees or 45 degrees. Oblique pictorials that are drawn at a 30-degree angle may be sketched using

Your Turn
Drawing an Isometric Sketch

Using isometric grid paper (if you need to, you can find an isometric grid pattern in PDF format on the Web and print it out), sketch an object found in a kitchen. Find an object that has a simple 3-D form, such as an electric can opener or toaster oven.

isometric grid paper. Oblique pictorials that are sketched at a 45-degree angle are easier to sketch on standard quadrille-ruled graph paper. Viewing an object straight on to a face would mean that it would be impossible to see any of the plane's 90 degrees to that face, so the angles are distorted.

Two common types of oblique projections include cavalier and cabinet. **Cavalier projection** uses full width, height, and depth measurements (see Figure 8-38). **Cabinet oblique drawing** scale the depth to one-half of the true distance, making the object appear more realistic than a cavalier oblique. An example of cabinet oblique is shown in Figure 8-39. These pictorials significantly distort one or more planes of the 3-D object. The advantage of these projections is that one face has no distortion, so circles and other features are pictured accurately and are easier and faster to draw. Figure 8-40 shows an application of oblique pictorial drawings.

cabinet oblique drawing: A form of oblique drawing in which the receding lines are drawn at half scale, and usually at a 45-degree angle from horizontal.

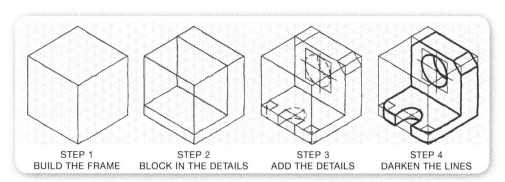

STEP 1 BUILD THE FRAME
STEP 2 BLOCK IN THE DETAILS
STEP 3 ADD THE DETAILS
STEP 4 DARKEN THE LINES

Figure 8-37: *An example of "crating" used to construct an object in isometric projection.*

REPRINTED WITH PERMISSION FROM *INTERPRETING ENGINEERING DRAWINGS*, SEVENTH EDITION, BY CECIL H. JENSEN AND JAY D. HELSEL. COPYRIGHT © 2007, 2002 DELMAR CENGAGE LEARNING.

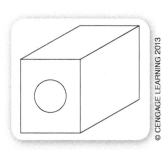

Figure 8-38: *Oblique projection view: Cavalier projection uses full distance in the depth direction.*

Figure 8-39: *Oblique projection view: Cabinet projection uses one-half distance in depth direction.*

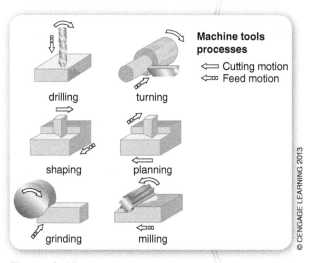

Figure 8-40: *A drawing using oblique projection.*

ORTHOGRAPHIC DRAWING AND SKETCHING

Orthographic projection is still the standard for most technical drawings, as described at the beginning of this chapter. To understand orthographic drawing, we will look at a number of important conventions based on the ANSI standards.

Arrangement of Views

There are two arrangements of views that are common in multiview drawings: first-angle projection and third-angle projection. The angle of projection refers to the placement of views in an orthographic drawing. In Figure 8-41, you see the camera views oriented in first-angle projection. Note the placement of the front view (the view with the most detail), and the one below it. If you were to rotate the top of the camera toward you, you would see the view below the front view. If you were to rotate the left side of the camera in the front view toward you, you would see the view to the right of the front view. First-angle projection is used in Europe and most Asian countries, and is generally considered an ISO standard. It is not generally used in the United States.

The same camera is pictured in third-angle projection in Figure 8-42. Note the placement of the top- and right-side views in third-angle projection. If you were to rotate the top of the camera in the front view toward you, you would see the view above the front view. This is the top view. If you were to rotate the right side of the camera toward you, you would see the view on the right of the front view. This is called the right-side view. Third-angle projection is used in the United States and Canada, and is considered an ANSI standard.

Envisioning an Object in Three Views

Certain occupations, including those in engineering, architecture, the natural sciences, computer science, and mathematics, rely on one's cognitive (thinking and reasoning) and perceptual abilities with space and shapes. For example, imagine you are looking at a side view of a complex object, such as a car. Can you visualize what it looks like from another angle, such as from the rear? This ability is called *spatial ability* or *spatial relations*, and the car example is just one aspect of this ability.

Some researchers think that engaging in sports activities and playing computer games may increase or reinforce a person's spatial ability. This is because these activities require the mind to constantly analyze three-dimensional space, as in the

Figure 8-41: *Orthographic drawing with first-angle projection.*

Figure 8-42: *Orthographic drawing with third-angle projection.*

case of catching a fly ball in baseball or making a basketball shot. Envisioning what a 3-D object will look like in an orthographic view depends on your spatial ability, and like sports, you can get better with practice.

For thousands of years, artists have wrestled with the problem of representing the 3-D world on a 2-D plane. It was not critical, however, that the artist's picture of the world involve exact **proportion** or scale. In art, the interpretation of a scene or person is what makes the art unique. In multiview drawings, precision is very important, so the picture must reflect true scale and proportion of the object drawn.

To achieve the precision required in multiview drawings, the drawing plane is located parallel to one of the faces on the object that is being drawn. In Figure 8-43, lines are projected from each edge and corner on the rear face of the object onto the drawing plane. The projected lines occur perpendicular (90 degrees) to the drawing plane. Because the drawing plane is parallel to the rear view of the object, the true size and shape of the object is represented, with each line showing true length. If the viewing plane were turned so that it is no longer parallel to the rear view, the length of the lines of the object would distort, resulting in an image that does not show true size or true shape.

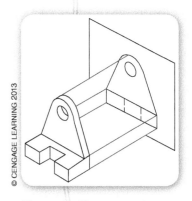

Figure 8-43: **Lines of projection from an object to the drawing plane.**

In Figure 8-44, a 3-D object is pictured inside a glass box. The views are projected from the object onto all six sides of the box, each side being parallel to one of the six principle views. When the glass box is unfolded, the six orthographic projections form a six-view multiview drawing (see Figures 8-45 and 8-46).

Unless otherwise specified, it is generally accepted that the front view is the object face that shows the most detail. The remaining views are developed as needed, but usually include a top- and right-side view (resulting in the common three-view drawing). Left side and bottom views are sometimes required if the object contains a lot of features that cannot be represented by the standard top-, front-, and right-side views (see Figure 8-47).

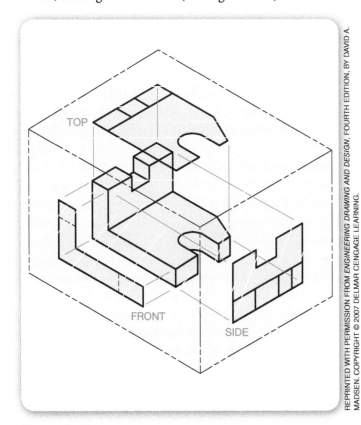

Figure 8-44: **The glass box principle.**

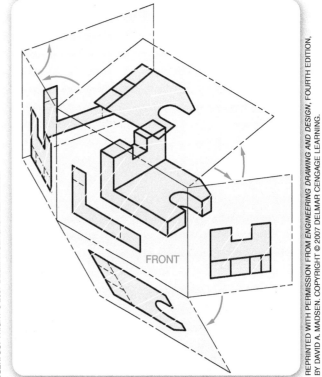

Figure 8-45: **Unfolding the glass box at hinge lines, also called fold lines.**

226 Part I: The Engineering Design Process

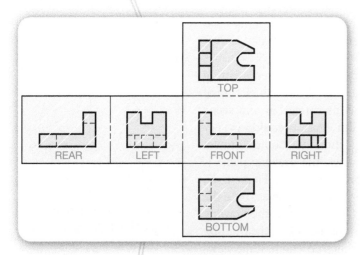

Figure 8-46: Glass box unfolded. Typically, not all views are used.

REPRINTED WITH PERMISSION FROM *ENGINEERING DRAWING AND DESIGN*, FOURTH EDITION, BY DAVID A. MADSEN. COPYRIGHT © 2007 DELMAR CENGAGE LEARNING.

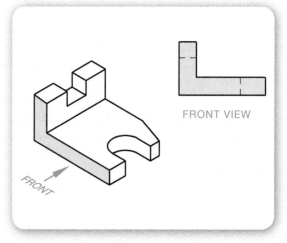

Figure 8-47: Front view selection.

REPRINTED WITH PERMISSION FROM *ENGINEERING DRAWING AND DESIGN*, FOURTH EDITION, BY DAVID A. MADSEN. COPYRIGHT © 2007 DELMAR CENGAGE LEARNING.

Your Turn
Selecting the Front View

Figure 8-48: Select the best front views that correspond to the pictorial drawings at the left. You may make a first and second choice, if you wish.

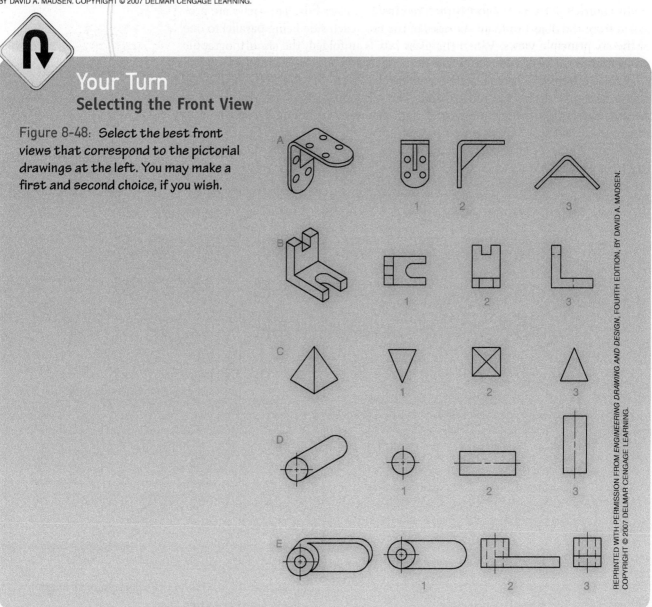

Your Turn

Create an Orthographic Sketch

Using the isometric sketch you created in the previous Your Turn (or another object of your choosing), develop a three-view drawing in third-angle projection that includes the front, top, and right-side view. You do not have to use a straight edge or ruler, but you can if you desire. Use 8½- by 11-inch or A4 size paper.

Spacing between Views

There are no standards that apply to the spaces between views. However, dimensioning guidelines dictate that dimensions should be placed, whenever possible, between the views. Therefore, the space between the views must be large enough to allow room for the object's dimensions.

Scale

Although some objects can be drawn true to size, many of the things we want to represent on paper or on the computer screen are too small or too large and must be scaled to fit within the space that is available. Hobby models are created to scale, such as in Figure 8-49. Scaling is done to permit the representation of these under- or oversized objects.

Figure 8-49: A scale model.

As noted earlier in this chapter, standard scales are used in architecture and mechanical drawings. In architecture building design, most units are in feet, so a scale of that unit is required to represent the building on paper. Common scales for residential design are 1/8 inch equals 1 foot and 1/4 inch equals 1 foot. In mechanical design, scale is a ratio of the full size. Common scales used include full, or 1:1; half, or 1:2; quarter (QTR), or 1:4; and double (DBL), or 2:1.

It is important that the scale be clearly marked in the drawing title block, which is the space reserved for the drawing title, date, scale, and other information about the drawing. Anyone reading the drawing may need to know how the object is represented.

The term *scale* is also applied to the tool used to measure drawing distances. Scales allow engineers or other designers to efficiently measure distances without resorting to a calculator or manually converting true distances to the scale distances used in the drawing (see Figure 8-50).

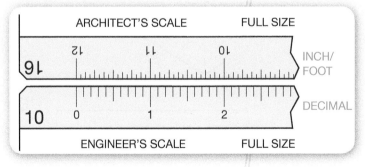

Figure 8-50: Comparison of architect's scale (16) and full engineer's scale (10). Note that in the engineer's scale, an inch is divided into 10 parts.

REPRINTED WITH PERMISSION FROM *ENGINEERING DRAWING AND DESIGN*, FOURTH EDITION, BY DAVID A. MADSEN. COPYRIGHT © 2007 DELMAR CENGAGE LEARNING.

LINE CONVENTIONS

There are rules for lines in orthographic drawings. These rules standardize how different parts of views are represented in the language of technical drawing, just as parts of speech are standardized in spoken or written language. These rules are called line conventions, and deal with line weight (thickness) and whether the line is unbroken or dashed.

Object Lines

The visible lines that outline and detail an object's shape are called object lines. Object lines are bold (0.024 inch thick). With a few exceptions, all other drawing lines are lighter in weight. A no. 2 or 2H pencil is typically used to draw object lines.

Construction Lines

Construction lines are used to lay out drawings and sketches. They are very lightweight and are not intended to show when the drawing is reproduced. Often, the entire drawing will be carefully laid out in construction lines before object lines are drawn.

Hidden Lines. The object in Figure 8-51 is pictured in three views, with an isometric view in the upper right.

The problem is that some features are not clear, such as the notch on the left side of the object and the hole in the front. Does the hole go all the way through? Is the notch the same all the way to the back of the object? There are ways to show these features more clearly.

A hidden line is a dashed line that is used to represent an edge of a surface that cannot be seen from a particular view. When drawn, the thickness of a hidden line is half the thickness of an object line. Typically, a 4H pencil is used for hidden lines, but a sharp no. 2 pencil with a light touch will substitute.

In Figure 8-52, the same object is shown but now it is easy to see that the hole goes all the way through the object, because the hidden lines in the top view show that the edges of the hole extend from the front face to the rear face. Another hidden line, to the left of the hole's hidden lines in the top view, shows the inside corner of the notch. The hidden lines in the right-side view confirm that both features occur from the front face to the rear face. For clarity, hidden lines are sometimes included in isometric views, though are not required.

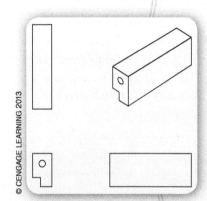

Figure 8-51: Three-view drawing of an object with an isometric view included.

Centerlines

Another type of line that helps clarify a drawing is the centerline. Centerlines have several purposes, but are most commonly used to identify the location of a hole or arc center. Once located, the centerlines may serve as extension lines for dimensions. Centerlines are made with a series of short and long dashes. When locating a hole and arc center, the short dashes from two centerlines will intersect

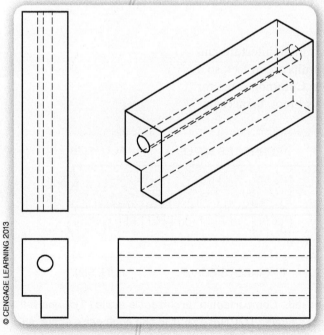

Figure 8-52: The same object as in Figure 8-51, with hidden lines included.

at the center of the arc or circle at 90 degrees. When drawn, the thickness of a centerline is equal to that of a hidden line. Figure 8-53 shows an object in which centerlines have been used to locate a hole.

Extension Lines

Extension lines are used to extend the edges of an object so they may be located with dimension lines. They are half the thickness of an object line, and used so that dimension lines can be located some distance away from the object.

Extension lines are collinear to (in line with), but never touch, their respective object lines. A gap always exists between an extension line and its respective object line, so that the two line types can be distinguished from one another. A typical gap is 0.06 inch (1.5 mm). The extension line extends beyond the arrow point of its respective dimension line about 0.125 inch. It is permissible for extension lines to cross object lines, hidden lines, centerlines, and other extension lines, but they may not cross dimension lines. Figure 8-54 shows a simple example of the use of extension lines.

Dimension Lines

Dimension lines are used to indicate an object's size, as well as the sizes and the locations of its features. In Figure 8-56, the object is dimensioned to show the height, width, and depth, as well as the location and size of the hole and arc. Dimension lines have arrows at each end. The tips of these arrows point toward their respective extension lines. A dimension line has the same thickness as a centerline and hidden line.

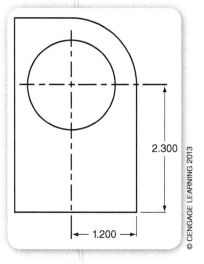

Figure 8-53: **Centerlines are used to locate the center of a hole, arc, or circle feature.**

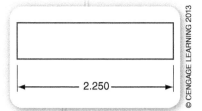

Figure 8-54: **Use of extension lines to create a dimension.**

Point of Interest

Figure 8-55: **Examples of line types and weights used in orthographic drawings.**

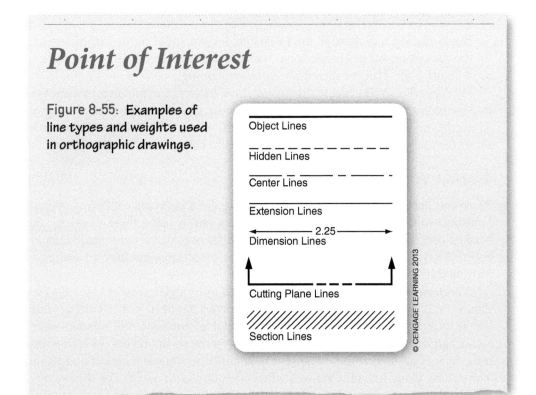

230   Part I: The Engineering Design Process

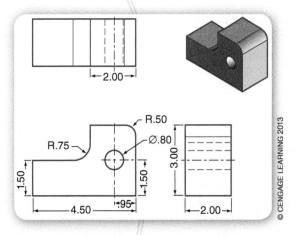

Figure 8-56: Example of a simple object with dimensions.

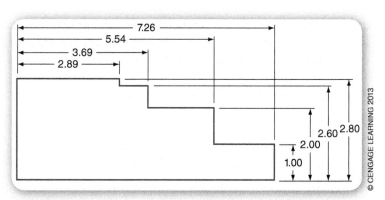

Figure 8-57: Orienting the dimension numbers to read in the same direction.

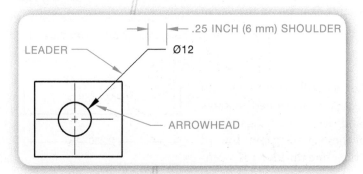

Figure 8-58: Leader line used to dimension the diameter or radius of an arc or circle.

REPRINTED WITH PERMISSION FROM *ENGINEERING DRAWING AND DESIGN*, FOURTH EDITION, BY DAVID A. MADSEN. COPYRIGHT © 2007 DELMAR CENGAGE LEARNING.

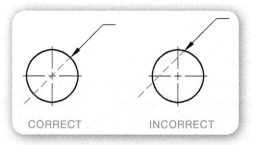

Figure 8-59: Circle to leader line relationship. The path of the leader should pass through the center.

REPRINTED WITH PERMISSION FROM *ENGINEERING DRAWING AND DESIGN*, FOURTH EDITION, BY DAVID A. MADSEN. COPYRIGHT © 2007 DELMAR CENGAGE LEARNING.

When placing a dimension on a drawing, leave a space in the dimension line for the dimension value. Dimensions should all be oriented in the same direction, as in Figure 8-57. This is called **aligned dimensioning**.

Leaders are also considered dimension lines. Leaders are thin lines that are used to dimension a hole diameter or arc radius. Typically, leaders are drawn at a 30-degree, 45-degree, or 60-degree angle, but they may be drawn at any angle. Figure 8-58 shows the use of a leader line. Figure 8-59 shows the correct way to draw a leader.

Section Views

As objects become more complex, representing them becomes a bigger challenge. Drawings can become filled with hidden lines, and reading these drawings can become confusing. When objects have interior features that are not readily apparent in the typical views already discussed, then a section view may be needed to provide clarity.

A **sectional view** provides a view of an object as though it were cut by a saw (see Figure 8-60). To represent the place where the object has been "cut," a **cutting plane line** is located in the top view. A cutting plane line represents the location where a cutting plane (similar to a saw blade) passes through the object. This line type is as thick as an object line, because it is intended to draw the viewer's attention. The cutting plane line tells viewers where a section is to occur. The arrows on a cutting plane line indicate what part of the object is to be removed.

Note the thin, diagonal parallel lines that occur inside the profile of the object's edges. These are called **section lines**, and they appear where an unfinished surface occurs as a result of the cutting plane. A section line has a thickness that is equal to an extension line, and must never be drawn parallel or perpendicular to an object line. Each section view in an engineering drawing is tied to a specific cutting plane line. The section view in Figure 8-60 is titled Section A-A. This corresponds to the letter A, which appears next to each arrow on the cutting plane line in the front view. If another section view occurred in the drawing, its designation would be Section B-B, thus corresponding to a cutting plane line that is labeled with the letter B next to its arrows.

The cutting plane does not have to be a straight continuous plane. In Figure 8-61, the cutting plane jogs to intersect the center of several holes so that a clear representation of the interior detail is shown.

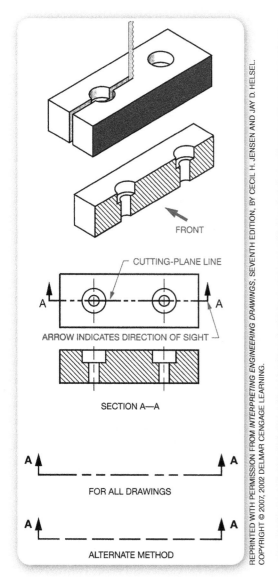

Figure 8-60: A section view is like cutting the object with a saw.

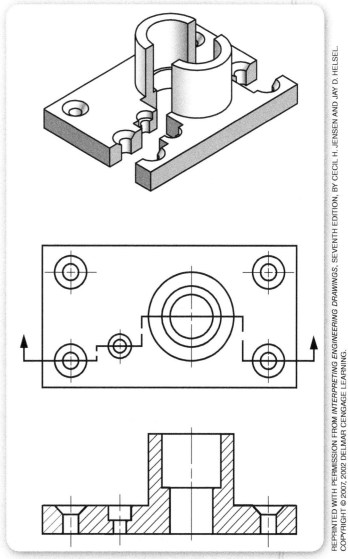

Figure 8-61: Section planes do not have to cut through an object in a straight line, although they must be parallel to the viewing plane.

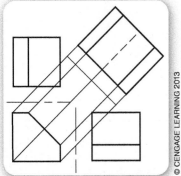

Figure 8-62: An auxiliary view at an angle to the front view.

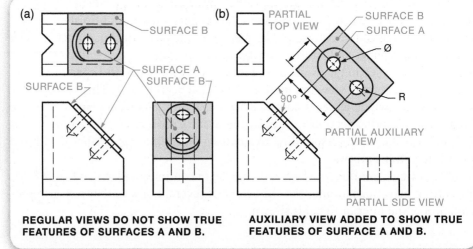

Figure 8-63: How an auxiliary view allows true distances to be represented.

REPRINTED WITH PERMISSION FROM *INTERPRETING ENGINEERING DRAWINGS*, SEVENTH EDITION, BY CECIL H. JENSEN AND JAY D. HELSEL. COPYRIGHT © 2007, 2002 DELMAR CENGAGE LEARNING.

Auxiliary Views

When an object has an inclined or angled surface that contains important geometric features, it is sometimes necessary to create an auxiliary view. An auxiliary view is an orthographic view that is used to show a surface that is not parallel to any of the principal view planes.

In Figure 8-62, the faces representing the object's angled surface in the top- and right-side views are foreshortened. A foreshortened face will show edges that are shorter than their true lengths, and will also distort any shapes associated with that angled surface. As you know, a head-on view of an object's surface will show the true size and shape of that surface, and is made through a process called orthographic projection. In this process, projection lines extend out from the object to a two-dimensional viewing plane that is parallel to the respective surface (see Figure 8-43).

The projection lines used to construct an auxiliary view will originate from one of the principal views (top, front, or side) in which the inclined surface appears as an angled edge. The projection lines will extend outward, perpendicular to the angled surface. The projection lines originate from the front view, and extend outward, perpendicular to the front view's angled edge. The depth of the auxiliary view was determined by the depth of object as measured on the top- or right-side view.

Figure 8-63 provides an example of how an auxiliary view helps accurately describe an object. The auxiliary view provides an opportunity to clearly dimension the features on angled surfaces, instead of providing the distorted view. The development of auxiliary views is often an involved process, depending on the complexity of the object you are drawing.

DIMENSIONING

In engineering drawings, dimensions are as important as the shapes that are drawn. To accurately manufacture an object, the craftsperson, machinist, or manufacturing engineer must have a complete account of the object's geometric information, which includes the object's overall width, depth, and height, along with the locations and sizes of its features.

projection line: A horizontal or vertical line that can be used to locate entities in an adjacent view.

There are two common methods for dimensioning an object. The first is called **chain dimensioning** (see Figure 8-64). When the relationship between features is important, this method applies. This is, however, a drawback in some situations. Because each dimension has a tolerance, each dimension can vary within certain acceptable limits. Chain dimensioning gives equal importance to each extension line, which means that dimensional variation can exist in either direction. When an object is made, each real dimension can vary slightly from what is specified on the drawing. This leads to accumulation of inaccuracies across the length of an object. However, if accuracy in the distance between features is important, such as the distance between two holes, then chain dimensioning helps achieve that accuracy.

Note the omission of the last dimension in the horizontal and vertical directions and, in both directions, an overall length dimension is given. This last dimension is redundant, and could be confusing when laying out the actual object. The allowable variation (tolerance) of each dimension would mean that this last distance may be slightly different than the "ideal" distance you specify, so the accepted standard is to leave it out.

Another common dimension practice that is used is called **datum dimensioning**. In datum dimensioning, sometimes called **baseline dimensioning**, a common point (which occurs where three surfaces meet) is used as a reference for the placement of overall size and location dimensions. A datum surface can be easily recognized by an extension line that shares two or more arrows from dimension lines, as in Figure 8-65.

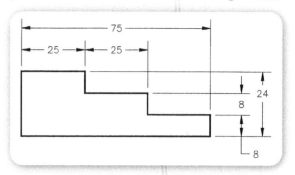

Figure 8-64: **Chain dimensioning.**

REPRINTED WITH PERMISSION FROM *ENGINEERING DRAWING AND DESIGN*, FOURTH EDITION, BY DAVID A. MADSEN. COPYRIGHT © 2007 DELMAR CENGAGE LEARNING.

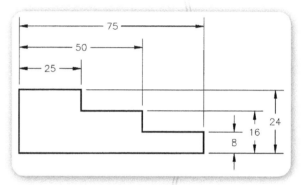

Figure 8-65: **Datum dimensioning from a common surface.**

REPRINTED WITH PERMISSION FROM *ENGINEERING DRAWING AND DESIGN*, FOURTH EDITION, BY DAVID A. MADSEN. COPYRIGHT © 2007 DELMAR CENGAGE LEARNING.

Point of Interest
Basic rules of dimensioning:

- ▶ Only place dimensions on edges and features that appear as true size and true shape. Do not dimension surfaces or features that are foreshortened.
- ▶ Do not place dimension lines inside an object view. Create a perimeter of space that immediately surrounds the object view's profile, wherein no dimension lines will occur.
- ▶ If possible, avoid sending extension lines through an object view. If this is not possible, place the dimension on the side where the extension line will extend into the object the shortest distance.
- ▶ Dimensions should be arranged so that the smallest dimension occurs closest to the object view and the larger dimensions occur farther from the view.
- ▶ Where appropriate, establish a common reference point (datum), and use datum dimensioning to avoid compounding tolerances.
- ▶ Use radius dimensions for arcs and diameter dimensions for circles.
- ▶ Avoid redundant dimensions; do not dimension the same feature in different views.
- ▶ Allow sufficient space between dimensions.

Dimension Precision

When designing an object, the precision that is indicated by the object's dimensions should be determined through careful consideration of a material's physical properties and the manufacturing methods that will be used to make the part. For example, if you were to design and build a cedar bench for a garden, you would need to create technical drawings that indicate the size and shape of the wood components that make up the bench.

Let us assume that you intend to use a handheld circular saw to cut the pieces. Even the most careful woodworker would be hard-pressed to cut wood to an accuracy of 0.01 inch (this means that the part geometry must not vary in either direction from the target dimension more than five times the diameter of an average human's hair). If this could be accomplished, what purpose would it serve? Wood is a material that experiences relatively large amounts of expansion and contraction due to changes in temperature, humidity, and barometric pressure (which result from changes in seasons). Given the manufacturing method and the material's properties, it would make little sense to specify overly precise dimensions on the bench's component part drawings.

Many of the manufacturing tools used to manipulate "industrial materials" are outfitted with precise adjustment controls. Such tools are capable of producing parts to very high dimensional accuracies. Parts made from metal and plastic usually have engineering drawings that are dimensioned with decimal values. Though not nearly as common in industry, drawings that are used to make wooden objects through the use of traditional craft tools may incorporate fractional dimensions. Knowing the material from which the object will be made should give you an idea if fractional or decimal dimensions are appropriate.

Dimension Tolerances

```
X.X    = ±.020 inch
X.XX   = ±.010 inch
X.XXX  = ±.005 inch
```

Figure 8-66: Unless otherwise specified, the number of decimal places used in a dimension determines the tolerance value.
© CENGAGE LEARNING 2013

Dimensional variation, even if it is very small, exists in all manufactured objects. Dimension tolerances are important for two main reasons. First, they tell the manufacturer how much dimensional variation can exist and still have the part be considered acceptable. Second, a tolerance may identify a dimensional range above and below, at which the part will no longer function. Tolerances may appear with a dimension value, or the number of decimal places that are shown in the dimension value may infer them.

General **tolerances** are applied to dimensions according to the degree of accuracy exhibited by a dimension's value. They take the form of **bilateral tolerances**, in which equal variation is permitted in both directions from the specified dimension. A bilateral tolerance uses a plus/minus (±) symbol.

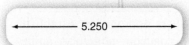

Figure 8-67: Example of a dimension with ±.005 inch.
© CENGAGE LEARNING 2013

Figure 8-66 shows the format for general tolerances. General tolerances tell the manufacturer that there are differences between dimensions that read 5.0 inches, 5.00 inches, and 5.000 inches. If an object's overall width dimension is 5.0 inches, then the manufacturer is free to produce a part that has an overall width value between 4.980 and 5.020 inches. If the overall dimension is 5.00 inches, then the manufacturer must produce a part that has an overall width between 4.990 and 5.08 inches. If the overall dimension is 5.000, then the manufacturer must produce a part that has an overall width between 4.995 and 5.005 inches. Bilateral tolerances are not solely reserved for title blocks, and may appear next to dimensions (see Figure 8-67).

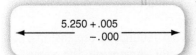

Figure 8-68: Unilateral tolerance varies in only one direction from the specified dimension.
© CENGAGE LEARNING 2013

A **unilateral tolerance** also uses a plus (+) and minus (−) symbol, but in a way that is different from a bilateral tolerance. Unilateral tolerances allow variation in only one direction from the specified dimension (see Figure 8-68).

Another type of tolerance, called **limit dimensions**, eliminates the need to calculate the acceptable dimension range by showing two dimensions stacked on top of each other. The value on top represents the upper dimensional limit, and the value on the bottom represents the lower dimensional limit (see Figure 8-69).

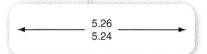

Figure 8-69: **Limit tolerance dimensioning.**
© CENGAGE LEARNING 2013

Good engineers understand that overly precise dimensions and excessively tight tolerances will make an object more difficult to produce, and result in increased manufacturing costs. It is a rule of thumb that you should only specify dimensions and tolerances that are necessary for the object to carry out its intended function.

How do engineers know how to apply tolerances to dimensions? The answer has to do with the engineers' ability to recognize the amount of dimensional variation that can exist without adversely affecting the design's ability to carry out its function. For example, tight dimensional tolerances are not needed to ensure that a simple plastic drink coaster will keep a cup elevated off of a table surface. Therefore, specifying a tolerance of $\pm.005$ inch on any of the coaster's dimensions would not only be pointless, but it would unnecessarily increase the cost of production. On the opposite end of the complexity scale, tight tolerances are needed to ensure that the mating components on an astronaut's suit will not allow atmosphere to escape during a space walk. This is one of the reasons why an astronaut's suit costs over $1,000,000. There are more rules for specifying dimension tolerances, so you may need to investigate further if your particular situation requires it.

Though CAD solid modeling programs automatically generate technical drawing views, accurately communicating dimensions and tolerances is still a manual process that designers conduct. Locating and applying tolerances to dimensions requires a knowledge and understanding of both the object's function and the manufacturing methods that will be used to produce the object. Though many CAD programs possess automatic dimensioning features, computers do not yet possess the intelligence needed to accurately and effectively apply dimensions to anything but the simplest technical drawings. Therefore, you will have to apply many of the dimensions in a complex CAD drawing, using the standards and rules stated previously.

Dimensioning Features

In addition to the overall length, height, and width, most objects have features that must be dimensioned. Holes, notches, arcs, angles, and other features have required dimensions that will locate them on the object and provide details about the feature itself. For example, a hole has both a diameter and a location. The location is typically detailed by a distance from the center of the hole to an edge or other feature in two axes (see Figure 8-70).

A **fillet** is an inside radius between two intersecting planes (see Figure 8-71). In production, fillets are used to provide strength to avoid a fracture where two planes meet. A **round** is applied to an outside corner to avoid a sharp edge and for appearance (see Figure 8-72).

A **chamfer** is also often used on outside intersecting planes by removing material at an angle to one of the planes (see Figure 8-73). Rounds and chamfers avoid sharp corners for safety and handling, and make it easier to produce the object because it is difficult to efficiently move material into a sharp corner when casting. To dimension a fillet or round, you need to provide a radius value, as in Figure 8-74. There are two common ways to dimension a chamfer shown in Figures 8-75 and 8-76.

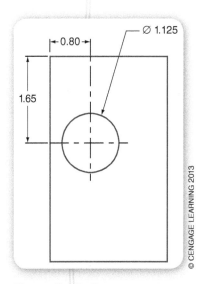

Figure 8-70: **Dimensioning and locating a hole.**

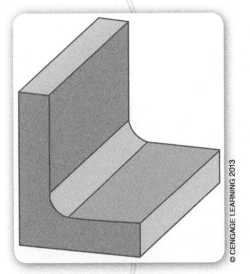

Figure 8-71: A fillet adds strength to an inside corner.

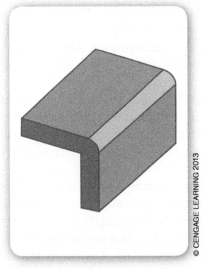

Figure 8-72: A round smoothes an edge or corner.

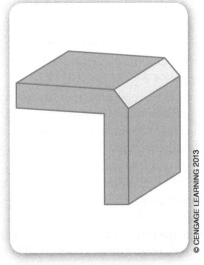

Figure 8-73: A chamfer breaks an edge or corner.

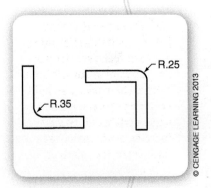

Figure 8-74: Dimensioning fillets and rounds.

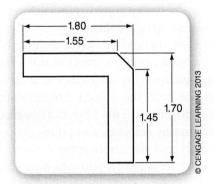

Figure 8-75: Dimensioning a chamfer—method 1.

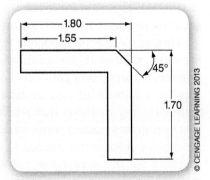

Figure 8-76: Dimensioning a chamfer—method 2.

Fun Facts
Rounds and Chamfers

The use of rounds and chamfers in production is common to avoid sharp corners. In fact, it's difficult to find products that do not use these features, except in the case of products that are intended to slice or cut. Corners that are sharp are dangerous, so rounds and chamfers are used for safety. Corners that are sharp also damage easily, that is, they are subject to rapid wear so they look sloppy after a short time, so rounds and chamfers are also used for appearance. It is difficult to produce components with sharp edges, so if rounds or chamfers are not used, corners may be inconsistent, requiring another production step and increasing the cost to produce the part. Rounds and chamfers provide a number of important advantages in production.

COMPUTER-AIDED DESIGN

As discussed throughout this text, the introduction of computers into the design process has brought about significant changes in how products are designed and the jobs of those involved in the design process. Two major types of CAD 3-D modeling software are used for the design of products: freeform surface modeling and solid modeling.

The principal reason that industry has so widely adopted CAD is because it has decreased the time from idea conception to manufacture. In business, time is money, and the time that is saved by using CAD reflects on a company's "bottom line." Computer-aided design has streamlined the product-development process, allowing design teams to communicate more efficiently. Coupled with the power of the Internet, CAD allows team members to work together from all over the world. Design teams may use standard parts files from different manufacturers and subcontractors, with provisions for updating these files when companies make changes to their products.

With CAD, design revisions can be made more quickly without the need to start over again with new drawings and specifications. **Parametric modeling** design software, pioneered by Parametric Technologies Corporation (PTC), allows solid model features to be changed and updated because of the way in which the program stores information about the model. Examples of **parameters** are dimensions, material density, and formulas that describe curves, sweeps, and lofts between sketches.

Another important aspect of many CAD programs is called **feature-based modeling**. In feature-based modeling, for example, a hole is an entity that carries instructions on how the hole behaves when other dimensions change. When a designer specifies a hole that goes all the way through an object, the hole will go all the way through, even if the designer later changes the thickness of that object. In this way, the program has captured "design intent," so it "knows" that the designer intended that the hole should go all the way through. Feature-based modeling is a very powerful tool and often found in parametric modeling programs.

Freeform surface modeling is a technique that allows designers to create a "skin" over a component, creating curves in almost any shape desired. Surface modeling uses control points (sometimes called poles) to curve and distort a surface. Figure 8-77a shows a flat surface that has been curved using control points. As the surface is stretched and squeezed by clicking and dragging control points, a 3-D form is created. Surface modeling is used extensively in animation, especially to create lifelike figures (see Figure 8-77b).

> **parametric modeling:** A CAD modeling method where each feature, such as a length of a side or radius of a fillet, uses a parameter to define the size and geometry of that feature and to create relationships between features. Because the software keeps a history of how the model was built, changing the parameter value updates all related features of the model at once when the model is regenerated.

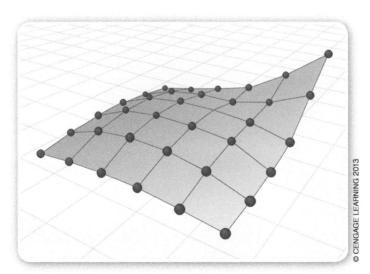

Figure 8-77a: **Control points on a surface. By dragging control points, a surface can be reshaped.**

Figure 8-77b: **Surface modeling of complex curves.**

> **OFF-ROAD EXPLORATION**
> **Which CAD Software?**
> There are many powerful CAD programs being used by different industries. Using the Web, find the most common CAD programs used in the following industries:
> - Aerospace
> - Architecture
> - Automotive
> - Industrial design
> - Landscape design
> - Yacht or ship design
>
> Develop a chart that lists the names of the most common CAD programs down the left side of the chart and the industries across the top of the chart. Indicate the two or three most common programs for each industry.

Many freeform surface modeling systems use nonuniform rational basis splines (NURBS) mathematics to define surface curves, although other mathematical techniques have been developed. NURBS software is used extensively for creating models for animation, jewelry, and a variety of other products where complex surface features are important.

Solid modeling is a method that involves mathematically describing both the exterior and interior of the object. To create a 3-D solid model in most programs, you begin with a 2-D sketch of a closed shape, as shown in Figure 8-78a. These 2-D sketches are then extruded or revolved to form the solid, as shown in Figure 8-78b. Other features are used to round or chamfer edges and add or subtract additional material. Complex solids are built up from these basic forms. From there, these solids are brought together into highly complex assemblies, as in Figure 8-79. Solid modeling is used for much of engineering design and mechanical component development.

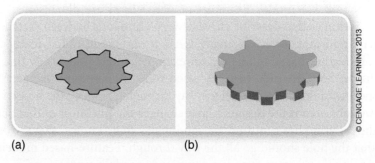

(a) (b)

Figure 8-78: The 2-D sketch on the left (a) is extruded to create the 3-D solid on the right (b).

Figure 8-79: A CAD model of an underwater scooter.

Creating Sketches in Solid Modeling Software

Parametric modeling is based on the creation of features that are developed from 2-D sketches. Developing skill in the creation of 2-D sketches is essential to the mastery of parametric modeling software. Some programs allow the development of these sketches without dimensional constraints; therefore, 3-D solids may be created and "sketched" in 3-D before they are constrained. Other programs require sketches to be dimensioned as they are developed.

Generally, sketches are created with only a handful of drawing tools. In programs like Autodesk Inventor and PTC's Pro/DESKTOP or Pro/ENGINEER (and others), there are tools to create lines, circles, arcs, rectangles, and other polygons (see Figure 8-80). There are also tools to delete line segments and move or rotate sketch lines. These tools are used to develop 2-D sketches that represent a profile of a desired 3-D part.

Sketches are turned into 3-D models by *features,* a term often used to describe a number of functions, such as extrusion, lofting, revolving, sweeping, and shelling. Figure 8-81 shows these features created from their associated sketches.

The future of CAD software programs is in combining surface and solid modeling, although there are still some issues of compatibility with the mathematics involved in the two systems. Some current products do combine solid modeling and surface modeling to allow complex surface curve development required by more and more consumer products, but as the systems evolve, most industry CAD software will incorporate both systems.

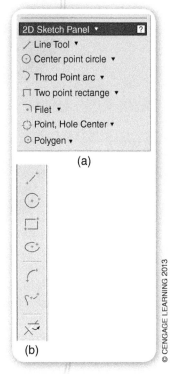

Figure 8-80: AutoDesk Inventor (a) and PTC's Pro/DESKTOP (b) sketch tools. These tools are used to create 2-D sketches that are then used by features to develop solids.

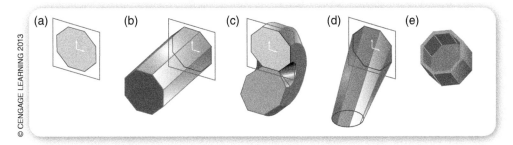

Figure 8-81: These 2-D sketches were used to create 3-D models using "features": (a) sketch, (b) extrusion, (c) revolve (220 degrees), (d) two sketches on perpendicular work planes with a loft created between them, and (e) shell feature applied to a solid.

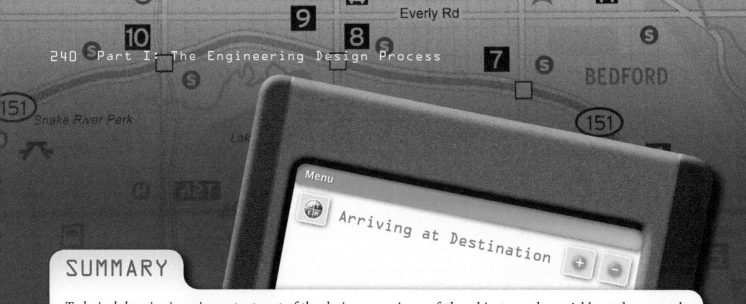

SUMMARY

Technical drawing is an important part of the design process and is a unique language used by industry to design and manufacture new products. As ideas develop from simple sketches to possible solutions, engineers and other design professionals need to see those solutions in more detail. Technical drawings are often based on orthographic projection, although other kinds of drawings are also important. Orthographic drawings show the object in true shape and size by looking down on the object. Drawings may be scaled so that very small or very large objects can be represented on a single sheet of paper and be displayed on a computer screen. When drawings are approved, the information can be translated into a machine language and parts can be made on CNC machines. Accuracy and precision are required in both dimensions and measurements. Designers must understand and be able to use the principle of variability. A vernier caliper allows designers and technicians to measure to thousandths of an inch or finer. Architectural scales allow measurements to be easily scaled larger or smaller. Measurement systems are important to all societies and become more accurate as the society develops technologically. Designers in the United States must be able to read and work in both imperial and metric measurements.

The introduction of computer-aided design (CAD) evolved the way in which designers develop designs and use technical drawings. Designers can now create objects in 3-D, called solid models, and CAD stores the geometry of the solids so that other views of the object can be quickly and accurately created. While draftspeople took many hours to make orthographic drawings from sketches, CAD allows designers to produce orthographic drawings in seconds and create actual 3-D models in a matter of hours.

Engineers and technicians use technical drawings to produce parts and products, and consumers use drawings to assemble products. Hobbyists use drawings to build items such as furniture and model aircraft. Quality-assurance personnel use technical drawings to inspect parts, making sure what has been manufactured is acceptable.

Orthographic multiview drawings are an important aspect of communication among engineers and other design professionals. These drawings are created to standards established by the American National Standards Institute (ANSI), followed in the United States, Canada, and several other countries, and the International Organization for Standardization (ISO), which is followed in most European and Asian countries. Standards have been created for views, dimensions, and drawing sizes, as well as many other aspects of orthographic drawing. These standards make it possible for anyone familiar with the standards to read and interpret drawings created by someone else.

Successful multiview drawing requires practice and attention to detail. Dimensioning is a critical part of the orthographic drawing process, and clearly placed dimensions with attention to appropriate tolerance are important for production.

BRING IT HOME—

OBSERVATION/ANALYSIS/SYNTHESIS

1. Using isometric grid paper, sketch the objects shown in Figures 8-82 through 8-86 in isometric projection. When dimensions are not specified, estimate proportional sizes in your drawing.

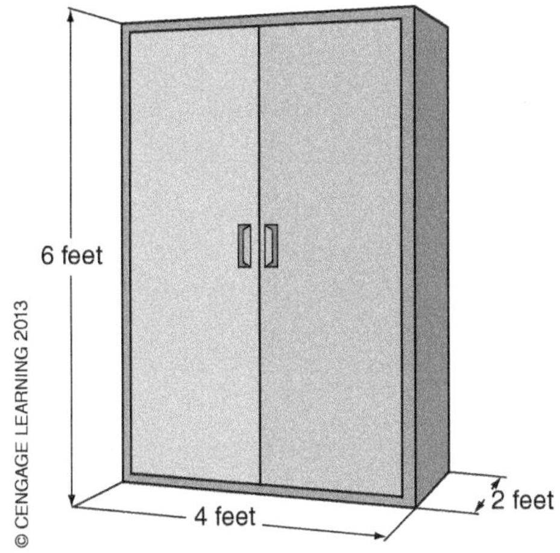

Figure 8-82

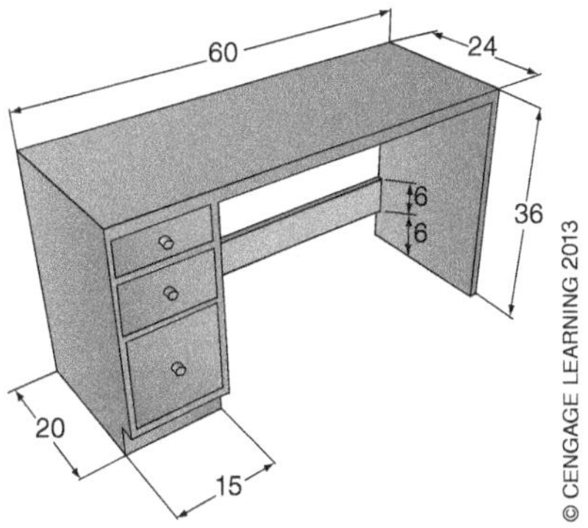

Figure 8-83

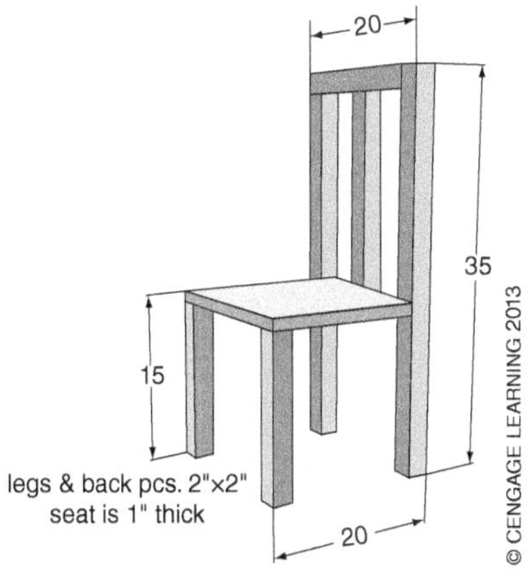

Figure 8-84

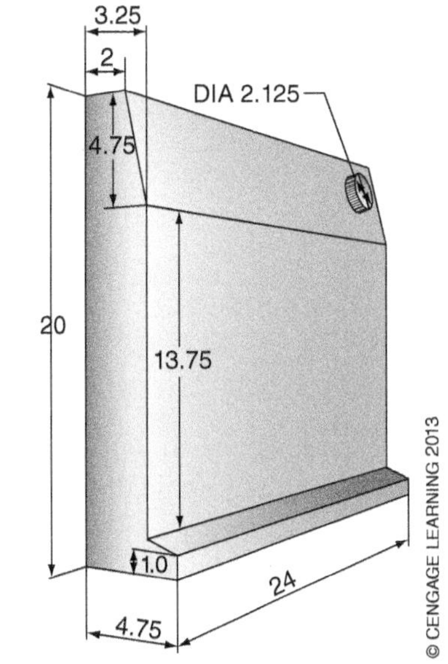

Figure 8-85

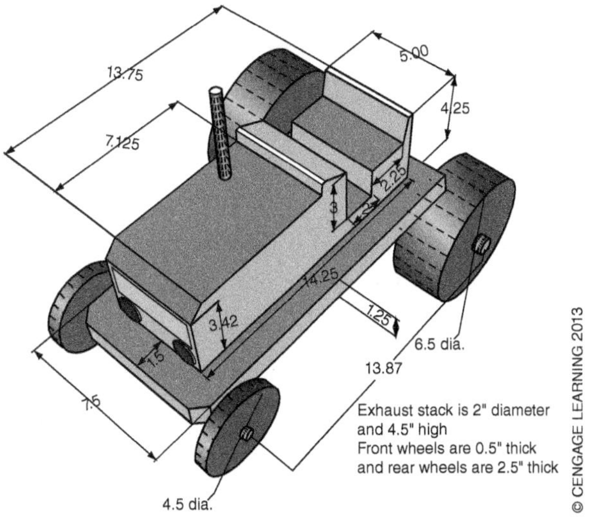

Figure 8-86

Figure 8-87

2. Sketch a three-view orthographic drawing of each of the objects in number 1.
3. Develop hand-sketched, three-view orthographic drawings based on the isometric drawings in Figures 8-87 through 8-90. Dimension the drawings.

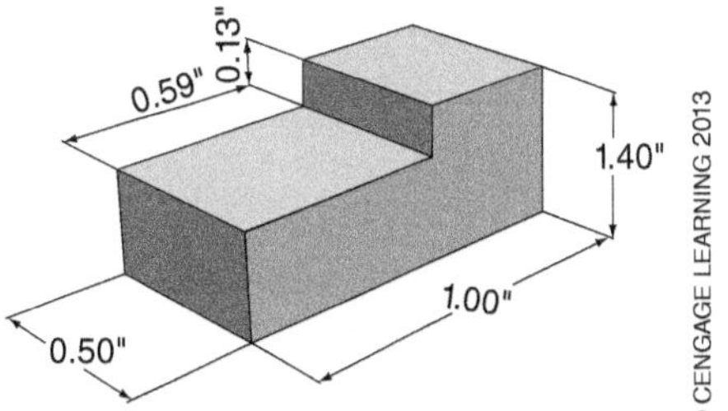

BRING IT HOME

(continued)

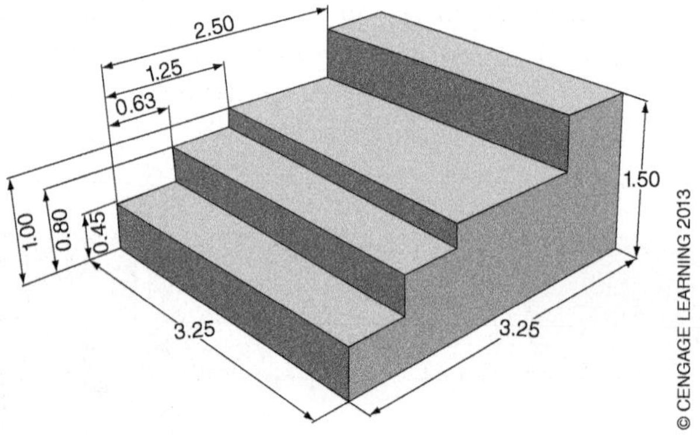

Figure 8-88

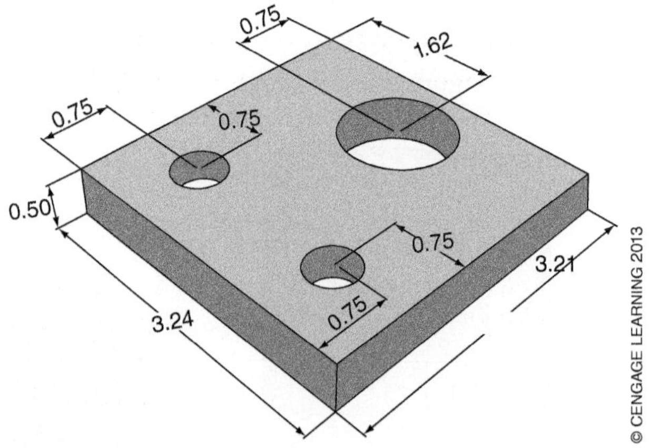

Figure 8-89

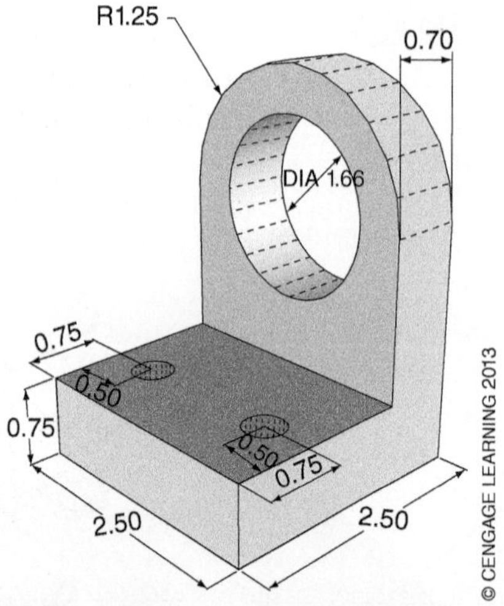

Figure 8-90

4. Draw Figures 8-91 and 8-92 in orthographic projection with dimensions. When dimensions are not specified, estimate the proportional sizes and include those dimensions in your finished orthographic drawing.

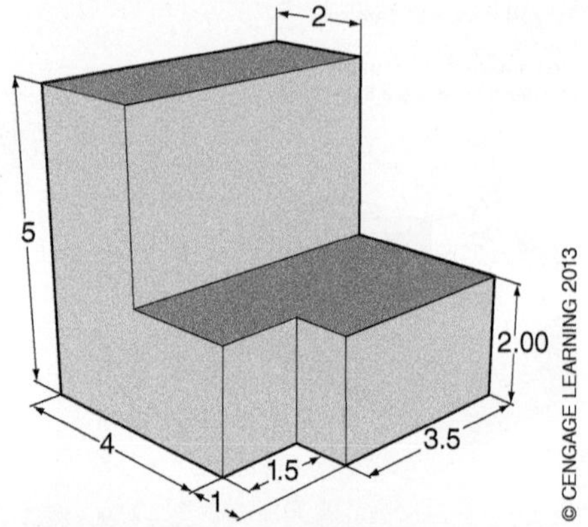

Figure 8-91

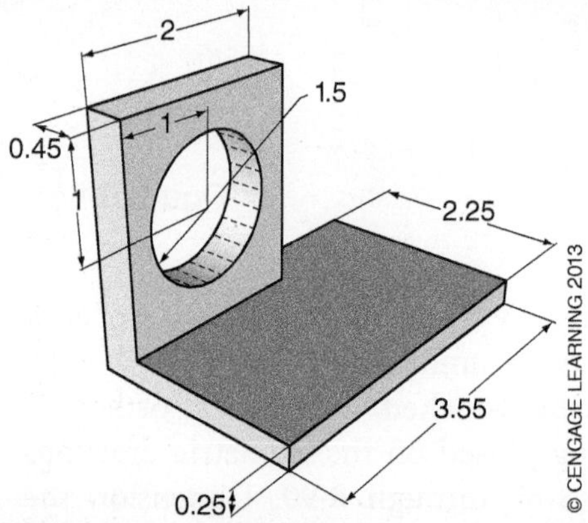

Figure 8-92

BRING IT HOME--

(continued)

5. Show step by step how to make a peanut butter and jelly sandwich using only a series of isometric sketches (no words). Each step should be in its own box with the step number.
6. Find a simple object in your home or school lab and carefully sketch it in isometric projection (use isometric grid paper if available). Measure as much detail as you can, writing down all the measurements using calipers and an engineer's scale, if available. Create an orthographic multiview drawing of the object to scale. Dimension the object following the guidelines in this chapter.

EXTRA MILE

DETAILED ORTHOGRAPHIC DRAWING

- Study the following isometric sketches and develop an orthographic drawing for each. Drawings should be created with drawing instruments (T square or equivalent, triangles, compass, and so on) to appropriate scale, and fully dimensioned.

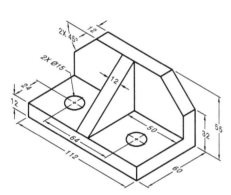

Figure 8-93

REPRINTED WITH PERMISSION FROM *ENGINEERING DRAWING AND DESIGN*, FOURTH EDITION, BY DAVID A. MADSEN. COPYRIGHT © 2007 DELMAR CENGAGE LEARNING.

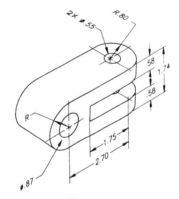

Figure 8-94

REPRINTED WITH PERMISSION FROM *ENGINEERING DRAWING AND DESIGN*, FOURTH EDITION, BY DAVID A. MADSEN. COPYRIGHT © 2007 DELMAR CENGAGE LEARNING.

Figure 8-95

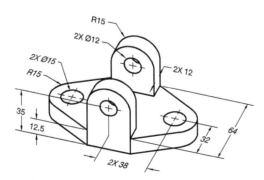

REPRINTED WITH PERMISSION FROM *ENGINEERING DRAWING AND DESIGN*, FOURTH EDITION, BY DAVID A. MADSEN. COPYRIGHT © 2007 DELMAR CENGAGE LEARNING.

CHAPTER 9
Testing and Evaluating

Menu

 Before You Begin
Think about these questions as you study the concepts in this chapter:

1. What kinds of tests would you develop to assess the success of your design work?

2. What kinds of tests do engineers use to test product materials for tensile strength, fatigue, and hardness?

3. Why do you think that product testing typically includes looking at aesthetics, ergonomics, safety, and durability/reliability?

4. How can assessing the way in which you went about your work help you evaluate your own design work?

INTRODUCTION

Testing is done as part of the process of product development for many reasons, but the main reason is to determine if the goals or criteria of the product (or project) have been met. The criteria could be a technical specification like "lift a weight of 275 pounds 10 feet in less than 18 seconds," or perhaps the criteria could be more directly related to the financial success of a product, like "reduce piece part costs for the product by 35 percent within three months." For design projects, whether they are in industry or as part of a school project, testing is crucial step for assessing the level of success. In business, testing is crucial for the economic health of the company.

Testing goes to the heart of designing and problem solving. Did the design or solution meet the criteria and specifications? For many products, a number of different tests may determine if the solution meets the specifications and requirements of the design. For example, evaluating an automobile prototype may involve testing, among other things, if the engine overheats in hot weather while idling. For this test, the prototype may be placed in an environmental chamber to simulate desertlike conditions with the engine idling for several hours (see Figure 9-1). Engine temperature and other physical parameters will be monitored and recorded. This may be only one of hundreds of tests to determine if the prototype is ready for production. Other tests may include acceleration performance, steering and handling, crash safety, exhaust pollution, and even paint and finish durability. These are examples of tests on complete products or systems. Testing of individual components or parts is also typically required and may take place during development and again when the product is assembled. Additionally, it may also be necessary to test the raw material that goes into a component. For example, perhaps to ensure that a plastic molding process for certain car parts works well, one may determine that the ~1-millimeter-diameter plastic spheres used as raw material should be pretested for density and color.

This chapter will look at strategies for testing design solutions as well as evaluating and presenting the results of these tests.

At some point in the process of your own designing, you must ask yourself how well you have done. It is not easy to answer that question. Seldom does a design fulfill each and every requirement of a problem, and you must always deal with the trade-offs and risks of technology. In addition, you have probably been working within limitations imposed by the amount of time allowed for your work and the resources available to you.

The first question you will want to address deals with how well your design and final product have solved the problems you identified in the early stages of the process: Does it solve the problem? Although this may seem to be a simple question, it is not.

Figure 9-1: Automotive testing in an environmental chamber.

DEVELOPING APPROPRIATE TESTS

Developing appropriate tests for your design work is an important stage of the design process. It will require that you define exactly what it is you want to test, investigate testing possibilities, devise a number of possible tests, choose the tests you decide are important, and implement them. Sound familiar? This is another trip through the design process, only the efforts are focused on testing as a design problem. Too often for students, testing and evaluation are just steps to go through as swiftly and easily as possible, but for business, this phase is extremely important and can take an enormous amount of time and resources. It is very often the case that initial decisions of what is to be tested and how the testing will occur are made near the beginning or the middle of the product-design process, while the actual testing occurs near the end of the process.

Testing is as important as any other aspect of design. In the world of consumer goods, organizations have been founded that do almost nothing but test products and publish the results. Consumers Union is a well-known example of a product-testing organization (see Figure 9-2). Each month, Consumers Union publishes a magazine, *Consumer Reports*, with the results of its tests on food, TVs, cars, tools, appliances, and many other products and services. For each product, standardized tests need to be devised that will provide useful information for consumers.

Another well-known product testing organization is Underwriters Laboratories (UL). The purpose of this organization is to evaluate products or components of products for compliance with over 800 safety standards. Most consumer electrical devices in the United States have a tag that states that the device is "UL Approved," which means that the manufacturer has submitted the device to Underwriters Laboratories for testing. In addition to electrical product testing, Underwriters Laboratories also provides standards for products and systems in the areas of environment and public health, fire and building safety, plumbing, and others.

In science, you have a hypothesis or a guess about what you think will happen. Devising and conducting the tests to determine if your hypothesis will hold up is the creative work of the scientist. In engineering and technology, testing is similarly used to validate a design.

At several points throughout this book, the design of a toy is mentioned. The toy industry is large, with gross sales in the many billions of dollars a year. Testing a toy is absolutely critical for a number of reasons, including liability and the safety of the child, and to prevent product returns and customer dissatisfaction, which could damage the reputation of the company. These types of tests could be categorized as product-level testing because they are primarily concerned with the performance of the product. But another kind of test should be completed early in the design process before the toy goes into production. That test would be to determine if the toy would sell, that is, if children would want to play with the toy and if parents (or others) would purchase the toy for a child. This type of testing could be referred to as market testing (or market analysis).

Testing a Toy

If you undertook the design and development of a toy, you probably developed a design brief and specifications that provided you with direction, such as the intended age of the child and other considerations. If this activity was designed to meet some other need, such as helping you develop your knowledge of simple mechanical systems, then the design brief or specifications may have required the use of computer-aided manufacturing (CAM) in the solution.

Figure 9-2: Front cover of a *Consumer Reports* magazine.

© 2011 BY CONSUMERS UNION OF U.S., INC. YONKERS, NY 10703-1057, A NONPROFIT ORGANIZATION. REPRINTED WITH PERMISSION FROM THE APRIL 2011 ISSUE OF *CONSUMER REPORTS* FOR EDUCATIONAL PURPOSES ONLY. NO COMMERCIAL USE OR REPRODUCTION PERMITTED. WWW.CONSUMERREPORTS.ORG.

Let us assume for a moment that you began with a problem situation you identified in your younger sister's nursery school. You had noticed that the children seem to quickly break many of the mechanical toys that you recognized as being more interesting and educational. From this observation, you developed a design brief that stated that you intended to develop a toy that would be (1) educational, (2) hold the interest of preschoolers, and (3) be more durable. You have designed and developed a toy prototype, and you are now being asked to test and evaluate it. What do you do?

To begin with, you will need to return to your design brief and specifications to find the standards against which you will measure the success of the toy. Going back to the design brief, you find that you intended the toy to be educational. Great! But how do you measure this? Second, you find that you promised that your toy would be interesting to preschool children. How will you measure this? Third, you specified that the toy would be durable. How can durability be measured?

All three of these aspects of the toy can be measured. There are likely multiple ways of measuring each of these three characteristics of the newly designed toy. The testing process is at least as creative as the idea creation (ideation) you went through in the generating and developing ideas stage of the design process. Let's look at some possible testing strategies for your toy and see if these will give you the information you need.

Measuring the educational value of a toy may not be too difficult if you have a clear idea in your mind of what the toy will teach. For example, suppose you have created a toy that will help a child learn to tell time with an analog clock. First, you could let a child play with the toy for a certain length of time and then ask the child to look at a clock and tell you what time it is. Do you think this is a good test? Might it have been better to learn if the child could tell time before he or she used the toy? Why? Might the time a child plays with the toy be a factor in whether the test will show the skill of telling time?

Suppose your toy did not seem to help the child tell time after the child played with it for an hour. Does that mean that it is not educational? Do some concepts take longer to learn than others? Do children of preschool age need longer to learn some concepts than others? Is it possible that telling time is beyond the reach of some preschoolers? These are valid questions that have a bearing on the measure of success for the toy. Can you see why some of these questions would have been very useful to consider during the product design process, and not simply after the product design was completed?

There are alternative schemes for evaluating the educational value of a toy, such as asking the opinions of a person who is an "expert" on both children of this age and educational concepts. This individual may provide you with some insights into the educational value of your toy. Can you think of other testing strategies?

In addition to the educational value of the toy, your design brief stated that it should hold the interest of the child. What does this mean? Does it mean that the child will play with this toy to the exclusion of all others? Does it mean that the child will play with it for 10 minutes, 20 minutes, or perhaps 30 minutes? What is a fair measure of the interest the child has in the toy?

Most children play with one toy for a while and then go on to another toy. If there are many toys from which to choose, they may not pick up the same toy again for quite some time, if at all. The other toys from which the child has to choose will determine whether your toy gets attention. It would hardly be a fair test to give a child only your toy, or your toy and some beat-up old broken toys to see if he or she played with your toy. Can you think of a fair test?

The last area of testing mentioned for your toy is durability. Suppose you gave the toy to your younger sister and let her play with the toy for a day. At the end of the day, the toy was still intact. Could you say with confidence that the toy is durable? Certainly, the conditions under which your sister played with the toy and

Figure 9-3: Testing of toys is a critical step in product development, as both product market success and safety issues will impact the financial success of the toy.

the kind of treatment the toy received at the nursery school are very different. What are some of the ways in which you might test the durability of a toy? Here are some suggestions:

- Study the kinds of things that happen to toys, and the kinds of stresses to which they are subjected. Find ways to simulate these events (see Figure 9-3).
- Research how toy companies test toys, and what they expect and require.
- Find out if there are mandated safety features for toys. Are there government regulations or consumer group guidelines?

The testing of the toy has been used as an example of a general product-testing process, but the testing for any problem solution would follow similar lines. Even though some testing procedures can be defined toward the end of a design process, it is very often the case that the specific testing procedures are determined near the beginning of the design process. The main reason for defining testing procedures early in the process is to make sure the design team is clear about the expectations for the project and product.

Your Turn
Creating a Test for a Product

Devise a strategy for evaluating the effectiveness of several bicycle warning lights for night riding. These lights are not intended to illuminate the road in front of the rider, but rather to alert automobile drivers and others of the presence of the bicycle rider. Your evaluation strategy should take the form of one or two tests, with descriptions of their purpose and how each test will be conducted.

TESTING AN ENGINEERING SOLUTION

In the previous example about developing a new toy, the two criteria of educational value and interest level were examples of criteria that are more focused on how the customer perceives the product, and less so on the mechanics of how the toy works. Some tests are more "engineering-oriented," having a closer association with the application of scientific and mathematics principles.

Engineering testing may be done on individual components as they are developed, as well as on the product or system as a whole once these components are assembled. For example, the development of a competition robot will require testing of both the individual components and system as a whole. You may ask if the chassis is strong enough to support the batteries, motors, controllers, and subsystem mechanisms such as those used to pick up or throw a ball. You may ask if the motors are powerful enough to withstand the heavy loads they will have to endure during competition. These questions may involve structural integrity (Will a component withstand the forces?) and performance (Will the remote control system perform effectively and reliably?). Following are a few examples of some common types of engineering testing.

Materials Testing

In engineering, materials testing is done to determine the properties of materials and to find their suitability for applications involving structures and other designs that require these materials to stand up to loads and forces. Although extensive data already exist on a wide range of materials, it is often necessary to test specific samples of materials that are intended for use in a particular design.

Typically, materials are tested using a standard size sample. However, it would not be unusual for tests to be conducted on special sizes and shapes, such as bent tubing or corrugated material.

Stress Versus Strain. There are a few terms that you will need to know to understand what materials testing is all about. The first term is **stress**. Stress is the force applied per unit area to a material. For example, if a 1,000-lb force is applied to one end of an aluminum rod, with the other end fixed, and if the rod is one square inch in cross-section, then the stress on that material is 1,000 lb per square inch or 1,000 lb/in^2 (see Figure 9-4). In an example using metric units, if a force of 300 newtons were being applied over a surface that has an area of 2 square centimeters, then the applied stress would be 150 N/cm^2.

A second term is **strain**. Strain is the actual change in length of the material that results from the stress. For example, if an aluminum rod stretches 0.02 inch as a result of a force being applied, then the strain is 0.02 inch. Figure 9-5 presents a chart of the relationship of stress compared to strain. It is a graphic display of the changes that occur in the length of a material as greater and greater force is applied.

Tensile Testing. The most basic material test is known as the **tensile test**, where material is pulled apart until failure. When forces are being applied in a way that compresses the material, like when someone pushing in on both ends of a rod, the material is said to be under "compression." When the opposite happens, forces being applied that tend to pull a material apart (like pulling on both ends of a rope), the materials are said to be under "tension." As material is put under tension, it goes through a number of predictable stages. First, the material stretches in proportion to the force applied. Yes, under tension the material actually gets a little bit longer. The larger the tension, the longer the material stretches. During this stage, the elastic stage, if the tension is released, the material reverts back to its original shape. A good example of this stage is a spring. Pull a spring apart by a little bit and it stretches. But release the tension and the spring returns to its exact original shape.

However, if you keep stretching a material, it will eventually get to a point where it will no longer return to its exact original shape. This happens because some permanent deformations have occurred within the material, which will not allow the material to recover to its original shape. This point at which permanent deformations occur is called the **elastic limit** or **yield strength**. If further force is applied, the material enters its *plastic* stage, and the material is permanently deformed.

Keep stretching the material and eventually it will fail. This is the **rupture point** or **failure point** of the material. The maximum force a material can withstand is called its **ultimate strength**. Figure 9-6 shows a chart of the stress versus strain of a material. Starting in the lower left corner, stress is zero, so there is no stretching of the material; therefore, strain is also zero.

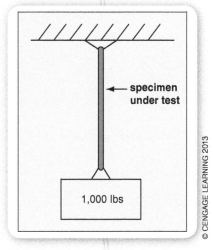

Figure 9-4: This diagram shows a material under tension. One end is fixed and the other has a load attached.

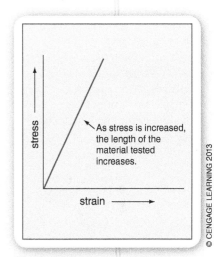

Figure 9-5: Materials under tension will elongate. The more force that is applied (stress), the more the material stretches (strain).

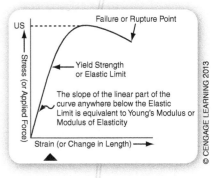

Figure 9-6: Typical graph of stress versus strain on a material loaded to its rupture point.

250 Part I: The Engineering Design Process

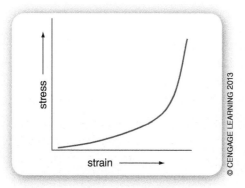

Figure 9-7: Stress-strain graph of a rubber band. Note that it stretches easily at first and then, at a certain point, more force is necessary to stretch it further.

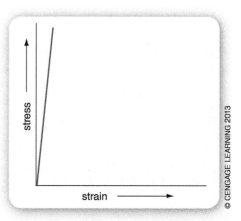

Figure 9-8: Stress-strain graph of glass. Because glass is brittle, little strain is observed until the material fractures.

Hooke's law:
The principle that stress applied to a material is proportional to the resulting strain (change in length). This is only true, however, within the elastic limit of the material.

As stress increases, the material enters its plastic stage at the *yield point*. From this point on, the material deforms more with the same increase in stress until the point of ultimate strength is reached. Further increase in stress stretches the material easily, and the curve turns down. The point of failure is finally reached.

All materials behave in a similar fashion, but the curves plotted on the stress-strain chart differ. Can you imagine the different plots that a rubber band and a piece of glass would make? The rubber band would show a great deal of strain (change in length) with only very little increase in stress. The chart would look something like Figure 9-7, where the plot would be close to the horizontal. Glass, on the other hand, would stretch very little and would fail after only very little strain (see Figure 9-8).

Hooke's Law. The elastic phase of a material is governed well by the scientific principle known as Hooke's law. Robert Hooke (1635–1703) was a scientist and mathematician who had a great deal of influence on the scientific revolution of the seventeenth century. Through experimentation, he found that there is a direct relationship between the amount a material deforms and the magnitude of the force applied.

As you can see by the charts (see Figures 9-5 through 9-8), the plot shows a straight line in the early stages of tensile testing, indicating that the relationship between the stress applied and the change of length of the material (strain) is linear. In this part of the plot, the relationship obeys Hooke's law:

$$\frac{\sigma}{\varepsilon} = E$$

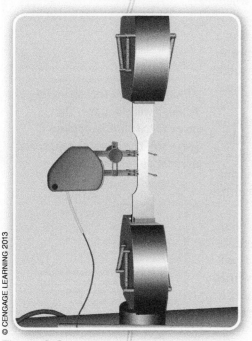

Figure 9-9: A metal sample is being placed under stress in a tensile tester. More and more force will be exerted until the sample fails.

where E is a constant as is the slope of the plot in this elastic region. In the previous equation, σ is stress and ε is the strain. This slope, or E, is called Young's modulus of elasticity or Young's modulus, and it is found by dividing stress by strain. By taking measurements at a number of different stress/strain points, a full plot can be created. Every material has its own modulus of elasticity. For example, aluminum has a modulus of 10 million pounds per square inch, or in metric units, $\sim 70 \times 10^9$ pascals or 70 Gigapascals (or 70 GPa). Steel is much stronger and has a modulus of 200 GPa (or $\sim$30 million lb/in^2).

Your Turn
Steel versus Aluminum

Steel (modulus of ~30 million psi) is substantially stronger than aluminum (modulus of ~10 million psi). However, can you think of several disadvantages of using steel instead of aluminum? Is there a metallic material that is "in between" that might be considered? In addition to considering common properties of these materials, are there substantial cost differences?

When a material stretches (strains) proportionally to the amount of force (stress), that material conforms to Hooke's law. Steel, for example, obeys Hooke's law right up to its yield point. Other materials, however, do not obey Hooke's law, such as rubber. Materials that do not obey Hooke's law are known as "non-Hookean" materials.

Special machines are used to test the tensile strength of materials. In Figure 9-9, a metal sample is put under tension and pulled apart. The information from these tests is used to help design parts and products so that they are strong enough to stand up to their intended use.

Science of Material Testing: Gathering scientific data on the properties of materials provides a baseline for testing. This is done in controlled situations with each material sample carefully sized so comparisons can be made among different materials.

Technology and Engineering of Material Testing: The design and development of material testing machines allow both researchers and product developers to test material samples. Using the results of material tests, engineers and other designers can calculate the appropriate size of parts that will be needed to resist loads and other forces.

Mathematics of Material Testing: Charts and graphs are developed for a wide range of materials that describe the effects of stretching (tensile), squeezing (compression), twisting (torsion), and bending on specific-sized material samples. This information is important for making decisions about the design of specific components for aircraft, bridges, automobile suspension systems, and many more products.

Fatigue. Have you ever taken a small piece of plastic or metal and bent it back and forth until it broke? If you have, you have performed a **fatigue** test. In many applications, it is important that a material should not fatigue quickly, such as in aircraft wing spars or landing gear components. Designers must know if these materials will stand up to the repeated application of forces. Wings provide lift for the aircraft but they are constantly being moved and bent due to wind forces in every direction.

Engineers must be assured that these components will not form fatigue cracks that can lead to failure. Of course, all materials would eventually fail under these conditions, so pilots keep careful aircraft logs of the hours an aircraft is in the air.

fatigue:
In engineering, the fracture that occurs when a material is subjected to repeated or fluctuating stress that has maximum values less than the tensile strength of the material.

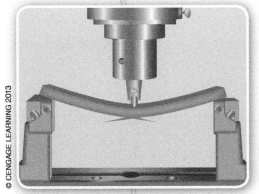

Figure 9-10: **Fatigue testing by repeatedly deforming a material until failure.**

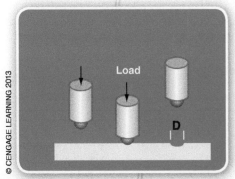

Figure 9-11: **Hardness test for materials. The diameter of the dent D is an indication of the material's hardness.**

There are design limits set for the "airframe," which means that periodic inspections of critical components must be made to grant an "airworthiness" certificate for the aircraft.

Of course, other products are also subjected to forces that cause material fatigue. The decks of skateboards and snowboards are constantly flexing, and cracks or delamination can occur over time.

Machines similar to that in Figure 9-10 are used to test materials for fatigue. Raising and lowering the point in the center forces the sample of material to constantly flex. Engineers inspect the sample periodically for cracks and other signs of fatigue.

Hardness Testing. By applying force to a material in a very small area, the hardness of that material can be found. Many hardness-testing machines use a hardened steel or diamond point that is pushed into the surface of a material with a predetermined force, thus carefully measuring the resulting diameter of the dent (see Figure 9-11).

Generally, the harder the surface of a material, the greater its resistance to wear. In applications where materials are subject to wear, such as gears, bearings, and other similar components, hardness is important.

A standard measurement of hardness is the Rockwell hardness test. In this test, a diamond or hardened steel point is pushed into a material with a small (minor) force. When the point has made its deepest impression (dent), a much higher force (major force) is used to further push the point into the material. When the maximum depth is reached, the higher force is withdrawn, but the minor force remains. This allows the material to partially recover and rebound a bit. The testing machine keeps track of the distance the point has penetrated, and the hardness is a result of the increase in the impression in the material. The degree of (Rockwell) hardness is indicated by a letter scale: A, B, C, . . . and so on.

Other Engineering Tests. A number of other tests may be performed on materials and components, such as finite element analyses (FEA) that can predict thermal and mechanical stresses. These tests are often performed as simulations, as opposed to actual physical tests on materials. Some of these tests are described elsewhere in this text. More about structural and mechanical systems is presented in Chapter 11, Designing Structural Systems, and Chapter 12, Designing Mechanical Systems.

TESTING AND EVALUATING YOUR OWN DESIGN WORK

There can be many important attributes in your design work. Some examples of important attributes include aesthetics, ergonomics, performance/functionality, durability, costs, and overall impact (to the environment or society). The following reviews each of these common attributes.

Aesthetics

We have all been told that aesthetics is just a matter of taste, that how good something looks is all in the eye of the beholder. But if that were true, we would not recognize great artists or great product design. For example, Apple, Inc. is recognized almost universally as a leader in product design. Through the work of Jonathan Ive, Senior Vice President of Industrial Design, Apple holds international acclaim for its product design and design styles it has pioneered (see Figure 9-12). Both experts and consumers recognize good design.

If you keep trying to create good-looking products and get better at it with practice and time, it means that the work you did earlier was inferior to the work you are doing now, and this means that your earlier tastes were not just less developed, but they were worse than your tastes are now. The only conclusion you can draw is that aesthetics is not just a matter of taste. There are standards for good aesthetics.

Aesthetics refers to the pleasurable response you get to an object or experience. A large percentage of this pleasure has to do with appearance—with form/shape, scale, color, and so on coming together as a whole. An object that has good aesthetics is pleasing to look at (however, if that object is a set of stereo speakers, sound quality is another aesthetic factor).

If you want to determine if what you have designed is aesthetically pleasing, it would be appropriate to get the opinions of a number of people for this test. A survey or a questionnaire of how well people like the appearance of the product might give you more useful results if you asked several questions dealing with such things as color, shape, style, and use of design elements such as texture or line. In this category, the overall quality of the finished product is emphasized, including evidence of careful crafting.

Figure 9-12: **The Apple iPhone. Apple has become one of the most recognized companies in the world for its innovative product design.**

Ergonomics

Ergonomics, or "human factors engineering," has to do with designing products and environments that work well with people.

The test of ergonomics often involves the product "feeling" right. Ease of use or comfort while using the product can indicate that the device is appropriately sized, weighted, balanced, and that the controls are in an appropriate place (see Figure 9-13). Because of the broad nature of ergonomics, you will need to think about the tests necessary to measure success in this area. Chapter 15, Human Factors in Design and Engineering, will provide you with more information about this topic.

ergonomics: The study of workplace equipment design or how to arrange and design devices, machines, or workspace so that people and things interact safely and most efficiently.

Performance/Functionality

Performance refers to whether the product does what it is supposed to do. The actual functioning of the product or system determines if it performs as intended. You may need to consider simulations or controlled trials if there is a potential for accident or injury. If you are testing a toy for a child, you may need to first try out the product yourself to make certain that no injury to the child could result. You should also have your teacher look over the toy carefully so that any potential dangers can be identified and corrected, such as sharp edges or corners, small parts that could be swallowed, or unsafe materials.

Durability/Reliability

Durability refers to a product's ability to remain functional over its expected lifetime. Something that remains useful for less than three years is usually considered a nondurable good while products that remain useful for more than three years are considered durable

Figure 9-13: **A device for determining driver and passenger posture and driving position.**

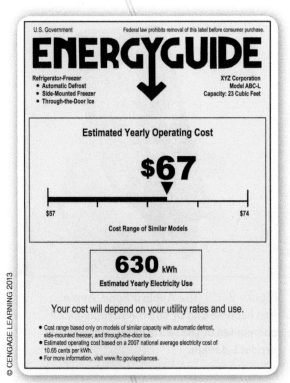

Figure 9-14: The EnergyGuide sticker on appliances provides information about the efficiency of a particular appliance compared to other similar appliances. As a comparison, a yearly cost of energy use estimate is provided.

profit:
In economics, profit is the difference between a company's total revenue, which includes all sources of income such as the sale of products, and total costs (direct costs such as salaries and materials and indirect costs such as rent on facilities, equipment purchases, energy, and other costs that cannot be associated with any one project).

life-cycle cost:
The total cost of a product that includes the amount spent on energy to run it. Also included may be the cost of recycling and/or the environmental impact cost to produce it.

goods. As an example, you may expect a high-definition TV (HDTV) to last at least three years. So, durability testing would be tests run on populations of HDTVs to ensure that they would remain functional over a three-year period. *Reliability* is a term often used in industry and is also related to lifetime expectations. Reliability refers to how long a population of a product remains functional, either under accelerated or nonaccelerated conditions, resulting in a measure of mean lifetime for the product. For example, reliability testing on the same population of HDTVs would be more open-ended in that a large population of HDTVs would be tested over a long period of time, keeping track of the mean time of failure, resulting in a statistically precise measure of mean HDTV lifetime. More about durable/nondurable goods and planned obsolescence is covered in Chapter 11, Designing Structural Systems.

A solution may function as intended but only for a very short time. Does this mean that the solution is a good one? When we purchase a product, we expect that it will last for some length of time. To determine if your solution is reliable or durable enough, you may want to test it a number of times or simulate such conditions.

Cost

How much does it cost to manufacture a product? How much money might a customer spend to buy a product? These are very important questions to consider. If you cannot manufacture a product for less than what a customer will pay for it, then the product cannot be successful. All companies must make a profit to stay in business.

Increasingly, *cost* is becoming a term that means more than just the purchase price. For example, the Federal Trade Commission of the U.S. government requires that all new appliances have an *EnergyGuide* label (see Figure 9-14). This label is intended to help consumers make intelligent purchase choices using information about the energy requirements of that particular appliance. A less expensive but less energy-efficient appliance may not be such a good deal when the cost to operate the appliance is considered. Figure 9-14 shows an example of the life-cycle cost of just electricity for two refrigerators.

By the term life-cycle cost, we include the money to clean up poor industrial practices, such as pollution to the air or groundwater as a result of producing that product. It may also include the cost to remove contaminates such as heavy metals from the factory site after the business moves. Some environmental groups are suggesting that the health care costs related to products that are polluting our air, water, and land should be considered when looking at the cost of a product. When these factors are considered, the real cost of a product may be many times higher than the purchase price. In calculating this figure, the cost of maintaining and disposing of the product is taken into account along with the purchase price.

Impacts

Impacts describe many positive and negative qualities of an object, including its purpose, function, the materials from which it was made, and the processes used in its manufacture. There may be *environmental impacts, personal impacts, social impacts,* or *legal impacts*.

Your Turn

Find three similar products (such as air conditioners with the same BTU rating or refrigerators of the same interior volume) with different EnergyGuide ratings, and calculate energy life-cycle costs (in this case, the purchase price and electricity operating cost) for each over 10 years. Present your results in a chart or graph form.

Example Energy Use as a Part of Appliance Cost

Compare two refrigerators: Refrigerator A costs $500; Refrigerator B costs $650. Which is the cheaper refrigerator? Not so fast! Refrigerator A has a lower efficiency rating than B, which means that it uses more electricity than B. If A uses $0.70 per day ($255.50 per year) and B uses $0.45 of electricity per day ($165.25 per year), and the life expectancy of both refrigerators is 10 years, then the cost of owning refrigerator A is

$500 + (255.50/year)(10 years) = $500 + $2,555 = $3,055.00

The cost of owning refrigerator B is

$650 + (164.25/year)(10 years) = $650 + $1,642.50 = $2,292.50

The life-cycle cost of refrigerator B is actually cheaper than A.

You will need to consider a number of questions related to impacts in evaluating your work:

▶ What impact does a product have on the environment? For example, what happens to the product when its useful life is over? If the object is designed and made with little or no regard to its impact when it is discarded, then the object cannot be considered "good design." Are the materials from which it is made or the chemicals used in its manufacture environmentally hazardous? Many substances have no environmentally sound method of disposal (see Figure 9-15).

▶ What are the personal impacts of the product? Will the product change your daily life? For example, do you use pay phones less now that you have a cell phone?

▶ What are the social impacts of the product? Has the introduction of Web blogs resulted in fewer people reading books and newspapers? Are people more aware of current events because of Web blogs or are they simply reading blogs that reinforce their own point of view or prejudices?

Figure 9-15: **Materials used in industrial production can have a disastrous impact on the environment if responsible disposal procedures are not followed.**

OFF-ROAD EXPLORATION

To learn more about the impacts that technology can have, read *The Victorian Internet: The Remarkable Story of the Telegraph and the Nineteenth Century's On-Line Pioneers* by Tom Standage (ISBN-13: 978-0802713421). The telegraph had an enormous impact on society because it was the first time that people could communicate effectively over long distances without actually going there in person.

OFF-ROAD EXPLORATION

For more information about good design, see *The Design of Everyday Things* by Donald A. Norman (ISBN-13: 978-0465067107).

▶ What are the legal impacts of the product? Did you infringe on a patent or copyright? Are there potential dangers with the product that could result in legal action?

▶ What are the ethical responsibilities on the part of designers? Could you apply the Golden Rule, "Do unto others (including the environment) as you would have them do unto you"?

The criteria of good design (aesthetics, ergonomics, performance durability, cost, impacts, and so on) are all interrelated, and compromises are often made to assure that a product is worth making and marketing. At the same time, other values must be at work in the design and development of products and systems besides profit. If profit is the only value, what will happen to the environment and your personal safety? Profit makes business and industry possible, but other values must be considered if we want a healthy planet on which to live.

PRESENTING TEST RESULTS

Your design portfolio is a record of your design work. To communicate your efforts in testing and evaluation, you have a number of possibilities. You can use descriptions, numbers, checklists, and testimonials in presenting your results.

Descriptions

Describing a test and its results can be done through a written description or a combination of graphic and written description. Graphical descriptions may include an illustration of a testing apparatus and a graphical summary of the test results.

Numbers

Test results most often take the form of numbers, such as the number of minutes a child plays with a toy or the amount of force a component or product can withstand without crushing or other failure, or the percentage of people rating the product "very good" and "excellent." Presenting this information in a narrative or description can sometimes be tedious. However, graphic presentation through charts, graphs, diagrams, and tables are often more effective techniques for conveying important information (see Figure 9-16). Be certain that figures are well labeled and clear for readers. Chapter 16, Math and Science Applications, gives examples of various graphic options. Charts and graphs can be developed in most presentation software. These programs create a professional and colorful display of number data.

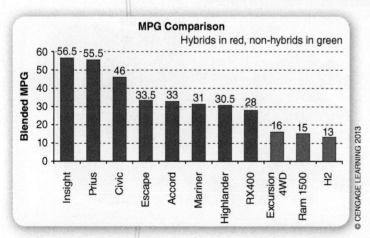

Figure 9-16: **A graph showing results of mileage testing on eleven vehicles.**

Checklists

An effective way of testing a solution against the design brief and specifications is to develop a checklist (see Figure 9-17). The checklist provides a visual summary of the extent to which the solution meets the design requirements.

Chapter 9: Testing and Evaluating 257

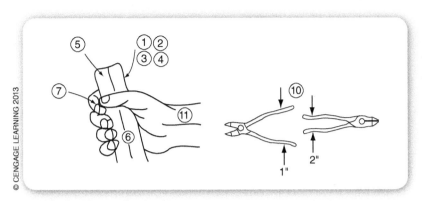

Figure 9-17: Checklists are a good way of comparing data. In this chart, three hand tools are compared against a list of specific ergonomic criteria.

Item	Ergonomic Feature	Yes	NA	No	Score
1	Grip surface is nonslippery.	10		0	
2	Grip surface does not have sharp edges, undercuts, deep ribs, or finger grooves.	10		0	
3	Grip surface is electrically insulated; tool handle is either made of wood or coated with rubber or soft plastic.	10		0	
4	Grip surface is thermally insulated; it will not get hot or cold quickly when working in a hot or cold environment.	4		0	
5	Handle is made of wood, or grip surface is coated with semipliable material, not too hard and not too soft, similar to the rubber used in the soles of sport shoes.	10		0	
6	Grip length is 4 to 6 inches; handle does not end inside the palm of the hand.	10		0	
7	For one-handle tools: Size of handle cross-section is not too small or too large. The index finger and the thumb are allowed to overlap by 3.8 inches when gripping (for hammers and hammerlike tools, overlap of 1 inch is acceptable).	8*	0	0	
8	For one-handle tools other than screwdrivers: Shape of handle cross-section is oval or rounded-edge rectangular.	10*	0	0	
9	For screwdrivers: The basic shape of handle cross-section is circular, hexagonal, square, or triangular.	10*	0	0	
10	For two-handle plier-like tools: Grip span is greater than or equal to 2 inches when fully closed and less than or equal to 3.5 inches when fully open.	10*	0	0	
11	Angle of the handle is formed so that the work can be done keeping a straight wrist.	10		0	
12	The tool weight is less than 5 pounds.	10		0	
13	The tool can be used with either hand.	10		0	
14	The tool can be used with the worker's dominant hand.	10		0	
15	The tool will allow a two-handled operation (using both hands at the same time).	2		0	
16	The tool and accessories are clearly marked and/or coded so they are easy to identify; colors are bright and tool contrasts with the surroundings of the work area.	5		0	
	Total Score of the Tool				

"Yes," "No," or "NA" (not applicable). Place the score that corresponds to your response in the "Score" column. Add the scores of all items to get the total score of the tool.
(*Items 7, 8, 9, and 10 are not applicable for all tools.)

Careers in the Designed World

How Things Work

When Patty Ratchford was seven years old, she took her bicycle apart and put it back together. "I just wanted to know how it worked," says Patty. Her early interest in mechanics eventually put her on a career path in mechanical engineering. Patty is now in her tenth year as a manufacturing methods engineer with the German-based automobile company BMW.

Patty Ratchford: BMW Manufacturing Co.

On the Job

At BMW, Patty has been an assembly planner and senior manufacturing methods engineer. Now her title is Section Manager for Metrology and Geometric Analysis. Patty's job is to make sure that the design of a new vehicle can actually be built on the assembly line.

Patty continues to learn from the unique challenges of working in a global manufacturing company like BMW. Most of the designers for BMW speak German as their first language, and the communication of technical information can challenge the design team. Software is available for translating technical terminology, but when it came to car parts, it didn't work very well. Patty often uses pictures with arrows pointing to parts of interest, and videos of mechanical design problems along with brief, written highlights of the design issues. This more visual approach to communication works better on Patty's team than long, written explanations.

For example, Patty recalls working on a wire harness placement on the BMW Z4 convertible. The harness included 85 pounds of two-inch bundled wire (1,500 wires), and it needed to fit in back and under the glove compartment. When the designer in Germany created the design on the CAD system, it was drawn as a pipe with 90-degree bends. In reality, this two-inch wire bundle would not bend 90 degrees. To communicate the problem to the German-speaking designer, Patty sent a video of the space in question and pictures of the bends in the wire bundle. The designer was able to actually see the problem and adjusted his design to work in the given space.

Inspirations

Before working at BMW, Patty worked at Pratt & Whitney, where she was an engineer in the Joining (Welding) Technology Development group. From there, Patty went to work at Performance Friction as a process engineer, and at FACSO Controls as a development engineer before moving to BMW.

Patty's work at Pratt & Whitney showed her how important it is to know and understand manufacturing processes for a design to be successful. Patty spent valuable time on the shop floor with the technicians welding test plates to understand the limitations and process for joining two pieces of metal. "It doesn't really matter how many books you have read about a topic," says Patty. "You can't truly understand it or be good at it unless you have done it yourself. Plus, people will have more respect for you if you are willing to try things yourself."

Education

Patty chose math and science course work while in high school. After high school, she went on to Clarkson University, where she earned a bachelor's of science and master's of science degrees in mechanical engineering.

Advice for Students

In a global-manufacturing environment, Patty values speaking and writing skills. "Don't ignore writing and speaking skills," she cautions, as they can help engineers build successful careers. "Public speaking doesn't always come naturally to a person in engineering," says Patty. "But there are many times you have to present your work or communicate a design problem. These skills are very important to your success on the job."

Bodywork in progress at BMW's Spartanburg, South Carolina, plant.

Testimonials

Testimonials are opinions of those who have tried your product and will endorse it. They may like it because of its overall qualities or because of one quality, such as aesthetics or ergonomics. These individuals have given permission to have their names associated with their comments about the product. Testimonials are presented through direct quotes (see Figure 9-18).

EVALUATING YOUR DESIGN SKILLS

To this point, you have critically looked at the product or system you developed or perhaps the customer or market for which your solution was designed. This was the "testing" portion of your project. However, there is an assessment aspect of your project that has less to do with your product or market and more to do with your or your team's performance or perceptions. This is called "evaluation." In your evaluation, you must look at your own role, or your team's role, in the designing and making of the solution.

You had the responsibility for the decisions made and the management of time and resources that went into the final results. You now need to look critically at how well you did these things. Here are some questions you may use as a guide in the evaluation of your work:

Figure 9-18: Testimonials are often used in advertising.

- ▶ Did the design brief provide a solid direction for the project? Were specifications too vague or too restrictive?
- ▶ Did you collect appropriate information about the problem? Did you apply this information to the problem? Did you find out about how others had solved similar problems? Did you return to this step as problems came up in other steps of the process?
- ▶ Did you use the first idea that you thought of? Were the alternatives presented workable and well thought out? Were creative strategies used to help develop solution ideas? Did you combine attributes of several ideas?
- ▶ Can you defend the solution chosen in terms of "good design"? Do you have reasons you can list for choosing one solution over the other?
- ▶ Did you work out structural, mechanical, electronic, or pneumatic problems before you developed working drawings? Were working drawings developed? Did the working drawings leave unanswered problems in the design?
- ▶ Was the model and/or prototype well crafted? Were appropriate materials, adhesives, fasteners, and finishes used?
- ▶ Were the tests that were developed appropriate? Were the results of tests presented clearly and honestly? Were your evaluations critical and honest? Did you give yourself credit when it was due, or self-criticism when you deserved it?
- ▶ Did you use the time you had to work on this project appropriately? Did you attempt a project that was too difficult or too simple for the time allotted? Did you adequately plan the time you had to complete the project and follow through with the plan?

SUMMARY

Testing and evaluating your ideas and solutions is an important step in the design process. It is not always apparent, however, how to evaluate the success of your work. You need to develop appropriate tests that can assess the effectiveness of a solution in terms of the design brief and specifications you developed in the early stages of your design work. It is often necessary to apply indirect tests for certain aspects of a solution, such as the specifications that a toy be educational. Often, the types of testing are clearly defined near the beginning of a design process.

Materials testing can include tensile, fatigue, hardness, and others. These tests are normally performed using specialized equipment. The results of these tests can be used in the design of a component or product. In presenting test results, you may choose to provide descriptive, numerical, checklist, and testimonial results. The presentation of these results then becomes an important part of a major design project portfolio.

Finally, the evaluation of your work should take two forms: evaluation of your design work, usually the end product; and evaluation of your design skills, including the planning, thought, and effort you put into the project.

BRING IT HOME

OBSERVATION/ANALYSIS/SYNTHESIS

1. Develop a way to test the solution to a problem based on these design briefs:
   a. Design and make a game for 7- to 8-year-olds. The children should learn some important historical facts while playing the game.
   b. Design and make a product that helps a person with arthritis to remove the lids on jars.
   c. Design and make an emergency light for a person who has fallen overboard at sea at night.
2. Make a list of the kinds of stresses a toy for a 2-year-old has to withstand. Devise a test for each one.
3. Find a commercial product, and analyze it for "good design," using the criteria outlined in this chapter.
4. Discuss in detail how a chosen product rates on the cost and ethics criteria for "good design."
5. Conduct a product analysis on three similar products from different manufacturers, detailing how each rates on the "good design" criteria. Present your results in a graphic form.
6. Find two examples of the same product from two different manufacturers. Come up with at least five criteria for evaluating their use. Develop a test for each criterion and describe it. Evaluate these products and present the results.

EXTRA MILE

- Conduct simple engineering tests on a product category, such as paper grocery bags or cardboard boxes. Develop a testing apparatus that will hold the bags and apply a gradually increasing load, such as pouring sand into the bag until failure. For boxes, create a device that will apply a measured impact to a corner or side and increase the force until the box is damaged.

- The system you design should provide a way to compare a number of items to determine which carries the most load before failure, or which withstands the most force until damage is observed.

CHAPTER 10
Manufacturing

Menu

Before You Begin
Think about these questions as you study the concepts in this chapter:

1. What are some of the fundamental changes the Industrial Revolution brought about in how the majority of people in developed countries live, work, and play?

2. What role did the steam engine play in the early Industrial Revolution?

3. What impact did the development of interchangeable parts make on the production of goods?

4. What does "mass production" mean and how has this relatively recent practice influenced product development and distribution?

5. How are materials processed to change their shape or physical characteristics?

6. How have computers impacted modern manufacturing?

INTRODUCTION

The design and engineering principles presented in the preceding chapters primarily addressed the design of new things: products, if you will. However, a very important portion of a product lifecycle, and the design process, is the manufacture of the product. Manufacturing is so critical that it is very often the case that key manufacturing criteria are included in the design criteria. One example given in Chapter 2 was the design of a new very high gas mileage hybrid sport utility vehicle (SUV), where one of the criteria for the new SUV design was to be 100 percent compatible with existing mass-manufacturing platforms. Cost is an important reason for having such a strong connection between design and manufacturing. For example, for automobiles, the manufacturing equipment platforms can cost hundreds of millions of dollars, so it is a clear advantage to use this expensive manufacturing equipment as much, and as long, as possible. The genius of human invention and innovation is only partly seen through the design of new products. The other half of technological genius can be seen through the organizations and systems we create to manufacture these new designs. Organizations, in the form of managed production systems, are necessary to produce products effectively. The resources needed by these production organizations are large and include:

- Capital
- Energy
- People
- Tools and machines
- Knowledge processes
- Materials

We have all heard that we live in a global community and that we are in an Information Age. Although this is true, it is the production of goods that fuels our economy. The purpose of this chapter is to introduce how industry and production activities affect the individual, the world community, and the environment.

We often take for granted the idea that goods are mass-produced. The fact that we can go online and buy an incredible range of items, from toothbrushes to MP3 players, and that next year these products will likely include even more functionality or cost less, is something we barely think about. In the industrialized nations of the world, we have access to more products and services than could ever be imagined just a hundred years ago.

The key to providing the goods we consume is mass-production manufacturing. Mass-production manufacturing has only been in existence for a little over a century. But it all started with what was one of the biggest events in human history—an event that changed everything: the Industrial Revolution.

THE INDUSTRIAL REVOLUTION

Technology significantly changes human existence. The early development of tools, the discovery and use of bronze and iron, and the development of agriculture are all examples of technology. Most of these innovations happened in the distant past, but what occurred in the mid-1700s fundamentally changed the way we work, live, travel, and communicate today. As mentioned in Chapter 1, the Industrial Revolution did not occur overnight, but rather began slowly and picked up momentum over the course of 100 years. The revolution began in England and spread across Europe and the North American continent in a wave that still affects us daily.

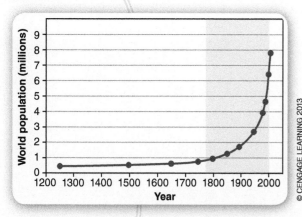

Figure 10-1: With the increased food supply made possible by new agricultural techniques and the introduction of mechanized farming, world population levels have increased dramatically since the start of the Industrial Revolution.

The Agrarian Revolution

One of the most impressive influences of the Industrial Revolution is the impact on human population size. Population levels in industrialized nations began increasing at astonishing rates in the early 1900s, and the trends continue today (see Figure 10-1). Food supply has always been a limiting factor on population size. Historical and current hunter-gatherer populations have been limited in size to approximately 20 individuals. In these groups, each person requires about 10 square miles of land in which to gather edible plants and hunt animals for food and other resources. Therefore, a group of 20 people requires approximately 200 square miles, which is also about the limit that the group can cover on foot. Industrialization caused a mass movement of people into cities for work in factories. This influx in population required significant food-production capacity.

When humans developed agriculture, the amount of land required per person decreased to approximately three square miles. This did not happen overnight, but rather took thousands of years as dry agriculture was gradually replaced with controlled irrigation methods. This allowed more people to live in smaller areas and towns, and later, cities grew up around the agricultural lands.

The development of agriculture, or the Agrarian Revolution, set the stage for the population growth of the Industrial Revolution. The new methods of farming, including crop rotation, improved livestock breeding, and improved land management allowed a sufficient food supply to feed the rapidly increasing population. These agricultural improvements began in the early 1700s, and continued with the invention of mechanized farming in the 1860s. A key to mechanized farming, as well as industrialization, was the steam engine.

The Steam Engine

The idea of a steam engine is thousands of years old. Hero of Alexandria, a Greek inventor who lived in the first century B.C., developed a simple steam turbine (see Figure 10-2). It was only a novelty, however, and no one at the time seemed to recognize the potential the device had to provide powerful energy. It would be another 1,800 years before someone else would successfully develop a working steam engine.

The first working steam pump was developed in 1698, by Captain Thomas Savery. The purpose of this device was to pump water out of mines in southern England. Although not a true steam engine, it led the way to further steam technology.

The first true steam engine was developed by Thomas Newcomen in 1712 (see Figure 10-3). Newcomen was a blacksmith with sharp technical skills. He designed and constructed this revolutionary invention that literally changed the world.

Prior to the steam engine, small factories and grain-grinding mills had to be located next to a moving river or stream for waterpower. Of course there were wind-driven mills, but these produced less power and were less reliable due to the unpredictability of the wind. Therefore, waterpower was the preferred energy source.

With the introduction of steam power, factories could be located anywhere. Even in cities bordering on water sources, factories could be spread

Figure 10-2: Hero's steam turbine was called the aeolipile. Heated water that had turned into steam entered the sphere and escaped out of the two angled tubes. The escaping steam rotated the sphere.

out some distance from the river or stream. Small steam engines allowed the development of the railroad, steam-powered boats and ships, farm equipment, and many other devices. The significance of the steam engine cannot be overstated.

Principles of the Steam Engine

Steam engines have a piston that slides back and forth in a cylinder when the pressure of steam is applied. Valves control the movement of steam into and out of the cylinder. In **double-acting steam engines**, steam is used to push the piston in both directions.

Many of the early steam engines, called **atmospheric engines**, operated on the principle of creating a vacuum by condensation. Steam pressure pushed a piston to the top of a vertical cylinder, but the piston was pushed down by injecting a spray of cold water into the cylinder that condensed the steam, creating a partial vacuum under the piston. Atmospheric pressure would then push the piston down to the bottom of the cylinder (see Figure 10-4).

The Industrial Revolution set the stage for our modern society, and the steam engine drove this Industrial Age. As factories sprang up across England, Europe, and North America during the 1800s, people moved into the cities to take the factory jobs. Cities expanded as populations boomed.

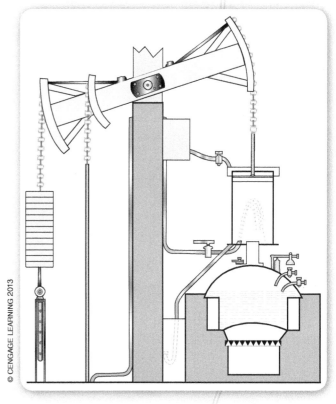

Figure 10-3: Thomas Newcomen's beam steam engine of 1712.

> **STEM Connection:** The Industrial Revolution came about when knowledge in math and science began to merge with experience in engineering and technology. Here are some of the important connections that helped to advance the emerging field of engineering during that time:
>
> **Science:** Isaac Newton's work with forces and motion revolutionized mathematics and our understanding of the universe. The "scientific method" and many of the basic principles of physics were established.
>
> **Technology:** Techniques used in large-scale production of iron and the development of early steel production were critical to the creation of the steam engine and factory machinery.
>
> **Engineering:** Math and science applications made it possible to design and improve the steam engine and power transmission machinery. As inventions proved successful, and sometimes when they did not, the knowledge and principles of the young field of engineering began to emerge.
>
> **Mathematics:** Isaac Newton also made the most significant advancement in mathematics: calculus. Gottfried Leibniz also developed calculus independently of Newton. More advanced mathematics knowledge led directly to better designs and production techniques.

The Industrial Revolution began with the invention of textile machinery and the development of the iron-making industries in England. This led to the manufacture of machinery for other industries. Eventually, the steam engine was adapted to create the railroad and then fitted to power ships.

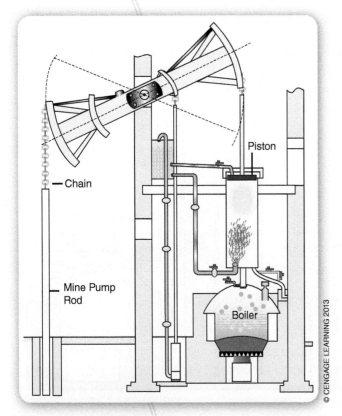

Figure 10-4: Atmospheric steam engine. Water is injected into a cylinder to condense steam when the piston is at the top of its travel. The partial vacuum that is created allows atmospheric pressure to push the piston to the bottom of the cylinder.

Interchangeable Parts

When parts are made by hand, it is difficult to keep tight tolerances. Tolerance is a term used to describe the accuracy of the dimensions of a part. A higher-tolerance part means that there is less variation in the dimensions among several examples of the same part. If you want parts to fit together, the accuracy of the part's dimensions is important.

For the thousands of years that things were made by hand, fitting pieces together to make a product, such as a chair, was a matter of carefully shaping pieces one at a time and hand-fitting them. But hand-fitting is time-consuming and costly and does not lend itself well to high-volume production.

To avoid hand-fitting, parts need to be made to a specific size with little variation. This kind of precision is difficult when making things by hand. The use of machines, however, can assist operators in working more accurately. Machines were developed with the purpose of increasing the precision with which parts could be made.

A Frenchman and gunsmith, Honoré Blanc, started producing firearms with interchangeable parts around 1778. As the story goes, he produced nearly a thousand muskets and put all their parts on a table and assembled rifles by picking parts at random. He did this in front of a committee of politicians, scientists, and representatives of the military. Almost 20 years later, in 1798, Eli Whitney used the same strategy when he assembled 10 muskets with interchangeable parts in front of the United States

Point of Interest
Part tolerance

In Figure 10-5, there are two parts (A and B). Part A is a circular pin while part B is a circular hole. The purpose of these two parts is to have the pin fit snuggly into the hole. If the diameter of the pin is 70 mm and the tolerance of that part is plus or minus 1 mm, then the diameter of any versions of part A could be as small as 69 mm, or as large as 71 mm. If the same dimension and tolerance (70 ± 1 mm) were used for the hole, then significant problems would arise. For example, if the hole is on the larger end of the tolerance and the pin is on the smaller end, then the extra space between the two parts may be as much as 2 mm. Conversely, if the hole is on the smaller end of the tolerance and the pin on the larger end, then the pin would simply not fit into the hole, clearly an undesirable result. This is the importance of tolerance. The tolerance, or allowed "variability," in dimensions is very important to consider for all dimensions of a design.

Figure 10-5

Congress. Congress, impressed with Whitney's demonstration, created a standard for military firearms. However, Whitney had created all the rifle parts by hand and had not really developed a system to make interchangeable parts. When he received a contract to deliver 4,000 rifles, it took him eight years to produce them and even then the parts were not really interchangeable. But others went on to make the method work.

tolerance:
The total permissible variation in a size or location dimension.

THE ASSEMBLY LINE

The assembly line has been around for over 200 years. Marc Isambard Brunel developed it in 1801, to make blocks (the part that holds the pulley in a block and tackle) for the English navy (see Figure 10-6). With a single frigate ship using approximately 1,500 blocks at the time, the English navy was buying about 100,000 a year. There was a need to produce blocks in great numbers.

Making a block by hand required several operations. Brunel developed a series of machines, each performing a single, simple operation. This sped production and did not require skilled crafts workers. With 44 machines, 10 unskilled workers could do the work of 110 skilled workers. This saved the English navy a lot of money, but put those skilled crafts workers out of jobs.

Labor is almost always a significant portion of the cost to manufacture a product, so business is always looking to cut labor costs. There is, of course, a tension between reducing labor costs and putting people out of work. The people who do not have a job cannot afford to buy the products you produce, so demand soon decreases. In a society where many people are out of work, the economy suffers and business profits fall.

Many people believe that Henry Ford developed the assembly line, but Ford adapted the work of Brunel and Ransom Eli Olds. Olds, who is credited with the first automobile assembly line, started the "Olds Motor Works" in the city of Detroit. He produced 425 low-priced "Olds Curved Dash" in 1901, his first year in production.

But it was Henry Ford who designed and built whole factories around the assembly line concept and refined it to the point where a Model T car was coming off the assembly line every three minutes (see Figure 10-7).

assembly line:
A manufacturing process for the mass production of a product that involves adding interchangeable parts in sequence.

Figure 10-6: An antique block. The block serves as the housing for the pulley and keeps a rope passed across the pulley centered while moving.

MATERIAL PROCESSING

Most design solutions call for materials to be changed so that they will function in some desired manner. For example, pieces of steel and wood are shaped and assembled into a hammer. Your goal in design and engineering should be to develop an understanding of basic industrial material-processing techniques so that you can apply them to your design decisions. As an example of this principle, let us consider the design of a body or shell for a new camera. To make a decision about materials, designers would need to consider the material characteristics as well as the potential for efficient production if that material were selected. Understanding the basic industrial material-processing techniques will give designers the confidence to know that most plastics could be easily formed into the desired camera shape at a relatively

Figure 10-7: Assembly line at an early Ford automobile plant.

low cost, but may have drawbacks in the durability of the product over time. There are many different kinds of plastics, each with its own characteristics.

If a metal were chosen for the camera body material, then production processes would be different and production costs higher. But metal is generally quite durable and lends a product a certain quality. With an understanding of material processes, and material characteristics, both design and production decisions are more informed.

Material Production Cycle

All materials begin a production cycle in their basic form. **Raw materials**, the most basic form of materials, can be extracted from the earth, water, or air, or they can be grown as biological products. **Recycled materials** are increasingly used in manufacture, but these materials must typically be processed to return them to a raw material state or such a condition that they can be effectively used in production.

Silica, iron ore, timber, petroleum, and cotton are examples of raw materials. Most raw materials are refined or processed into **standard industry materials** by primary industries. Primary industries in textiles, metals, paper, petroleum, chemical, lumber, and energy produce products that other industries, in turn, use as resources to produce most consumer goods. The lumber industry, for example, processes trees to make dimensional stock for construction, as well as plywood and a number of other composite materials. Trees, as renewable resources, are also processed by papermaking industries to produce newsprint, book papers, and cardboard. These industries also produce paper towels and facial tissues. Some of these products must be processed by other industries before they become useful products (for example, newsprint used by a local newspaper). Other products, such as facial tissue, can be used immediately by consumers.

Most products that primary industries make are used as standard stock and become one of the inputs into the **manufacturing process**. Manufacturing industries produce components that are to be assembled into a final product. Components such as automobile tires, gears, belts, ball bearings, glass bottles, and cardboard boxes may be used to make many different types of products. A ceramics coffee mug, on the other hand, is an example of a manufactured component that is also the end product. Most complex products require many components (see Figure 10-8).

The combining of many components is called product **assembling**. Recently, consumers have learned that it is difficult to determine where a complex product, such as an automobile, is actually made because the components in most cars are produced in many countries. Usually, product-assembling industries require workers with lower skills than industries in which primary component parts are manufactured.

> **standard industry materials:**
> Raw materials that have been processed into standard size, shape, or composition to be used in production of products, such as dimensional lumber, plywood, paper, cloth, and so on.

> **manufacturing process:**
> The transformation of raw material into finished goods through one or more of the following: casting and molding, shaping and reshaping for forming, shearing, pulverizing, machining for material removal, or joining by transforming heat or chemical reaction to bond materials.

Figure 10-8: **Many different components are needed to make a complex product.**

Your Turn

Researching Standard Industrial Materials

Identify at least 10 different standard industrial materials. Create a list with the raw materials from which your 10 standard industrial materials are made. Finally, give an example of 10 products that use each of the standard industrial materials in their production.

The Importance of Materials

The role that material choice plays in the manufacturing of a product cannot be overemphasized. In Chapters 2 and 4, the effect of the introduction of new materials on design was briefly discussed. New materials that had unique characteristics provided designers with new options for the design of products.

Many fields of engineering emphasize an understanding of materials and their properties. The properties of materials are so important that they have their own field of studies: materials science and materials engineering. Understanding material properties informs the design and decision-making processes and helps ensure safety, durability, cost, and a host of other factors that lead to a successful product. Most mechanical and industry-related engineering programs also have courses that emphasize materials selection and manufacturing engineering in their core.

Forming Materials

Changing the shape of a material is called **forming**. During the Stone Age, a number of materials were shaped into useful forms, including wood, bone, and stone. Because stone is much stronger than either bone or wood, early tools were made by striking one stone with another, resulting in sharp edges that could be used to cut and rip. These tools were also useful for forming implements from the other materials. This technique of striking one stone with another is sometimes called **flint knapping**. There are even examples of early mortise and tenon joints made in stone.

The blacksmith during American colonial times shaped metal using a hammer and anvil. Considerable skill was necessary, and each item was given a unique and distinctive shape. Products also took a long time to make. Although blacksmithing is still practiced today, modern processes have been developed to shape many different types of materials. Two primary means of forming materials are by (1) compressing or stretching and (2) **casting**.

Bending. Bending involves both compressing and stretching the material, and is a relatively simple forming operation. When a material is bent, the material at the inner side of the bend is compressed while the material on the outer side of the bend is stretched (in "tension"). Most materials are bent cold, while some, such as thermoplastics and glass, are heated prior to bending. Bending is changing the shape of the material in one direction. Cardboard is bent at right angles to make a package. Metal can be bent using a metal brake (see Figure 10-9). Thermoplastics can also be bent but must be heated and softened on a strip heater

Figure 10-9: **Sheet metal brake for bending light gauge sheet metal.**

Careers in the Designed World

Thinking from the Outside In

Most people don't think of Estée Lauder lotions and industrial engineering in the same sentence. Most people who shop for cosmetics usually don't ask what it takes on the production line to fill a tube with lotion. They are just thinking about the lotion.

But when Molly Hawthorne went to work as an industrial engineer for Estée Lauder Companies, she thought about cosmetics from the outside in. Molly's position was to set up and design the production process for creating tubes and dispensers for various cosmetic lines. As an engineer, she knew that the tubes needed to hold the lotions, and they had to be filled at a certain rate and quantity. It was Molly's job to monitor and evaluate the production process for the final filling of the tubes.

On the Job

Most people don't realize that a design will not work if the production process is not created and managed to fit the product design. "I closely monitored the production lines and looked for any improvements that could be made. I routinely worked with the mechanics, manufacturing engineers, and other industrial engineers to brainstorm new ideas for process improvements. I was also closely involved in the start-up of a new cosmetic line," says Molly.

When setting up a new production line, Molly worked with line mechanics and product managers to ensure that product delivery deadlines were met. Molly performed test runs to check the speed of the machines that produced the product tubes and to decide how many people were needed to keep the production line running at the desired speed. "Once a product line was running smoothly, I would write up the operating procedure, and it would become the standard procedure for that production line," says Molly.

Today, Molly is a product manager for Saint-Gobain Performance Plastics, a company that provides plastics for a wide variety of applications including automotive and aerospace. Molly is responsible for a $17 million global product line. She participates in new product design, marketing, strategic planning, business development, and product pricing. "I interact with plant management, customer service, materials management, quality control, production management, and engineering management on a daily basis. I now oversee many engineering projects," says Molly. She says her educational and work experience have prepared her for her current position. "I have reached my goal!"

Molly Hawthorne: Industrial Engineer, St. Gobain Performance Plastics

Inspirations

Molly's love of engineering comes from her inquisitive nature and interest in math and science. When she was a junior in high school, Molly's CAD instructor suggested that she take a course in principles of engineering. Molly learned about the different fields of engineering and met with practicing engineers. One of those engineers recommended that Molly attend engineering school.

Education

Molly took her teacher's advice and entered Clarkson University, where she earned a degree in interdisciplinary engineering and management. She joined the Society of Women Engineers (SWE) and the American Society of Mechanical Engineers (ASME), completed two internships, and worked on several design teams as part of her training. These experiences taught Molly the value of technical training, but also helped her focus her interest in business and management. Soon, Molly pursued a master's degree in business administration (MBA) with a concentration in innovations and new venture management.

Advice for Students

Molly recommends that young people spend as much time with practicing engineers as possible. Look for opportunities to intern, co-op, or job shadow, she suggests. "By seeing what an engineer does on a day-to-day basis and the different types of opportunities out there for engineers," says Molly, "you may be able to determine what a good fit for you will be."

before they can be bent. Wood and glass require more effort to bend, while textiles are obviously bent very easily. Bending usually distorts the material, and this must be accounted for in the design.

When a material is put under a bending force, part of the material is under compression (squeezed) and part of the material is under tension (stretched) (see Figure 10-10). Because of these forces, material on the inside of the bend is compressed while the material on the outside of the bend can sometimes pull apart. These distortions are usually visible, although not overly objectionable, unless the material begins to tear. Pipe or other tubular materials require bending with a special tool that keeps the material from collapsing at the bend. Scoring heavy paper and cardboard helps make the bend smoother. Wood and paper are best bent with the grain, while metals are bent across the grain.

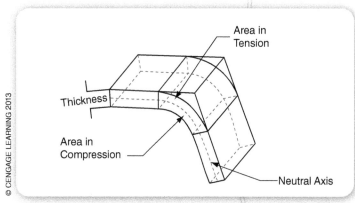

Figure 10-10: When a material is bent, the outer part of the bend is under tension and the inner part of the bend is under compression.

Pressing. Another way to compress and stretch materials is by pressing. This usually involves processes that are more complex than bending. Pressing uses male and female molds, sometimes called dies. Making the molds is expensive and contributes to the fixed costs of the product. Design specifications must state the mold dimensional tolerance and surface finish. Most molds are custom-produced using special metal alloys and are hardened to withstand the wear of production.

Most metals are pressed cold, but materials such as thermoplastics and glass must be pressed hot while the material is soft. In production, pressing cycles must be adjusted to the material being processed. Because metals are pressed cold, the production cycle is more rapid. Pressing thermoplastics and glass requires a cool-down cycle. Most composites require a slower cycle to give the resin time to cure. Ceramics and paper need cycling time for water used in the process to be removed.

A simple punch and die illustrate the making of a simple V-bend in metal (see Figure 10-11). Also, this figure shows holes punched, which is a shearing process we will discuss later in this chapter. Extensive use of formed sheet metal throughout the past century led to the development of large power presses. Many products, such as automobile body parts and appliance parts, utilize complex three-dimensional (3-D) shapes to give relative lightweight sheet metal parts

die:
A term used for a variety of tools used in production to create a shape or 3D form.

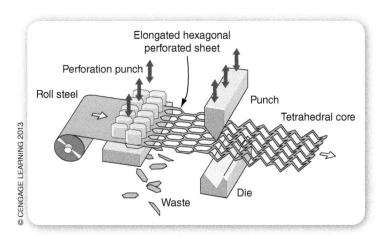

Figure 10-11: A punch and die is used to make a series of corrugations in metal.

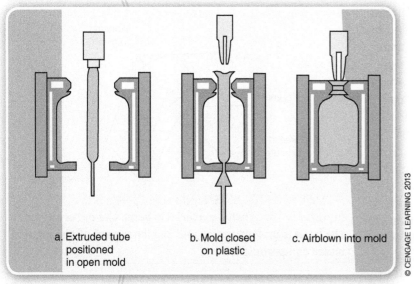

Figure 10-12: **Plastic bottles are typically formed by blow molding.**

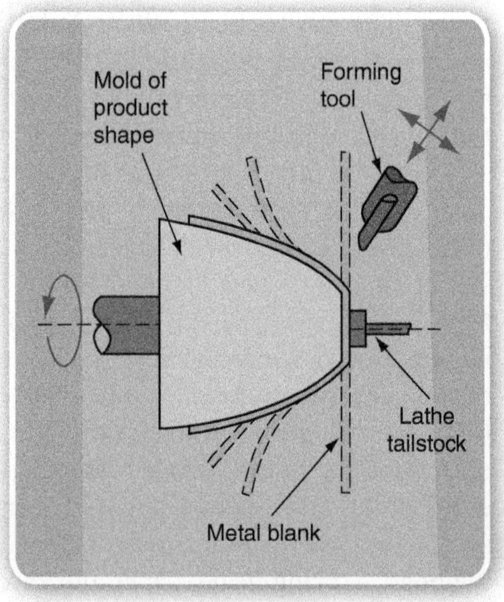

Figure 10-13: **Forming by spinning.**

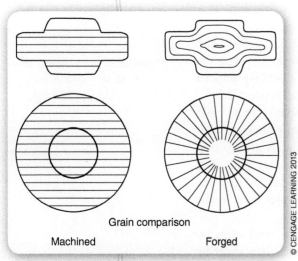

Figure 10-14: **Forging causes the grain structure to bend, making the part stronger.**

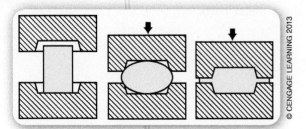

Figure 10-15: **Impression-die forging. Two dies are brought together to change the shape of a metal billet.**

good stiffness qualities. When the part to be formed has considerable depth, multiple forming stages will be used, each of which stretches the material to a practical limit until the desired shape is achieved.

Engineers use many forming techniques for a variety of materials. Vacuum, drape, and blow forming are used to form plastic, glass, and some special metal alloys (see Figure 10-12). Some forming processes, such as the deep-draw technique used to make aluminum cans, use both male and female dies, while other processes use only one die or forming mold.

Explosive hydroforming uses only a female die and the pressure wave of the explosion forces the forming material to take the shape of the mold. Rotational spinning, shear, and flow forming use only a male die (see Figure 10-13).

Forging. Although pressing usually involves sheet material, forging takes a solid blank and processes it into a desired shape using very high pressure. High strength requires cold forging, such as automotive steering and suspension components. Generally speaking, products formed by this process have high dimensional tolerance and poor finish. Forged products are strong because of the unique grain structure created as the material is formed into the desired shape (see Figure 10-14). Engineers use this process on warm (up to several hundred degrees) metal.

Impression-die forging, sometimes called closed-die forging, forces hot metal to change shape between two dies (see Figure 10-15). This manner of forging can shape components from several ounces up to 25 tons into integrate forms.

Open-die forging squeezes metal between two flat dies, producing no specific shapes. Open-die forging is typically used to shape very large components (up to 200,000 pounds and 80 feet in length).

Extruding. An extruding process causes the shape of toothpaste as it comes out of the tube. **Extruding** is the process of forcing a material through a die that imparts a predetermined shape (see Figure 10-16). Extruding pushes the material through the die. This process can create many shapes using aluminum, copper, cast iron, thermoplastics, and clay. Metal windows and door frames, as well as tubing and plastic-coated wire are examples of extruding products.

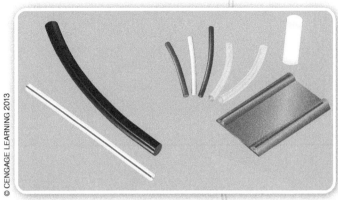

Figure 10-16: **Sample extruded products.**

Drawing. **Drawing** is the opposite of extruding in that the material is pulled instead of pushed through the die. Designers make products such as metal or plastic rod and wire by this process. Drawing usually forms the product in a series of steps. Electrical wire is formed with this process, with thinner wire requiring more drawing steps. For some applications, such as making pipe, the material is initially formed by extrusion and then finished by drawing.

Casting. When the material to be formed is in a liquid state, casting operations are used. In casting, liquid material is poured into a mold (sometimes referred to as molding). Skilled pattern makers produce a pattern that represents the shape to be taken by the cast material. Patterns are made to exacting specifications, including the molding technique to be used and shrinkage factors. Usually, when a material is formed as a liquid, it shrinks as it solidifies. The pattern is used to make a cavity in the molding material and usually contains two or more parts so the finished part can be removed (see Figure 10-17). With few exceptions, the pattern is removed before the liquid material is poured into the mold. Although most people know that iron and steel are cast, any material that has a liquid state can be formed by casting.

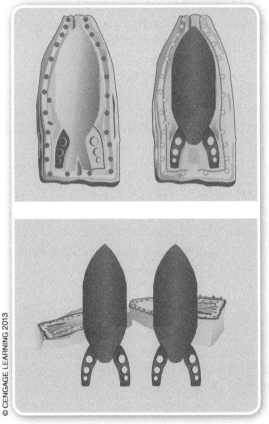

Figure 10-17: **Resin casting of a model spaceship. The two-piece mold is clamped together while the liquid resin is poured and cures. The mold is taken apart so the ship can be removed.**

Figure 10-18: Creating a "tree" of wax jewelry rings to be used as a mold for investment casting.

Figure 10-19: Die casting mold with part release mechanisms.

Types of Molds. The cheapest types of molds are called one-shot molds, and are called this because they are only used once. Sand casting, shell mold casting, and investment casting (also called *lost-wax casting*) are common one-shot molds (see Figure 10-18). Manufacturers often use this process to create a part in large quantities.

Designers first create the original part in a durable material such as metal. They next make a latex mold of the part and then make wax patterns from that mold. Individual parts, as in the case of jewelry, are often connected together to cast many pieces at once. Wax patterns are used to create new molds from plaster (called the *investment*) or other similar material that will withstand the heat of the molten material that will make the part. The plaster mold is heated to a high temperature to burn out the wax, leaving the complex form of the mold. When the wax has been removed, molten metal is poured into the mold. When cooled, the plaster (investment) is broken apart to remove the cast parts. When the product is to contain hollow region(s), such as in the case of a brass spigot or pipe fitting, or even an engine cylinder block, a *core* must be placed in the mold.

Unlike one-shot molds, *permanent* molds can be reused. Die casting, transfer molding, centrifugal casting, injection molding, and lay-up forming represent types of permanent molds. Permanent molds are typically used to form thermoplastics and ceramics (see Figure 10-19).

Casting creates products with distinguishing characteristics. The molding technique determines the finish of the product. In the case of sand casting, the sand from the mold imparts a rough texture to the finished product. Because a path is needed to allow the liquid to flow into the mold and, in some instances, from one part of the mold to another, marks from these paths, or gates, often remain on the finished product. The advantage to casting is that it produces fairly complex shapes at a relatively low cost. This is especially true of one-shot molding practices, which require little tooling up for production. The major disadvantage of one-shot casting is the slowness of production, although sand-molding operations currently employ numerous forms of automation techniques.

The designers must carefully consider the nature of the material in all forming processes. In pressing, they must consider the degree of elasticity, as the material will tend to return to its original shape after being processed. If the material has low ductility, it may need to be heated prior to processing. Designers also must know the amount of material shrinkage in molds and dies. Special material characteristics such as grain structure must be properly oriented before being processed. Although wood has an obvious grain structure, other materials, including paper and metal, can also have grain. Pressing and casting are important material-processing techniques because complex forms can be made cheaply and with little or no material waste.

An important technique for creating detail in a finished product is called **centrifugal casting** (sometimes called *rotational casting*). Metal or plastic resin material is poured into a revolving mold. The rotation forces the molding material to the outside of the mold, filling in all the details. Jewelry makers use centrifugal casting extensively, but even large objects can be made using this technique (see Figure 10-20).

Separating Materials

Separating processes are used to take something away from the material being processed. Unlike forming processes, in which little waste is created, **separating** has the potential of creating considerable waste. Although some of this material can be recycled as **preconsumer waste**, it is important to minimize the amount of wasted or removed material. Designers can help to minimize waste by planning products around standard stock sizes.

Removing materials can be expensive, because it typically requires extensive time and machine processing. Separating techniques can create nearly any shape; however, these processes usually are not as efficient or productive as forming techniques. For this reason, many finished products use separating processes in combination with forming processes. For example, the automobile engine block is formed by casting to produce its basic shape, and then it is processed using material separation techniques to its final specifications (see Figure 10-21). The casting process was employed to reduce the amount of material that needed to be removed to produce the final product.

Separating techniques vary from a simple operation, like cutting 2- by 4-inch lumber to length, or drilling a hole in a part for future assembly, to sophisticated laser, chemical, and ultrasonic processing. Separating techniques can be used on all materials.

Figure 10-20: Two-piece mold for centrifugal casting of a small boat.

Figure 10-21: Cast engine block being machined.

Designers must solve three problems in planning for separating techniques:

▶ How is the material going to be removed?

▶ How is the material going to be held during removal?

▶ How is the cutting tool going to be controlled?

Mechanical, electrical, chemical, and thermal processes all achieve material separating. Mechanical separating is the most common process.

Figure 10-22: **Separating by shearing.**

Figure 10-23: **Die-cut gaskets made from paper and rubber.**

Mechanical Separating Techniques. Previously, we learned how thin sheet metal and other materials can be shaped to yield a product with a relatively high degree of stiffness. Product demand for sheet materials forced development of many specialized mechanical separating techniques. Most of these processes combine a cutting action that causes plastic deformation at the tool edge and eventual material failure by shearing (see Figure 10-22).

Anyone who has used scissors has separated materials by shearing. Shearing is one of the basic forces acting on materials. Product designers must select materials that will withstand an expected shearing force. In material processing, designers must create shearing machines that will exert enough shearing force to separate the material. Properly sheared edges will show a slightly burnished surface opposite the two outside surfaces, with a somewhat duller center area. Shearing operations are commonly used to trim excess material from finished products, such as flashing resulting from pressing and casting operations.

Piercing operations, which also incorporate shearing principles, are used to make small holes in thin materials. Common drilling operations, which will be discussed in the section on chip removal, are not appropriate for making holes in thin stock. Blanking operations are similar to piercing except that the material being removed is the desired product. The common washer is made by combining both operations.

In addition to metal, paper and cardboard products are commonly sheared by several operations, including die cutting, slitting, and perforating. All special shapes, such as envelopes, packaging materials, and cardboard displays, use die-cutting processes (see Figure 10-23). Because economy of materials is always important for both cost and environmental reasons, shearing operations must be carefully planned. Parts to be made by blanking operations should be oriented to minimize waste while still taking into consideration material characteristics such as grain.

Heavy machining is often associated with separating by chip removal. Unlike shearing, by which the product was removed from a stock sheet, machining removes the unwanted materials and leaves the product. Common hand tools such as the chisel, drill, and saw use chip-removing techniques as a way of separating materials. Most chip removal, however, is done by heavy machines, including the drill press, the machine lathe, the power saw, and the milling machine (see Figure 10-24). The wedge-shaped cutting tool is the basis for most chip-removing processes.

Chapter 10: Manufacturing   277

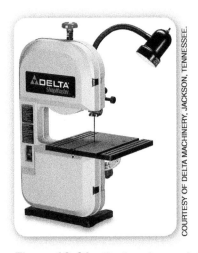

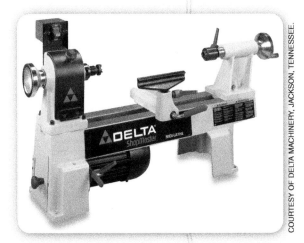

Figure 10-24: **The band saw, drill press, and lathe are examples of chip-removing machines.**

The essential features of the wedge-shaped cutting tool are the clearance angle and rake angles (see Figure 10-25). The **clearance angle** is necessary to prevent the cutting tool from rubbing against the material being cut. The **rake angle**, which actually controls the amount of cutting action, must be carefully designed. Using the proper rake angle determines the efficiency and quality of the cut. The material and the process of separation determine the angle. The rake angle will affect the amount of energy used and the length of the cutting tool's life.

When a particular shape is planned, a **form-cutting tool** can be used. Router bits are good examples of specialized form-cutting tools (see Figure 10-26). Metal, plastic, wood, and other composites are best suited for chip-removing processes.

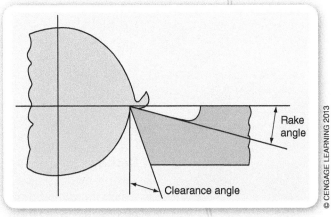

Figure 10-25: **Principle of the wedge-shaped cutting tool.**

Common products use other material-separating processes. The most important of these processes use abrasives as the basis of the chip-removing process. *Grinding* and *ultrasonic machines* use abrasives to separate and smooth materials. *Electrical* and *chemical* methods are used to erode material from the workpiece. *Thermal* methods are used to remove materials by melting. A hot-wire plastic cutter, oxyacetylene setup, and laser are examples of thermal material-separating processes.

To separate materials, it is necessary to hold both the material being processed and to guide the tool doing the cutting. Machine tools use both linear and rotary cutting actions. In a lathe operation, the tool is stationary while the workpiece moves. In a milling operation, the workpiece is stationary while the tool moves. Both the tool and the material move in other operations. A variety of clamping devices have been developed to aid in holding the workpiece.

Fixtures consist of a broad range of devices that hold the workpiece (see Figure 10-27). **Jigs** are devices designed to guide the tool in production. Typically, fixtures are more permanent and are attached to the machine during the operation, where jigs perform a similar function but are temporary in nature. In modern production, a computer controls the tool. In production settings, material-separating processes have been highly automated, with both transfer and holding of the workpiece, and automatic tool operation. We discuss modern industrial organization methods later in this chapter. Industrial engineers often design new jigs and fixtures for new production lines.

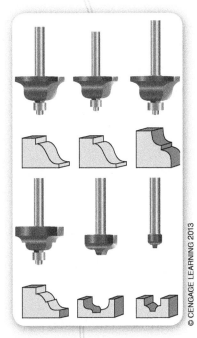

Figure 10-26: **Router bits and some of the interesting shapes that can be created.**

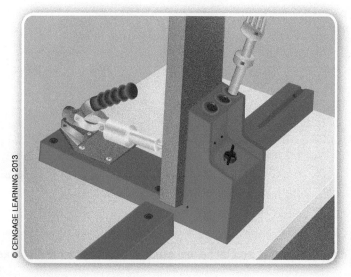

Figure 10-27: A drilling jig accurately drills holes at an angle.

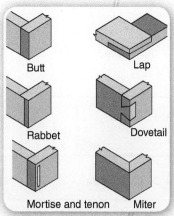

Figure 10-28: Common wood joints.

Combining Materials

It is sometimes necessary in production to combine components made by forming or separating processes. Combining techniques are often associated with product assembly or fabrication. Joining components can be done by mechanical or chemical techniques. In addition to assembling components, combining techniques are used to coat products with paint or to put a printed image on paper or some other substrate. In general, assembly and the finished product require combining materials.

Mechanical Fastening. One of the oldest methods of fastening materials together was by making various and ingenious joint configurations. Archaeological evidence shows that prehistoric humans used mortise and tenon joints. These and other wood joints form a basis for many modern material joints (see Figure 10-28). Some joints, such as the dovetail, are self-locking but require more skill to cut.

To simplify the locking of joints between two components, craftspeople developed other types of mechanical fasteners (see Figure 10-29). Machine screws and bolts are designed with standardized thread types and sizes, so any screw of a

fixtures:
Devices used to hold, clamp, align, or space during fabrication or a machining operation.

jig:
Devices that help guide a tool for machining operation.

Figure 10-29: An array of mechanical fastening devices.

certain size will mate with a nut or other threaded fastener of the same size. The United States standardized nails into specific sizes based on the "penny" or *d* for short (most of the rest of the world uses inches or centimeters). The use of the penny dates from quite a long time ago and was the price for buying 100 of a particular-sized nail.

Many products are fastened with threaded devices. Threaded, or tapped, devices come in an almost infinite variety of shapes and sizes (see Figure 10-30). Threaded devices are often used when the product may need to be disassembled, such as removing the tire from an automobile or bicycle.

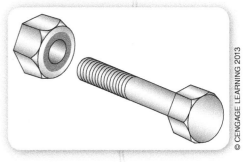

Figure 10-30: Threaded fastener.

Less expensive mechanical fastening devices include nails, staples, sewing, and lacing. Nails are mostly restricted to construction and other applications where wood is used. Some specialized nails can be used with masonry products. The printing industry uses staples in a variety of applications, and staples have become common in surgical procedures as well.

Sewing, because of its application to the textile and shoe-making industries, is the most used mechanical fastening technique. In addition to these techniques, a large variety of spring clips, rings, pins, rivets, spiral and comb binders, and other special mechanical devices have been invented. Modern material developments have provided designers with additional fastening techniques.

Chemical Fastening. Soldering, brazing, and welding use thermal processes to fasten components together. Thermal bonding joins two materials together by melting adjacent areas, thus allowing a covalent bond to form when the material cools. Thermal bonding provides a very strong bond but also has the potential to distort the material due to uneven heating. Thermal fastening is restricted to metal and plastic materials (see Figure 10-31).

Adhesives also employ chemical-bonding principles. Although natural adhesives have been used for centuries, modern developments led to the invention of very strong synthetic adhesives. These adhesives are used to glue ceramic tile on the space shuttle, the rearview mirror on automobile windshields, assembled furniture, and utility pipes. Some materials, such as thermoset plastics, can only be fastened by adhesives.

Modern adhesives are used to fasten metal, plastics, ceramics, leather, paper and cardboard, wood, and nearly any other material. Adhesives may prove to be an excellent material-fastening technique for many design applications. They are often stronger than the material itself and, if applied carefully, will provide a very clean joint.

The selection of the type of combining technique is determined by analyzing questions of application, including the type of material, anticipated stress, cost, and aesthetics:

▶ How permanent will the joint need to be?

▶ What are the characteristics of the joining materials? (Often dissimilar materials are difficult to join with adhesives.)

▶ How strong does the joint need to be?

▶ Is the joint to be stiff or flexible?

▶ What forces will be present at the joint (such as vibration, heat, or corrosives)?

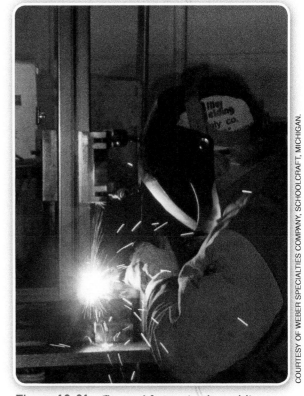

Figure 10-31: Thermal fastening by welding.

- What are the aesthetic considerations? Can the joint be seen?
- What are the environmental impacts of the joining material on the worker, consumer, and environment?

Organizing for Production

The transition from craft or custom production to the factory systems of the twentieth century caused fundamental shifts in society. As people moved into cities, they became better informed. For example, in this period of time, workers formed labor unions, resulting in improved working conditions. Leisure time also increased. With more efficient production systems, productivity increased and the standard of living improved for the middle-class workers. To continue increasing efficiency, new, more complex machines replaced older, less productive machines. These new machines did more of the work previously performed by a laborer. Complex, automated machines controlled the tool and workpiece to a greater degree. Automation usually has the effect of replacing the need for skilled machine operators. Therefore, the nature of the workplace and production in a technological society is constantly changing.

Computer-Integrated Manufacturing

Computer-integrated manufacturing (CIM) is an umbrella term used to describe a full range of contemporary manufacturing techniques. CIM (pronounced *SIM*) encompasses a broad range of modern processes, including computer-aided design (CAD), computer-aided manufacturing (CAM), automated material handling (AMH), total quality control (TQC), robotics, and others. In actuality, it is management and the involvement of everyone in the organization that makes the system successful (see Figure 10-32). The computer is the technological tool that drives CIM. The CIM technique has the potential to provide all the necessary information essentially *immediately* (so-called "real time"), which optimizes all production operations. Areas of concern for a production system include:

- The impact of product design decisions on production.
- Converting design ideas into production drawings.
- Translating specifications into parts lists and production routing.
- Preparing schedules.
- Controlling machine operations (timing and yield).
- Providing current cost of materials.
- Providing real-time information to managers.
- Performing routine data-recording tasks.

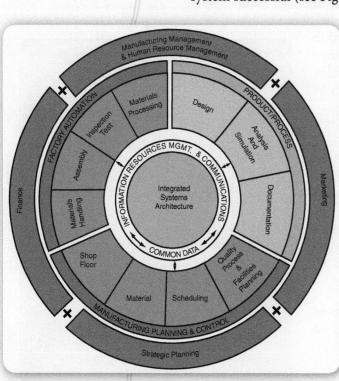

Figure 10-32: Critical CIM links prepared by the society of manufacturing engineers.

CIM techniques are intended to increase productivity and quality. By carefully planning, organizing, and controlling production, wasted time and materials can be reduced. These savings reduce raw material and labor costs. They also reduce the time a product is in the production cycle.

A CIM technique called group technology (GT) is being incorporated into traditional batch production to increase productivity. As much as 75 percent of production is done by batch processing. Under GT principles, management groups together parts or components that can share common machine operations and setups. Through careful planning, savings can be achieved by sharing expensive jigs and fixtures, by reducing setup time, by needing fewer machines, and by improving the flow of parts. One of the newest developments in GT is the introduction of the manufacturing cell. The cell is a collection of tools and material-handling equipment grouped together to produce a family of parts (see Figure 10-33).

Figure 10-33: CIM cell.

Just-In-Time. Just-in-time (JIT) is a CIM technique largely associated with Japanese manufacturing. With this technique, material handling, inspections, and storage are considered wasteful, leaving only production processes as a means of adding value to the product. Over a five-year period, Japanese companies reported that JIT was responsible for a 30 percent increase in productivity, a reduction in inventories by 60 percent, and a 90 percent lower quality rejection rate.

Inventories are materials that the company stores as either stock for production or as finished components or products. Simply, the storage of the inventories needed for production can be relatively expensive. So reducing inventories is a critical CIM function for JIT production. To reduce inventories, the company must work closely with their material and parts suppliers because they will be ordering smaller quantities of materials more frequently. One of the disadvantages of JIT is the inability of the system to react quickly to demand shifts. Companies that use JIT techniques also develop quality circles and forge new relationships between management and the labor force.

> **CAD/CAM:**
> The integration of computer-aided design with computer-aided manufacturing to improve efficiency.

CAD/CAM. One area of development that affects nearly every industry is the introduction of computer-aided design and computer-aided manufacturing (CAD/CAM) (see Figure 10-34). CAD/CAM is successfully reducing the amount of time needed to bring new products to the marketplace. These CIM techniques often reduce duplication of effort in preparing prototype models and product specifications. With CAD/CAM, information from the design stage can be directly input into prototype and production equipment, saving considerable time and money.

Flexible Manufacturing System (FMS). Similar to the cell concept, this new CIM technology may be the factory of the future. The flexible manufacturing system (FMS) incorporates multiple workstations into an integrated manufacturing

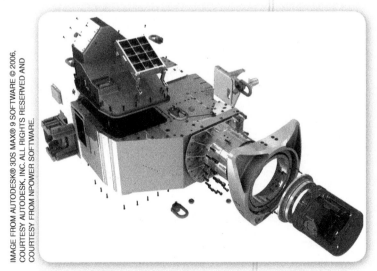

Figure 10-34: CAD/CAM has revolutionized the design and production of products. Image from Autodesk 3ds Max 9 Software.

> **OFF-ROAD EXPLORATION**
>
> For more information about robotics, see *Robot Building for Beginners* by David Cook (ISBN-13: 978-1893115446), or *Absolute Beginner's Guide to Building Robots* by Gareth Branwyn (ISBN-13: 978-0789729712). These books will provide clear guidance for the construction of basic hobby robots and a resource for the design of more complex robotic projects.

environment. In FMS, all scheduling, material handling, process and tool management, quality control, and data keeping are integrated into the system. Robotics and other forms of automation are key elements of flexible manufacturing.

Robots were first used, and continue to be used, to do many monotonous or dangerous jobs in industry. With increased use of microprocessors, robotics is moving into more sophisticated operations, including material processing, assembly operations, finishing, material handling, and inspection operations with such techniques as machine vision. Currently, 90 percent of all robots are used in the automobile industry.

With robotic systems, or more generally called "automated manufacturing systems" costing between $100,000 and several million dollars each, companies must look at both the short- and long-term implications. Some advantages of using robotic or automated systems include consistency, reliability, and, most important, cost-effectiveness. The safety factor of using robots in dangerous and unsafe environments is important to consider. Although robots are incorporated into flexible manufacturing systems, the major advantages of FMS are realized in the reduction of material-handling time and inventory reductions.

DESIGNING FOR MANUFACTURE

With the advent of computers, it is easier to design parts to be manufacturing-ready. A key reason for this is the use of computer-driven rapid prototyping techniques.

Rapid Prototyping

Watching the old *Star Trek* TV shows, you often see *Enterprise* crewmembers going to the "replicator" to get food or to instantly produce a tool or weapon. On the starship, the replicator created the thing the crew wanted just by pressing a few buttons. Although modern rapid prototyping has not reached that level of sophistication (and probably never will), seeing the results of a computer file taking three-dimensional form for the first time is an amazing experience.

> **rapid prototyping:**
> The process of creating a solid model in a computer-controlled machine using a CAD design file.

Rapid prototyping machines typically use the geometry files from CAD programs and then, with their own software, mathematically "slice them up" into narrow slivers like you would slice a loaf of bread, but much thinner, in the range of approximately 0.005 inch. The result is a series of mathematical cross-section views of the part. The thinner the cross-sections, the more detail the final prototype will have.

There are several different types of rapid prototyping machines, and they continue to evolve in sophistication, accuracy, and speed. **Stereolithography** was the name given to the early rapid prototyping process that used lasers to harden a liquid plastic solution held in a tank. The plastic solution reacts photochemically, which means that the light from the laser changes the physical characteristics of the liquid. A platform sits just below the surface of the liquid and a laser beam moves back and forth to solidify the plastic solution near the surface (see Figure 10-35). When the cross-section of the part is completed, the platform moves down and the process is repeated until the entire part is built. In the early versions of this process, the liquid solution was expensive and the parts produced by this method were very costly. Today, this process is refined,

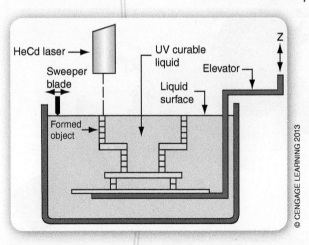

Figure 10-35: **Stereolithography prototyping uses a laser to cure one thin layer of resin at a time.**

and several manufacturers produce stereolithography rapid prototyping equipment, although this process is still expensive.

Using the same concept of mathematically slicing a part into thin sections, a number of companies produced machines that would cut heavyweight paper into cross-sections of a CAD component. The paper slices were then stacked to create a 3-D model (see Figure 10-36). This process was quite inexpensive but required a careful hand to build an accurate model.

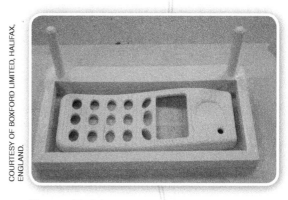

Figure 10-36: Stacking thin card stock creates a prototype of a cell phone case.

The next generation of rapid prototyping machines produced components in plaster or starch, one thin layer at a time. A thin layer of powered plaster is laid down on a small movable platen (table) and then a thin layer of binder (glue) is sprayed on top in the shape of the cross-section of the part, very much like an inkjet printer sprays inks on paper in the shape of letters or images.

The platen moves down a small distance and then another thin layer of plaster is spread across the platen. Again, a thin layer of glue is sprayed in the shape of the cross-section of the part. This process is repeated until the part is completed. The finished prototype is then carefully lifted out of the unglued plaster powder (see Figure 10-37).

Today, rapid prototyping machines that build components in plastic are common, creating usable and durable parts. They use the same layer-by-layer buildup process but typically use acrylonitrile butadiene styrene (ABS) plastic to create the part. The process is called fused deposition modeling (FDM), but because of the time involved to make each piece, it is not a substitute for mass production. However, the ability to create usable parts to develop prototypes and models has a significant impact on the manufacturing industry and the way products are designed and developed. Figure 10-38 shows a variety of parts that have been made using the FDM process.

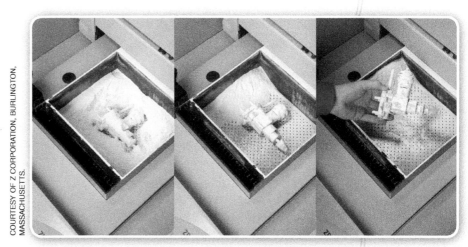

Figure 10-37: Model made from powdered plaster in a "3-D" printer.

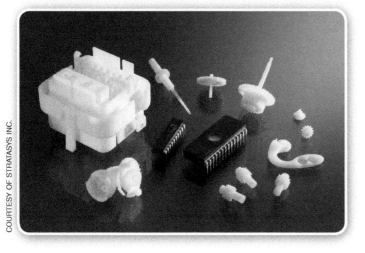

Figure 10-38: Strong and detailed components can be created with the FDM process. The integrated circuits are included for scale.

Case Study

Anytime™ Chair

Richard Douglas Rose, Industrial Designer

Richard Douglas Rose is a freelance industrial designer and the designer of the Anytime multi-purpose chair manufactured and marketed by SitOnIt Seating (see Figure 10-39). He has been interested in design since high school. He noted that as a child, "I was continually fascinated with manufactured objects—disassembling them, creating new objects from parts, building things, fixing things."

Figure 10-39: *Anytime™ Chair*.

After receiving his bachelor of fine art (BFA) in furniture design from the Rochester Institute of Technology and working in the contract furniture industry, Richard established Richard Douglas Design. A product design specialist, Richard focuses primarily on solutions for the office furniture industry. He draws from a diverse set of interests including all things mechanical (especially aviation), extreme sports, travel, art, architecture, and ergonomics to form the foundation of his design principles: performance, practicality, value, and aesthetics. With over 50 individual seating items in production, Richard remarks, "I am extremely grateful and gratified that my work has been so rapidly accepted by the industry." Richard Douglas Design is based in Bucks County, Pennsylvania.

Concerning furniture design, he noted that some very large companies have a design department, but that it is typical for companies to hire freelance designers as well. When he works as a freelancer, he signs a royalty contract and is paid a percentage of the gross income from the sales of the product he designs for the company. Projects may begin as an idea Richard has for a new product, or a company such as SitOnIt Seating may contact him with a design brief. Although most companies consider design briefs confidential, Richard indicated that a brief typically involves a description of the specific product, marketing, and business targets such as:

1. Product goal outlining application, function, and usage.
2. Specific product requirements often including model configurations, materials, finishes, and options.
3. Meeting government and industry standards.
4. Target list pricing and target profit margins.
5. Competition during development and analysis of rival products.
6. Estimated annual sales, development and marketing costs, and project milestone dates such as prototype sign-off, engineering testing, and first order shipment.

Sometimes companies have a "bake-off" where they send out design briefs to industrial designers and ask them to submit a design proposal. To participate in a bake-off, designers must create drawings or renderings. Sometimes the company even requires the designers to develop a prototype. When extensive work is involved, a fee may be paid to each participating designer. Design projects usually take Richard a year from beginning to manufactured product!

Working from a design brief, Richard usually prepares three proposals. A proposal involves considerable work to create a photo-realistic rendering. Richard uses various design and engineering software for 3-D modeling and rendering such as Rhino3D, Autodesk Alias Studio, and PTC ProEngineer, as well 2-D design software such as Adobe Creative Suite. He indicated that designers who do not know how to use these high-tech software packages are at a decidedly competitive disadvantage (see Figure 10-40).

He begins his projects by working on visual appearance and function, but he must consider manufacturing and cost. As Richard noted, he must be mostly right from the beginning. What he meant by that is that if he designs a great-looking chair that cannot be

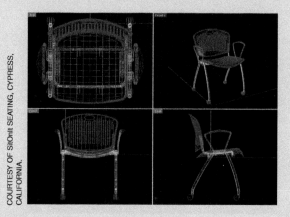

Figure 10-40: *Computer models of the Anytime chair created by Richard Rose.*

Case Study

(continued)

manufactured for a marketable price, he is out of business. He does research on new materials and processes. Most of his chair designs involve injection-molded plastic parts, molded plywood, and steel. In addition to the business and marketing side of product development, he must understand material science, manufacturing, and tooling for production; part production and product assembly; and distribution. On the Anytime chair, the frame was manufactured in China, the injection mold for the seat was made in Taiwan, and the seat was injection-molded in polypropylene plastic in California—typical in today's global economy. SitOnIt Seating performs final assembly, upholstery, and other details at the company headquarters in California and then ships the final product to customers worldwide. This process is part of SitOnIt Seating's "Mass Customization" business model, which combines operational excellence with build-to-order manufacturing. This approach enables the company to provide highly customized, high-quality products to customers quickly and affordably.

Richard sends his proposal renderings to the company for consideration by the management, marketing, and production personnel. If they sign off or approve the proposal, Richard begins to develop or engineer the component parts. To do this, he makes detailed 3-D solid CAD models and 2-D drawings of the parts used for analysis and production. Although he has considerable knowledge of materials and processes, he will need to work with engineers in various locations. For example, he worked with plastic engineers to develop the Anytime seat, and company engineers on the welded steel components. The cost of materials is important. For example, a typical seat made in polypropylene plastic may cost $2.00, while a seat made in a more expensive plastic might cost $10.00. That makes a big difference in the final cost and the marketability or success of the product. His developmental work leads to a prototype model (see Figure 10-41). Rich resends that model to the company for management approval. Rich uses custom modeling for some things such as the metal frame, and rapid prototyping machines such as FDM for the plastic seat component. Upon model approval, called "prototype sign-off," he begins preparing for product manufacturing.

Richard tests the product design to meet ANSI/BIFMA standards. Not all furniture manufacturers

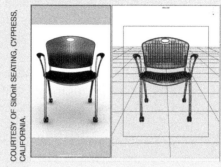

Figure 10-41: *Prototype model created by Richard Rose.*

meet these standards. If a chair, for example, is not designed for these standards, it may not be durable or strong enough for the continued use in a business or education environment. Richard was very proud of the fact that the arm for his Anytime chair was tested at over 500 pounds—far in excess of the 300-pound ANSI/BIFMA standard.

As a freelance designer, Richard Douglas Rose gets involved in all aspects of the design and manufacturing of a product. For the Anytime chair, he worked with SitOnIt Seating to enhance the company's adherence to the four elements most important to lean manufacturing: labor efficiency, cost of quality, safety, and complete and on-time delivery. He writes design briefs (first step), does all the design work including engineering components, prepares for the manufacturing of parts, and works with the manufacturer to help market the product. With the introduction of a new product to the market, Richard attends trade shows to talk with potential customers about the design. He draws a parallel between the textile industry in the United States in the mid-twentieth century and the furniture industry today. He believes that design is entrepreneurial and one of the strongest parts of the American economy. He noted that although the United States is manufacturing fewer garments, much of it is still designed domestically, and everyone still needs and wants attractive clothing. Most importantly, he finds industrial design to be fun and rewarding.

He says, "Design is about problem solving and it is about creating products that people desire. There is a symbiotic link between practicality, durability, function, value, performance, aesthetics, and desirability, especially with products we use in our daily lives."

FUTURE IMPACTS

It is extremely important for the health of the economy in a free enterprise system that companies remain viable and competitive in producing goods for the global marketplace. Competition begins with appropriate design solutions—design solutions that are also capable of being efficiently manufactured. An efficient production system is one of the key elements in maintaining a viable economy.

It is estimated that over 70,000 different natural and synthetic materials are available for production. As a society, we must be concerned about the depletion of needed natural resources. Critical resources can be saved by finding or designing alternative materials or, more important, by conserving those that already exist. The current trend to design new products using fewer materials (source reduction) is one way to save valuable resources. Recycling and waste reduction are others.

In the quest to remain competitive, society will need to consider important environmental questions. In 1970, the U.S. Environmental Protection Agency (EPA) established standards for air and water quality. These regulations have been subsequently revised, and many other countries have adopted similar standards. The current concern over the depletion of the ozone layer and global climate change points to the need of all citizens of the planet to be involved in the decision-making process in balancing the trade-offs and risks of technological progress. Good design always considers the impacts of production and consumption on the individual, the society, and the environment.

SUMMARY

Manufacturing products involves a managed production system that combines the input of capital, people, knowledge, materials, energy, tools and machines, and processes. Products begin their life cycle as a design to a perceived need and end the cycle as discarded waste. In between, the product impacts the individual, the society, and the environment. Manufacturing is an important part of our economic system. Planning production is important so that the best method of production can be found. Materials are processed in the system by forming, separating, and combining. New techniques for managing production involve computer-integrated manufacturing (CIM) technologies. The way in which our society identifies and solves technological problems will continue to have a major impact on individuals, societies, and the environment.

BRING IT HOME

OBSERVATION/ANALYSIS/SYNTHESIS

1. Select one of the products found in the scene in Figure 10-42, and list as many of the materials and processes you can that are needed to produce the product.
2. With the product you chose from #1, list as many of the raw materials needed for both the product itself and for producing the product.
3. With the product you chose from #1, ask 10 typical users of the product what they think would be the most difficult aspects about manufacturing the product. Compare their answers to your teacher's answers to the same question.
4. Design a 6-inch cube made of 6 flat pieces of ½-inch thick plywood. There is no constraint on how these 6 pieces of plywood would be fastened together to form a cube. The purpose of the 6-inch cube is simply to robustly support the weight of a much larger cube (resting flat surface to flat surface) that weighs 1,000 pounds. Also, design the processes used to make 20 of these cubes. Outline the ways your design conforms to standard materials and processes. How would designs differ for the following constraints: (a) use of no fastening devices (screws, nails, and so on), and (b) wood screws required as a fastening device.

Figure 10-42: **A variety of manufactured products.**

COURTESY OF MEMORY PROTECTION DEVICES, INC.

BRING IT HOME

(continued)

5. Select one of the least efficient processes necessary to produce your 6-inch cube product, and suggest alternative, more efficient processes.
6. Evaluate your cube design in #4 for each of the following criteria:

   a. Manufacturing viability
   b. Durability
   c. Maintenance
   d. Aesthetics
   e. Material source reduction
   f. Environmental impact
   g. Social impact

EXTRA MILE

MASS-PRODUCTION LINE

- Design a mass-production line to assemble a simple structure you have designed from LEGO parts or other construction kit parts. Organize at least five people to perform a single task each in the assembly and to pass the product to the next assembly station. Document the assembly line during production with photos (including captions) or a short narrated video.

PART II
Resources for Engineering Design

Chapter 11 Designing Structural Systems
Chapter 12 Designing Mechanical Systems
Chapter 13 Designing Electrical Systems
Chapter 14 Designing Pneumatic Systems
Chapter 15 Human Factors in Design and Engineering
Chapter 16 Math and Science Applications
Chapter 17 Design Styles
Chapter 18 Graphics and Presentation

© SHUTTERSTOCK IMAGES, LLC/MC_PP

CHAPTER 11
Designing Structural Systems

Menu

 Before You Begin
Think about these questions as you study the concepts in this chapter:

1. What are structures and how are they different than other human-designed things?
2. Why do civil engineers and architects design and build structures?
3. Why do structures fail?
4. What must be understood to design a structure?
5. How do loads and natural forces affect structures?
6. What types of components are used to design and build structures?
7. What are common types of bridge structures?
8. How do you calculate loads on structures?

INTRODUCTION

Early humans built structures to provide shelter and protection. The first human structures were made of materials indigenous to the region and designed and made by the people living there. For example, Native Americans living throughout the continental United States built very different housing. In the Northwest and Northeast where an abundance of forests existed, these early designers used wood to construct their homes. On the plains, where animal skins from buffalo were a staple of life, the tepee made of long wooden poles covered with a tent of skins provided protection from the elements and allowed for the homes to be moved as the buffalo roamed (see Figure 11-1). In the Southwest, people made homes of adobe bricks made from clay and straw that were dried in the sun. These adobe bricks were especially good for the hot, dry desert climate. Interestingly, they constructed apartment-style homes and simply added rooms as families grew (see Figure 11-2). Early humans built successful structures without fully understanding the principles that we will cover in this chapter. These early designers observed nature well and improved on existing technique as they built their technological environment.

Today we build major structural systems from many types of materials. Although natural materials are still appropriate, humans increasingly use synthetic or engineered materials for their improved strength, reduced weight, resistance to natural forces, and reduced cost. A large variety of materials are available to designers or engineers for design application in structures such as residential and commercial buildings, bridges, dams, and tunnels. Examples of other structures are manufactured pipelines, utility towers, furniture, transportation vehicles, and even medical equipment. Civil engineers, structural engineers, or architects design most structures.

An example of a modern skyscraper is the Taipei 101 Tower designed by C. Y. Lee and Partners, Architects (see Figure 11-3). The massive 2.34-million-square-foot office and retail space building completed in 2004 visually dominates downtown Taiwan. The 60-foot spire was inspired by native bamboo plants and borrows from Chinese culture in both exterior and interior motif. In the tower design, the architects and engineers needed to take into consideration the possibility of earthquake and typhoon activity.

Figure 11-1: The tepee was the home of the plains Indians.

Figure 11-2: American Indians of the Southwest built Adobe apartment-style homes.

Figure 11-3: Built in 2004, the Taipei 101 Tower is one of the world's tallest buildings at 508 meters (1,667 feet).

STRUCTURAL SYSTEMS

structure:
A structure is a body that supports a load and resists external forces without changing its shape, except for that due to the elasticity of the material(s) used in construction.

In a properly designed structure, failure can only occur if a joint or the material itself fails. Unfortunately, designers today must consider one additional factor that can potentially cause structural failure—sabotage. Today the design of most new commercial structures considers ways to protect the structure from terrorist attack or to be able to resist the force of an attack.

In the design of a new structure, engineers and architects must use scientific and mathematical principles and concepts. Designers must also meet a given set of design specifications and satisfy the client's intended purpose for the structure. Safety considerations include not only standard safety factors, such as the maximum loading for a bridge or building, but also considerations to promote safe use by humans. Structures must be designed for ease of maintenance and service. Finally, it is important that the design uses materials and construction processes efficiently to minimize costs. Entry-level engineers often are asked to design beams, columns, or other basic structural subsystems. To design these components, engineers must accurately calculate loads and understand forces that will be applied to the structural component (see Figure 11-4).

As a young teenager, my son said he wanted to build a table for his room and he wanted to do it by himself. After hearing much cutting and hammering in our basement workshop, he appeared with a table consisting of a top and four legs nailed through the top. Although it looked like a table (a type of structure), it lacked all the elements necessary to give it stability. Figure 11-6a illustrates how something can look like a structure but not function as one. In this figure, four bars are pinned to form a rectangle. If a force is applied, as in Figure 11-6b, the bars move and the shape becomes deformed, which is just what happened to my son's table. Even if we replaced the pins with strong, tightened bolts and nuts, the ability to resist the force would only be in the increased friction caused by the tightness

Figure 11-4: **Engineered lumber.**

Fun Facts

The Eiffel Tower is a famous and elegant structure (see Figure 11-5). The French civil engineer Gustave Eiffel (1832–1923) designed the iron tower. It is 320 meters (1,050 feet) tall and stands on the Champ de Mars beside the River Seine in Paris, France. It is the tallest structure in Paris and one of the most recognized structures in the world. Eiffel also designed the Statue of Liberty located on Liberty Island, New York. The French government gave the statue to the people of the United States in 1876, in celebration of the centennial of the United States.

Figure 11-5: **Eiffel Tower, Paris, France.**

of the bolts, not in the redesign of the object itself. If a fifth member were added diagonally to the object, forming two triangles, as in Figure 11-6c, then its shape would have the ability to resist much higher forces. This same concept, of forming triangular shapes, could also have been used in the table design, resulting in a much stronger structure. The reengineered shape is now a structure, because the material of the bars or the joints would have to fail for the structure to fail. The triangle shape is very central in many structural designs.

A structural subcomponent using the triangle shape is used frequently in the Eiffel Tower.

As illustrated in Figure 11-7, structures occur in both the natural and the technological world. Structures must support a load and resist various forces by holding each structural element in a relative position to other parts. Most important, structures should not fail prematurely! Natural structures can teach us a lot. A natural structure with which we are most familiar is our endoskeleton. The human endoskeleton gives our body form, carries the load of our organs and other internal parts, withstands external forces, and provides an ability to move. Many of the capabilities of the human are discussed in Chapter 15, Human Factors in Design and Engineering. The body's ability to withstand external loads such as those created by standing, sitting, walking, lifting, or occasionally even running into an object, is necessary for us to function (see Figure 11-8).

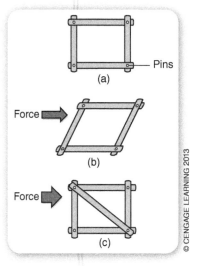

Figure 11-6: (a), (b), and (c): What makes something a structure?

Figure 11-7: Examples of natural and technological structures.

Figure 11-8: The human body withstands external loads and acts as a structure.

Your Turn

Can you think of "everyday" structures that use triangle-shaped components?

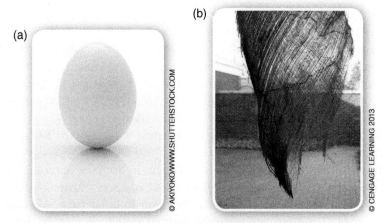

Figure 11-9: Two examples of natural structures: (a) Buckminster "Bucky" Fuller described the egg as a perfect structure, and (b) the fabric-like pattern of the palm frond.

Skeletons consist of components that include bones, joints, and muscles. Joints are capable of certain movement and are held together by ligaments and cartilage. Muscles provide the power to move body parts and are connected to the bones by tendons. The body is also capable of sensing internal and external stimuli, processing information, and producing responses. The human body is an extremely complex and marvelous creation. By looking at the structure and abilities of our bodies and other natural structures, much can be learned and applied to solving technological problems. We can also look at other natural structures for ideas to design new synthetic or technological products (see Figure 11-9).

TECHNOLOGICAL STRUCTURES

Just as nature provides thousands of unique natural structures in the plant and animal kingdoms, humans have designed thousands of unique technological structures. Both natural and human-designed structures are essential to our modern life. One obvious way to look at structures is by the purpose they serve. Structures provide housing, containment, and transportation as well as serve other purposes (see Figure 11-10). Even your cell phone, a communications device, uses structural designs.

Civil engineers and architects must consider structural purpose and other important issues when designing successful structures. All structures, regardless of their purpose, must support loads caused by internal and external forces and, very importantly, must not fail prematurely.

Structural Failure

It should be assumed that all well-designed technological structures will fail at some point due to natural forces over time or misuse. According to the U.S. Department of Transportation, Federal Highway Administration, more than 590,000 bridges exist in the National Bridge Inventory (NBI). Most of these bridges were designed for a safe, useful life of 50 years. Today, more than 40 percent of the NBI bridges are over 40 years old. Bridges are inspected for safety and are repaired or rebuilt as needed. All buildings, bridges, dams, tunnels, towers, and other technological structures are designed for a safe and useful life.

Careers in the Designed World

A BALANCING ACT: STEPHEN DOUGLAS, CIVIL ENGINEER, HIGH COUNTRY ENGINEERING

The engineering design process is often a balancing act. No one knows this better than civil engineer Stephen Douglas. In his work for High Country Engineering in Denver, Colorado, Stephen often weighs the cost of a good design that meets local codes against the client's wish to spend as little money and time as possible. Despite these constraints, the client still wants the same quality product.

The best way to balance the equation, Stephen says, is to maintain open lines of communication between the client, the reviewing agencies, and the engineering team. Stephen tries to present everyone with a wide range of design alternatives so everyone has as much information as possible to make the best decision.

On the Job

High Country Engineering in Denver, Colorado, is a civil engineering and land surveying firm that designs land development, municipal infrastructure, and road and highway systems. Stephen works on projects from start to finish, reporting directly to a senior project manager as he prepares designs and final documents for a new land development project. Stephen manages a team of project engineers and CAD designers, who will complete the project. Together, this team will develop the design concept, listen to the client to fine-tune the design concept, control the project budget, and work with the city or county agencies that set the design standards and guidelines.

Inspirations

Stephen enjoys the challenges of problem solving. One of Stephen's most rewarding projects to date was the design of a small steep plot of land for a single-family home in the mountains. The steep, mountainous terrain controlled every aspect of the design, from grading to placement of the septic tank. The challenge was making the human-made structure fit within the difficult land formation.

Education

Stephen joined the U.S. Navy's nuclear training program right after high school. His work with the electrical systems on a nuclear submarine sparked Stephen's interest in pursuing a degree in engineering. Stephen received his civil engineering degree from Colorado State University, in Fort Collins, Colorado. In college, he learned the importance of attention to detail and the base of knowledge that he applies now to his career path in civil engineering.

Advice for Students

"Listen to the project manager that has been doing this for years," says Stephen. "That person is a wellspring of knowledge and experience that can help you tremendously. Also, keep a notebook with you at all times to write down directions that have been given to you. This will act as a reminder and a guide as you step yourself through the design process." Stephen has learned that keeping a record of design changes and important information can be of great value for both the clients and the engineering company during the many design meetings that are held to evaluate the progress of the design and development of the new land project.

"Working in a design team contributes to the success of a project because you have the benefit of everyone's knowledge and there are more people to get the job done. If you enjoy challenging yourself and at the same time improving the world you live in," says Stephen, "a career in engineering just might be for you."

Figure 11-10: Structures serve many useful purposes: (a) Housing, (b) containment, (c) communication, and (d) transportation.

planned obsolescence:
The conscious decision on the part of a designer to produce a product that will become obsolete in a defined time frame.

Figure 11-11: *1968 Chevrolet Corvette.*

durable and nondurable goods:
A designation established by the U.S. Department of Commerce to describe the length of time a product is intended to be useful. Durable goods are those that are intended to last more than three years while nondurable goods are designed to be useful for less than three years.

The concept of **planned obsolescence** was first used in the 1920s and 1930s, but the term did not become popular until 1954 when Brooks Stevens, an American industrial designer, used it in a number of presentations. In the United Kingdom, the concept is known as *built-in obsolescence*. Planned obsolescence can be either functional or aesthetic (style). Producers benefit from planned obsolescence because consumers will be required to or will desire to buy their product repeatedly. Alfred P. Sloan, president and chairman of the board (1923–1956) for General Motors, hired Harley Earl to head a new design department. Earl, who greatly influenced the style of all GM automobiles, is considered the father of the Chevrolet Corvette (see Figure 11-11). Under Earl's management, the stylizing of new models reflected an application of planned obsolescence in the automobile industry. Most people are concerned with both the functional qualities of their car, with warranties between 50,000 and 100,000 miles, and with the style because they want the look of the latest style, or model.

The U.S. Department of Commerce uses the terms **durable goods** and **nondurable goods** to describe the length of time a product is intended to be useful. Durable goods are those that are intended to last more than three years while nondurable goods are designed to be useful for less than three years. Sometimes poor design due to insufficient knowledge about the structure, material, or system can cause a technological system to prematurely fail.

An excellent example of early failure is the Tacoma Narrows Bridge, which failed in 1940 due to resonant oscillation, or vibrations, created by naturally occurring wind passing by the bridge deck (see Figure 11-12). Today, these oscillations are better understood and are controlled by using updated bridge design rules, including stiffer bridge decks. The failure of the Tacoma Narrows Bridge represents something called a *Type IV technological impact* as it was unexpected (and certainly undesired!). Some systems fail because the design incorporates materials, components, and processes incorrectly. Have you ever purchased a cheap pair of scissors or other tool that never worked properly or broke because of poor design or poor choice of material(s)?

Systems can fail for expected reasons as well. Because all systems are exposed to such external forces as moisture and vibrations, these external forces begin to weaken the elements of a system over time. Many of these forces can be controlled. The paint on a metal swing set, which resists rust forming from moisture in the environment, illustrates how designers and engineers properly considered the impact of external forces on the product. In addition to the expected forces, systems fail when they are used improperly. For example, all products are rated for maximum loading. If the system is overloaded, it will fail at some point. Designers and engineers will calculate and use a planned safety factor in setting load limits.

Figure 11-12: **Tacoma Narrows Bridge.**

Safety Factor

A safety factor determines how much a product, or an element within a product, is overbuilt. The safety factor is usually determined early in the project. Knowledge of how materials and components or systems will work is important in determining a safety factor. Safety factors (S.F.) should be greater than 1 (S.F. > 1.0). A race car, for example, is designed to minimize its weight to attain higher accelerations. The designer attempts to make the car as light as possible so that it will go as fast as possible. Therefore, the safety factor for a racing car is often very close to 1.0.

We expect that most products will work properly and safely, even though we may be using them to full capacity. A two-ton jack (1 ton equals 2,000 pounds) is not expected to fail if 4,001 pounds are lifted. A jack typically has a safety factor of around 1.5, meaning that a two-ton jack will lift up to 6,000 pounds safely—2,000 pounds above its rated capacity.

safety factor: A measure of how much the product is overbuilt and the ratio of the ultimate stress (breaking point) and the working stress (maximum expected load).

> S.F. = Ultimate stress (breaking point)/Working stress (max. expected load)
> S.F. = 6,000 pounds (designed capability)/4,000 pounds (max. load rating)
> S.F. = 1.5

NEWTONIAN MECHANICS

Newton's laws of motion form the basis for classical Newtonian mechanics. These principles consist of three physical laws that provide relationships between the forces acting on a body and the motion of the body. They were first formulated by Sir Isaac Newton and published in 1687. Design professionals will need to develop a full understanding of Newton's laws of motion to be successful engineers and architects.

> *First law:* An object at rest remains at rest and an object in motion continues to move in a straight line with a constant speed unless and until acted upon by an external unbalanced force.
> *Second law:* The rate of change of momentum of a body is directly proportional to the force acting upon it and is in the same direction.
> *Third law:* For every force applied to an object, there is an equal, but opposite, reaction force.

Example: Applying Newton's Third Law

A practical example will help illustrate the relationship between science and technological activity. For scientific purpose, a load applied to an object is subject to Newton's third law: For every force applied to an object, there is an equal, but opposite, reaction force. In this example, the force is a person sitting on the chair. As a designer or engineer, knowing and being able to apply the appropriate scientific principles is essential to a successful design. The designer must know that if a 100-pound person sits on the chair, the chair will be subject to 100 pounds of compressive force, F = R (see Figure 11-13).

Figure 11-13: Applying necessary scientific principles to a technological design problem.

engineering design analysis: The process of applying mathematical and scientific understanding to a proposed design solution to determine if there are any reasons why the design will not function as expected when it is prototyped or ultimately produced.

The relationship between science and technology is well illustrated by looking at the effect of forces acting on technological systems. Remember that the purpose of science is to discover and describe, while the purpose of all design fields is to invent or innovate. Although an important symbiotic relationship does exist, for the purposes of this book, scientific principles will be applied without further explanation. For example, in discussing forces, the concept of weight will be used and expressed as either English (American) pound force or the Systeme International (SI) newton force [one newton of force accelerates a mass of 1 kg at 1 meter per second per second, $N = (1kg)(1m/s^2)$], commonly referred to as the metric system.

In design, technology, and engineering, known scientific principles cannot be violated in a new design (see also Chapter 16, Math and Science Applications). Sometimes design involves a natural principle that is not fully understood. Usually, designers document principles that are in play for new designs and use both scientific principles and mathematical understanding to analyze their design work. In engineering, the process of applying mathematical and scientific understanding to the design is called **engineering design analysis**. For engineers, this process helps to determine any reasons why the design will not function as expected when it is prototyped or ultimately produced.

In addition to the science, of equal importance is an understanding and correct application of all necessary technological principles to a successful design solution.

Although there is only one correct answer to the amount of force, many interesting technological challenges and opportunities must be answered in designing a chair. Designers need to fully understand the constraints, such as time and cost. Many questions will need to be answered, such as: What materials should be used? How will the chair look and what will its shape be (human factors)? How will it be produced and marketed? In addition, designers need to consider worker safety, product reliability, and the environmental impacts during the design process (see Figure 11-14).

Figure 11-14: An adjustable chair is a good design solution because it is comfortable for a wider range of people but will cost more to produce.

STRUCTURAL LOADS AND FORCES

Structural loads are classified as either live loads or dead loads. Live loads include the weight of all occupants in the building as well as things that can be moved, such as furniture. Dead loads include the weight of the materials in the structure itself and all major architectural or product features. Other loads also need to be considered. For example, natural forces created by wind, water, snow, or seismic and tsunami activities need to be considered in the design. Also, temporary forces occurring during construction like the weight of machines, equipment, construction materials, and personnel need to be considered.

Not all loads are equal. A person sitting quietly in a chair is considered a static load. Static loads are those associated with a load at rest. Static loads are easy to predict, because they are easily measured.

Dynamic loads are characterized by forces in motion. An example of a dynamic load is a person in the process of sitting down in a chair. You may have younger brothers or sisters who throw themselves into a chair rather than sitting down in a chair. In this example, the dynamic load created is considerably higher than the static load or weight of the person. A chair or any product must be designed for expected dynamic forces. All loads and natural forces must be counteracted by internal structural forces or the structure will fail.

Equilibrium

When an external load is applied to a structure such as a chair, internal forces are necessary to counteract each load force. Internal forces are generated by the molecular structure of the material in question to counter external forces. External forces are the loads that are applied to the object in question. In the example of the chair, the wood or metal components have a molecular structure that pushes back against the load (the weight of the person sitting in the chair) with a force that is equal to the load. This condition is known as equilibrium (see Figure 11-15).

Figure 11-15: *Forces in equilibrium.*

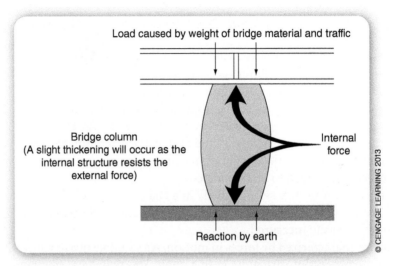

Figure 11-16: **Internal forces must equal external loads in a structure in equilibrium.**

If the load were greater than the force that could be generated by the molecular structure of the material in the chair, then the chair would fail or break. If the internal force of the chair were greater than the external force of the load (person), then the person would be shot into space.

The scientific principle of equilibrium will be applied when designing or engineering a technological structure. When a force or a load is applied to a structure, the internal forces help the material to resist being changed by the external load (see Figure 11-16). Although changes in the material may not be evident, slight thinning or thickening of all materials occurs under a load. A concrete highway may show little change, while slight pressure on a water balloon will show a significant physical change in the shape of the balloon.

Stress

Stress describes the quality or ability of a material to distribute internal forces when a load is applied. Consistent with Newton's third law, the material must have the ability to provide an equal but opposite force when a load is applied. Eventually, given sufficient force, all materials will fail or fracture. Stress is calculated by dividing F (force) by A (area), as shown in Figure 11-17.

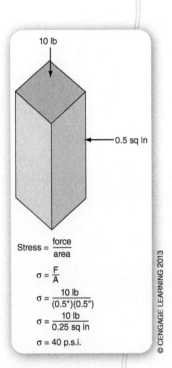

Figure 11-17: **Stress formula with sample problem.**

In SI measurement, stress is often expressed in meganewtons (MN) per square meter (MN/m^2). In the conventional English system, stress is expressed as pounds per square inch (psi). Stress is represented by s or the Greek letter σ (sigma). Stress (or ultimate stress) characterizes the strength of the material by quantifying how much force is needed to make the material fail or break.

Strain

Strain is used to describe the change in shape of a material caused by compression or tension forces (see Figure 11-18). While stress describes how much force is needed to make the material fail, strain describes how much the material itself stretches or compresses under a load just prior to the ultimate stress point. Strain is usually expressed as a percentage of change in the original length. Strain is represented by the Greek letter ε (epsilon).

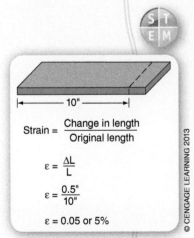

Figure 11-18: **Strain formula with sample problem.**

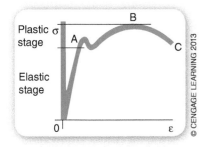

Figure 11-19: **Stress/strain graph used to determine the Young's modulus of elasticity for a material.**

Figure 11-20: **Young's modulus of elasticity, table of E values.**

Material	MN/m² (approximate values)	psi (approximate values)
Rubber	7	1,000
Shell membrane of egg	8	1,100
Human cartilage	24	3,500
Human tendon	600	80,000
Wallboard	1,400	200,000
Unreinforced plastics	1,400	200,000
Plywood	7,000	1,000,000
Wood (along grain)	14,000	2,000,000
Concrete	17,000	2,400,000
Magnesium metal	42,000	6,000,000
Ordinary glasses	70,000	10,000,000
Aluminum alloys	73,000	10,400,000
Brasses and bronzes	120,000	17,000,000
Iron and steel	210,000	30,000,000
Aluminum oxide	420,000	60,000,000
Diamond	1,200,000	170,000,000

By combining the measures of stress and strain, a measure of the strength of a material is represented as **Young's modulus of elasticity, or E** (Thomas Young, 1773–1829), as shown in Figure 11-19. Three points are important on the stress/strain graph. Each point represents a different stage in the molecular structure of the material. From point "0" to point "A," the material is said to be in an elastic stage. During the **elastic stage** a material under force will change its shape; however, when the force is removed, the material will return to its original shape. In this elastic region, the shape of the curve is linear. A fiberglass pole used by a pole-vaulter is a good example of a material that bends under force but returns to its original shape after a vault. The next stage occurs between points "A" and "B" and is known as the **plastic stage**. During this stage, the material under load changes shape permanently and will not return to its original shape after the load has been removed. If the fiberglass pole remained curved after the vault, it would no longer be useful. The final stage, occurring at point "C," identifies **material failure or breaking point**. Typically, no material should be used outside its elastic stage.

From a design and engineering perspective, the E (modulus of elasticity) value is useful in the selection of appropriate material for a particular application (see Figure 11-20). A material with a low modulus of elasticity, such as concrete, should not be used where it would be exposed to tension or bending forces, even though concrete is extremely strong (under compression). Mild steel or aluminum, on the other hand, combines both strength and flexibility.

COMMON FORCES

There are five forces that are commonly considered when designing structures: (1) compression, (2) tension, (3) bending, (4) shear, and (5) torsion. As previously noted, external forces are loads that are applied to a structure, while internal forces resist changes in the structural shape. These forces can be static (at rest) or dynamic (in motion). The concepts of stress (strength), strain (change in shape),

Figure 11-21: **Material under compression.**

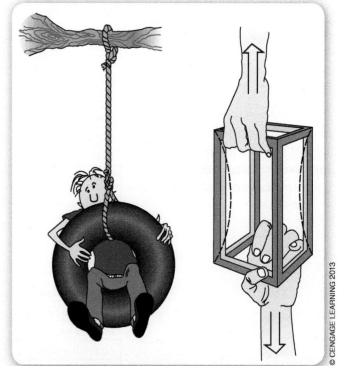

Figure 11-22: **Material under tension.**

and elasticity (change in shape without permanent deformation) describe the physical changes in the material under load. The five different ways in which forces or loads are exerted on a structure help determine how the structural element is to be designed or engineered.

Compression

Compression is a force that is pushing against something. A load placed on top of a structural element is typically under the influence of gravity and as such is pressing down on one end of the element (see Figure 11-21). In compression, the load is pressing down on one end of the element while the ground, or some other structural element, is pressing up on the other with an equal force. If the element is to be in equilibrium, it must have sufficient strength (stress) to withstand the compression force. Compression loads try to collapse the element. The legs of a chair or table are under compression.

Tension

Frequently, a load is applied to the element in a pulling action (see Figure 11-22). When this occurs, the element is said to be in tension. Tension forces try to pull the element apart such as the force on the rope in tug-of-war. To maintain equilibrium, the element must be able to resist the outward pulling forces. It is easiest to design for materials that are only under tension loading.

In both compression and tension, the loads are acting along the length of the element. Some loads, however, act across rather than along the element.

Bending

When an element, such as a bookshelf, is loaded with books, the bending force is applied across the material (see Figure 11-23). A neutral axis existing near the center of the element causes the material to be under both compression and tension forces.

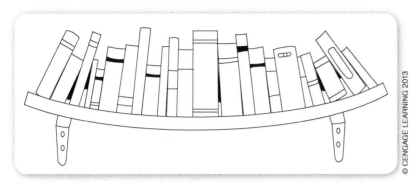

Figure 11-23: Material can bend under a load.

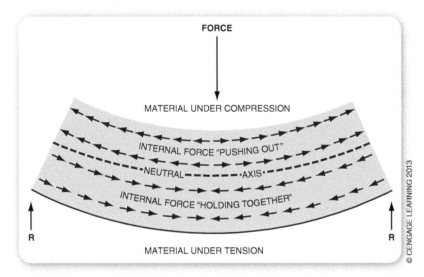

Figure 11-24: Material being bent is under both compression and tension forces.

Material on the inside of the bend is under compression; that is, the load is trying to force the molecules together. At the same time, the material on the outside of the bend is under tension. The bending action on this side of the material is trying to pull the molecules apart (see Figure 11-24).

Most materials used in shelving will bend to some degree under a load. Plywood, which is often used in shelving, is a composite material made from thin layers of wood glued together. Why would this material more effectively resist bending (consider that each layer has a neutral axis)? If the supports are too far apart or too many heavy objects are placed on the shelf, the shelf material could be strained beyond its elastic stage and fail.

Your Turn

The compression and tension of a bending action can be easily seen by taking a piece of thick foam rubber or similar material. A line representing the neutral axis is drawn along the center of one edge. A series of lines, approximately 2 centimeters apart, are drawn perpendicular to the original line. If the foam rubber is bent, you will observe how the material is under compression above the neutral axis and under tension below the neutral axis.

Figure 11-25: Material under shear.

Figure 11-26: Material under torsion.

Shear

When compression force is applied on a material in opposite directions across a perpendicular plane, the result is called a shear force (see Figure 11-25). In this figure, the weight of the person is pushing or compressing the board at the point perpendicular to the support that is pushing or compressing the board in the opposite direction. The opposing compression forces are attempting to shear the material on the perpendicular line. Scissors use shear force to separate materials from thin paper to heavy metals. Shear forces must be taken into consideration and can be a major reason why a structure fails. Most fasteners, such as bolts, nails, and welds, are under shear force.

Torsion

Torsion is a twisting force (see Figure 11-26). In applications such as an automobile drive shaft or simply spinning on a single rope swing seat, the structural element is under torsion. Torsion describes forces that try to twist, or rotate, the material apart. As with bending, there is a neutral axis and materials within the element are under both compression and tension forces. Torsion (also known as *torque*) is most associated with mechanical systems, such as motors, engines, and the related moving parts. Torsion also occurs in many other systems, including airplanes, bridges, and nearly all other technological systems.

The five types of forces—compression, tension, bending, shear, and torsion—typically occur in varying combinations with one another in structures. For example, a chair or a table leg may be primarily under compression but must also resist bending and torsion forces. When you are designing structural elements, it is necessary for you to determine the type of force and the amount of load at each point in the structure. Different types of structural components must be selected to maintain structural stability or equilibrium.

Figure 11-27: (a) Early beams using natural materials; and (b) modern engineered beams.

STRUCTURAL COMPONENTS

The job of all structural components is to support a load, maintain structural stability, and not fail. The load on the structure, including all internal and external forces, is attempting to make the structure lose stability and ultimately fail. Designers and engineers must first calculate the strength and type of force (compression, and so on) that will be present on structural components. Based on this information, they may select an appropriate structural element or, if necessary, design a new structural element if a standard element is not available.

Beams

Horizontal structural components are called beams. **Beams** are primarily designed to resist bending forces but must also resist shear and torsion forces. Early beams were large, solid materials, such as large stones or timber logs, as shown in Figure 11-27a. Through innovation, new beam designs have incorporated less material and reduced weight while at the same time making the beam stronger. These new engineered beams have improved the beam weight-to-strength ratios. By using less material, natural resources are conserved and costs are reduced. New materials may be developed to overcome other problems, such as a reduction in material availability, material weight or strength, reduction in the impact of moisture or temperature, cost, environmental concerns, or a desire to change the aesthetics created by the material (see Figure 11-27b). In bridges, beams are used to make trusses and girders. In construction, beams are incorporated into all residential and commercial construction. Beams are used in nearly all structures.

BRIDGE STRUCTURES

Bridges meet important societal needs by spanning difficult terrain and allowing for the design of improved transportation systems. Although people and animals can travel up and down relatively steep hills, most cars, trucks, and trains have limited capability. When the great Union and Pacific Railroad was built between Nebraska and California after the Civil War, it needed to pass through the steep Sierra Nevada

Point of Interest

The Brooklyn Bridge is a treasured landmark and elegant structure. It is one of the oldest suspension bridges in the United States spanning 1,825 m (5,989 feet) across the East River, connecting the important New York financial district with the very desirable living environment in Brooklyn. The bridge was designed by an engineering firm owned by John Augustus Roebling from Trenton, New Jersey. The primary building materials for the structure are limestone, granite, Rosendale natural cement, and "wire rope." The architecture style is Gothic, with characteristic pointed arches above the passageways on each tower.

As construction began in 1870, Augustus was injured in a sudden accident and died. His son, Washington Roebling, succeeded him but was soon stricken with *caisson disease*. Although not understood at the time, the illness afflicted men working in the caisson. The caisson was a large wooden structure on which the towers for the bridge were being built in the river. As the heavy stones of the tower forced the caisson into the riverbed, men working in the caisson removed the mucky mud to allow the caisson to sink deeper into the riverbed and ultimately to the bedrock. Today, caisson's disease is understood to be a decompression sickness commonly called "the bends." With Augustus incapacitated, his wife Emily Warren Roebling took over the management for the completion of the bridge. To complete the construction, she learned engineering so that she could communicate with on-site managers. Emily Roebling was the first person to cross the bridge when it opened on May 24, 1883. The bridge cost $15 million to build and 27 workers lost their lives during its construction. Today, anyone visiting New York is encouraged to take a stroll across the bridge's popular walkway.

(a)

(b)

Figure 11-28: The Brooklyn Bridge.

Figure 11-29: (a) The Brooklyn Bridge (circa 1890), and (b) tower design.

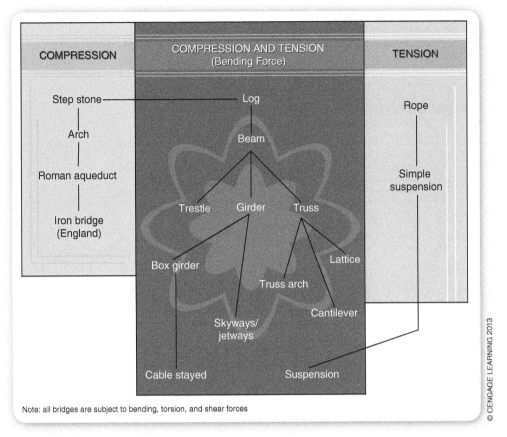

Figure 11-30: Bridge taxonomic chart by principle forces employed.

Mountains. To meet design specifications, all bridges and tunnels were designed and built to accommodate a maximum of a three-degree angle change in elevation. Now that was a tremendous surveying and engineering achievement for the 1860s!

The major difference between bridge types is their ability to span distances. The distance to be spanned depends on the distance between two supporting points. In bridge design, these points can be the opposing walls created by a canyon or other natural structure, or a column or tower built between the area to be spanned. Architects and engineers usually work with some combination of beam bridge, truss bridge, arch bridge, or suspension bridge design (see Figure 11-30).

Figure 11-31: A beam bridge.

Beam Bridges

The **beam bridge** is typically not used for distances over 60 meters (200 feet). Beam bridges are the oldest type of bridge. Someone laying a log across a small stream has created a type of beam bridge. Many of the nearly 600,000 bridges in the United States are beam designs (see Figure 11-31).

A beam bridge must resist a bending force. Any load consisting of people, cars, trucks, or a train, creates a bending force over the entire distance of the beam (see Figure 11-32). Beam bridge designs are placed on supports called piers, which must support the shear force at the pier of the dead load consisting of the weight

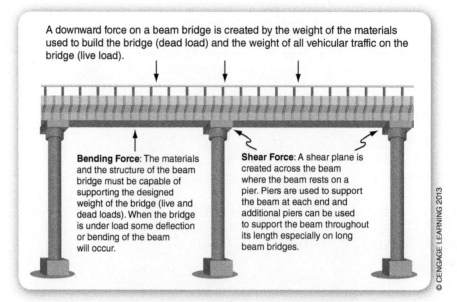

Figure 11-32: Bending and shear forces on a beam bridge design.

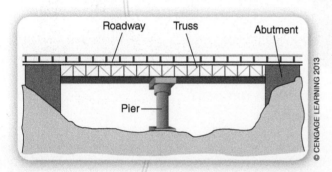

Figure 11-33: Basic components of a truss bridge.

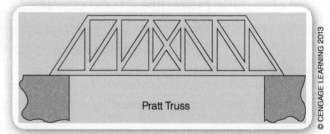

Figure 11-34: Pratt truss.

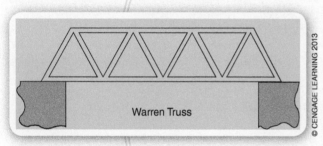

Figure 11-35: Warren truss.

of the bridge and the live load created by objects using the bridge. Engineers must use existing beam designs or create new ones capable of carrying the design load while remaining stable. Beam designs are used not only for roadway and railroad bridges but are commonly used for skyways for people crossing over busy streets or between buildings.

Truss Bridges

A **truss bridge** must resist compression and tension forces and is a more complex variation on the basic beam bridge. Although components in a beam bridge are solid structural members, trusses are made of a variety of components. These components are designed to use less material while maintaining truss stability, as shown in Figure 11-33.

The most common type of truss design is the Pratt truss. Named for its original designers Thomas and Caleb Pratt in 1844, the Pratt truss was originally made of wood. Today, many variations of Pratt trusses exist and are characterized by a diagonal member creating triangular shapes between vertical components (see Figure 11-34). Many types of Pratt trusses were developed by engineers of the Pennsylvania Railroad in the 1870s.

A Warren truss, patented by British engineers James Warren and Willoughby Monzoni in 1848, can be identified by the equilateral or isosceles triangles formed by the structural members connecting the top and bottom chords (see Figure 11-35).

A Howe truss has diagonal members creating triangular shapes between vertical components like the Pratt truss. Patented by William Howe in 1840, the Howe truss causes tension and compression forces to be opposite those of the Pratt truss (see Figure 11-36). The Howe truss was used in early railroad bridges and was commonly made of wood. Because the vertical elements were under tension force, thinner iron tension rods could be connected to the thicker wooden beams that were under compression.

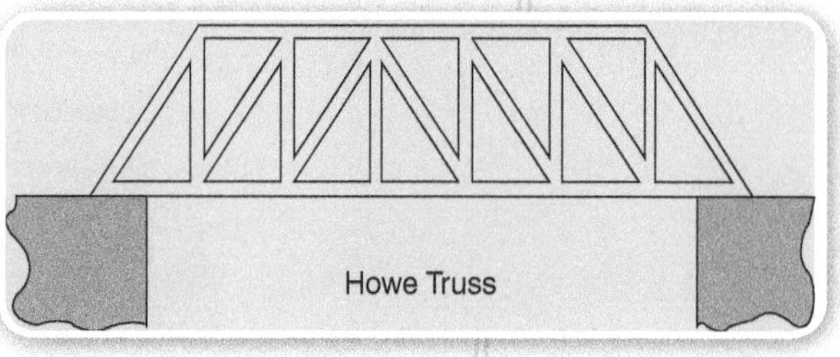

Figure 11-36: **Howe truss.**

Truss designs often employ a cantilever structural member. A **cantilever** is a structural member that projects beyond its support and is supported at only one end. Cantilevers can be seen in overhanging house decks and porches. Many building designs incorporate cantilevers, and most roofs are cantilevered well beyond the sides of the structure.

In bridges, the cantilever truss design allows for a greater span to be achieved. In the case of the Firth of Forth Bridge between South and North Queensferry, Scotland, the cantilever style bridge has a 350-foot center span, as shown in Figure 11-37.

Arch Bridges

The **arch bridge** has its design roots in Roman engineering. Roman arches were typically made of stone

Figure 11-37: **Firth of Forth cantilever truss railroad bridge.**

and were used for buildings, roadway bridges, and aqueducts to carry water from mountain reservoirs to the cities. Original Roman arches still can be seen in many parts of Europe and the Middle East (see Figure 11-38). Arch structures are very strong and transfer the weight of the load as a compressive force through the arch. The base or abutment prevents the ends of the arch from moving apart. Modern arch bridges can span a distance up to 300 meters (1,000 feet).

Suspension Bridges

Modern **suspension bridge** designs are capable of spanning the greatest distance of all bridge designs. The Akashi-Kaikyo Bridge completed in 1998 has a main center span of 1,991 meters (6,529 feet), as shown in Figure 11-39.

(a)

(b)
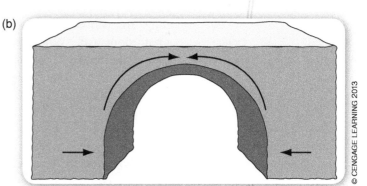

Figure 11-38: **Ancient Roman aqueduct.**

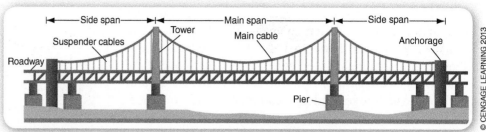

Figure 11-40: Structural elements of a typical suspension bridge.

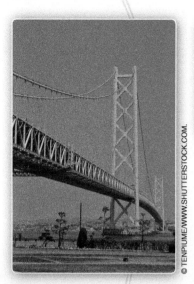

Figure 11-39: The Akashi-Kaikyo Bridge in Japan is the world's longest bridge.

Suspension bridges transfer dead and live loads through massive steel cables like those used by Roebling in the Brooklyn Bridge. The cables, which are under tension, are hung over towers and secured to both ends of the bridge by anchorages. The towers must support most of the load and are under compression from the load transferred by the cables. The bridge surface, hung on the cables, is usually a truss structure to give stability to the roadway (see Figure 11-40).

Structural components must resist all forces. If a component buckles (bends) or snaps (breaks), the bridge will fail. Bridge designs either dissipate a force by spreading it out to all components or transfer the force away from a lesser element that would be unable to carry the full load. Arch designs dissipate the load while suspension designs transfer the load. To resist compression forces, structural components called struts are used. Struts are used to make bridge piers and columns in construction. Ties are structural components that must resist tension. Ties can be steel cables on a suspension- or cable-stayed bridge, or they may be more rigid steel elements in truss bridges. All components within a structure must be fastened together with joints that are either mechanical such as rivets or bolts or chemical such as welds. Fasteners are subject to shear forces.

CALCULATING LOADS ON STRUCTURES

To be able to create an effective design, designers must calculate the load and type of force acting on the structure. With this knowledge, the proper material and type of component (beam, and so on) can be selected. If unexpected forces are applied to the structure, or if incorrect components or materials are used, it may fail.

Simple Testing of Loads

Basic load measurements can be made through simple empirical testing and modeling. You can determine typical dynamic and static loads by placing a chair on a bathroom scale. When you sit down, does the scale briefly measure more than your actual weight? (See Figure 11-41.) You can also determine the force applied on a hanging lamp or swing seat using a spring scale (see Figure 11-42). You can also use spring scales to predict the force of a 25-mph wind on a tent rope support. A simple cardboard and string or rubber band model can be used to determine which elements in a bridge or a truss are under compression and tension (see Figure 11-43). In this test, components believed to be under compression are made of cardboard, while components believed to be under tension are made with string. If a string component collapses under load, the component is really under compression and must be replaced with cardboard. All elements that are not obviously

Chapter 11: Designing Structural Systems 311

Figure 11-41: Simple testing of loads using a bathroom scale.

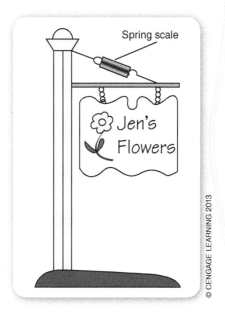

Figure 11-42: Simple testing of loads using a spring scale.

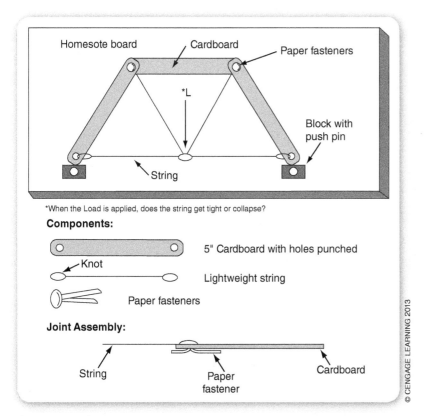

Figure 11-43: Determining compression and tension forces by modeling with string and cardboard.

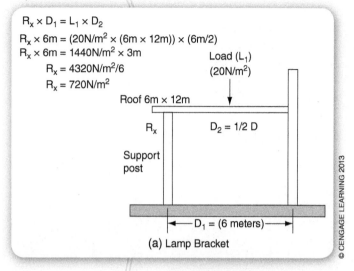

Figure 11-44: Calculation of a UDL load.

under compression should be tested with string. Even though they can be very useful, it is not always appropriate to use these types of physical modeling techniques. When appropriate, you should use mathematical or graphical models.

Using Mathematics to Calculate Loads

Predicting loads in structures usually involves mathematical or computer modeling. A simple **moment** calculation demonstrates this point. Calculation of a moment can determine the effect of a force applied to a door handle or a box wrench. Moment calculations are important measurements when a mechanism such as a lever handle for a tool is being designed. The force of the turning moment continues until equilibrium is achieved by tightening the nut. The force acting on a structural component must be counteracted by a reaction force if the structure is to remain in equilibrium. The load in this example is called a point load. Many structures, such as a roof, have loads distributed over an entire surface. When a calculation is to include the weight of the material making up the structure, as well as loads that may be distributed over the entire surface, a **uniform distributed load (UDL)** measure is used (see Figure 11-44).

You can determine the amount and type of load applied to a wall bracket structure by using vector analysis (incorporating algebraic and trigonometric functions). As evident, even the mathematical method involves some graphical analysis (see Figure 11-45).

Figure 11-45: Determining force vectors using trigonometric functions.

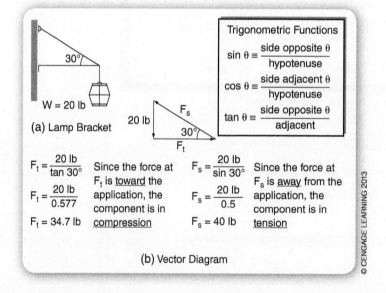

Your Turn

Additional questions about this structure can be answered using UDL measures. What is the shear force at the wall? What is the maximum expected load, assuming a 30 cm (12-inch) wet snow load? What is the load if a new 25 pound shingle roof is put over the existing roof?

Chapter 11: Designing Structural Systems

In physics and in calculus, a vector is a concept used to describe both magnitude and direction of a force. Vectors function in Euclidean space and are drawn from an initial point to a terminal point. Magnitude is represented by the length of an arrow and direction of the force by the direction of the arrow. As such, the arrow carries the essential information. In fact, *vector* in Latin means "one who carries."

Using Graphical Analysis

Vector polygons form the basis for answering questions using graphical analysis. **Vectors** are lines that describe both the direction and magnitude of a force being applied to the system. Polygons are figures with more than four angles. A bicycle frame will serve as the illustration for this graphical analysis (see Figure 11-46). Remember that, to solve a technological problem, it is necessary to know the magnitude of the force on the component within the system, as well as the type (compression) of force it is.

Figure 11-46: Analyzing the forces on a bicycle.

Bow's Notation

Bow's notation is a method of identifying forces within a structure. It is a graphical method in that accurate drawings of the structure must be made to scale, as shown in Figure 11-47a.

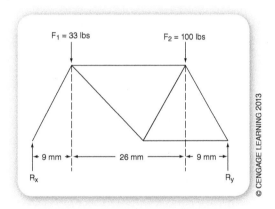

Figure 11-47: (a) Determining R_x and R_y.

>>> STEP 1

Calculate the moment reaction of F_1 and F_2 at R_x and R_y. The distance between R_x and R_y = 44 mm. The distance between R_x and F_1 = 9 mm, and the distance between F_2 and R_y = 9 mm.

$$(R_x)(44 \text{ mm}) = (9 \text{ mm})(100 \text{ lb}) + (35 \text{ mm})(33 \text{ lb})$$
$$(R_x)(44 \text{ mm}) = 2{,}055 \text{ mm-lb}$$
$$R_x = 2{,}055 \text{ mm-lb}/44 \text{ mm}$$
$$R_x = 46.7 \text{ lbs.}$$
$$(R_y)(44 \text{ mm}) = (9 \text{ mm})(33 \text{ lb}) + (35 \text{ mm})(100 \text{ lb})$$
$$(R_y)(44) = 3{,}797 \text{ mm-lb}$$
$$R_y = 3{,}797 \text{ mm-lb}/44 \text{ mm}$$
$$R_y = 86.3 \text{ lb}$$

STEP 2

Label the scale drawing with Bow's notation, as shown in Figure 11-47b. Begin in the lower left corner and work around the outside of the structure in a clockwise direction. Letter each external and internal space. Label all known forces and number each joint. Note that each line can be identified by two letters (for example, AE describes the bicycle front fork).

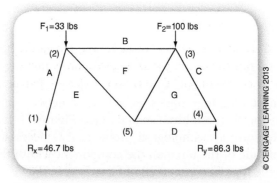

Figure 11-47: (b) Applying Bow's notation.

STEP 3

The actual calculations begin by making a vector diagram drawing of all forces. The drawing must be made to a chosen scale (we have chosen 0.5 mm = 1 lb), as shown in Figure 11-47c. Note that some forces are directed downward (the weight of the rider), while other forces are directed upward (the reaction of the ground through the tires). The up and down forces must be in equilibrium ($F_1 + F_2 = R_x + R_y$).

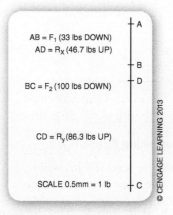 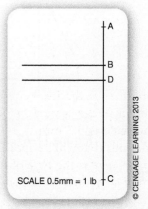

Figure 11-47: (c) Vector diagram (known vertical forces).

Figure 11-47: (d) Vector diagram (unknown horizontal forces).

STEP 4

Add all horizontal vectors to the vertical forces shown on the vector diagram. In this example, lines BF and DG are horizontal vectors with unknown force, so the actual length of the line cannot be determined, as shown in Figure 11-47d. The basic vector diagram will be used to measure the magnitude and type of force at each joint.

STEP 5

Draw to scale a free body diagram of the joint in question. Start in the lower left corner with joint #1. The free body diagram will be used to determine the direction of the forces (compression or tension), Figure 11-47e. Notice that we are only concerned with force R_x and line AE. If there were a connecting component between joints #1 and #5, then line ED would also be considered.

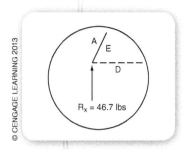

Figure 11-47: (e) Free body diagram for joint #1.

 STEP 6

Draw on the vector diagram, as shown in Figure 11-47c, a vector representing line AE. Begin at point A and draw the line in the same direction or angle as line AE. Extend the line until it intersects the horizontal line at point D, as shown in Figure 11-47f.

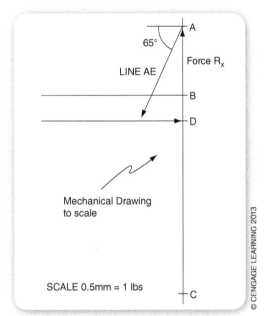

Figure 11-47: (f) Vector of line AE drawn on vector diagram.

Interpreting the Graphical Analysis

The length of line AE on the vector diagram is equal to the force on that component in the scale drawing. Because line AE is 26.5 mm long and the scale is 0.5 mm = 1 pound, then line AE represents a force of 53 pounds that will be applied to the front fork of the bicycle. Of equal importance is the type of force that the load (53 lb) will be exerting on the structure. Note that the arrows have been drawn on the vector diagram in the direction of the force being applied to the system. Because the analysis is of joint #1, it is necessary to follow the force at R_x through the components in the free body diagram. Force R_x is being applied up the vertical line of the vector diagram. At point A, the force follows down line AE to point E, where it turns back toward point D. It is important that the direction of the force identified in the vector diagram be transferred to the free body diagram, as shown in Figure 11-47g. If the arrow (direction of the force) is pointing toward the application point (joint #1 in this example), then the component is under compression. If, on the other hand, the arrow is pointing away from the application point, then the component is under tension.

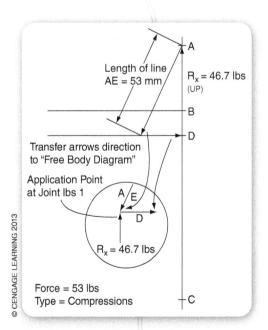

Figure 11-47: (g) Determining force direction and magnitude.

STEP 7

Repeat Steps 5 and 6 for each of the remaining joints:

Joint #2, lines EF and BF, Figure 11-47h;

Joint #3, lines FG and CG, Figure 11-47i; and

Joint #4, line DG, Figure 11-47j.

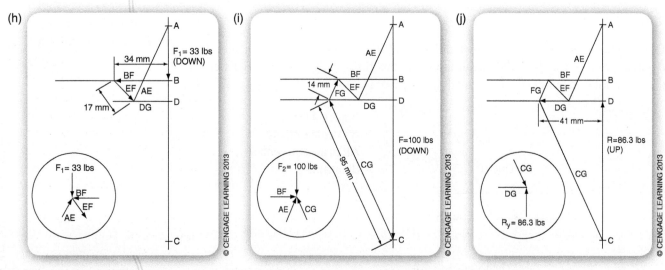

Figure 11-47: (h), (i), and (j) Combined diagrams showing force magnitude and direction for joints #2, #3, and #4.

STEP 8

Make a chart of all components showing magnitude and type of force (see Figure 11-48).

Figure 11-48: Graph of bicycle components showing force magnitude and type of force.

Component	Force	Type
AE	53 lb	Compression
BF	34 lb	Compression
EF	17 lb	Tension
FG	14 lb	Compression
CG	95 lb	Compression
DG	41 lb	Tension

Your Turn

If we consider the bicycle structure and observe that no component ties joint #1 with joint #5, what conclusion can you make about the forces at joint #2? You can form other new questions by asking, "What magnitude and type of forces applied to the bicycle structure can I consider?"

SUMMARY

Early humans built structures to provide shelter and protection and made them out of materials indigenous to the region. Today we use many types of materials to create major structural systems. Although natural materials are still appropriate, engineers increasingly use synthetic or engineered materials for their improved strength, reduced weight, resistance to natural forces, and reduced cost. All materials are available to designers or engineers for design application in residential and commercial buildings, bridges, dams, and tunnels, among others. Most major structures are designed by civil engineers, structural engineers, or architects.

A structure is a body that supports a load and resists external forces without changing its shape, except for that due to the elasticity of the material(s) used in construction. In a properly designed structure, failure can only occur if a joint or the material itself fails. One obvious way to look at structures is by the purpose they serve. Structures serve many purposes, including housing, containment, communication, and transportation.

In the design of a new structure, engineers and architects must use scientific and mathematical principles and concepts. Designers must also meet a given set of design specifications, consider all live and dead loads to be supported by the structure, and satisfy the client's intended purpose for the structure. Safety considerations include not only standard safety factors, such as the maximum loading for a bridge or building, but also considerations to promote safe use by humans. Structures must be designed for ease of servicing.

We may assume that all well-designed technological structures will fail at some point due to natural forces over time, misuse, or possible terrorist attack. All building, bridges, dams, tunnels, towers, and other technological structures have a designed safe and useful life. Planned obsolescence is the conscious decision on the part of the designer to produce a product that will become obsolete in a defined time frame. A safety factor determines how much a product, or an element within a product, is overbuilt. Safety factors are usually greater than 1 (S.F. > 1.0). Newton's laws of motion form the basics for structural design. These principles consist of three physical laws that provide relationships between the forces acting on a body and motion of the body.

When an external load is applied to a structure, internal forces are necessary to counteract each load force. Internal forces are generated by the molecular structure of the material in question to counter external forces. There are five forces that are commonly considered when designing structures: (1) compression, (2) tension, (3) bending, (4) shearing, and (5) torsion. External forces are loads that are applied to a structure, while internal forces resist changes in the structural shape. These forces can be static (at rest) or dynamic (in motion). The concepts of stress (strength), strain (change in shape), and elasticity (change in shape without permanent deformation) describe the physical changes in the material under load. The five different ways in which forces or loads are exerted on a structure help determine how the structural element is to be designed or engineered.

To create an effective design, designers must calculate the load and type of force acting on the structure. Loads can be calculated using simple empirical testing and modeling, mathematical formulas, or graphical analysis.

BRING IT HOME

OBSERVATION/ANALYSIS/SYNTHESIS

1. Make a simple one-view sketch of a structure in your community that uses triangular shapes to create inherent stability.
2. Make a simple one-view sketch of at least five well-designed structures in your community, including a residential home, commercial building, bridge, dam, and tunnel, if possible. If any historical site is available, it may be appropriate to consider as a subject.
3. Look up information about the history and/or designer for one of the following structures: Brooklyn Bridge, Eiffel Tower, Ferris Wheel, Petrona Twin Towers, Akashi-Kaikyo Bridge, Hoover Dam, Taipei 101 Tower, Palm Islands (Dubai, United Arab Emirates), Millau Viaduct, Deltawerken (Netherlands), Eurotunnel, or the Millennium Dome. Prepare a 10-slide PowerPoint presentation.
4. Select a bridge type from this chapter or choose a bridge located in your community. Do a simple paper analysis by using cardboard for all components that you believe will be in compression and string for all components that you believe will be in tension. Tape all joints. Apply a load, and evaluate your model.
5. Design a beam using one ounce of wooden stirring sticks. The beam should span a 30-cm open space. The beam must be designed to resist bending, shear, and torsion forces while being supported only at the two ends. Load the beam until there is a 1.0 mm deflection, and record the weight.
6. Participate in the West Point Bridge Design Contest. Go to http://bridgecontest.usma.edu/tutorial.htm to download the program.

Safety Note
Be sure to get permission from your teacher and a parent before selecting an object in the community to sketch.

EXTRA MILE

ENGINEERING DESIGN ANALYSIS CHALLENGE

- You are working as an apprentice to a licensed structural engineer and a client has hired the engineer to sign off on a set of drawings for a bathroom addition to be placed over an existing single-story family room (see Figure 11-49). Two sides of the addition are on load-bearing walls so no additional support will be needed there. Because the southeast corner of the addition will not be able to be placed on an existing load-bearing wall, a new beam must be added to the east side of the addition and a post (pier) must be placed under the beam on the southeast corner.

- The engineer has asked you to review the drawing and recommend an appropriate beam to carry the load of the new bathroom and the roof from over the existing family room. Using appropriate formulas, calculate the live and dead loads for both the shed roof and bathroom addition. Because the house is located in the Northeast U.S., consider a possible six-foot snowdrift on the roof. Once the UDL loading is determined for the roof and bathroom, use published tables for manufactured floor beams and select the appropriate beam size. Information about Weyerhaeuser's TrusJoist Parallam beams can be found at http://www.ilevel.com/literature/TJ-9000.pdf. The engineer will want to see all your calculations so that your work can be checked. If you are correct, the engineer will fill in the recommended beam and sign off on the project.

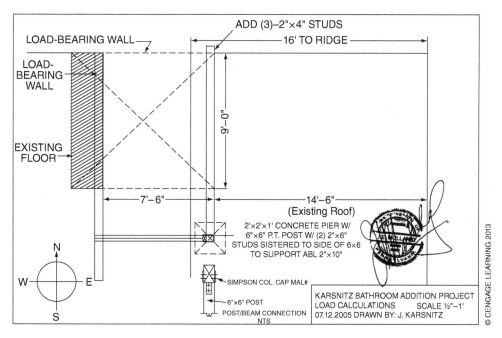

Figure 11-49: **Drawing of bathroom addition.**

CHAPTER 12
Designing Mechanical Systems

Menu

 Before You Begin
Think about these questions as you study the concepts in this chapter:

1. Why do mechanical engineers and other design professionals design and build mechanisms?
2. What are mechanisms and how are they different than other human-designed objects?
3. What STEM principles must be understood to design a mechanism?
4. How are desired output motions created with mechanisms?
5. How are levers and linkages used to design mechanisms?
6. How are rotary motions created with mechanisms?
7. How can mechanical designs be modeled?

INTRODUCTION

Mechanisms and mechanical systems can be found in many products and devices. The Greatest Engineering Achievements of the 20th Century website identifies the automobile, airplane, agricultural mechanization, air conditioning and refrigeration, spacecraft, household appliances, and health technologies among those achievements—none of which would be possible without an understanding of modern principles of mechanics (see Figure 12-1). Mechanisms can help extend human capability by creating a desired output of motion or force. Consider how pliers can help a technician grip a nut or assist a surgeon who needs to hold or clamp an artery. This chapter introduces mechanical concepts and the scientific and mathematical principles (STEM) used to describe how mechanical systems function.

Figure 12-1: Examples of modern products that use mechanical systems.

Throughout history, people have used mechanisms to help solve technological problems. Leonardo da Vinci (1452–1519), the Italian Renaissance painter of the famous *Mona Lisa* painting and many other important artworks, was also a designer and inventor of many mechanisms. Over 4,200 sketches still exist today of mechanisms he designed (see Figure 12-2). Most of da Vinci's designs could not be made, because the materials and production practices necessary to execute the designs were not developed until many years after his death. Some of his ideas included an experimental flying machine. Four hundred years later, the Wright brothers made the first successful flight in 1903. His idea for a helicopter was not able to be produced until the 1940s. Da Vinci's designs for a movable bridge, parachute, and a construction crane were done without a full understanding of the underlying scientific principles; rather, he was an excellent observer of the natural world.

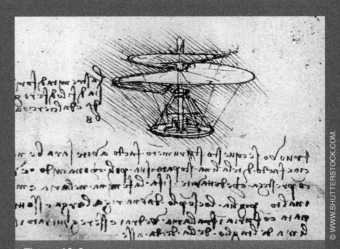

Figure 12-2: Leonardo da Vinci's rotating airscrew design is the basis of the modern helicopter.

Fun Facts
The helicopter

Sharing aviation history with Wilbur and Orville Wright, Igor Sikorsky is best known as the inventor of the helicopter. Born in Russia, Sikorsky (1889–1972) was a scientist and engineer who immigrated to the United States in 1891. His work in fixed and rotary wing aircraft led to many inventions including the VS-300, the first practical helicopter, as shown in Figure 12-3.

He came to America from Russia to build aircraft. He taught mathematics and aviation to earn a living. Encouraged by friends to start an aeronautical enterprise, he opened the Sikorsky Aero Engineering Corporation in 1923. Sikorsky's company has a proud and rich history in aircraft design and construction and is considered one of America's pioneering aerospace companies.

Sikorsky had been working on helicopters since 1909, before he finally succeeded in building a working aircraft. That is 30 years of failures before a success! He continued to work on improving the VS-300 until he achieved the first working production helicopter, the VS-316 (R-4) in 1943. The R-4 saw its first military application in the Pacific during WWII. Because much of the Pacific front involved "island hopping," the R-4 was ideally suited for delivering supplies, especially for parts to service critical military machinery including island-based aircraft.

The helicopter creates lift with the rotary wing or rotor. Controlling the flight of the aircraft is not easy, especially landing where the pilot must navigate the slipstream of self-induced vortices (vortex rings). The tail rotor is critical to the control of the aircraft. Forward flight is created by tilting the rotor to create a forward thrust.

The first commercial helicopter to enter service was the Bell 47 model in 1946.

Figure 12-3: Piloted by Sikorsky, the first successful flight of the VS-300 was made on May 13, 1940.

OFF-ROAD EXPLORATION
Information about Igor Sikorsky's life work as an aviation pioneer:
http://www.sikorskyarchives.com

OFF-ROAD EXPLORATION
Information about how helicopters work:
http://www.helicopterpage.com/

Chapter 12: Designing Mechanical Systems

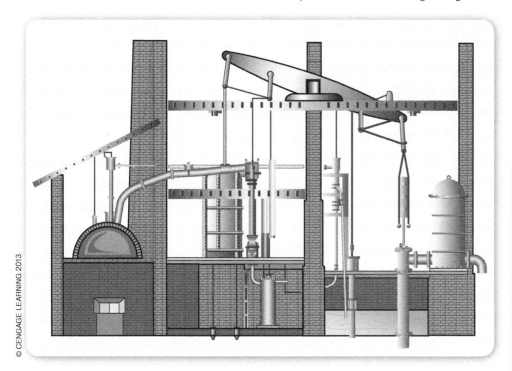

Figure 12-4: James Watt's steam engine circa 1769.

machine:
A device that is capable of transforming energy to accomplish a task. Machines can be mechanical such as a power tool or an automobile, or electrical such as a computer. The simple machines studied in science including the lever, wheel and axle, pulley, inclined plane, wedge, and screw, only transform the direction or magnitude of a force. For mechanical engineering design purposes, students will need to understand the role of the lever and crank, wheel and gear, cam, screw, things that transmit tension or compression, and things that provide intermittent motion to possible design solutions.

Mechanisms played an important role in shaping society. In 1769, James Watt (1736–1819) patented the principle of the separate condenser for a steam engine. A Scottish engineer, Watt is best known for his improvement of the steam engine (see Figure 12-4). The steam engine is one of the primary forces behind the Industrial Revolution in Europe during the eighteenth century and in the United States during the early nineteenth century. Other innovations by James Watt include crank movements (so that the steam engine could turn wheels), a throttle valve, a governor, and many other innovations to make the steam engine more useful.

MECHANISMS AND MACHINES

Mechanisms and machines are similar in that they both are a combination of moving parts that are capable of defined movement. Machines are different from mechanisms in that they are also capable of transforming energy to do work. We usually think of a machine as being more complex than a tool and more independent of the human user. The Watt steam engine shown in Figure 12-4 was a **machine** (transformed energy) connected to a **mechanism** (capable of designed movement). Together, the machine and the attached mechanisms consisted of a series of parts that were capable of doing work. Early steam engines were connected to pumps to remove water from mines or were placed on wheeled platforms to serve as train locomotives. A dentist drill, lawn mower, washing machine, electric hand drill, food processor, and printing press are all considered machines. Some machines, such as the steam engine, diesel or gasoline engine, and hydroelectric turbine, are also considered "prime movers" because they are intended to convert energy directly into mechanical motion.

mechanism:
An assembly of moving parts involved in or responsible for taking an input motion or force and transforming it to an output motion or force. Mechanisms are governed by physical principles and laws.

KINEMATICS

kinematics:
A branch of engineering mechanics that studies motion without regard to the force or mass of the things being moved.

In the late nineteenth century, a German mechanical engineer by the name of Franz Reuleaux (1829–1905) developed the fundamental concepts of modern mechanisms known as kinematics. Extremely famous in his time and one of the first honorary members of the American Society of Mechanical Engineers (ASME), Reuleaux is considered the "father of kinematics." Kinematics is a branch of solid mechanics that deals with relative motion of elements or pairs in machine and mechanical systems. Kinematics and kinetics (the action of forces on bodies) together form the basic elements of engineering dynamics and machine design theory (see Figure 12-5).

Figure 12-5: Reuleaux's model for a planetary gear chain. You can view still and moving pictures of Rouleaux's models in Cornell University's collection at http://kmoddl.library.cornell.edu.

OFF-ROAD EXPLORATION
Digital models of six collections of mechanisms, including models by Franz Reuleaux: http://kmoddl.library.cornell.edu/

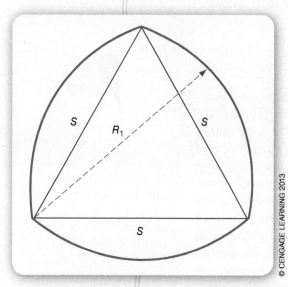

Figure 12-6: Reuleaux's triangle: Each side of the triangle is replaced by an arc with a radius equal to the original side.

Reuleaux's six mechanical elements of machine design (see Figure 12-7) and symbolic notation are important to modern mechanical engineering design and mechanical design theory. In addition to understanding these mechanical elements, mechanical engineering students study machine dynamics and vibrations, methods of multibody dynamics, and analysis of differential equations and finite element methods in developing necessary design skills. One of his best-known inventions is the "Reuleaux triangle," a type of equilateral triangle in which each side is an arc whose radius is equal to the length of one side of the inscribed equilateral triangle (see Figure 12-6). The mechanism forms the basis of the modern Wankel rotary internal combustion engine. In addition to his contributions to kinematics, his ideas about design synthesis, optimization, and aesthetics in design are part of engineering education. Also important in modern machine design are dynamics, stress analysis, and materials science.

Reuleaux's Six Mechanical Elements

Examples of how Reuleaux's six mechanical elements are used in today's engineered products are shown in Figure 12-7:

- ▶ Levers and cranks (Figure 12-7a)
- ▶ Wheels and gears (Figure 12-7b)
- ▶ Cams (Figure 12-7c)
- ▶ Screws (Figure 12-7d)
- ▶ Things to transmit tension or compression (Figure 12-7e)
- ▶ Things that transmit intermittent motion (Figure 12-7f)

(a) Lever and crank used to open a window.

(b) Wheels and gears used to make a transformer.

(c) Cams open and close valves in automobile engines.

(d) Screw acts as a worm gear in a LEGO transmission.

(e) Things that transmit tension or compression, such as belts or chains used to drive the wheel of a motorcycle.

(f) Things that transmit intermittent motion, such as the ratchet used in a clock mechanism.

Figure 12-7: Reuleaux's six mechanical elements used to design mechanical systems.

326 Part II: Resources for Engineering Design

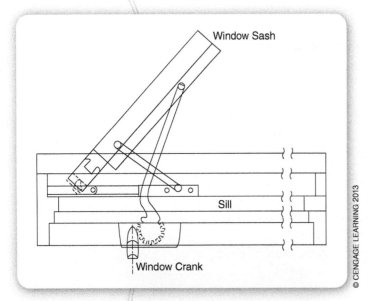

Figure 12-8: The operation of the mechanism is not effectively shown in the mechanical drawing of a casement window.

OFF-ROAD EXPLORATION
Source of paper animation kits of mechanisms: http://www.robives.com

Kinematics can give us information about the movement within mechanical systems and provides us with an organized way to study, record, and design these systems. For example, in the casement window in Figure 12-8, the designer of the opening mechanism used a kinematics diagram to determine if the window could be fully opened and closed by the rotary motion of the handle.

Kinematic Diagrams

As your ideas begin to take shape, it is important that you be able to estimate how the mechanism will function. Because of the complex nature of a mechanical drawing, such as the one shown in Figure 12-8 for a casement window, it is difficult to visualize and analyze the critical movements of the mechanism.

To solve this problem, designers and engineers use kinematics or skeleton diagrams (see Figure 12-9). These simple diagrams show only the essential elements of the mechanism.

Kinematic diagrams are referenced, beginning with all fixed or ground links (see Figure 12-9, #1). All parts or elements of the mechanism or machine must be referenced. Two elements in relative motion and in contact with each other are known as **pairs**. Pairs in Figure 12-9 have been referenced by letters. Elements that join pairs together are known as **links**. For example, pair AC is a link connected to the crank (A) and the slider (C). A group of elements including pairs and links that are joined together are called a **kinematic chain** and when one link of the kinematic chain is fixed, the system is considered a mechanism. The casement window has a fixed element at the crank (1) with a kinematic chain formed by the slider linkages (AC) and (DE) and the pivoting link (BD). "Input" motion and "output" motion are also noted.

Kinematic diagrams allow preliminary visualization. Figure 12-10 shows some common kinematic symbols. When a more detailed analysis of the system is needed, a kinematic diagram is made to scale. From this scale diagram, an analysis of pivot locations, fixed angles, link length(s), motion(s), displacement, force, and other factors can be calculated.

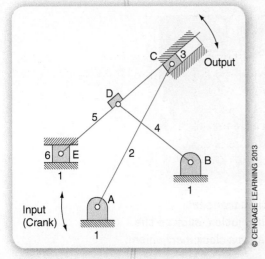

Figure 12-9: A kinematic or skeleton diagram of a casement window opening mechanism.

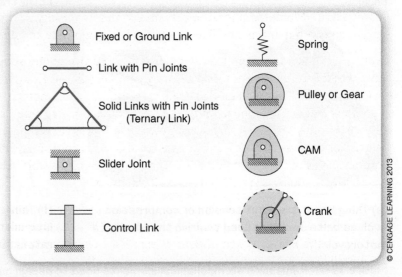

Figure 12-10: Common kinematic diagram symbols.

Kinematic Members

All mechanical elements such as levers, cranks, wheels, gears, cams, screws, belts, chains, and ratchets can be represented as kinematic members. As noted in Figure 12-9, kinematic elements forming pairs (elements with relative motion) and links (connecting pairs) become kinematic chains and must be referenced in the drawing. Links may be solid elements such as those used in the casement window or nonrigid members, such as cable or chain. Nonrigid elements must be used in tension. The casement window and all kinematic chains with a fixed element are considered mechanisms, as shown in Figure 12-11(a). As previously stated, a mechanism is a device that takes an input motion or force and creates a desired output motion or force. If the combination of elements, after being fixed, is not capable of movement, it is no longer considered to be a mechanism, but is more correctly called a structure, as shown in Figure 12-11(b).

When the machine or mechanism rotates, the effect of the force is equal to the product of the force (strength) and the distance perpendicular to the line or axis of the rotation. In Figure 12-13, the force applied to a wrench (a lever) creates torque around the lever's fulcrum. The torque or moment force is determined by the distance (D) the force is applied from the fulcrum × the amount of force (F) and is expressed in newton or pound force. Torque is a vector that points along the axis of the rotation and is an arc measured from the fulcrum point to the point of force on the lever. An example might be helpful.

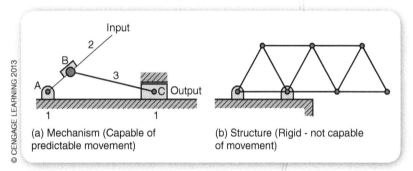

Figure 12-11: Kinematic chain showing the difference between a mechanism (a) and a structural system (b).

Suppose the wrench in Figure 12-13 is 20 cm (0.2 m) long and the force applied by the user is 20 newtons. The applied torque is then (0.2 m)(20 N) = 4 N-m.

Motion. Motion describes a change in position of an object that can occur in a straight line or some other shape in space. When considering a change in position of an object over time, velocity and time must also be measured frequently enough to achieve the desired accuracy. Velocity is the measurement of an object's displacement per unit of time. Acceleration can also be an important measurement related to motion and is the variation of the velocity of the object being displaced per unit time.

Momentum. Momentum is an object's mass times its velocity. Because velocity is a vector, so is momentum, and like all vectors has magnitude and direction.

Work. Work is a force applied over a distance that overcomes some resistance as noted in Newton's laws. The units of work are (force) (distance), so this is foot-pound in the English system, and newton-meters (N-m) in the SI (metric) system. Work is also units of energy. In the metric system, the units of energy are joules (1 joule = (1 N) × (1 m) = 1 N-m). As in the example, in the Watt steam engine, steam entering a cylinder does work when it expands and moves the piston against the resisting forces such as the drive wheels of a locomotive. People do work when they lift an object where the resisting force is that of gravity.

Power. Power is the rate at which work is done and has the units of energy per unit time. In SI units, power is measured in watts (W) and is equal to one joule per second. In the British system, power is expressed as horsepower, which is a term rarely used except in association with the automobile.

Energy. Energy is the ability to do work and is measured in joules. All objects possess energy that may be in the form of thermal, chemical, electrical, radiant, or nuclear. Mechanical systems have potential energy, kinetic energy, or both. Potential energy is sometimes thought of as stored energy, and as with money, it can be held for later use. Potential energy can create forces capable of acting on an object. Kinetic energy is energy in action and is caused by the motion of a body. A hammer lifted in space has the ability to do work when it is dropped on a nail head, and is an example of potential energy. A fly wheel connected to an automobile engine uses the kinetic energy it receives from the engine to drive the attached transmission more smoothly.

Physical Principles

The action of machines or mechanisms is governed by physical principles described in Newton's three laws of motion. Because mechanisms are responsible for any action or reaction by transmitting forces through a series of pre-determined motions, these components are associated with dynamic forces. A force causes a body (mechanism) to accelerate (move). Forces have vector qualities that cause an object to be lifted, pushed, or pulled at a predetermined magnitude (strength) and direction.

Forces are measured in either newtons (SI system of units) or pounds (English). A newton is the amount of force required to accelerate a mass of one kilogram at a rate of one meter per second per second:

$$1 \text{ N} = (1 \text{ kg})(1 \text{m/s}^2)$$

A vector is used to represent physical quantities that have both magnitude and direction. The magnitude corresponds to the strength of magnitude of the force while the direction of the vector represents the direction of the force. A vector can be represented using a Cartesian coordinate system to determine a starting and ending point and a direction on a surface (x and y axis) or in space (x, y, and z axis). In its simplest form, a vector is a measure based in geometry. Vectors are usually indicated in boldface, and are shown in graphical form using arrows (see Figure 12-12).

In this illustration, point A is the starting or origin, while point B is the ending or destination. The length of the arrow, or vector, represents the magnitude of the force while the direction of the arrow represents the force's direction. A graphical analysis problem using vectors was used in Chapter 11, Designing Structural Systems.

Torque, also known as a *moment*, can be thought of as "rotational or angular force" that causes a change in rotational motion. Represented by the Greek letter tau "τ," torque is defined as the linear force multiplied by a radius where the force is being applied (see Figure 12-13).

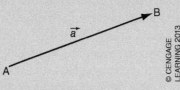

Figure 12-12: **Representing a vector force by indicating magnitude and direction.**

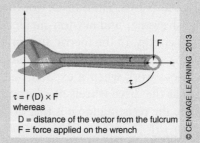

Figure 12-13: **Calculation of torque (moment) applied to the end of an adjustable wrench.**

Your Turn

Many toys incorporate mechanisms to create interesting motion and often serve as an excellent problem-solving situation for students (see Figure 12-14).

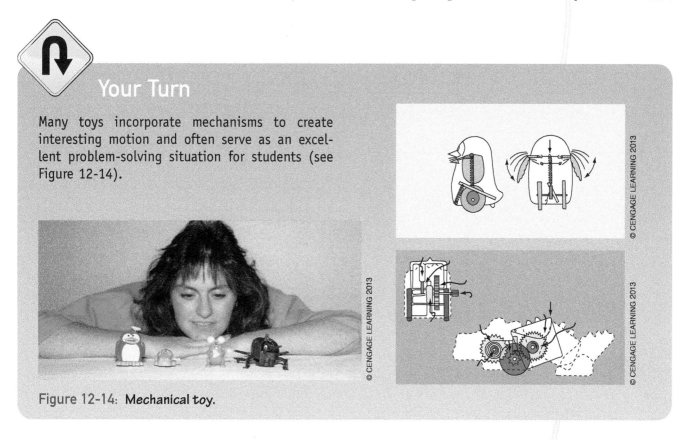

Figure 12-14: **Mechanical toy.**

CREATING MOTION IN A MECHANISM

A mechanism takes an input motion or force and creates a desired output motion or force. A mechanism may consist of a single lever, crank, wheel, gear, cam, screw, belt, or ratchet, or more typically, a combination of these elements. The analog clock is considered a mechanism, because its purpose is to create a desired output motion (movement of its hands). Most mechanisms serve unique purposes, such as clamping, lifting, locating, opening, coupling, making fine adjustments, or folding (see Figure 12-15).

Because mechanisms deal with motion, it is important to understand the different kinds of motion possible. Most mechanisms use a rotary input motion often from an electric motor to create the desired output motion. The four common types of motion used in a mechanism are linear, reciprocal, rotary, and oscillating (see Figure 12-16).

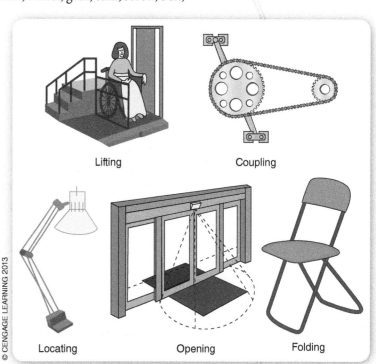

Figure 12-15: **Sample mechanisms.**

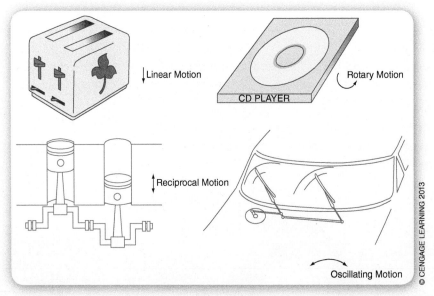

Figure 12-16: **Four common types of motion used in mechanisms.**

OFF-ROAD EXPLORATION

Useful information, tables, schedules, and formula related to mechanical engineering and engineering materials: *http://www.roy-mech.co.uk/Useful_Tables/Drawing/Drawing.html*.

Linear Motion. Linear motion is a straight-line motion that occurs in one direction. The toaster switch and the drawer that slides out to hold the CD in a compact disk player are examples of linear motion. Linear motion can be returned to a starting position.

Reciprocal Motion. Reciprocal motion is back and forth linear motion. The piston of an internal combustion engine and the needle of a sewing machine both have a reciprocal motion.

Rotary Motion. Rotary motion occurs around an axle or center point, such as the rotation movement of the steering wheel of a car or the handlebar of a bicycle. Rotary motion is very efficient and is the most common form of mechanical motion generation by a prime mover. Rotary motion is often used in machinery, such as the huge printing presses used to print daily newspapers or this book.

Oscillating Motion. Oscillating motion involves a back and forth movement in an arc. A clock pendulum and the agitator of a top-loading washing machine move with an oscillation motion.

Mechanisms use one form of input motion and create a modified or different output motion. The output motion can be continuous or nearly continuous. Many mechanisms create an output motion that changes relative speed throughout the motion cycle, which can cause other problems. Mechanisms are also capable of intermittent motion.

Your Turn

How much force would be needed to lift the gate if the total length of the gate is 20 feet long and weighs 70 pounds? The fulcrum or post is placed 3 feet in from one end. To lift the gate, the operator must push down on the short 3-foot end. How would the amount of force needed change if the fulcrum was moved to 5 feet from one end? What is the advantage of having the long side of the gate (output motion) operated by the short side of the gate (input motion)?

LEVERS AND LINKAGES

The lever is one of the basic machines of science and is a simple mechanism that has been used to extend human capability throughout history. Lever mechanisms are grouped by class. The lever class is determined by the placement of the fulcrum, or pivot point, relative to the effort and load (see Figure 12-17).

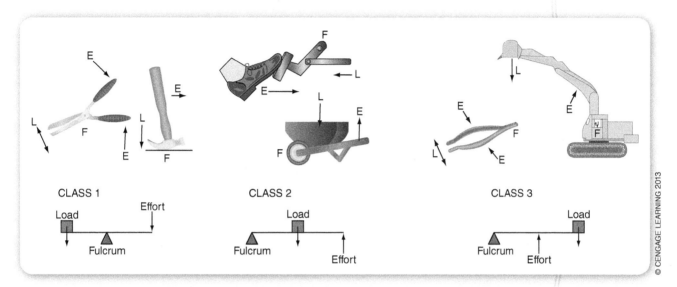

Figure 12-17: Examples of Class 1, 2, and 3 levers.

A Class 1 lever has the fulcrum between the effort and load. A playground seesaw, pliers, and scissors are all examples of Class 1 levers. A Class 2 lever has the load placed between the fulcrum and the effort. A wheelbarrow, nutcracker, and an appliance hand truck are all examples of Class 2 levers. A Class 3 lever has the effort exerted between the fulcrum and the load. The garden shovel and bathroom tweezers are Class 3 levers.

Changing the relationships between the load, effort, and the fulcrum allows the lever to be used to gain mechanical advantage or to gain motion (see Figure 12-18). Levers are in equilibrium when the effort moment $(F_E)(D_E)$ is equal to the load moment $(F_L)(D_L)$. Remember that a mechanism is used to create an output motion and force. When the lever is used to increase output force, then the amount of input motion will need to be increased. When the lever is used to increase output motion, then the amount of input force will need to be increased. These trade-offs can be calculated mathematically.

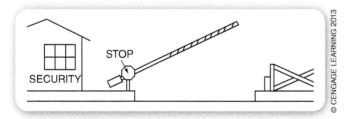

Figure 12-18: Gaining motion by moving the fulcrum closer to the effort.

Mechanical Advantage

The measure for mechanical advantage (MA) is the ratio between load and effort (see Figure 12-19). A mechanical advantage greater than 1 (MA > 1) means that a gain in output force has been achieved by an increase in input motion. A mechanical advantage less than 1 (MA < 1) means that a gain in output motion has been achieved by an increase in input force. Because Class 3 levers always have a MA less than 1, they would not be used to gain output force. Class 3 levers are typically used where a gain in motion is desired or to conserve space.

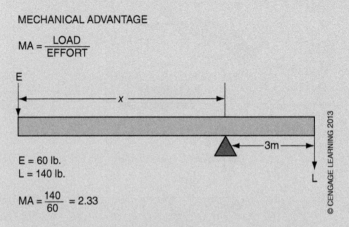

MECHANICAL ADVANTAGE

$$MA = \frac{LOAD}{EFFORT}$$

E = 60 lb.
L = 140 lb.

$$MA = \frac{140}{60} = 2.33$$

Figure 12-19: **Determining mechanical advantage of a lever.**

Velocity Ratio

For each gain in mechanical advantage, efficiency in motion is lost. In science, as in technology, many trade-offs exist. **Velocity ratio (VR)** is the relationship between the distance moved by the effort and load as shown in Figure 12-20. To gain mechanical advantage, the effort must be moved over a greater distance. An example of this is when a hoist is used to lift a heavy load. By using a number of pulleys, a hoist can have a very high mechanical advantage. The trade-off, however, is that you may have to pull the hoist chain 25 feet to lift the load 1 foot. A VR greater than 1 (VR > 1) means that the effort is moving a greater distance than the load.

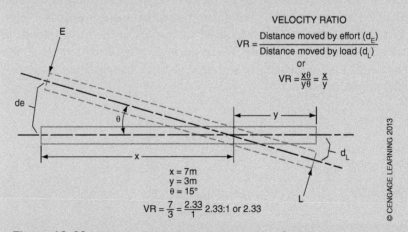

VELOCITY RATIO

$$VR = \frac{\text{Distance moved by effort } (d_E)}{\text{Distance moved by load } (d_L)}$$

or

$$VR = \frac{x\theta}{y\theta} = \frac{x}{y}$$

x = 7m
y = 3m
θ = 15°

$$VR = \frac{7}{3} = \frac{2.33}{1} \quad 2.33:1 \text{ or } 2.33$$

Figure 12-20: **Determining the velocity ratio of a lever.**

Linkages

Linkages are very important to mechanisms, because they transmit (link) the input motion or force to the desired output location. Linkages can change the direction of the force, change the length of motion of the force, or split the motion and force over multiple paths. Linkages can make things move in just about any way desired.

Bell Crank and Reversing Linkages. The bell crank is a basic linkage that got its name because it was used to ring bells (see Figure 12-21). The bell crank linkage changes the direction of force 90 degrees. Double bell cranks are also used. A bell crank linkage is used as part of the braking system on many bicycles as shown in Figure 12-22.

Reversing linkages changes the direction of the force 180 degrees (see Figure 12-23). By changing the length of the linkage and the relative position of the fulcrum point, the force and motion can be altered. Linkages work on the same principle as levers. Mechanical advantage and velocity ratio can be calculated for each linkage used in a system.

Figure 12-21: **An example of a bell crank linkage.**

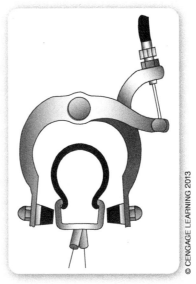

Figure 12-22: **Bicycle braking system using a bell crank linkage.**

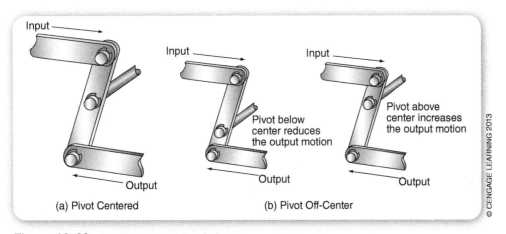

Figure 12-23: **Motion-reversing linkage.**

Your Turn

Compare the mechanical advantages of the braking mechanisms of a 10-speed bike and a mountain bike. How are they different?

Parallel Linkages. A variety of fascinating mechanisms incorporate parallel linkages. Parallel linkages are based on the geometric parallelogram. These linkages allow component parts to move while maintaining a parallel relationship. Parallel linkages are used in scissor-type gates and tables, and the toolbox shown in Figure 12-24.

Treadle Linkage. When it is desirable to change an oscillating motion to a rotary motion, a treadle linkage is incorporated. The treadle linkage got its name from the foot-operated treadle used to turn sewing machines, wood lathes, and other similar devices prior to the twentieth century (see Figure 12-25). Today, the treadle linkage is used in such common areas as the operation of a car's windshield wipers. In this instance, the mechanism is reversed, causing a rotary motion to output an oscillating motion.

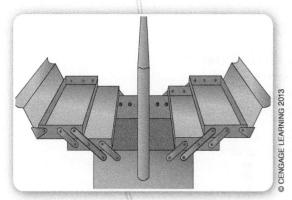

Figure 12-24: **A toolbox design using parallel linkages.**

Toggle Linkages. A toggle linkage is considered a clamping linkage, because it generates tremendous force as semivertical angle θ nears 90 degrees as shown in Figures 12-26 and 12-27. Many mechanisms use toggle linkages, including clamps and locking pliers. Toggle linkages are also used in general consumer products, such as card tables, folding chairs, and collapsible baby strollers. Notice that, if the toggle linkage is taken past straight and against a solid member, the linkage is locked in place (see Figure 12-27). Additional force will need to be exerted to take the linkage into the open position.

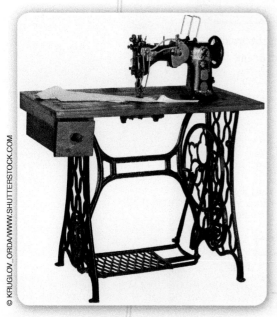

Figure 12-25: **Treadle-operated sewing machine from the eighteenth century**

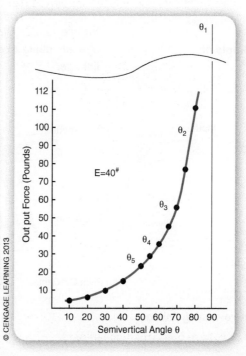

Figure 12-26: **Chart showing the changing force generated by a toggle mechanism at different angles.**

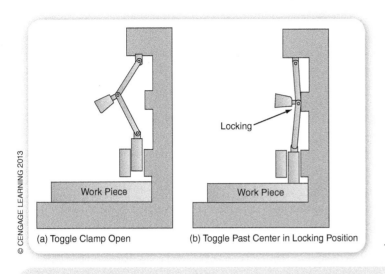

Figure 12-27: **Locking a toggle mechanism.**

The Greek letter theta "θ" is used to represent an angle measurement. Angles are measured by drawing an arc centered at the vertex of the angle. The angle, in radians, is determined by dividing the length of the arc s by the radius r (see Figure 12-28).

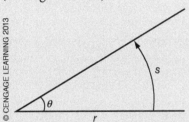

Figure 12-28: **Calculating an angle measurement.**

ROTARY MECHANISMS

Gears, pulleys, cams, and other related mechanisms are considered higher-order kinematic pairs. These rotary mechanisms transfer or change an input rotational motion and force to an output motion and force. The output can be either rotary or reciprocating motion, depending on the rotary mechanism employed (see Figure 12-29).

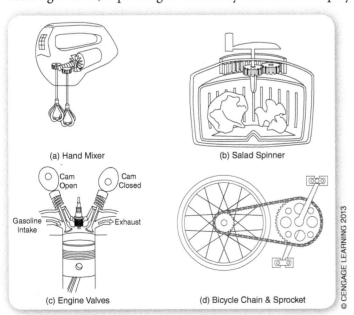

Figure 12-29: **Examples of rotary and cam mechanisms.**

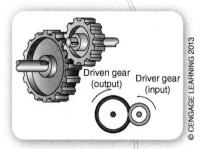

Figure 12-30: Driver and driven gears.

Gears

Gears are toothed wheels fixed to an axle. Remember that the wheel and axle is one of the basic machines. One gear fixed to an input axle is called the *driver gear*, and is the gear where the input force or energy is being applied. The other gear fixed to an output axle is called the *driven gear* (see Figure 12-30), and is the gear where the output force or energy is going. When a number of gears are connected together, the system is known as a *gear train*.

The relationship between input motion and force and output motion and force for gears is the same as for levers, and can be calculated using the following formulas (see Figure 12-31), where gear ratios are used instead of velocity ratios.

Gear Ratio

$$GR = \frac{\text{Number of driven teeth (output)}}{\text{Number of driver teeth (input)}}$$

$$GR = \frac{16}{12} = 1.333 : 1$$

(Gear ratio for gear train in Figure 12-30)

Figure 12-31: **Determining gear ratios.**

When two gears mesh, the motion of the driver gear turns the driven gear in the opposite direction. When both the input (driver gear) and output (driven gear) are required to turn in the same direction, an idler gear is needed (see Figure 12-32).

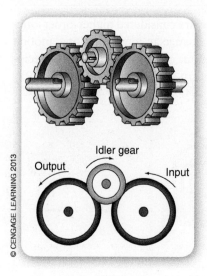

Figure 12-32: **Idler gears allow input and output motion to occur in the same direction.**

Sometimes large speed or force changes are needed. In this situation, where a high- or a low-velocity ratio is required, compound gear trains are used. Compound gear trains have two gears of different sizes on one shaft, one of which acts as an input or driven gear and the other as a driver gear to the next gear in the train. This system makes compact gearing mechanisms possible. To determine the input to output gear ratio in compound gear trains, the gear ratio for each pair of gears must be determined (see Figure 12-33).

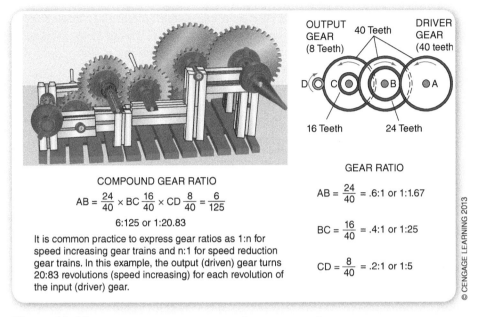

Figure 12-33: **Determining gear ratios for compound gear trains.**

The most basic type of gear is called the spur gear. This gear is relatively easy to manufacture and is one of the earliest forms of gears (see Figure 12-34). The helical gear is a more technologically advanced gear design (see Figure 12-35). This gear is cut at an angle, and its shape is part of a helix. These gears are designed to run quieter at high speeds and to transfer more torque than other gear designs. This is because helical gears have more teeth area in contact. In transmission systems, where there is a gear train, helical gears generally shift more easily than spur gears.

Both the worm and bevel gear are used to change motion direction, typically at 90 degrees. The worm gear looks something like a screw but is really a single-gear tooth wrapped around a driver axle. This gear system is capable of making large speed reductions. Figure 12-36 shows a worm gear system with a 30:1 velocity ratio.

Figure 12-34: **Spur gear from Stanserhorn cable car, Switzerland.**

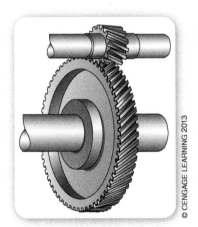

Figure 12-35: **Helical gears.**

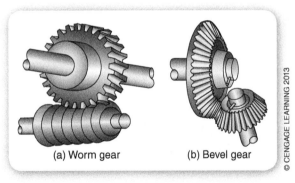

Figure 12-36: **Worm and bevel gears change the rotation axis.**

In the bevel gear, VR changes are made by changing the size of the driver and driven gears. The bevel gear system is used for smaller velocity ratio changes than the worm gear system. A form of bevel gear system is used in the rear axle of rear-wheel drive cars and trucks to change both the speed of rotation and the direction of the rotary motion of the drive shaft. These systems generally have a ratio of 3.00:1 to 4.10:1.

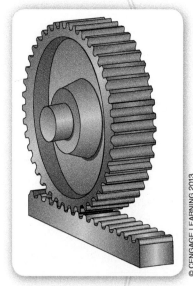

Figure 12-37: **Rack and pinion gear system.**

The rack and pinion gear system changes a rotary input motion to a linear output motion (see Figure 12-37). In this gear system, one gear is produced as a flat strip. Rack and pinion gears can be found in many automotive steering systems and some machines in the technology laboratory such as a drill press.

Gears are made of a variety of materials. Although most gears are manufactured in steel, other metals such as brass are used. Plastic gears are also common. Plastic gears are not as strong as their metal counterparts but are often less expensive, run more quietly, and do not require as much lubrication.

Other Rotary Systems. Pulley and sprocket systems use a variety of belts and chains to convert a rotary input motion and force to a desired output motion and force (see Figure 12-38).

These systems work under the same measure of mechanical advantage and velocity ratio as the gear. Pulley and belt systems are quiet to operate, and they are capable of changing both the speed and direction of the output motion. In most pulley and belt systems, adjustments must be possible to hold the belt in proper tension. Sometimes a spring-loaded or adjustable idler wheel is used (see Figure 12-39).

Sprocket and chain systems are not as quiet as pulley and belt systems but they are capable of transmitting greater force, as shown in Figure 12-40. Chains must be kept well lubricated, or rapid wear occurs. In addition to changing the output speed of rotary motion such as in a bicycle gear mechanism, sprocket and chains can be used to change rotary to linear motion. A conveyor system is a good example of rotary to linear motion.

Figure 12-38: **Examples of pulley and sprocket systems.**

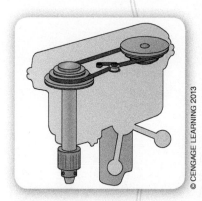

Figure 12-39: **The idler pulley puts tension on a belt to prevent slippage.**

Figure 12-40: **Sprocket and chain system.**

Cam and Crank Slider Mechanisms

Cams are like smooth gears but are not always round in shape. Crank slider mechanisms may be driven by cams and convert rotary input motion to a desired reciprocating or oscillation output motion. An eccentric is the simplest cam shape and provides a smooth reciprocating motion to the cam follower. The follower can be blunt or a shaft with a wheel (see Figure 12-41).

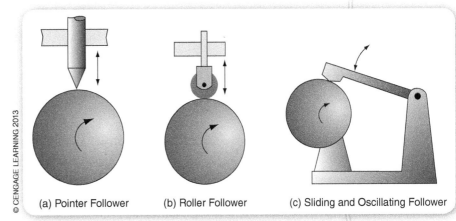

Figure 12-41: Eccentric cams with different types of followers.

The crank and slider mechanism is used to change between rotary and reciprocating motions. In a VCR, the crank and slider mechanism converts the rotary input motion from a small electric motor to the linear action that accepts or ejects the tape cassette. A prime mover, such as a diesel or gasoline engine, uses a crank and slider mechanism to convert a reciprocating motion caused by the engine pistons moving back and forth in a cylinder into a rotary output motion. This action is similar to the treadle linkage mechanism discussed earlier in the chapter.

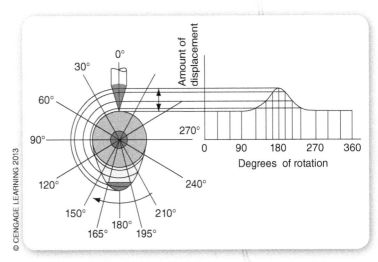

Figure 12-42: This figure illustrates the displacement motion of the follower created by the cam shape.

Cam Motion. The input and output motions of a cam are represented by a displacement diagram (see Figure 12-42). The eccentric cam, a circular disk with the axle placed off-center, gives a smooth, harmonic, linear output motion to the follower. The pear-shaped cam creates an intermittent linear motion with a set dwell time. Dwell is the time, expressed in degrees of rotation, when the rotating motion of the cam does not cause a change in follower motion. For example, in an automobile engine, a cam opens and closes valves. The shape of the cam, however, causes a dwell that allows a valve to remain closed for a period of time and to remain open for a period of time, instead of simply snapping open and shut.

Your Turn

Cams can be used to obtain all kinds of interesting motion, such as in a child's toy. A heart-shaped cam is another commonly used cam. Can you determine its motion and think of an application for this cam design (see Figure 12-43)?

Figure 12-43: Crank and slider mechanism used to create motion for a simple toy.

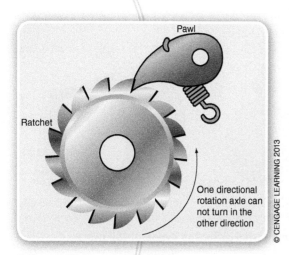

Figure 12-44: Pawl ratchet.

Ratchet Mechanisms

Ratchets are the oldest form of mechanism to create intermittent motion (see Figure 12-44). Leonardo da Vinci used ratchet mechanisms on many of his designs. When a mechanism must be designed to turn in only one direction, a ratchet mechanism is used. A ratchet mechanism can be used to prevent an axle from turning in the wrong direction. This is often desired when a load is being lifted, such as when an engine is being removed from a car. Without the ratchet, the engine could crash back into the car if the mechanic's hands slipped on the chain. Many different types of automobile jacks also use a ratchet mechanism to protect the operator.

A ratchet mechanism can also be used to convert linear motion to a rotary motion. In this design, the pawl is used to push against the teeth of the ratchet. The amount of rotary motion is determined by the amount the pawl is moved.

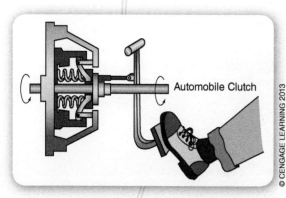

Figure 12-45: Clutch mechanisms are used to connect rotating shafts.

Clutches and Brakes

When the motion created by a mechanical system is intended to be momentarily disrupted, a clutch mechanism is used. When the motion must be stopped, a brake is used (see Figure 12-45). A clutch is a form of coupling that can be easily connected and disconnected. A friction clutch can be engaged or disengaged while either the input and/or output shaft is moving. A centrifugal clutch used in a lawn mower or a moped is a type of friction clutch that engages when the input shaft is rotating at a predetermined speed. A positive clutch interlocks the input and output shaft and can only be used if both shafts are stopped or are turning at the same speed.

Traditionally, brakes use friction to reduce the speed of a mechanism, as shown in Figure 12-46. Both drum and disc brakes are commonly used. Mountain bikes and most modern automobiles use a form of disc brake. Disc brakes generate greater braking force and dissipate heat more efficiently than drum brakes. Some automobiles use drum brakes on the back wheels because they are easier to connect mechanically to an emergency brake. Less braking force is needed on the back brakes of a vehicle as the load shifts forward during braking.

You may not think about innovation in a braking system as very important but that has been one of the major design breakthroughs in hybrid cars. Regenerative braking systems designed by automotive engineers use the car's electric motor to slow or stop the vehicle. A moving car has considerable kinetic energy and, in traditional braking systems, a brake pad is pressed against a brake drum causing friction. Because energy cannot be created or destroyed, braking in a conventional car system produces heat that is of no practical value. The

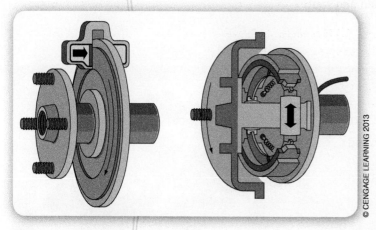

Figure 12-46: Brake mechanisms.

genius of the hybrid braking system is that the electric motor is put in reverse, making the motor a generator. A generator under load produces electricity and drag for stopping. The energy that is normally lost as heat is now transformed as electricity that can be stored in the battery system and used to power the car. A conventional braking system still exists as a backup in the hybrid braking system.

MODELING MECHANICAL DESIGNS

Many educational kits are available for modeling mechanical designs (see Figure 12-47). Some kits combine mechanical and electrical components to make an operating robotic or vehicle system. Modeling kits can be very complex and are actually used for industrial training simulations.

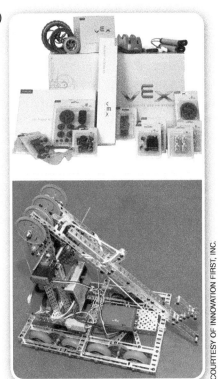

Figure 12-47: Kits for modeling mechanical systems. (a) Fischertechnik systems are used to model mechanical and structural systems and subsystems. Profi kits are available that utilize solar energy, pneumatic power, electronic sensors and logic gates, and even automotive technology with gearboxes and differentials. (b) Vex Robotic kits are used to model robotic and autonomous systems. Kits can include a programmable microcontroller, variable speed motors, gears and wheels, and chassis components.

Careers in the Designed World

NICHOLAS BARNICK: MANUFACTURING ENGINEER, DEPUY ORTHOPAEDICS, INC., A JOHNSON & JOHNSON COMPANY

It Takes a Team to Manufacture a Design

Rarely will you find Nick Barnick at his desk at DePuy Orthopaedics, a Johnson & Johnson company. Nick is the team leader for a group of manufacturing engineers who support the manufacture of orthopaedic implants. He spends most of his time on the manufacturing floor or in team meetings checking on the progress of any one of the projects he is responsible for at DePuy.

When you think of the Johnson & Johnson company, you might think of baby powder or shampoo, but Johnson & Johnson has other companies that produce medical devices. One such company is DePuy Orthopaedics, Inc. Nick's team of manufacturing engineers develops processes for the manufacturing of orthopaedic medical devices. In his primary role as team leader, Nick ensures that projects are resourced appropriately and aligned with the company's strategic initiatives.

On the Job

To develop manufacturing processes for new orthopedic implants, Nick works with cross-functional teams. These types of teams include members from different departments. The team uses each member's knowledge and skills to develop manufacturing processes for new products. Nick explains, "I work with team members from other groups that are knowledgeable about product development, quality engineering practices, microbiology, regulatory requirements, marketing, and production practices. Together we review designs for manufacturability. Then we choose the manufacturing processes and identify and purchase the necessary equipment to produce this new design."

Nick explains how his group works closely with design engineers at DePuy to develop new product ideas. "Design engineers in our company use specialized design software to generate electronic models of our products. From these electronic models, drawings are generated. From these drawings, manufacturing processes are developed."

To create these new processes, Nick's team will develop tooling and fixtures to produce new products. From the electronic models and drawings, they will develop computer numerical control (CNC) programs for machining parts, and design their own inspection techniques. Finally, they will determine manufacturing costs and create a bill of materials. From the beginning of a new product design to the final packaging and labeling of the product, the cross-functional team works together to problem-solve the development of new medical devices.

"Because our products are designed and manufactured around the world," says Nick, "it is important that our communication be efficient and clear. Communication is critical to the success of our product and process development. We use various technologies such as email, video conferencing, Web conferencing, and online e-rooms to ensure clear and efficient communication."

Inspirations

Nick has always been inspired by how things work. At first, he followed his father's career path to an engineering job in the automotive industry. As time went on, Nick was inspired by the medical devices that enhanced the quality of life for his disabled brother. "After some reflection," says Nick, "I knew that I wanted to work in an engineering field where I could improve the quality of life for people." Today, Nick draws inspiration from watching the development of DePuy's products go from concept to a real live product, usable by people.

Education

Nick earned an associate's degree in engineering science from Hudson Valley Community College in Troy, New York. He completed his bachelor of science in mechanical engineering at Syracuse University. During his college summers, Nick picked up some hands-on experience by working in a prototyping shop for a manufacturer of outdoor power equipment. He obtained a master's degree in mechanical engineering from Rensselaer Polytechnic Institute.

Advice for Students

"My advice for a young person is to keep an open mind with respect to the ideas and opinions of others. Many of the most important engineering solutions that I have seen have not been from independent efforts but from the collaborative efforts of a cross-functional team."

SUMMARY

Mechanisms and mechanical systems can be found in many products and devices such as the automobile, airplane, agricultural mechanization, air conditioning and refrigeration, spacecraft, household appliances, and health technologies. Human history is rich with mechanical designs. Leonardo da Vinci designed many mechanisms including a flying machine, but it took over 400 years for technological developments in tools, materials, and processes for the ideas to be successfully manufactured. Igor Sikorsky built and piloted the first successful flight of the VS-300 helicopter in 1940.

The action of machines or mechanisms is governed by physical principles described in Newton's three laws of motion. Because mechanisms are responsible for any action or reaction by transmitting forces through a series of predetermined motions, these components are associated with dynamic forces. A force causes a body (mechanism) to accelerate (move). Forces have vector qualities causing an object to be lifted, pushed, or pulled at a predetermined magnitude (strength) and direction. Forces are measured in either newton (SI) or pound (English). Torque, also known as a moment, can be thought of as "rotational or angular force" that causes a change in rotational motion. Motion describes a change in position of an object, which can occur in a straight line or in space. Momentum depends on the mass and velocity of an object moving on a plane or in space. It is a vector force and like all vectors has magnitude and direction. Work is a force applied over a distance that overcomes some resistance as noted in Newton's Laws. Power is the rate at which work is done. In SI units, power is measured in watts (W) and is equal to joules per second. Energy is the ability to do work and is measured in joules. All objects possess energy that may be in the form of thermal, chemical, electrical, radiant, or nuclear. Mechanical systems have potential and kinetic energy.

Mechanisms and machines are similar in that they both are a combination of rigid bodies and are capable of defined movement. Machines are different from mechanisms in that they are capable of transforming energy to do work. In the late nineteenth century, a German mechanical engineer by the name of Franz Reuleaux (1829–1905) developed the fundamental concepts of modern mechanisms known as kinematics. Kinematics is a branch of solid mechanics that deals with relative motion of elements or pairs in machine and mechanical systems. Kinematics and kinetics (the action of forces on bodies) together form the basic elements of engineering dynamics and machine design theory. Reuleaux identified six mechanical elements used in mechanical design. They include:

- The lever and crank
- The wheel and gears
- The cam
- The screw
- Things that transmit tension or compression, such as belts, chains, and hydraulic fluid lines
- Things that transmit intermittent motion, such as the ratchet

Because mechanisms deal with motion, it is important to understand the different kinds of motion possible. A mechanism was previously defined as a device that takes an input motion (and force) and creates a desired output motion (and force). Four common types of motion are typically controlled by a mechanism: linear motion, reciprocal motion, rotary motion, and oscillating motion. The measure for mechanical advantage (MA) is the ratio between load and effort. A mechanical advantage greater than 1 (MA > 1) means that a gain in output force has been achieved by an increase in input motion. A mechanical advantage less than 1 (MA < 1) means that a gain

SUMMARY
(continued)

in output motion has been achieved by an increase in input force. Because Class 3 levers always have a MA less than 1, they would not be used to gain output force. Class 3 levers are typically used where a gain in motion is desired or to conserve space. For each gain in mechanical advantage, efficiency in motion is lost. In designing mechanical systems, many trade-offs exist. Velocity ratio (VR) is the relationship between the distance moved by the effort and load. To gain mechanical advantage, the effort must be moved over a greater distance. An example of this is when a hoist is used to lift a heavy load. By using a number of pulleys, a hoist can have a very high mechanical advantage. The trade-off, however, is that you may have to pull the hoist chain 25 feet to lift the load 1 foot. A VR greater than 1 (VR > 1) means that the effort is moving a greater distance than the load.

Figure 12-47 provides examples of a wide variety of mechanisms.

BRING IT HOME

OBSERVATION/ANALYSIS/SYNTHESIS

1. Do an Internet search for "Leonardo da Vinci inventions." Leonardo da Vinci (1452–1519) made many detailed drawings of his inventions. Pick three of his inventions that you think are very close to modern devices.
2. Do an Internet search for "Igor Sikorsky biography." What designers did Igor Sikorsky study when he was young? Were Sikorsky's business ventures successful? Can you determine which famous musician first invested in Sikorsky's company, "Sikorsky Manufacturing Company?" Who now owns the Sikorsky helicopter business?
3. Make a kinematic diagram of a lower and higher kinematic form of mechanism. Use some existing product, such as a home device, tool, or machine as a model. Identify the input motion and force, as well as the output motion and force.
4. A type of modern bicycle has two gears in the front attached to the pedal (driver gears) and nine gears on the back wheel. The number of teeth on the two front gears is 50 and 34, while the number of teeth on the largest and smallest gears in the back is 27 and 11, respectively. Calculate its mechanical advantage and velocity ratio for the two conditions where it is "hardest" to pedal (for going fast down hills or on flat roads) and where it is "easiest" to pedal (for going up hills).
5. Search Wikipedia for "escapement" or "escapement clock." What is the purpose of the escapement designs? How many designs are shown? Which escapement design do you think costs the least to manufacture, and why? Digital clocks have no gears, so how do they extract time? Search www.howstuffworks.com.
6. Using a modeling system, such as Lego, Fischertechnik, or Meccano, design, construct, and test a mechanical system. Some suggestions are wind thread on a bobbin; simulate the motion of a sewing machine needle; and model the motion of a steam engine.
7. Design a cam that keeps a valve closed twice as long as it is open. The linear motion of the valve should be 1 cm.

EXTRA MILE

ENGINEERING DESIGN ANALYSIS CHALLENGE

IDENTIFICATION OF PROBLEMS: Americans accumulate over 240 million tons of solid waste each year. The waste is polluting the environment. Many communities require recycling of a number of items, including aluminum cans. Every time someone has a soda, that person can choose either to clutter the kitchen counter or to take the can to the recycling bin located outside. Opening the door to take the can outside lets a lot of cold air into the house and wastes energy, but the empty cans on the counter look messy.

DESIGN BRIEF: Design and build a hand-operated device that will reduce the volume of an aluminum can by at least 75 percent. The device must be attractive (salable), fit on or hang above a counter or on the back of a cabinet door, and must be usable by a person 12 years of age or older. The device must be safe to operate (does not require safety glasses, gloves, and so on) and must provide a relatively safe product (no extremely sharp or jagged edges). The design must represent an efficient and proper use of materials and production practices.

CHAPTER 13
Designing Electrical Systems

Menu

Before You Begin
Think about these questions as you study the concepts in this chapter:

1. What are six revolutionary events in electrical science and technology?

2. Over what time frame did electrical technology develop? How does this compare to the time frame for the development of the three important metallurgical technologies (copper, bronze, and iron)?

3. What is the name for the mathematical form for the equations of force for both gravity and electricity?

4. What is electricity?

5. What are some materials used as electrical conductors or insulators and what is the basic science of what makes them conductors and insulators?

6. What are the three basic electrical circuit components and what do they do?

7. What are several electrical circuit components used for each of the following portions of an electrical design: (a) input, (b) process, and (c) output?

8. What single electrical device has revolutionized modern human society?

INTRODUCTION

Of all human-designed technologies, it would be very difficult to find a technology that has had a larger impact on modern human existence than electronic technology. Constructing a time line of the major events in electronics can help summarize the impact of electronic technology. Major events in both the science and technology of electronics are shown in Figure 13-1. These events span a very brief two- to three-century period from the mid-1700s to 2000.

A time period not shown on the time line is the time span from ancient Greece to the mid-1700s. In this time span, electricity was used only as a curiosity or entertainment tool. For example, in the court of Louis XV in the mid-1700s, Abbe (Jean-Antoine) Nollet used electric shock to the delight of many. Today, a mere 200-plus years later, we see the mass proliferation of Internet-connected personal computers, cell phones, smart phones, and a worldwide human society dependent and interdependent on numerous electronic-based capabilities.

In the late 1700s, society was strictly supported by mechanical and metallurgical technologies, like horse- or oxen-drawn plows built with wood and metal-based processes or crafts. However, by the late 1800s, transportation, housing, food cultivation and preparation, health care, and communications had all been revolutionized by the use of electronics-enabled technologies. Some of these enabling technologies included motorized vehicles (internal combustion engines),

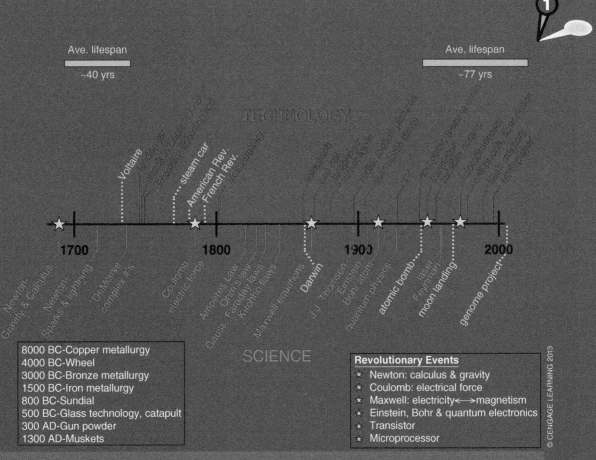

Figure 13-1: Time line of key events in both the science and technology of electronics. Also included for reference are a few politically or scientifically significant dates.

INTRODUCTION

lighting and heating equipment, telegraphs, radios, and telephones. This span of time was very brief—slightly greater than 100 years. The technological advancement from the late 1700s to the late 1800s was indeed very fast. However, in an additional 100-year span of time, from the late 1800s to 2000, electronics technology advanced even faster, and now electronics technology has truly become dominant in society. Today we enjoy the benefits of widespread electric power distribution throughout most of the world with the associated benefits for improved living conditions and industry. Automobiles, which use a high level of electronics technology, are widely prevalent, with commercially successful hybrid vehicles (shared fuel and electric power) achieving fuel efficiencies greater than 60 miles per gallon. In particular, the end of the twentieth century will likely be remembered as the age of the microprocessor. Fast, efficient, and inexpensive semiconductor-based microprocessors are abundant in many human tools. Computers, personal data devices, cell phones, automobiles, communications systems, heating and cooling systems, robotic manufacturing systems, and even electronic toothbrushes are microprocessor-controlled. Many devices often contain numerous microprocessors. Automobiles, for example, can contain over 50 microprocessor-based subsystems, including braking, fuel injection, and safety air-bag deployment. Automobiles can even communicate their health directly to the manufacturer via onboard wireless communications systems.

It could be argued that the beginning of the scientific revolution occurred in the late 1600s, with events surrounding Sir Isaac Newton's eloquent description of calculus and gravitational forces, appropriately termed "Newtonian physics." Newtonian physics accurately described the motion (position, velocity, and acceleration) of common everyday objects, objects that could be easily seen, touched, and observed. For example, Newtonian physics accurately predicted the planetary observations that Tycho Brahe made in the late 1500s, verifying the revolutionary, and dangerous, "sun-as-the-center" theory proposed by Nicolaus Copernicus. Closer to home, Newtonian physics also accurately predicted projectile motion, something very useful for military purposes both then and now.

The unique difficulty with describing the physics of electricity is that the basic driver of electric force, charge, cannot be easily touched, seen, or safely observed. Many of the early tinkerers of electricity were routinely shocked, and some even died. Discoveries in electrical science can be very difficult. For example, it took 50 to 100 years after other key discoveries in electrical science to prove the existence of the electron, one of two key charged particles in nature (J. J. Thomson in 1897; see Figure 13-1).

Careers in the Designed World

KRISTIN WEARY: ELECTRICAL SYSTEMS ENGINEER, LOCKHEED MARTIN

Documentation: A Link from the Past to the Future

Kristin Weary sees her position in electrical engineering as a critical link to the future. As a systems engineer for Lockheed Martin, Kristin often works on equipment and systems that were designed many years ago. Well-documented systems allow technology upgrades to be implemented more easily. Kristin says, "Documentation of your work is crucial, because twenty years from now you won't be around to explain what you did, and someone will want to know why you did it the way you did."

On the Job

Kristin supports the design, documentation, and troubleshooting of complex electrical systems in remote locations. "I write documentation for very sophisticated electrical systems," explains Kristin. She starts with complicated information and writes it in a simple manner so that someone without an engineering degree can comprehend and safely operate the systems.

"Most of the time I can't see the equipment when issues come up, and detailed information is limited," says Kristin. "When there is a problem with a system, I work with a team to develop strategies to resolve the problem."

A computer model is used to simulate the actual complex systems Kristin works with. She works in a lab where she has designed a simulator to model a smaller system that has been integrated with the computer model. Here are the steps that Kristin used to set up and use these computer models:

- ▶ Understand the problem
- ▶ Understand the overall system
- ▶ Conceptually frame out a solution
- ▶ Design a solution in the software
- ▶ Test/debug the software on a computer simulator and with a test box
- ▶ Merge the new design with the rest of the equipment and computer model

Kristin uses communication technology such as email to communicate with off-site team members and customers.

Inspirations

Kristin loves to solve logic problems and come up with creative solutions. She realized early on that an engineering career would let her apply this skill in real life.

For Kristin, fun is being able to debug a complex system by breaking it into smaller areas to find the problem. Being able to identify and fix problems provides satisfaction for Kristin.

Education

Kristin earned a bachelor's of science degree in electrical engineering from Virginia Tech. In addition, she completed two summer internships with General Electric Corporation. During one of the internships, she was challenged to fix a problem with a complex database that someone else designed. This experience set the stage for Kristin's future employment as a systems engineer at Lockheed Martin.

Advice for Students

Kristin says that students who plan to study engineering already know that math and science are important parts of their preparation. She also advises students to develop their writing skills. "A lot of engineering is the documentation of the design and also the ability to communicate in writing about the design to other people," says Kristin.

"Don't be afraid to ask for help," adds Kristin. "If you can't solve a problem or find the answer and you have put an honest effort into doing so, ask for help, because most likely there is someone who knows the answer and is more than willing to mentor you."

THE SCIENCE OF ELECTRICITY

OFF-ROAD EXPLORATION

http://spectrum.ieee.org/geek-life/profiles/special-report-dream-jobs-2011
http://www.amasci.com/amateur/elehob.html
http://www.technologystudent.com
http://www.williamson-labs.com
http://electronicdesign.com/

The basic building blocks in nature are the more than 100 elements in the periodic table. Atoms of elements are made up of neutrons, protons, and electrons. Varying numbers of protons and neutrons make up the nucleus, or center, of atoms while the electrons are in constant planetary-like motion around the nucleus. Copper, a familiar and commonly used metal in both industry and electronics, is an element and has an atomic number of 29. This means that copper atoms have exactly 29 protons in the nucleus and exactly 29 electrons in motion about the nucleus. As the names suggest, protons have a *positive* electric charge and neutrons are neutral, having no electric charge.

In opposition to the protons, electrons have a negative charge. Additionally, the magnitude, or amount, of the negative charge of an electron is exactly identical to magnitude, or amount, of the positive charge of the proton. So, for example, if one were to observe a copper atom from far away, the equal numbers of positively charged protons and the negatively charged electrons would cancel in charge, resulting in a copper atom that is a completely neutral entity. This net charge neutrality of atoms is a very important attribute of matter in general, as we will discover next when discussing the size of electric forces.

The equations of force for gravity (F_g) and electricity (F_e) between two objects are given in Equations 13.1 and 13.2. The subscripts "g" and "e" remind us that we are talking about either gravitational or electrical forces. The metric unit of force is a *newton*, named in honor of Sir Isaac Newton. The equation for gravitational force determines the size of force between any two masses while the equation for electric force determines the size of force between any two charges. In the equation for gravitational force, M_1 and M_2 are the masses of the two objects, r is the distance between the masses, and G is a constant (the "universal gravitational constant"). In the metric system, mass is typically measured in kilograms (kg), r is typically measured in meters (m), and G has the value of 6.67×10^{-11} newton-m^2-kg^{-2}. In the equation for electrical force, Q_1 and Q_2 are the charges of the two objects, r is the distance between the charged objects, and k_o is a constant. In the metric system, charge is typically measured in coulombs (C), r is typically measured in meters, and k_o has the value of 9×10^9 newton-m^2-C^{-2}. The electron and proton are the prevalent charge-carrying particles in nature, with the electron having a charge of -1.6×10^{-19} C and the proton having a charge of $+1.6 \times 10^{-19}$ C.

$$F_g = G\frac{M_1 M_2}{r^2} \qquad \text{(Eq. 13.1)}$$

$$F_g = k_o \frac{Q_1 Q_2}{r^2} \qquad \text{(Eq. 13.2)}$$

Newton accurately described the gravitational force in the late 1600s, while Coulomb did not accurately describe the electrical force until the late 1700s (see Figure 13-1). The similarities between the force equations for gravity and electricity are striking. Indeed, the equations are completely identical in form. Can you imagine how history might have changed if Newton, in 1690, had realized that electric force, first described accurately by Charles Coulomb in 1784, had the same form as his famous gravitational force? We might have had MP3 players 100 years earlier, just in time for use in the Ford's first Model T motorcars or on the battlefield in WWI!

If the distance between two charges is doubled, by what factor does the force decrease?

A very important similarity between gravitational and electric forces is that they are both governed by an **inverse square law** (an r^2 appears in the denominator). Another similarity between electrical and gravitational forces is that the direction of the force is along a line formed directly between the two objects. However, this is where the similarities stop, and they stop dramatically. The first big difference is that gravitational forces are always attractive; objects with mass always attract other objects with mass. However, because there are two types of charge, positive and negative, electric forces can either be attractive or repulsive. Like charges repel each other while opposite charges attract each other.

A second substantial difference between electrical and gravitational forces is the magnitude, or strength, of the forces. In reality, it is a bit awkward to compare electric forces to gravitational forces because, after all, gravity acts strictly on masses whereas electrical forces act strictly on charges. However, as an example, let's look at the earth–moon orbiting system. The distance between the earth and moon is approximately 400,000 km, the mass of the earth is 6×10^{24} kg, and the mass of the moon is 7×10^{22} kg, resulting in an attractive gravitational force between the moon and the earth of about 10^{20} newtons. Not surprisingly, this is a very large force because the earth and moon have quite large masses. However, if the earth and moon were each replaced with only electrons, having the same net mass, then the force (which would now be repulsive) would be 10^{63} newtons, a factor of 10^{43} larger! Indeed, electric forces can be tremendously large. It is a very good thing that matter on earth, in general, is neutral, being made up of equal amounts of positive and negative charge. Otherwise, matter would literally be ripped apart.

A third substantial difference between electrical and gravitational forces is that charges can be quite mobile, an attribute due to both the large size of electrical forces and the tremendously small mass of charges (an electron only weighs approximately 10^{-30} kg). We have all observed the highly mobile nature of charge when we see (and feel) sparks when we sometimes reach for a doorknob. The mobility of charge cannot only be painful but also makes analyzing charged systems quite difficult.

A summary of the similarities and differences between gravitational and electrical forces is given in Figure 13-2.

inverse square law: Any physical law stating that some physical quantity is inversely proportional to the square of the independent variable (often distance).

Electrical Conductors and Insulators

Even though electrical forces can be large and electrical charge can be catastrophically mobile at times, we can easily control most electric forces and the flow of electric charge by using electrical **conductors** and electrical **insulators**. Most people understand that electrical devices use wires to transport electric power, and that the wires are typically made up of a type of metal, the conductor, coated by a plastic-like material, the insulator (or insulation).

The physics of conductors is relatively straightforward. A common material used as a conductor of electricity is copper. As mentioned earlier, a stand-alone copper atom has 29 electrons that orbit around its nucleus. When a large number of copper atoms are joined tightly together to form a metal, each copper atom is

conductor: A material that allows the flow of charge (typically, but not always, electrons).

insulator: A material that does not allow the flow of charge (typically, but not always, electrons).

Figure 13-2: Similarities and differences between gravitational and electrical forces.

Similarities	Differences
Governed by inverse square law	Attractive or repulsive
Force between two bodies	Size of electric forces can be huge
Force is along a line connecting the two bodies	Charge can be very mobile

Point of Interest
Powers of 10

In several engineering disciplines, it is routine to use scientific notation to represent numbers. Scientific notation is when a number is expressed using powers of 10. This makes sense given that our numbering system is a base-10 numbering system. (A base-10 numbering system is when 10 distinct numbers are used for counting, like 0 through 9.) Scientific notation is very useful when working with really big and really small numbers. For example, the U.S. government deficit in the year 2007 was about $8,900,000,000,000, which can be written as 8.9×10^{12}. It is definitely easier to write this large number, and carry through a calculation, when it is written in scientific notation. The same holds true for really small numbers. Can you imagine how difficult it would be to carry through a calculation using the mass of an electron, 9.1×10^{-31} kg, by writing the number in longhand, including all 30 zeros?

Even though scientific notation is very useful, powers of 10 can also be quite deceiving. For example, to the human mind it is very easy to comprehend any number between, say -35 and $+26$. We encounter many items on a daily basis between these two numbers. We might get paid $8 per hour for a job, or we might buy a new T-shirt for $22 or we might need to drive 13 miles to school. However, if we use these same numbers as the powers of 10 in scientific notation, we obtain quite incomprehensible numbers. Let's limit our discussion to length, in meters, and talk about the two lengths 10^{-35} and 10^{+26} meters. The number 10^{-35} meters is approximately the value of a "Planck length," the smallest distance even possible in modern physics. It does not even make physical sense to talk about a length smaller than a Planck length. To put a Planck length into perspective, the size of an electron—which is a really, really small entity itself—is approximately 10^{-15} meters in diameter, but is a length that is much larger than a Planck length ($\sim 10^{-35}$ meters). Taking the large end of our numbers, a length of 10^{+26} meters is the size of the whole visible universe! Therefore, the lengths between 10^{-35} and 10^{+26} meters, as summarized in Figure 13-3, span all known sizes of all things. So, while you are using scientific notation to make your calculations easier, try not to lose sight of how truly big, or small, the powers of 10 you are working with really are.

Figure 13-3: Several important lengths, expressed as powers of ten. The span of lengths from 10^{-35} to 10^{+26} meters includes the sizes of all known things.

Length (m)	Physical Quantity
10^{-35}	A Planck length
10^{-15}	Approximate diameter of an electron
10^{-9}	Diameter of DNA helix
10^{-7}	Wavelength of visible light & gate width of transistor (in year 2007)
10^{-4}	Diameter of a human hair
10^{-2}	One inch
10^{2}	Tallest tree
10^{5}	Length of 1 degree of latitude on earth
10^{7}	Diameter of earth
10^{10}	1 light-minute (distance light travels in 1 minute)
10^{15}	1 light-year (distance light travels in 1 year)
10^{21}	Diameter of Milky Way galaxy
10^{26}	Diameter of the visible universe

exposed to the large electrical forces of the 5 to 10 surrounding copper atoms. The closeness of all these other atoms literally frees one, and only one, of the outer orbiting electrons in each copper atom. For copper, and metals in general, there is about one free electron per atom, thus making metals essentially a gas of electrons free to move about within a matrix of interconnected metal atoms. Are there a lot of these free electrons? In a typical six-foot-long power extension cord, there are about 10^{24} copper atoms and because there is one free electron per copper atom, there are 10^{24} electrons that are completely free to move about within the metal conductor. This is about the same number of O_2 molecules in a cubic foot of air, a volume that is substantially larger than the copper inside a six-foot-long power cord. Therefore, it is indeed very appropriate to think of free electrons in conductive metal as a gas, a relatively dense gas. It is the flow of these free charges, electrons in this case, which is "electricity." By definition, electronic devices are devices that use electricity to achieve their functionality. (By the way, copper metal is definitely neutral because the total number of electrons, some of which are free and some of which are confined near the nucleus, is equal to the total number of protons. So, we do not have to worry about large electrical forces pulling apart the copper wire!)

When a material has a relatively large number of free electrons, it is a good conductor of electricity, and is therefore referred to as a conductor. Figure 13-4 shows a list of materials that are good conductors. Silver is the best conductor, but is rather rare and expensive so it is not often used as a conductor. Copper is also a very good conductor and is relatively abundant and inexpensive so it is commonly used. The wire inside the walls of your house is routinely made of copper. Copper also conducts heat very well so it does not easily heat up, which is a great side benefit.

In the mid-1900s, housing wire was made of aluminum, which is extremely inexpensive and is a good conductor, but also oxidizes (rusts) easily, causing local areas of electrical insulation resulting in heat buildup and fire. Once this fault was verified, building codes were updated, disallowing the use of aluminum as a conductor. Aluminum is still used in microelectronics where the oxidation process can be better controlled. Also shown in Figure 13-4 is the numerical value of electrical conductivity for the conductors. The higher the conductivity, the better the conductor. The electrical resistivity is also shown in Figure 13-4. The electrical resistivity is the mathematical inverse of electrical conductivity. The higher the resistivity, the more resistive the material is to the flow of charge (electrons). Both conductivity and resistivity are commonly used to describe important materials used in electronics, so Figure 13-4 gives both values.

Other types of materials are found to be bad conductors of electric charge, or rather, they insulate against the flow of electricity, and are therefore called insulators.

Figure 13-4: **Common conductor materials and their conductive and resistive properties.**

Material	Conductivity, σ (Ω^{-1}-m^{-1})	Resistivity, ρ (Ω-m) [$\rho = 1/\sigma$]
Silver	63×10^6	16×10^{-9}
Copper	60×10^6	17×10^{-9}
Gold	45×10^6	22×10^{-9}
Aluminum	38×10^6	26×10^{-9}
Nickel	15×10^6	69×10^{-9}
Iron	10×10^6	96×10^{-9}
Sea water	5	0.2
De-ionized water	5.5×10^{-6}	2×10^5

Figure 13-5: **Common materials used as electric insulators.**

Material	Resistivity, ρ (Ω-m)
Glass (silicon dioxide)	$10^{10} - 10^{14}$
Polyvinyl chloride (PVC) [rubber]	10^{13}
Sulfur	10^{15}
Teflon	$10^{22} - 10^{24}$

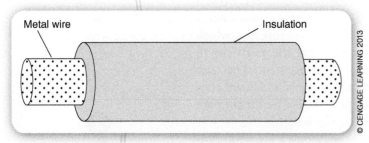

Figure 13-6: **Typical electrical wire design: solid or stranded metal wire (full of free electrons) surrounded by PVC insulation.**

Insulators have no free electrons; all the electrons get trapped close to the nucleus. Because insulators are poor conductors, they are better described by their resistivity. Figure 13-5 shows a list of materials commonly used as insulators, along with their resistivity values. The larger the resistivity, the better the insulator. The resistivity values of insulators are over 20 orders of magnitude (10^{20}) larger than conductors, dramatically designating these materials as poor conductors, and therefore excellent insulators. Paper is not shown in Figure 13-5, but is also a reasonable insulator. There is a very important class of materials appropriately called *semiconductors* that have resistivity values halfway between those of good conductors and good insulators. Semiconductors, like silicon, play a crucial role in modern microelectronics and will be discussed later in the chapter.

Why is the prefix *semi* used to describe semiconductors?

Most electrical wires are coated with a plastic-like material, PVC (polyvinyl chloride), a very pliable insulating material. Glass and various ceramics are routinely used as insulators in high-voltage power lines because they are reliable over the extreme fluctuations of temperature, heat, and moisture occurring outdoors. Oxides (silicon dioxide and aluminum oxide) are often used as insulators in microelectronic and micro-optical chips. Figure 13-6 shows the geometry of a typical insulated wire. It shows the gas of electrons, restricted to the (metal) conductor, surrounded by a layer of PVC. The outer PVC layer keeps the conducting wire from contacting other conductors, providing a well-controlled, and safe, flow of electricity. Even though this wire design appears quite simple to us in the 21st century, Ben Franklin and other early experimenters of electricity would have paid dearly for insulated wire during their numerous experiments with electricity.

Electrical Resistance

No material conducts electric charges perfectly, so even good conductors exhibit some resistance to the flow of charge. The total electrical **resistance** of a piece of material can be calculated using Equation 13.3. In Equation 13.3, R is resistance, ρ is resistivity, L is the length of the material, and A is the material's cross-sectional area. In Equation 13.3, the factor (ρ) holds all of the necessary information about the material while the fraction (L/A) contains all of the geometrical dependence. An example of a resistance calculation for a gold contact wire on a microchip is shown in Figure 13-7. A rectangular gold wire is 0.5 cm long, 2 μm high, and 10 μm wide. These are typical dimensions of a contact wire deposited on a microelectronic

resistance:
A measure of the resistance a material or component has to the flow of electricity (charge). The higher the electrical resistance, the more resistant the material or component is to the flow of charge (current). The unit of electrical resistance is volt-seconds per coulomb. These units of electrical resistance, being a bit complicated, are often referred to as an ohm. One ohm is 1 volt-second per 1 coulomb. The symbol omega (Ω) is often used to represent an ohm. For example, a resistance of 12 ohms is the same as writing 12 Ω.

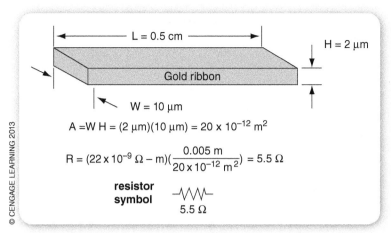

Figure 13-7: Example of a resistance calculation using a wire of rectangular cross-section (or wire "ribbon"). Ribbon wire of this approximate size is not uncommon for certain integrated circuits. Even though gold is a very good conductor, this calculation shows that even a good conductor can have appreciable resistance (5.5 ohms) if the cross-sectional area is small enough. Also shown is the circuit schematic symbol for a resistor.

circuit. (As an example of dimensions, 1 μm = 1 micrometer [μ stands for micro] = 1×10^{-6} m. For reference, a human hair is about 80 μm in diameter.) The calculation shows that the resistance is approximately 5 ohms, which is very appreciable. For comparison, the resistance of a standard six-foot long power cord is approximately 0.1 ohm.

$$R = (\rho)\left(\frac{L}{A}\right) \qquad (Eq.\ 13.3)$$

Also shown in Figure 13.7 is the symbol for a resistor: a triangle corrugated line. Resistors are, by far, the most simple and widely used electrical component.

> If the length of a resistor were doubled, would the resistance increase or decrease? By what factor would the resistance increase or decrease? If the cross-sectional area of a resistor were doubled, would the resistance increase or decrease?

Ohm's Law

One of the first practical and very useful laws of electricity was Ohm's law. This law was first proposed and proven by Georg Ohm around the year 1827. Ohm's law states that between any two points, the rate of flow of charge is equal to the voltage difference divided by the total electrical resistance. The important words here are "between any two points." The two points can be anywhere: two points in an electrical circuit, two points in space, two points in the atmosphere, or two points in your front yard. In mathematical form, Ohm's law is

$$I = \frac{V}{R} \qquad (Eq.\ 13.4)$$

or, with the terms rearranged (cross-multiplied), and in the more popular form

$$V = IR \qquad (Eq.\ 13.5)$$

Current. In Ohm's law, the variable "I" represents the rate of flow of charge, which is called current. The term *current* is very appropriate here because it represents a flow, just like current in a river or the flow of gas through a pipe. Remember, it can be helpful to think of the electrons in a conductor as a flowing gas. Being a flow rate, electrical current is measured in charge per second or coulombs per second (C/s) because a coulomb is the measure of charge. The unit of coulombs per second is most often abbreviated by the use of the term *amperes*, in honor of André-Marie Ampère, another famous scientist of electricity. Amperes are often further abbreviated as amps or A. In summary, current is a flow of electrical charge and a flow of 1 coulomb per second equals 1 C/s, which equals 1 ampere or 1 amp or 1 A.

voltage: A measure of energy that can be imparted to a charge. The higher the electrical voltage, the more energy a charge will have. The unit of electrical voltage is volts, which is often abbreviated with a V. For example, a 12-volt battery is equally referred to as a 12-V battery.

current: A measure of the flow rate of charge. Charge is measured in coulombs, and a typical flow rate is in coulombs per second or abbreviated C/s. Another name for C/s is an ampere, often abbreviated amp or the letter A.

Point of Interest

Because we cannot see what is actually happening in electrical circuits, comparisons with gravity can help us better understand some important principles of electricity. For example, to better understand electrical resistance, imagine three tall cylinders, say 1,000 meters high, and at the top of each cylinder is a large steel ball with a cross-sectional area the same as an average human (see Figure 13-8). The reason for this constraint on area will become clear soon. Let there be a vacuum in the first cylinder. That is, all air has been pumped out, leaving absolutely nothing in the way of the steel ball. The second and third cylinders are filled with air and water, respectively. Now let's examine what happens when the different steel balls are dropped.

In the vacuum, the ball will fall with an ever-increasing velocity. More specifically, the velocity of the steel ball in the vacuum will increase linearly with time over the entire fall, finally hitting the ground with a speed of 140 meters per second (m/s), or 313 mph.

As any skydiver would tell you, dropping the steel ball in the air-filled cylinder has a very different result. Due to the friction of air ("air resistance"), after falling for about 15 seconds, the steel ball's speed stops increasing, or "terminates," at about 53.6 m/s (120 mph). This speed is called the object's terminal velocity. The terminal velocity depends on, among other things, the cross-sectional area of the object and the density of the medium that the object is falling through. (We chose the cross-sectional area of the steel ball to be the same as a human skydiver so that the terminal velocity of the ball will be the same as that of a skydiver.) The terminal velocity of 120 mph is substantially slower than the case with a vacuum (313 mph), but the skydiver still needs a way to slow down before he or she contacts the ground! A parachute will accomplish this nicely by presenting a much larger cross-sectional area. A typical parachute reduces the terminal velocity to about 5 mph, a relatively safe landing speed for a human.

Being denser than air, water provides substantially more friction than air ("water resistance"), resulting in a terminal velocity of the steel ball of only 4 mph, without the use of a parachute.

Electrical resistance in materials is very similar to the friction in gravitational systems with air or water as the medium; the total resistance depends on material properties (density of the air or water) and geometrical factors like cross-sectional area.

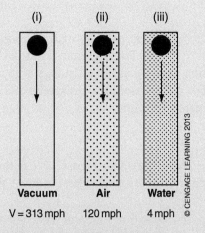

Figure 13-8: Air and water resistance experienced within a gravitational force is very much like electrical resistance. Shown here are three columns filled with different materials: (i) nothing (vacuum), (ii) air, and (iii) water. The resulting speed of the steel ball depends dramatically on the material within the column. This is like electrical current; the amount of electrons (current) depends on the material the electrons are passing through and on the geometry of the conductor.

 Using only information within the Point of Interest, and a calculator if needed, convert 60 mph to m/s.

An example of current might be useful. A flow rate of 1 ampere through the power cord of your computer means that 1 coulomb of charge is flowing through the wire every second. By the way, 1 amp of current corresponds to the flow of a lot of electrons, about 10^{19} electrons per second. A typical appliance in your home

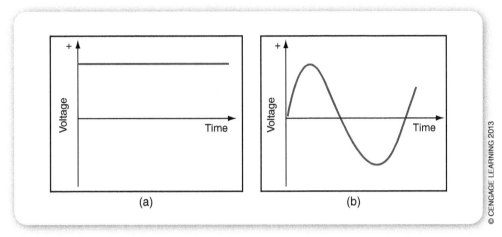

Figure 13-9: Graphs depicting typical behavior of (a) direct current (DC), and (b) alternating current (AC).

might flow between 1 and 10 amps of current. A digital output of the microprocessor in your MP3 player or computer would flow a much smaller current of approximately 0.02 amps of current, which can also be referred to as 0.02 A or 20 mA (or milliamps where milli refers to 10^{-3}), or a flow of 20 millicoulombs (mC) per second.

There are two types of current that flow in circuits: direct current and alternating current. **Direct current (DC)** is electricity that flows in only one direction and often, but not always, varies little with time. Batteries, photovoltaic cells (solar cells), and some types of generators produce direct current. Figure 13-9a shows that the voltage polarity in direct current remains the same over time.

Alternating current (AC) is electricity that is constantly switching the direction of flow. Alternating current, or voltage, is often sinusoidal in shape; that is, it has a shape like a sine wave in time. Figure 13-9b shows an example of AC voltage, where the sign of the voltage reverses over time. Per Ohm's law, when the sign of the voltage reverses, the direction of current reverses also. The wall receptacles found in buildings provide sinusoidal AC current, and most appliances, lights, and power tools operate on AC power. Although the basic principles of electricity apply to both DC and AC current, this chapter only deals with direct current.

Resistance. The variable R in Ohm's law represents the electrical resistance and is measured in ohms. The symbol omega (Ω) is often used to represent an ohm. Note that in Ohm's law (Eq. 13.4), the current varies inversely with R. When resistance increases, the flow of current goes down, which is just what happens with our gravitational analogy (see "Point of Interest"). With gravity, an object falling in a substance with higher resistance (water versus air resistance, for example) has a slower terminal velocity, which can be equated with a lower current (flow rate).

Voltage. The third variable in Ohm's law, V, refers to voltage. Voltage is measured in volts, often abbreviated V. The term *volt* is in honor of Alessandro Volta, an Italian scientist that first developed electric batteries. Being a key differentiator between various types of batteries, the term *volt* is fairly commonplace in everyday life. Batteries are typically purchased with a desired voltage. Also, power converters for MP3 players or laptop computers output a specific voltage (about 20 V).

direct current (DC): Electricity that flows in only one direction and typically, but not always, varies very little with time.

alternating current (AC): Electricity that is constantly switching the direction of flow. It is also often the case that the alternating current is sinusoidal in form; that is, it has a shape like a sine wave in time.

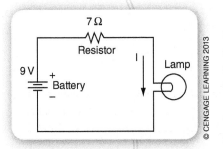

Figure 13-10: A simple battery-powered circuit that lights an incandescent bulb. A resistor is used to limit the current.

A detailed mathematical description of voltage requires Newton's calculus techniques and is not appropriate for this book. However, a gravitational analogy should prove useful to help understand voltage better. Voltage in electrical systems is like height in gravity; it measures the amount of potential energy that can be imparted to the charge. With gravity, an object like a steel ball that is 1,000 m above the ground has twice the potential energy as the same steel ball that is only 500 m above the ground. And because the steel ball that is 1,000 m above the ground has more potential energy, it will obtain a higher speed if dropped (in a vacuum). The gravitational analogy is so good that physicists and electrical engineers equally refer to the "voltage difference" between two points as the "potential difference" between two points, where "potential" is referring to potential energy. In electrical systems, the higher the voltage, the higher the energy that can be imparted to the charge. For example, in Ohm's law (Eq. 13.4), one can see that as voltage increases, the amount of the charge, or current, goes up as well. Also, it is important to note that, like potential energy in gravitation systems, *voltage* is a relative term—a voltage is always a relative measure between two points.

In summary, Ohm's law states that 1 ampere of current will flow through a 1-ohm resistor that has 1 volt of potential across it (one side of the resistor has a potential that is 1 V higher than the other side of the resistor). A few examples of Ohm's law may help clarify its use.

A simple electrical circuit diagram containing a battery connected to a resistor and a lightbulb is shown in Figure 13-10. Batteries and lamps are used frequently in electronics. So, standard symbols have been defined for these components and are appropriately labeled. Let's assume that the battery is a 9 volt (V) battery, the resistor has a resistance of 7 ohms, or 7 Ω, and the lightbulb provides negligible resistance. Ohm's law tells us that the current through the circuit is simply the voltage divided by the resistance, as shown in Equation 13.6, giving a value of 1.29 A (amperes), or 1,290 milliamperes (mA), where *milli* refers to 1/1,000 or 10^{-3}.

$$I = \frac{V}{R} = \frac{9V}{7 \text{ ohms}} = \frac{9V}{7\Omega} = 1.29 \text{A} \qquad \text{(Eq. 13.6)}$$

A resistor is very useful because it limits current, or voltage, in circuits. For example, in the previous example, if a resistor of 3 Ω was mistakenly used, the current would have increased substantially to 3 A, as shown in Equation 13.7.

$$I = \frac{V}{R} = \frac{9V}{3\Omega} = 3\text{A} \qquad \text{(Eq. 13.7)}$$

This value of current may be higher than the lightbulb can handle, thus burning out the lightbulb. Similarly, if a resistor of 100 Ω were mistakenly used, then the current would have been only 0.09 A, or 90 mA, leaving the lightbulb dark. In summary, Ohm's law uses the parameters of resistance and voltage to describe the flow of current between any two points. The key words again are "between any two points." Resistors, the most common electrical component, are very useful for setting the appropriate level of current.

Kirchhoff's Laws

Two very powerful techniques used in designing and analyzing almost all electrical circuits are called the Kirchhoff's laws of electricity. Kirchhoff's laws are often referred to as the conservation laws, simply because they are the electrical equivalent to the conservation laws of Newtonian physics. For example, in Newton's laws, mass is conserved: Mass cannot be created or destroyed, so mass is referred to as a "conserved" quantity. The first law is Kirchhoff's current law. It states that the total flow of current into a node must equal the total flow of current out of a

Fun Facts

In the mid-1700s, Benjamin Franklin wrote, "We say **B** (and other bodies alike circumstanced) are electrised positively; **A** negatively. Or rather **B** is electrised plus and **A** minus." In this statement, he became the first person to coin the terms *negative* and *positive* when referring to electricity. At the time, Franklin was recognized as one of the leading scientists in the world studying the properties of electricity (see Figure 13-11).

This attribute of charge as innately having "opposites" is very important. The opposite nature of charge not only leads to a correct understanding of electrical forces but also introduces the concept of polarity. With electric fields and forces, there is a definite direction to the forces and therefore a definite direction to charge flow. This is referred to as polarity. For example, when a component, like a resistor, is used in a circuit, one side of the component typically has a higher potential (voltage) than the other, which is sometimes noted with + (positive) and − (negative) signs written at the ends of the component. The positive side of the resistor has a higher voltage than the negative side. The component can be said to be polarized, or have a certain polarity. The polarity of a component is important because it sets the direction of current flow; a positive charge will flow from the positive side (higher potential) to the negative side (lower potential). Polarity is also like gravity; its pulls in a certain direction. But unlike gravity, electrical systems can be made to pull in either direction, not just down.

A resistor is symmetric, in that it can be polarized in either direction. That is, charge can be made to flow just as easily in either direction through a resistor. However, this is not true in general for all electrical components, so care must be taken to connect electrical components in the appropriate direction, or with the appropriate polarity. Examples of components that require specific polarizations are diodes, light-emitting diodes (LEDs), electrolytic capacitors, transistors, and DC motors (the polarity determines direction of rotation).

(a)

(b)

Figure 13-11: Benjamin Franklin was very interested in the science and application (technology) of electricity. This photo (a) shows his invention, the lightning rod, above the orange external fuel tank of the space shuttle *Atlantis* at the Kennedy Space Center in Florida. Back in the 1700s, especially in Europe, people were very skeptical of lighting rods. They thought they interrupted nature and were therefore unnatural and dangerous.

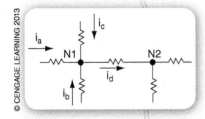

Figure 13-12: Portion of a circuit containing several resistors showing two nodes, N1 and N2.

node. A node is simply any single point in a circuit where more than two wires join together. Examples of nodes, labeled N1 and N2, are shown in Figure 13-12.

Kirchhoff's current law is a statement of charge conservation. Charge cannot be created or destroyed, so what flows in must flow out. This is similar to the flow of water or gas in a network of pipes; what flows in must flow out (assuming there are no holes in the pipes, which is usually a pretty good assumption). If we assign a positive sign to current flowing into a node and a negative sign to current flowing out of a node, then a more common version of Kirchhoff's current law states that the sum of currents passing through a node must be equal to zero. For example, in Figure 13-12, there are four wires that join together at node N1. Therefore, because Kirchhoff's current law states the sum of currents into a node must be zero, we have $i_a + i_b + i_c - i_d = 0$. In this particular case, the sign in front of i_d is negative because i_d is leaving the node while all the other currents will be positive because they are entering the node.

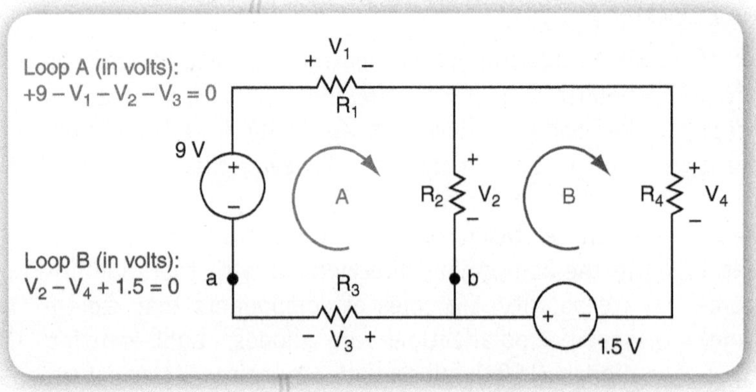

The second law, the Kirchhoff's voltage law, states that the sum of voltages around any closed-circuit path, or loop, must be equal to zero. This is the same as the conservation of work in a gravitational system. Say, for example, you live on the side of a large hill, and you go out for a walk, meandering up to the top of the hill, and then return back home. From a physics point of view, you do positive work walking up the hill, and perhaps a lot of positive work if the hill is tall. However, an equal amount of negative work is done when you walk down the hill. Walking up the hill is hard because you have to work against the pull of gravity. However, walking down the hill is quite easy because gravity is now working on you! Taken as a whole, walking up the hill and returning to the same place (home in this case), a net work of zero is accomplished. Examples of Kirchhoff's voltage law are shown in Figure 13-13. Two loops in a circuit are shown: Loop A and Loop B. Each loop is shown as a clockwise loop, although the direction of the loop does not matter. The equations to the side of Figure 13-13 indicate that the sum of all of the voltage drops in a loop must sum to zero. For ease, the voltage drop variable over each resistor is chosen to have the same subscript as the resistor subscript. Also, as a reminder, using Ohm's law, you can calculate the voltages. For example, V_1, the voltage over resistor R_1, is given by the equation $V_1 = (I_1)(R_1)$ where I_1 is the current through resistor R_1.

Figure 13-13: A circuit with two loops, one shown in green (A) and the second in red (B). These two loops are used to demonstrate Kirchhoff's voltage law: The sum of voltages around any closed loop is zero.

Magnetism

Magnets and magnetism are used frequently in electronic devices and systems. Magnetism is a commonly observed attribute of materials, and is actually a consequence of electricity. Magnetic forces are due to moving charges. In general, there are two types of magnets: (1) so-called permanent, or semipermanent, magnets; and (2) electromagnets. With permanent magnets, the material itself is magnetic. Materials like iron, called lodestone in ancient times, as well as nickel and copper can become permanent magnets. Permanent magnet materials are magnetic due to the rotation (that is, motion) of an inner-orbiting, non–free electron about its own axis (much like the earth rotates about its own axis). With materials made up of billions of atoms, when each of these tiny magnetic fields (a "rotating" electron) line up, a strong magnetic force is created: a so-called permanent, or semipermanent, magnet. Figure 13-14 gives a diagram showing a permanent magnet. This figure shows how

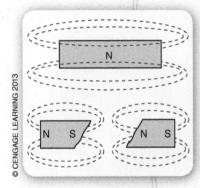

Figure 13-14: Magnetic fields around a permanent magnet are elliptical-like in shape, and extend from the south pole to the north pole. When a permanent magnet is split in two, the separate pieces are still magnets, each with south and north poles, but are only smaller.

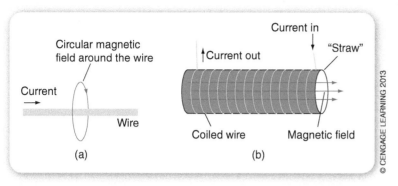

Figure 13-15: Magnetic fields are created by moving charge. Diagram (a) shows how a (circular) magnetic field is formed around a straight wire that carries a constant current. The principle shown in diagram (a) can be used to form a straight magnetic field by coiling the wire into a circular pattern (b), like wrapping a wire around a straw. With coiled wire, the magnetic field is formed "inside" the straw, parallel to the central axis of the straw.

the magnetic field starts at the south (S) pole and traverses to the north (N) pole. Also, as shown in the diagram, even if a permanent magnet is cut into sections, magnetism is maintained but only with smaller pieces.

The second type of magnet, the electromagnet, is constructed in its most simple form by wrapping wire in a circular fashion forming a cylindrical shape. Imagine wrapping wire around a straw, starting from one end of the straw and working toward the other end of the straw. This simple type of magnet is shown in Figure 13-15. When current is passed through the circular paths of wire, a magnetic field is created in a direction that is parallel to the axis of the cylinder (parallel to the "straw"). When a magnetic material, like iron, is placed in the center of the cylinder (the "hollow" portion of the straw), the magnet field is enhanced and results in a stronger magnetic field.

MAJOR CIRCUIT COMPONENTS

There are three basic electronic circuit elements: (1) resistors, (2) capacitors, and (3) inductors. Almost every circuit requires detailed knowledge of at least one of these elements. However, in many circuits, detailed knowledge of all three types of components is required. Briefly stated, resistors only consume energy but are extremely useful at controlling voltage and current levels, so are, by far, the most widely used component. Capacitors and inductors are unique in that they can store energy. The ability of a component to store energy leads to an important feature, energy being released over time.

For reference, a list of schematic symbols for several widely used electrical components is given at the end of this chapter (see Figure 13-53).

Resistors

Resistors are, by far, the most widely used electrical component. Resistors are generally constructed by either using carbon or by winding a wire of higher-resistance metal around a ceramic core. Carbon resistors are the most common and least expensive. Wire-wound resistors are generally used in higher-current conditions where the effect of the wound coil of wire will not adversely affect the circuit (see

Point of Interest
Resistors control voltage and current

Resistors are critical in controlling voltage and current. For example, certain devices may require less voltage than the available voltage source supplies. Fortunately, resistors provide a way to control the voltages, and currents, in circuits.

For example, assume you have a 10-V power supply but a relay that you want to use has a maximum voltage limit of 6 V. As shown in the "voltage divider" circuit in Figure 13-16a, an output voltage, V_{out}, of 6 V can be obtained by using resistor values of $R_1 = 6\ \Omega$ and $R_2 = 4\ \Omega$.

$$V_{out} = 10V\left(\frac{6\Omega}{6\Omega + 4\Omega}\right)$$

$$= 10V(0.6) = 6V \quad \text{(Eq. 13.8)}$$

The voltage divider circuit is also an example of a **series circuit** because there is only one path for current. In a series circuit, an example of which is given in Figure 13-16a, all of the current must flow through R_1 and R_2 before returning to the voltage supply (V_{in}).

Some devices require a certain level of current to operate properly. Fortunately, resistors also provide a way to control current. For example, assume you have a 2 A current supply but a device that you want to use has a maximum current limit of 0.5 A. One could certainly limit the current by using a single large resistor (in series), as demonstrated in Figure 13-10. However, this also limits the current to every other component in the (series) circuit. Another method is possible and is shown as the "current divider" circuit in Figure 13-16b. Using a current divider, a current of 0.5 A can be delivered through R_D, your device, by constructing a parallel circuit. A **parallel circuit** is one where current has more than one path to follow. A parallel circuit is like the water pipes in your house. One water pipe comes into your house (appropriately referred to as your "main" pipe). However, the water can flow in a variety of paths. For example, water can flow to one of several bathroom or kitchen sinks or to the washing machine. The water can flow in several directions simultaneously, so many of your water pipes are connected in parallel. In the electrical circuit shown in Figure 13-16b, the total current splits into two paths: one path that follows resistor R and a second path that follows resistor R_D. The currents through each of these paths are labeled I_R and I_D, with the subscripts referring to which path (resistor) the current takes. Assuming your device, R_D, has a resistance of 30 Ω, then choosing a value for R of 10 Ω will result in a current of 0.5 A through your 10 Ω device:

$$I_4 = 2A\left(\frac{10\Omega}{10\Omega + 30\Omega}\right)$$

$$= 2A(0.25)$$

$$= 0.5A \quad \text{(Eq. 13.9)}$$

Both voltage and current dividers are routinely used in circuit designs.

series circuit:
A circuit that has only one path for the current to flow. Components connected in such a manner are often referred to as "in series."

parallel circuit:
A circuit that has more than one path for the current to flow. Components connected in such a manner are often referred to as "in parallel".

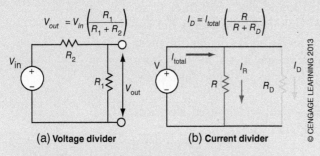

(a) Voltage divider (b) Current divider

Figure 13-16: Resistors are routinely used to control voltage and current levels in circuits. Here, simple series and parallel circuits are used to control voltage, or current, by forming "voltage and current dividers."

the section on inductors in this chapter). Resistors are manufactured to specific tolerances. You can purchase resistors that are made to an accuracy of ±20 percent, ±10 percent, ±5 percent, or ±1 percent of their stated value. Resistors that are manufactured more precisely are more expensive. Resistors are also made to operate within certain power conditions. Resistors in small consumer electronic products are typically rated at 1/8 or 1/4 watt, although resistors rated at much higher operating power are available. Electrical power is defined at the end of this section.

Everything about a circuit is known once all the voltages and currents are known. Hence, the relationship between voltage, V, and current, I, is extremely important. A resistor is the simplest component, because the voltage over a resistor is linearly related to current. That is, $V = IR$ (Ohm's law), where the proportionality constant is simply the resistance, R. Due to this simple linear relationship, resistors can be referred to as a linear component. Resistors cannot store energy. Resistors can only dissipate, or use up, energy.

Power Consumption of Resistors

Consumers of electronics equipment routinely demand three very important items: (1) low cost, (2) small size, and (3) low power consumption. Without substantial advances in cost, size, and power, the widespread use of consumer electronics—products like MP3 players, laptop computers, and cell phones—would simply not be possible. For example, would your cell phone or MP3 player be useful to you if you had to recharge it every hour? The power consumption, cost, and size of many electronic devices had to be reduced to achieve widespread acceptance. The amount of electrical power consumed or generated by electric components is indeed a critical consideration. The power consumed or created by any circuit or component is given by a very simple formula (see Equation 13-10). The power, P, consumed or generated by *any* device is simply the current through that device times the voltage over that device, with the direction of the current and voltage determining the sign of the power (whether power is being delivered or consumed). The units of power are watts. In the formula for power (Equation 13-10), if current is in units of amperes and voltage is in units of volts, then the calculated power will be in watts. For example, if an electrical device uses 2 A of current while being operated and the voltage over the device is 12 V, then the device is consuming (2 A)(12 V), which equals 24 watts, or 24 W:

$$\text{Power} = (\text{current})(\text{Voltage}) \quad \text{or}$$
$$P = IV \qquad \text{(Eq. 13.10)}$$

For resistors, Ohm's law also gives a simple relationship between voltage and current. So, we can write two very useful and simple formulas for the power a resistor consumes. Ohm's law tells us that $V = IR$, and equivalently that $I = V/R$. Inserting these two formulas into the general formula for power, $P = IV$, tells us that the power consumed by a resistor, P_r, is given by either of the two following formulas:

$$(i) \; P_r = IV = I(IR) = I^2R \quad \text{or}$$
$$(ii) \; P_r = IV = \left(\frac{V}{R}\right)V = \frac{V^2}{R} \qquad \text{(Eqs. 13.11 and 13.12)}$$

For example, for a known resistor, only knowledge of the voltage over the resistor or the current through the resistor is needed to calculate the power dissipation of the resistor. As an example, if the current through a 1,000 Ω resistor is 20 mA, then the power consumed by the resistor is $(20 \text{ mA})^2(1{,}000 \text{ Ω}) = (0.020\text{A})^2(1{,}000 \text{ Ω}) = 0.4$ watts = 0.4 W.

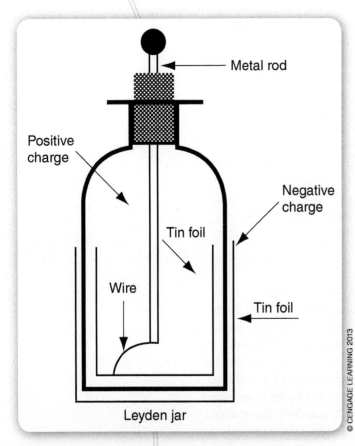

Figure 13-17: A typical Leyden jar design; an insulator (the glass jar in this case) surrounded by two sheets of (metal) conductor.

Figure 13-18: An array, or "battery," of Leyden jars can be used to increase the voltage (or current).

Let's say you only know the voltage over a resistor. You can still calculate the power dissipated (consumed) by the resistor. As an example, if the voltage over a 1,000 Ω resistor is 1 V, then the power consumed by the resistor is $(1\ V)^2/(1,000\ \Omega) = 0.001$ watts = 1 mWatt = 1 mW.

Capacitor

Capacitors are nonlinear components, in that the relationship between the voltage and current through a capacitor is not a simple one. Calculus is required to accurately describe this relationship. The important aspect of capacitors though is that they are capable of storing energy, a very useful capability. Even though the math for capacitors is more complicated, capacitors themselves are physically simple. A capacitor is made by placing a sheet of insulator between two sheets of conductor (metal). The first capacitor was the Leyden jar, named after Leiden University where it was first developed. This early capacitor was created using a thin foil of metal on the inside of the glass jar and another thin foil of metal wrapped around the outside of the jar (see Figure 13-17). In a Leyden jar, the glass walls of the jar serve as the insulator. On the inside, a metal rod is inserted through a rubber stopper and a flexible wire on the bottom end rests against the inside metal foil, making electrical contact. The metal foils on the inside and outside of the glass jar do not directly touch each other.

Benjamin Franklin utilized Leyden jars and referred to several jars hooked together as a battery (after a "battery" of cannon, which it looked like). Figure 13-18 shows a battery of four Leyden jars that could store a substantial charge. Several Leyden jars connected together (in series) could be used to obtain higher voltages, a useful tool in electricity experiments. The strength of a capacitor is given by its capacitance, C. The higher the value of C, the higher the capacitance, and the more energy that can be stored. Capacitance is measured in farads (named after Michael Faraday) and is abbreviated by the letter F. Capacitances are typically very small so are often measured in microfarads (1 microfarad = 1 μFarad = 1 μF = 1 $\times$ 10^{-6} Farads).

Typical operation of a capacitor relies on time-varying electric currents or voltages. This is perhaps apparent when considering the physical design of a capacitor. A capacitor is essentially two metal plates (like the foils surrounding the glass jar) separated by a thin insulating layer (the glass of the jar). Figure 13-19 shows a generic picture of a capacitor. With a DC voltage applied, a capacitor is equivalent to an infinite resistor because no current can flow through the insulator. Remember, Ohm's law states that $I = V/R$, and an insulator has a huge (infinite) resistance, so the current is zero. What is special about a capacitor is that when the voltage becomes time-varying, the effective resistance of the insulator reduces substantially, allowing time-varying current to flow easily through the capacitor. To be more specific, the effective resistance reduces as the frequency of the time-varying voltage increases: The faster the voltage signal changes with time, the lower the effective resistance, resulting in more current flow.

Inductor

Inductors are also nonlinear components, in that the relationship between the voltage and current through an inductor is not a simple one. Like a capacitor, calculus is required to accurately describe this relationship. The important aspect of inductors is that they, too, are capable of storing energy. An inductor is made by coiling wire into a circular fashion around a cylinder, like wrapping wire around a straw (see Figure 13-20). The wire is typically coated with a thin plastic so the metal in the coils do not touch one another. Very much like the electromagnet described earlier, when current flows in this configuration of wire, a magnetic field is created down the center of the cylinder. Central to the operation of an inductor is that the direction of the magnetic field opposes the flow of current that caused the magnetic field in the first place, thus reducing the current. Inductive elements are found in most circuits. A common application for an inductor is in the ignition system of cars. The strength of an inductor is given by its inductance, L. The larger the value of L, the higher the inductance. Inductance is measured in henrys (after Joseph Henry), abbreviated by the letter H. Like capacitors, the values of L are typically quite small and are routinely quoted in microhenrys (10^{-6} henrys).

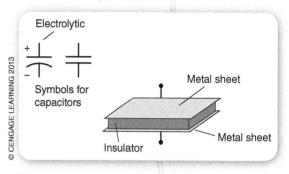

Figure 13-19: Capacitors are essentially two parallel conductors separated by an insulating material. The operation of a capacitor depends on time-varying electric fields. A capacitor is often used to store energy. The symbols for a capacitor are also shown.

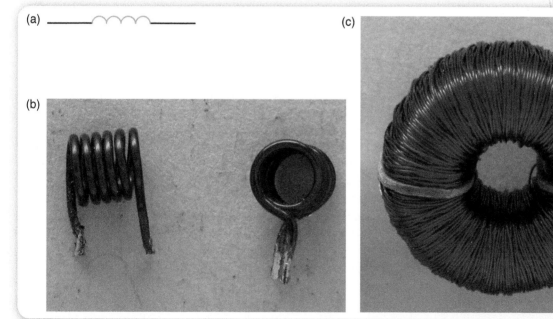

Figure 13-20: An inductor is typically a coil of wire, available in various shapes and sizes. To keep the metal coils from touching one another, the wire is typically coated with a very thin layer of insulation (plastic). The center may be filled with air or other materials, like iron, that will affect the magnitude of the inductance. The operation of an inductor depends on a self-induced magnetic effect. An inductor can store energy. The symbol for an inductor is also shown.

ELECTRONIC SYSTEMS DESIGN

Electronic systems are a series of electrical components that work together to control, monitor, or measure. As you learned in Chapter 1, a system consists of (1) inputs, (2) processes, and (3) outputs (also see Figure 13-21). Because electrical systems are routinely comprised of many subsystems, this three-part segmentation of input-process-output is particularly important. Of mission-critical importance at the beginning of any electronic design is a detailed definition of all inputs, processes, and outputs.

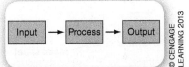

Figure 13-21: **An electronic system consists of inputs, processes, and outputs.**

Input. The input section of an electronic system often consists of one or more sensors. These sensors convert some physical phenomenon, such as heat, light, humidity, radiation, or magnetism, into an electrical signal, a signal that contains some kind of information that can be understood and processed by other parts of the circuit. The information may be a voltage level that varies with input, or something more complex like security-encoded information being sent to a bank's computer. Inputs may also be, and often are, outputs from other electrical subsystems. The connection of multiple subsystems is very common in electrical design so designers have developed a habit of saving and documenting designs that might be useful as a subsystem for future use. The use of clearly defined subsystems is also an excellent manufacturing strategy to achieve low cost because it is typically easier to manufacture many of one item as opposed to many different items.

Process. Modifying or conditioning ("processing") of the input signal is often required before the output. Devices that switch, time, compare, or amplify, or any combination of these, provide the control between the input and the output. Due to its extreme level of flexibility and low cost, a very effective processor of inputs is a semiconductor-based microprocessor. Microprocessors are discussed in more detail later in this chapter.

Output. Output devices take the processed electronic signal and convert it back into some usable form. Output devices may be transducers or **actuators** that convert electrical signals into physical phenomenon like sound (speakers and buzzers), light (lamps, displays), or movement. Actuators are devices that cause movement. Motors and solenoids are examples of actuators. Outputs may provide information for human users, or they may be simply electrical signals that are to be input into other electrical systems or devices. For example, when you type a website name into your computer and press <return>, this becomes an output signal that is encoded and sent out, eventually being received by a server. The server treats this signal as one of many inputs, processes it (decodes it), and responds by sending updated information (the new website page) back to your computer, resulting in an updated display.

A good example of an electronic system with clearly defined input, process, and output would be an audio system in a school or concert hall (see Figure 13-22). A microphone is used as an input. The microphone is a transducer that translates voice energy (air pressure variations) into a small, time-varying electrical signal. The microphone acts as the input. Because the voltage signal produced by the microphone is extremely small, an amplifier must be used to increase the signal strength. The electronic amplifier is the important process to be performed. A good amplifier will increase the signal level without adding too much noise (bothersome "hum"). Finally, the signal is distributed to a series of speakers that comprise the output to human ears. Speakers are another type of transducer that converts electrical signals back into pressure vibrations (sound).

OFF-ROAD EXPLORATION

For more information on how sound can be produced see:
http://en.wikipedia.org/wiki/Loudspeaker ("cone" speakers)
http://en.wikipedia.org/wiki/Subwoofer ("cone" speakers)

Chapter 13: Designing Electrical Systems   367

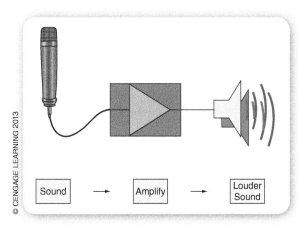

Figure 13-22: An electronic audio system, like that used in a concert hall or a school, is a good example of an electronic system that consists of an input (sound picked up by a microphone transducer), a process (audio amplification), and output (audible sound via speakers).

Figure 13-23: A bicycling computer is an example of a device that has multiple integrated sensors. Popular bicycling computers include sensors for measuring altitude, temperature, and power as well as position, speed, and time.

System Inputs

Electronic systems often need input from the real, physical world, as opposed to input from another electrical system. Sensors provide this doorway between the physical world and the electronic world by converting physical properties into electrical signals. Electronic sensors detect either the presence or absence of a physical property or the actual amount of the property. For example, a simple sensor may detect if the environment is either cold or hot. However, a more functional detector may go one step further and measure the actual temperature, in perhaps degrees Celsius. Some common input sensors include switches, variable resistors, or sensors designed to measure a unique property of the environment. Sensors that measure presence of light, heat, magnetism, humidity, motion, strain, acceleration, position, radioactivity, visual clarity, and numerous other physical properties are routinely used.

Modern microelectronics have revolutionized the sensitivity and size of input sensors. It is now common to see multiple "microsensors" in single compact devices. For example, small and lightweight bike computers are available that measure and display temperature, average and maximum speed, pedal cadence, time, date, elapsed time, and power output as well as altitude (see Figure 13-23). Compact global positioning system (GPS) function is also now routinely incorporated into compact bike computers.

Light. Light, a form of electromagnetic radiation, travels through tiny particles known as photons. Photons are theoretical in nature in that they have zero mass. To detect photons of light, a photovoltaic cell can be used as a sensor (see Figure 13-24). Photovoltaic ("solar") cells are devices often made out of silicon that produce a small amount of voltage and current when exposed to light. Solar cells are quite sensitive and are available in a wide variety of shapes, sizes, and outputs. Researchers have recently developed methods for making solar cells with the use of special polymers (plastics), which offer the use of flexible solar cell sheets, and may one day provide an extremely inexpensive conversion of light into electric energy.

Another device sensitive to light energy is the photoresistor, or light-dependent resistor (LDR). A typical LDR, shown in Figure 13-25, has a resistance of 150 megaohms (1 megaohm is 1×10^6 ohms) in the dark but drops substantially to only 2,000 ohms when the LDR is exposed to ambient room light. LDRs are widely available, very inexpensive, and are very handy for providing light-sensitive action. For example, if an LDR is connected in series with a voltage source and a light source, as shown in Figure 13-25a, the output current in the series circuit varies directly with

Figure 13-24: Photovoltaic cells, often made of silicon, convert light energy to electrical energy.

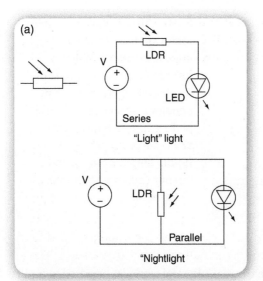
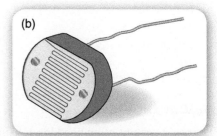

Figure 13-25: A photoresistor, also known as a light-dependent resistor (LDR), changes its resistance with light level. LDRs can be very useful for controlling a circuit with light levels (a). Shown here are a picture of an LDR (b) and the symbol for an LDR. Also shown are two circuits that turn on an LED depending on if it is either light or dark out (a night light).

REPRINTED WITH PERMISSION FROM GATES, *INTRODUCTION TO ELECTRONICS*, FIFTH EDITION. COPYRIGHT © 2007 DELMAR CENGAGE LEARNING.

the light hitting the sensor. In this series circuit, as light hits the LDR, the resistance of the LDR decreases, resulting in more current flowing to the light (via Ohm's law), thus making the light brighter. The opposite function can also be designed to obtain operation like a night light. That is, a light turns on when it's dark; otherwise, the light is off. To accomplish this function, an LDR is connected in parallel with an output light source, as shown in Figure 13-25b. A light-emitting diode (LED) is a very efficient and easy-to-use light source for such a circuit. As before, the resistance of the LDR decreases substantially when light hits the LDR. However, in this case, when the resistance of the LDR goes low, it enables most of the current to flow through the LDR, leaving very little current passing through the second path of the parallel circuit (the path with the LED), leaving the light turned off. Conversely, when it is dark, the resistance of the LDR is quite high, resulting in most of the current going through the LED, turning the LED on. To help understand the operation of this parallel circuit, it is again helpful to think of the circuit as a set of water pipes. Remember, electrons are like a gas, or a fluid. In a set of water pipes that split into two paths ("parallel" paths), more water flows through the path that presents the least resistance to the flow of the water. For example, perhaps one of the paths is mostly clogged with debris (more resistance) so most of the water will flow through the clear pipe, leaving only a trickle of water flowing in the clogged pipe. The same thing happens with the electrical system: More current flows through the path of least resistance.

Heat. Another form of electromagnetic radiation is heat. A thermocouple is a device made by "coupling together" (that is, putting into physical contact with each other) two dissimilar metals, like nickel-chrome (chromel) with nickel-aluminum (alumel). When heat is applied to the junction, a small voltage is generated across the metallic junction at the "couple." This small voltage changes in a predetermined way with temperature.

The thermistor is another device used for detecting temperature changes (see Figure 13-26). Similar to how an LDR changes resistance in the presence of light, a thermistor works by changing resistance as it is heated or cooled. Thermistors are most often used to detect the presence of temperature change because they are inexpensive and are widely available in a number of heat-sensitivity ranges.

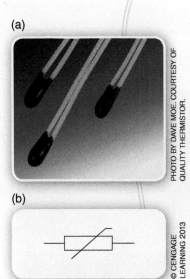

Figure 13-26: (a) A thermistor, shown here, changes resistance with temperature. (b) Also shown is the schematic symbol for a thermistor.

Temperature sensors can also be fabricated directly with semiconducting materials, resulting in very low cost. Because of their low cost, wide availability, and accuracy, these solid-state, semiconductor-based thermometers are used extensively.

Sound. Sound is a series of pressure changes traveling through air, high-pressure regions followed by low-pressure regions. A crystal microphone, shown in Figure 13-27, produces a small time-varying voltage when these waves of pressure change, causing a diaphragm to move and distort a quartz crystal. An effect in the quartz crystal called the piezoelectric effect is responsible for creating the voltage. A quartz crystal can produce a relatively high-voltage output and is therefore quite sensitive to low sound levels.

A type of microphone called a dynamic microphone produces a voltage when sound, striking a diaphragm, causes a coil to move through a magnetic field. In many ways, a dynamic microphone is like a common speaker cone, but in reverse.

A picture of a common wireless (Bluetooth) microphone used frequently in cell phones is shown in Figure 13-28. Like other sensors, recent advances in semiconductor technologies have revolutionized microphones. Figure 13-29 shows a picture of a

> **OFF-ROAD EXPLORATION**
>
> See the following websites for information on temperature sensors:
> http://en.wikipedia.org/wiki/Thermocouple
> http://en.wikipedia.org/wiki/Thermister

> **OFF-ROAD EXPLORATION**
>
> See the following websites for information on audio speakers and microphones:
> http://en.wikipedia.org/wiki/Loudspeaker
> http://en.wikipedia.org/wiki/Piezoelectric_speakers#Piezo_tweeter
> http://en.wikipedia.org/wiki/Microphones

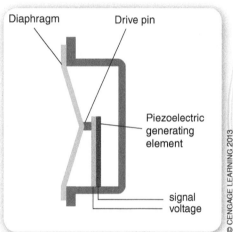

Figure 13-27: Various piezoelectric speakers and a diagram of the design of a piezoelectric speaker design.

Figure 13-28: Wireless (Bluetooth) microphones are often used with cellular telephones. These small headsets contain both a miniature speaker and microphone.

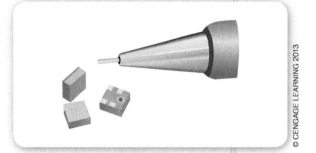

Figure 13-29: A really tiny microphone, measuring only a few hundred microns across, fabricated with semiconductor-based processes (an example of a microelectromechanical system [MEMS]). For reference, these MEMS microphones are shown next to the top of a mechanical pencil.

OFF-ROAD EXPLORATION
http://en.wikipedia.org/wiki/Microphone#MEMS_microphones

very small microphone made using silicon-based microelectromechanical systems (MEMS) technology. This tiny microphone only measures a few hundred microns across (1 micron = 1 micrometer = 0.0001 cm). For reference, a human hair is only about 80 microns in diameter.

Position. A position change can be detected by a sensor. A simple switch detects binary position—a switch is either in the "on" or "off" position, for a total of only two possible positions. A variety of different types of switches are useful in electronics, including push button (normally on or off), single-pole-single-throw, single-pole-double-throw, double-pole-double-throw, or multiple position switches (see Figure 13-30).

Even though switches are extremely useful, there are applications where knowledge of a continuous number of positions is required. For example, a volume control on a radio needs to detect a variety of possible positions, either straight-line position, as a slide control, or rotary position like with a knob. Variable resistors, also called potentiometers (*pots* for short), are useful for translating physical position into a change in resistance (see Figure 13-31). A potentiometer has a piece of carbon or other resistive material with an electrical connection on each end. In addition, a terminal on a movable wiper rubs on the carbon material. The wiper's position moves with the potentiometer's sliding or rotating arm, and this changes the amount of carbon material, and thus the resistance between the wiper and each end of the carbon piece. Potentiometers come in either linear or logarithmic versions. For a linear pot, the resistance is linearly related to the position, while the resistance for a logarithmic pot changes logarithmically with position. A logarithmic pot is typically used in audio systems because the human ear is itself a logarithmic sensor. So, for example, a logarithmic variable resistor is required to have it appear to a human ear that a setting of 5 on a volume knob sounds half as loud as a setting of 10 on a volume knob.

Figure 13-30: *There are many types of switches for a wide range of applications.*

Figure 13-31: *Potentiometers are variable resistors. They are available in rotating (shown here) or sliding versions.*

System Outputs

Output devices can be grouped into three categories: (1) displays, (2) actuators, and (3) transducers. Displays provide visual information, actuators make movements, and transducers convert an electrical signal into a physical property.

A lightbulb is a simple display component. It can be used to show that a device, such as a computer, is on when the light is lit. A motor is an actuator, because it provides rotational movement when electricity is applied. A speaker is a transducer, because it converts a voltage signal into sound, a physical property. Examples of some output devices are meters, lamps, LEDs, seven-segment LED displays, flat-panel LCDs (liquid crystal displays), cathode ray tubes ("televisions"), solenoids, motors (AC, DC, or stepper), electromagnets, relays, speakers, headphones, horns, buzzers, and heaters.

Light-Emitting Diodes. Light-emitting diodes (LEDs) are devices that very efficiently emit light, using very little electrical current. LEDs use such little power that most traffic lights and motor vehicles now use LEDs as opposed to power-hungry filament-based incandescent bulbs. Maintenance costs are also reduced substantially because LEDs have long lifetimes. LEDs come in a variety of colors: red, green, blue, amber, infrared (invisible to the human eye), and even white (many-colored). LEDs also come in a variety of shapes, sizes, and optical powers (see Figure 13-32). LEDs are typically connected to a direct current (DC) power supply in such a way that the negative terminal is attached to the negative lead of a battery or current source. The negative side of a source is also called the cathode. LEDs have a flat spot next to the cathode wire (see Figure 13-33). Noticing that the "flat" looks like a minus sign is an easy way to determine the polarity of an LED. The other lead, the positive wire, is called the anode, and it is connected to the positive terminal of the current source. LEDs are typically connected in series with a resistor to limit current (see Figure 13-33).

Seven LEDs can be arranged in a figure-eight pattern, forming a display called a seven-segment display. A seven-segment display is shown in Figure 13-34. Lighting various combinations of the seven LEDs allows the numbers between zero and nine to be displayed. It is also possible to form several letters. Seven-segment displays are available in two different types: common cathode and common anode. As the names suggest, common anode displays have all seven anode (+) leads connected together while common cathode displays have all seven cathode (−) leads connected together.

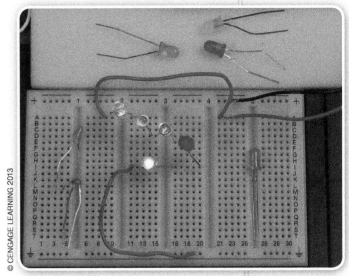

Figure 13-32: Light-emitting diodes come in a variety of shapes, sizes, colors, and powers.

Actuators. Output devices that cause movement are called actuators. A variety of devices fall into this category, but only a few of the most common will be discussed here.

Direct current motors are used to make rotary motion (see Figure 13-35). Direct current motors come in an almost endless variety of sizes and shapes, and are typically very inexpensive. Most small DC motors contain permanent magnets and electromagnets (see Figure 13-36). The DC voltage applied to the motor produces an electromagnetic field within one coil on the rotor. The magnetic field produced repels a permanent magnet fixed inside the shell of the motor, and this causes the shaft to rotate a partial turn. As the shaft turns, the electrical contacts,

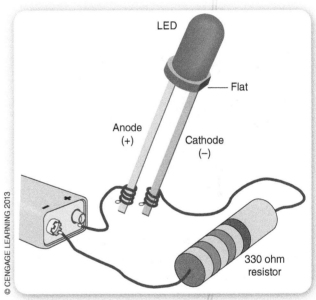

Figure 13-33: A resistor is typically used in an LED circuit to limit the current. An LED has a set polarity. Shown here is the flat portion of an LED, depicting the minus (−) side of an LED.

called the commutator, switch the DC voltage to another coil, which causes the electromagnetic field to be produced in a slightly different location. But, because the shaft has turned a bit, this location is back where the original field was produced when the motor was first connected to the voltage source. The like magnetic fields of the electromagnet and the fixed permanent magnets now repel again, and the rotor and shaft rotate a little more. This process repeats, causing a continuous rotation. A substantial disadvantage of DC motors is that they are either on or off. Therefore, the motor is either rotating at tens, hundreds, or thousands of rotations per minute, or it is not rotating at all. When rotating, the mechanical load placed on the motor shaft typically limits the speed of a DC motor.

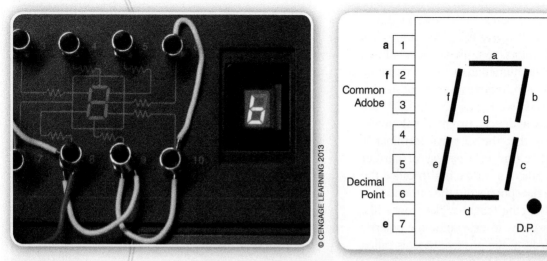

Figure 13-34: A seven-segment LED display is often used to display numbers and certain characters by illuminating a portion of seven LEDs that are prearranged in a figure-8 pattern. Shown here is a seven-segment display along with a schematic view of the same.

Figure 13-35: DC motors come in a variety of sizes, shapes, and powers.

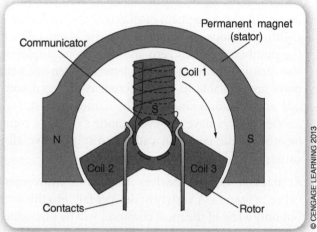

Figure 13-36: A DC motor uses magnetic fields, both from a permanent magnet and an electromagnet, to rotate a shaft.

In some applications, DC motors do not offer enough precision. Stepper motors can be used to achieve better control. Stepper motors also operate on DC voltages but use a special control circuit to feed either six or eight wires needed to control the motor (see Figure 13-37). A stepper motor moves only a tiny increment, sometimes less than one degree of rotation, each time it receives a pulse of current. To move again, a different combination of the feed wires must receive a pulse of current. A control circuit is necessary to coordinate the pulses to the correct input wires. It is possible for the stepper motor either to turn continuously or to turn to a specific spot, stop, and rotate again to another location. A computer interface is often used to control a stepper motor, although there are also integrated circuit chips designed specifically to control stepper motors.

Stepper motors are ideal in several applications such as moving the printing head on a printer, positioning the worktable on a computer numerical control (CNC) machine, or for accurate motion control in a robot. Although accurate, stepper motors are not capable of very high torque and are more expensive than standard DC motors.

Chapter 13: Designing Electrical Systems 373

Figure 13-37: A stepper motor is more difficult to control but also provides precise control of rotation. Some stepper motors have steps that are less than one degree.

Figure 13-38: Electric solenoids are commonly used to control the air inlet to a pneumatic cylinder.

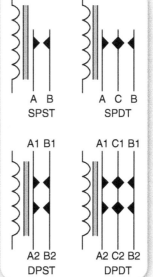

Figure 13-39: (a) Various electronic relays. (b) Circuit diagrams for several types of relays. A relay is a simple switch controlled by an electromagnet that connects several electrical contacts.

REPRINTED WITH PERMISSION FROM GATES, *INTRODUCTION TO ELECTRONICS*, FIFTH EDITION. COPYRIGHT © 2007 DELMAR CENGAGE LEARNING.

Solenoids are devices that produce rapid linear motion over a short distance, from about 1/8 inch to several inches (see Figure 13-38). Solenoids are used to engage mechanical mechanisms or gears, or to operate valves. Solenoids consist of a coil of wire wrapped around a hollow core, with a sliding magnetic armature within the core. When a magnetic field is formed by passing current through the windings of the coil, the armature moves. Some solenoids are capable of exerting a great deal of force. An example of a powerful solenoid is one that engages and disengages the starter motor gear to the flywheel gear of an automobile engine. Other types of solenoids are used for more precise movement and can be found in video recorders and multi-CD players.

A close cousin to the solenoid is the relay (see Figure 13-39). A relay acts as a switch, because electrical contacts are connected to the armature whose position is controlled by a magnetic field that is created when current flows through a coil of wire. A relay is often used as a remote switch to control one circuit with another. For example, a relay would allow a small switch carrying a small amount of current to switch a large amount of current to a device like a large, heavy-duty motor. The starter circuit in automobiles uses relays because the current requirements of most starter motors is over 300 amperes, a current much too dangerous to be running through the key circuit located near the steering wheel (and passenger).

Electromagnets are also actuators. Current running through a coiled wire causes a magnetic field to form down the center of the coil. The strength of the magnetic field depends on essentially three factors: (1) the core material; (2) the number of turns of wire around the core; and (3) the amount of current flowing through the wire. Electromagnets may be used for all sorts of interesting output devices. They may be used on a robotic arm to retrieve and release magnetic objects, to sort magnetic from nonmagnetic objects, or to cause motion in devices by attracting a magnetic object (like in a relay).

System Processors

By the term *process*, we mean that some intelligence is imposed on an input signal, causing a specific and desired output. There are two general classifications of processors: (1) analog and (2) digital.

Analog processors are processors that manipulate analog signals, or information. Analog information is information that varies continuously over time, changing by small amounts over small periods of time. An example of an analog process is how the temperature outdoors varies continuously as the day progresses. In the morning perhaps the temperature starts out at 60°F, and very slowly the temperature increases to 86° F by midday, and then decreases again as evening approaches (see Figure 13-40).

Digital processors can only process digital information. Digital information is comprised of only two types of signals, either a 1 or 0. So, voltage signals inside a digital processor are either on (1) or off (0). Examples of analog and digital signals are compared in Figure 13-40.

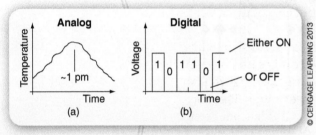

Figure 13-40: Examples of both analog (a) and digital signals (b). Analog signals vary continuously while digital signals are either on or off.

Analog Processors

Transistors. The ability to switch is one of the cornerstones processes in electronics. Transistors are semiconductor devices that can switch large currents on and off by using a small control current. This is similar to the function of a relay. Transistors have truly revolutionized electrical systems and human society in general. Transistors are widely used for both analog and digital applications. A typical processor in a digital computer holds close to several hundred million transistors on a single chip. Many of the transistors in a computer are used as switches, where a switch is either on or off (binary), but transistors are also often used as amplifiers. Only the switching capability will be reviewed here.

A circuit demonstrating the use of a transistor as a switch is shown in Figure 13-41. The control current may be from a sensor, such as the light sensor in a solar hot water system. A

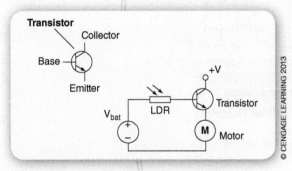

Figure 13-41: A control circuit that uses an LDR to turn on a transistor, which in turn activates a DC motor. Transistors are routinely used as switches; when the current injected into the base is large enough, then the resistance between the emitter and collector decreases quickly and substantially, allowing current to flow (in the direction of the arrow drawn on the transistor) between the emitter and collector connections.

light-dependent resistor (LDR), which changes resistance according to the amount of light falling on it, detects the presence of light. As the level of light changes, this device causes a small change in the current entering the base of a transistor. This change in current switches the transistor on or off. The transistor is then used to control the current going to the DC pump motor in the solar hot-water system. Figure 13-41 shows a simple control system that uses an LDR, a transistor, and a DC motor. If an AC motor were used, a relay would have to be added to the circuit, because a transistor cannot switch alternating current.

A bipolar transistor has three leads: the emitter (e), the base (b), and the collector (c) (see Figure 13-41). The larger current flows through the emitter/collector circuit. The smaller control current flows through the base-emitter circuit. The two types of bipolar transistors are NPN and PNP, which designate how the transistor is constructed and used.

555-Timer Chips. Another control need for electronics is to control "when" things happen (that is, to provide timing or clocking function). The 555 integrated circuit (IC), or "chip," is one of the most popular IC chips. It can be used to provide a pulse of variable duration, which means that it can be used to turn something on for a particular period of time. Using the 555 chip in this mode is called *one-shot*, or *monostable*, operation. The chip can also be used to provide continuous timed pulses, providing a clocking-like function. This clocking operation is called *multivibrator*, or *astable* operation. The 555-timer chip is usually in an eight-pin, dual in-line pin (DIP) package.

A monostable circuit is a timer that turns on just once for a specified amount of time and then turns off. Monostable operation is used by a clothes iron, which switches itself off after being unattended for 10 minutes. Each time the iron is moved, a small sensor resets the timer to zero to begin another count. The iron remains on as long as it is moved within the 10-minute limit. If it has not been moved, the circuit will return to its stable state and switch the iron off. Figure 13-42 is a schematic diagram of a circuit for building a monostable timer that switches on an LED. The use of an LED as a temporary output device is common to make sure the circuit works. Other output devices could replace the LED once the circuit operation has been validated.

Referring to Figure 13-42, a 555-timer chip uses two external components, a resistor (R1) and a capacitor (C1), to provide the required control that determines the monostable pulse. Changing the values of either or both of these components will change the output pulse duration. By changing the time it takes for the capacitor to charge, you can change the timing duration of the output of the 555-timer chip. The resistance R1 (in ohms) times capacitance (in farads) has the unit of time (seconds). So, if one wanted to increase the length of time of the pulse, then one would either increase R, or C, or both. Similarly, to reduce the length of the pulse, one would decrease the value of R, or C, or both.

A 555-timer chip can also be used to achieve astable operation, resulting in a continuous train of pulses. How rapidly the pulses occur is called the frequency, which is measured in hertz. One hertz is equal to one pulse per second; 1,000 hertz is equal to 1,000 pulses per second, or 1 kilohertz (1 KHz). A 555-timer chip is capable of providing output pulses up to 1 million times per second, or 1 megahertz, depending on the specific values of the control components (a few resistors and a capacitor). To make an LED appear like it is flashing to the human eye, a much slower pulse rate is required (a few Hz to tens of Hz).

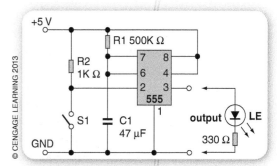

Figure 13-42: A schematic diagram of a monostable timing circuit that will turn on an LED for a period of time.

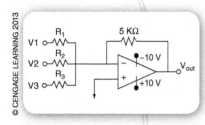

Figure 13-43: An operational amplifier (opamp) circuit that sums three voltages, V1, V2, and V3. In this circuit, as long as R1 = R2 = R3, the output, V_{out}, is proportional to V1 + V2 + V3. In general, the output is given by V_{out} = (R1)(V1) + (R2)(V2) + (R3)(V3).

Many circuits require the use of two 555-timer chips to achieve the desired results. For instance, you may want to have an LED flash at a perceptibly slow rate for a specific duration of time, say 15 seconds, each time the circuit is switched on. This is similar to a seat belt warning system in a car, where an LED and chime flash on and off for about 15 seconds. To do this, you would need a monostable circuit designed for one 15-second pulse. In addition, you would need an astable circuit to flash on and off the LED and chime. Conveniently, the 556 chip was designed for such purposes because it contains two 555-timer circuits.

Operational Amplifiers. Electrical circuits often need to perform mathematical operations, like adding and multiplying. A very useful device called an operational amplifier, or opamp for short, is used to provide mathematical functionality. The inner workings of an opamp are complicated but, fortunately, how to use one is straightforward. Referring to Figure 13-43, the symbol for an opamp, and an amplifier in general, is a triangle "laying on its side." Opamps are so named because they were designed to perform arithmetic *operations* such as addition and multiplication. Calculus-based operations of integration and differentiation can also be easily performed with an opamp. As an example, Figure 13-43 shows an "adder" opamp circuit that adds voltages, V1, V2, and V3. Similar circuits can be used to subtract, add, multiply, or divide.

Digital Processors

Digital processors, or microprocessors, have truly revolutionized modern society. Microprocessors are the brains of literally hundreds of billions of devices that humans use. Desktop and laptop computers, cell phones, smart phones, portable tablets, automobiles, airplanes, dishwashers, refrigerators, and cooling and heating systems all have microprocessor-controlled systems. A good example of the widespread use of microprocessors is the modern automobile. A single automobile typically contains over 50 microprocessors! Systems such as fuel injection, air-bag deployment, antilock brakes, and traction control are all controlled by microprocessors. Microprocessors have even been incorporated into toothbrushes. Billions of microprocessors are sold every year with the cost of lower-end microprocessors being less than 10 cents, truly making them "throwaway" commodities. Even the more expensive ($30 to $100) microprocessors used in computers are a quickly consumed commodity, being thrown away and replaced every two to five years. The driving force behind this phenomenon is a drive by scientists, mathematicians, engineers, and technologists around the world to construct smaller, faster, and semiconductor-based transistors that consume less power. Perhaps very surprisingly, this drive has followed a reproducible mathematical rule coined "Moore's law." Moore's law states that "the number of transistors on a single integrated circuit for minimum component cost doubles approximately every two years." Mathematically, this relationship is a power function ($y = 2^x$), which grows at an extremely fast rate. Over the past 10 to 30 years, humankind has benefited substantially from this astronomical growth rate.

Microprocessors are said to be digital because they only process digital or discrete information, information that is strictly represented by 1s and 0s. This digital system of numbering is called the binary numbering system. If a voltage state is high inside a microprocessor, it represents a "1-state," while if a voltage state is low (close to zero), it represents a "0-state." Knowledge of only two states is actually quite useful because knowledge of only two distinct states is all that is necessary to accomplish many important functions including: (1) conveying

Fun Facts
Powers of 2

As the following fable demonstrates, powers of two grow extremely fast. Two kings were playing chess and decided to make a bet. They agreed that the loser would pay the winner in grain, with the number of individual pieces of grain determined by the chessboard itself. A chessboard (an 8 × 8 matrix of squares) contains a total of 64 squares. The number of grains is determined quite simply by placing 1 piece of grain on the 1st square, 2 pieces of grain on the 2nd square, 4 pieces of grain on the 3rd square, 8 pieces of grain on the 4th square, 16 pieces of grain on the 5th square . . . and so on. The number of pieces of grain on each individual square is thus determined by the power function $y = 2^x$. The total number of grains is obtained by simply summing the number of grains on each square. The total number of grains turns out to be $(2^{64} - 1)$, which is quite accurately estimated at 2^{64} because 1 is very small compared to 2^{64}. But how big is this number 2^{64}? It is a good exercise for readers to calculate the weight, or volume, of this number of pieces of grain. The authors estimate that it would take approximately 1015, or 100 trillion, of the largest freight tankers to carry 2^{64} pieces of grain. This is clearly more grain than has been produced by mankind, ever! Moore's law (a power of two law) states, "the number of transistors on a single integrated circuit for minimum component cost doubles approximately every two years"— an almost unbelievable rate of technological development (see Figure 13-44).

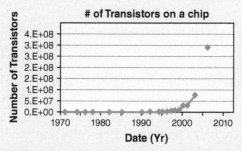 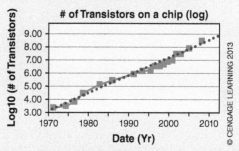

Figure 13-44: Plots of the number of transistors on a microprocessor chip as a function of time. The number of transistors integrated onto a single chip has increased dramatically over time, and follows a power-function (approximately 2^x) dependence (Moore's law). Graph (a) is a linear-linear plot, whereas graph (b) is a semi-log plot (the y-axis shows the logarithm of the y-value). In a semi-log plot, a power function should look like a line, which it does, indicating that Moore's law accurately describes the progression of semiconductor technology.

(code) information, (2) completing logical operations, and (3) performing mathematical computations, all of which are common in everyday life and within a microprocessor.

Coding. A unique series of 1s and 0s can be used as a code to represent a unique set of items. For example, as shown in Figure 13-45, a very delectable dessert can be constructed if one knows the digital code for certain items. Referring to Figure 13-45, if one wanted chocolate ice cream with only hot fudge and whipped cream, you would order a "0," a "100," and a "000." Knowledge of simple combinations of 1s and 0s are very useful to code information. Digital codes are prevalent. For example, on

Figure 13-45: Codes made up of binary numbers are easily constructed to represent information. In this case, different binary codes are used to represent type of ice cream, topping, or condiment.

a computer, when one types a lowercase "a," a code of "0110 0001" is sent to the microprocessor. However, if one makes a mistake and really wanted to type a capital "A," a code of "0100 0001" is sent. Computer hardware and software are so advanced now that if a lowercase "a" is incorrectly typed at the very beginning of a sentence, then the computer recognizes this as a grammatical mistake and automatically changes the lowercase "a" to an uppercase "A."

The digital code for representing letters and numbers is called the ASCII code (American Standard Code for Information Interchange). Codes also exist for all major written languages, so that information can be shared effectively between hundreds of cultures. Countless other binary codes are used for a variety of purposes, only limited by the designers for each microprocessor-controlled device.

Logic Operation. There are only three basic logic operations: INVERT, AND, and OR, each of which is performed by people every day. An INVERT function simply inverts the input. If the input is a "1" (high), then the output after an INVERTer is a "0" (low). If the input is a "0" (low), then the output after an INVERTer is a "1" (high). As the name implies, an AND function requires all inputs to be valid for the output to be valid. For example, for mammals, both a sperm and an egg cell are required to form a new living animal. It is a necessary and sufficient condition for a mammal that both a sperm cell *and* an egg cell come together. As another example, a history teacher may require two projects from students for a class, and the teacher may also require that students earn an A on both assignments to receive an A grade. Therefore, to get an A grade overall, students must earn an A on the first *and* the second project. An AND function is analogous to arithmetic multiplication.

An OR function is very different. For example, in the same history class, perhaps the teacher requires that the first report be on any war that the United States fought in after 1940. Therefore, a report on either WWII *or* the Vietnam War will satisfy the teacher's condition. Hence, an OR function requires that only one input to be valid for the output to be valid. An OR function is analogous to arithmetic addition.

The symbols and truth tables for the INVERT, AND, and OR functions are given in Figure 13-46. The truth tables simply define how each of the logic functions operates. For example, for an AND function, both inputs must be true ("1") for the output to be valid (a "1"). Similarly, for an OR function, if either of the inputs is valid (a "1"), then the output is valid (a "1").

Binary Numbers and Arithmetic. All digital electronic systems are based on the binary number system. The reason for this is simple: It is easy to represent **binary numbers**, 1s and 0s, in electrical circuits. If a wire has a high voltage, it is represented by a "1." If a wire has a voltage close to zero, it is represented by a "0." The binary system is called a "base-2" system because there are only a total of two numbers in the system. The numbering system we humans use every day

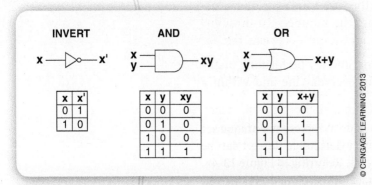

Figure 13-46: Logic function symbols and truth tables for INVERT, AND, and OR functions.

Fun Facts

Shown in Figure 13-47 is a representation of an operating room at Acme Hospital. The entrance hall to the operating room is actually comprised of two intermediate rooms with three doors. The purpose of the intermediate rooms is to keep dirt and germs out of the operating room, a highly desired attribute of an operating room. The entrance operates as follows. The inner Door C is enabled to be opened (actually opened automatically so no one needs to touch it) only when both Door A and Door B are fully closed (that is, no new germs/dirt can come in). Also, if the intermediate door (B) is open, then a high-speed bio-fan (fan plus bio-filter) must be turned on, ensuring that any leftover germs in the last intermediate room get sucked out. The operation of these doors is critical to maintaining a clean operating room.

The operation of this operating room "system" is correctly represented by a sequence of logic gates that represent the logic desired. A logic diagram solution to the operating room problem is shown in Figure 13-48. Once this solution logic diagram can be written down, it is a relatively quick process to design the appropriate circuit for controlling the operating room doors.

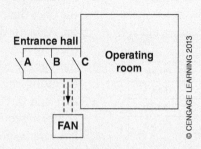

Figure 13-47: Blueprint diagram for an operating room that requires control over a series of three entrance doors to achieve an ultra-clean operating room.

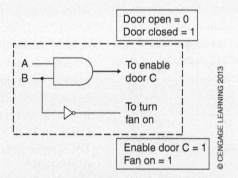

Figure 13-48: Logic diagram solution for the door control for the ultra-clean operating room.

is called the decimal numbering system. The decimal system is referred to as a base-10 system because it has a total of 10 numbers: 0 to 9. To help understand the binary system, let's first review how we count in the decimal, base-10, system. Using one digit, the "1s" digit, we count from 0 to 9 in sequential order. After we get to 9, we need to use one more digit, the "10s" digit. We increment the "10s" digit to a 1 (it was really a 0 before), and then we start over by incrementing the 1s digit again from 0 to 9, enumerating the numbers 10 through 19. At this point, we increment the 10s digit to 2 and then increment the 1s digit once again to form 0 to 9, generating the numbers 20 through 29. And this process continues, using as many digits as are needed. Let's look at a particular **decimal number**, say two hundred thirty-four (234). We can summarize what we have done with the placeholders in a very compact mathematical form. That is, $234 = \mathbf{4}(10)^0 + \mathbf{3}(10)^1 + \mathbf{2}(10)^2$, which is a sum of numbers with each number representing a placeholder digit. This form is simply adding the "1s" to the "10s" to the "100s"—something we do every day without thinking. Each term includes a power of 10, with the power corresponding to the exact placeholder, starting with a power of 0 for the first ("1s") placeholder. A power of 0 is used because $(10)^0 = 1$.

Figure 13-49: A chart showing the numbers 0 through 17 in both decimal and binary systems.

Decimal	Binary
0	00000
1	00001
2	00010
3	00011
4	00100
5	00101
6	00110
7	00111
8	01000
9	01001
10	01010
11	01011
12	01100
13	01101
14	01110
15	01111
16	10000
17	10001

Binary numbers are formed using the same process but with two important changes: (1) instead of 10 different numerals, we work with only two numbers (0 and 1); and (2) the counting can be summarized by powers of 2, not 10. Because the binary system can be summarized by powers of two, the digits are referred to as the ones digit (2^0), the twos digit (2^1), the fours digit (2^2), the eights digit (2^3), and so on. As in the decimal system, in binary we begin counting with zero followed by 1, but then we run out of numerals. Using the same principles that were used in the decimal numbering system, we now need to add a placeholder and make the next larger number using only 1s and 0s. So, the next number in our sequence is "10" a combination of 1 and 0. However, the number "10" in binary means two in decimal because it comes in sequence after the count of 1. Just as in the decimal system, each binary digit has a value according to its place in the number. To keep from confusing a number like 10 in decimal ("ten") and 10 in binary ("two") one usually puts a subscript on the number to indicate if it is binary or decimal. For example, eleven in decimal would be $(11)_{10}$ while two in binary would be $(10)_2$. A chart showing the numbers 0 through 17 in both binary and decimal systems is shown in Figure 13-49.

The conversion of numbers between binary and decimal is very important. After all, a computer, or microprocessor, "speaks" in binary, while humans "speak" in decimal. Fortunately, conversion between binary and decimal numbers is straightforward. For example, the binary number 101, or $(101)_2$, can be read using placeholders, just like we did earlier with the number 234 (234: 4 is in the 1s place; 3 is in the 10s place, and 2 is in the 100s place). The binary number $(101)_2$ has a 1 in the 1s place (2^0), a 0 in the 2s place (2^1), and a 1 in the 4s place (2^2). In the same compact form we used earlier for decimal numbers, it looks like this: $(101)_2 = \mathbf{1}(2)^0 + \mathbf{0}(2)^1 + \mathbf{1}(2)^2 = 1 + 0 + 4 = 5$, or $(5)_{10}$ or the 6th number in base-10. Another example may be useful: $(101110)_2 = \mathbf{0}(2)^0 + \mathbf{1}(2)^1 + \mathbf{1}(2)^2 + \mathbf{1}(2)^3 + \mathbf{0}(2)^4 + \mathbf{1}(2)^5 = 0 + 2 + 4 + 8 + 0 + 32 = (46)_{10}$ or 46 in decimal.

Binary numbers can be added as well, and this is extremely important in a computer or microprocessor. Adding binary numbers is also a very simple task, and an example is shown in Figure 13-50. In this example, the binary number 011 (decimal 3) is added to the binary number 100 (decimal 4) to get the binary number 111 (decimal 7), which makes sense because, after all, three plus four does equal seven! Key to adding binary numbers, just as adding decimal numbers, is the concept of carrying. For example, in binary, $0 + 0 = 0$, $0 + 1 = 1$, $1 + 0 = 1$, and $1 + 1 = 0$, but with a carry of a 1 to the next higher placeholder.

Microprocessors. Due to the extremely fast technological improvements of transistor design, microprocessors have changed quite a bit over the last 10 to 30 years. However, many of the basic operations remain the same. A typical functional

Figure 13-50: Two examples of binary addition. In binary addition $0 + 0 = 0$, $0 + 1 = 1$, $1 + 0 = 1$, and $1 + 1 = 0$, but with a carry of a 1 to the next higher digit (or placeholder). Some spreadsheet software programs are capable of performing binary addition.

Examples of binary addition									
0	1	1	(3)		1				
1	0	0	(4)		0	1	0	0	(4)
1	1	1	(7)		0	1	0	1	(5)
					1	0	0	1	(9)

architecture of a microprocessor is shown in Figure 13-51. The architecture consists of four main areas: (1) central processing unit (CPU), (2) arithmetic logic unit (ALU), (3) memory, and (4) input/output (I/O).

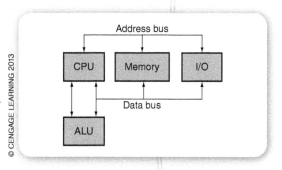

Figure 13-51: A typical microprocessor architecture showing the four major blocks (CPU, ALU, memory, and I/O) as well as the address and data buses.

CPU: The brain of a microprocessor. The CPU holds all of the basic logical instructions necessary for the microprocessor to work.

ALU: The ALU is exactly what the name implies. An ALU is where most, if not all, of the arithmetic in a microprocessor is carried out. This is much more efficient than requiring the CPU to do arithmetic because the CPU can be doing other tasks while the ALU is calculating.

Memory: When push comes to shove, microprocessors are all about managing memory (and data). Memory is required to store information that the microprocessor receives from the outside world as well as store internally generated results, even if it is temporary. The more memory that a microprocessor has, the more tasks it can do.

I/O: Microprocessors need to communicate to the outside world, so a microprocessor needs to be able to accept inputs and deliver outputs. It is very common for the I/O section of a microprocessor to contain both digital inputs and outputs as well as analog inputs and outputs. Digital I/O obviously transmits only digital information (1s and 0s), while the analog I/O lines send and receive analog (continuously varying) signals. As an example, many applications use microprocessors to input temperature (in degrees Fahrenheit or Celsius) and then use this information to control a process.

To facilitate communication between these different areas of the microprocessor, there are separate communication lines for data and addresses in a microprocessor. This is similar to a postal system—your postal delivery person needs to know your address to deliver and pick up your mail. What the postal person delivers (letters and packages) is the data that is useful to you.

Microprocessors vary greatly in capability, physical size, and cost. For example, relatively simple processors are used to manage relatively simple processes. These simple microprocessors are relatively small, measuring only a few millimeters on a side, and inexpensive (less than 10 cents in volume). For example, a processor included in a toothbrush has very little complexity so a very simple microprocessor would be used. A process that would likely require a bit more capability, especially speed, would be a processor that manages the songs on an MP3 player. Most MP3 players hold a substantial amount of memory, greater than several gigabytes, and your ears need the information (music) very quickly. A more challenging task in a microprocessor is interacting with the human user, you! Users are constantly pushing buttons: wanting to play a game, construct a new playlist, turn up the volume, or any of several hundred tasks. Microprocessors in computers used by humans need to be fast and handle a tremendous amount of data, hundreds of gigabytes of data. And as if this was not hard enough, a computer needs to handle immense amounts of data quickly for

Part II: Resources for Engineering Design

not just one application (program) but for many different types of applications, often at the same time (PowerPoint, Excel, Word, iTunes, Explorer, and so on). The chart in Figure 13-52 shows key attributes of a variety of microprocessors commonly used.

How many **bytes** are in 100 gigabytes? How many **bits** in 100 gigabytes?

(Hint: This is a bit of a tricky question. Search Wikipedia for "gigabyte" and "gibibyte.")

byte:
A unit of computer storage that contains eight bits.

bit:
A bit refers to something that holds just a single piece of binary information. In other words, a bit is in effect a single placeholder binary number. A bit is either a 0 or a 1. The binary number 11011 is comprised of five bits, while the binary number 101 is comprised of three bits.

Performance Level:	Low	Medium	High
Application(s):	Broad application (Control systems: automotive, motor control, heating & cooling, lighting, remote controls, human interface management . . . etc.)	Special applications (graphics, TV, HDTV, projection systems, MP3 players, audio equipment, memory management, navigation . . . etc.)	Broad application: "Computers" [Desktop, laptop, design stations (graphics/CAD), scientific calculations . . . etc.]
Size:	~ 1-to-4 sq. mm.	~ 10-to-25 sq. mm.	~ 100-to-400 sq. mm.
Speed:	1-to-40 MHz	50-to-500 MHz	0.4-to-3$^+$ GHz
Companies:	Zilog, Microchip	AMD, Intel	Intel, Motorola, IBM

Figure 13-52a: *This table compares microprocessors at different price points.*

OFF-ROAD EXPLORATION

The following websites are useful for learning how to program in C-language:
http://www.cprogramming.com
http://www.eskimo.com/~scs/cclass/notes/top.html

Figure 13-52b: *Photograph of a high-performance microprocessor commonly used in computers (actually using two processors ["Duo"] working closely together).*

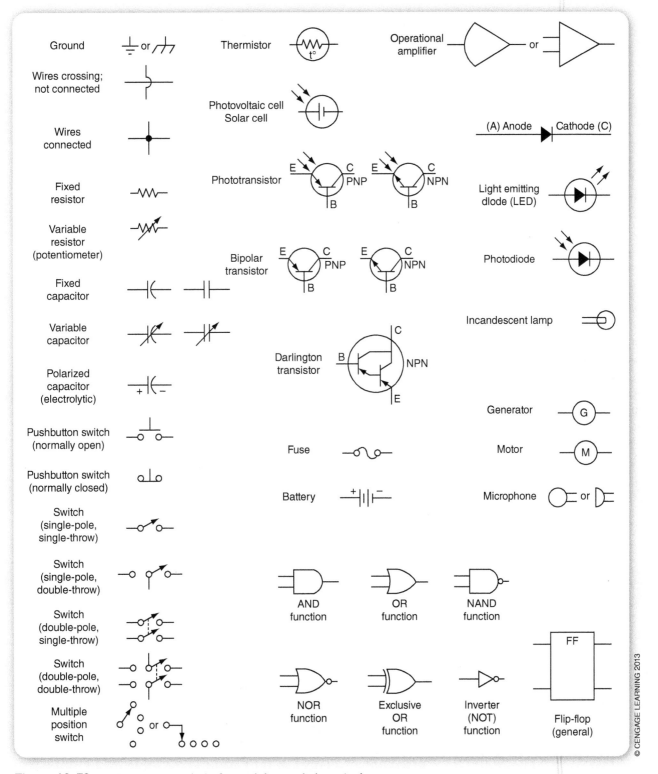

Figure 13-53: Schematic symbols for widely used electrical components.

SUMMARY

The science of electricity and the technology of electronics have developed extremely rapidly. Many of the basic principles of electricity were developed over just a few hundred years, a few human life spans. Semiconductor technologies have kept up this trend, and have developed at an even faster rate (a power-dependent rate). Over a span of about 50 years, a time frame substantially less than a human life span, semiconductor technologies have truly revolutionized the way we live, and how we use and impact earth's resources. No other technology has developed so fast, nor had such a large impact on human society, and perhaps the earth in general. Electronics is involved in almost every aspect of our lives on earth.

In this chapter, we have learned of charge having an attribute of opposites—charge can be positive or negative. Some of the basic equations governing electric forces are identical, in mathematical form at least, to Newton's equation for gravitational forces. However, the opposite nature and high mobility of charge combined with electrical forces being quite huge results in substantially unique behavior for charges. One of the first valuable laws of electricity was Ohm's law, which states that $V = IR$, a surprisingly simple linear relationship between V and I (with a proportionality constant of value R) given the complex nature of electric forces. We discussed that magnetism is a unique type of force that is due strictly to moving charge, and is therefore strongly linked to all of the equations of electricity.

There are only three basic electric components: resistors, capacitors, and inductors. All three of these basic electric components are widely used. Resistors are the most widely used component and are used for controlling voltage and current levels in circuits. We presented examples of both voltage dividers and current dividers, circuits routinely used to control voltage and current. Capacitors and inductors are unique in that they are each capable of storing electrical energy, making them very useful in time-varying applications. Capacitors are like batteries in that they can store charge directly. Inductors use a self-induced magnetic field to control time-varying electric currents.

In the last half of this chapter, we reviewed many specific electrical components and devices. This portion of the chapter was organized around the I-P-O concept of design—detailed knowledge of input, process, and output (I-P-O) should be clearly defined before attempting designs. This is true for many types of design work, but is especially true for electronic designs due to their complexity and the fact that many electrical components naturally fit into one of these three categories. Many input and output devices were described, including a variety of types of sensors and transducers. Processes were divided into two categories: analog and digital. Electrical designers commonly refer to analog and digital portions of designs, and can often combine both analog and digital designs within the same circuit. Microprocessors are perhaps the ultimate processor in that they are inexpensive and very flexible in their range of applications. Many microprocessors are so powerful that they can contain both digital and analog circuit capability, effectively bridging the digital world of a computer and the analog world of humans on a single chip.

BRING IT HOME

OBSERVATION/ANALYSIS/SYNTHESIS

1. What is the charge of a single electron? What is the charge of a single proton? What is the charge on a single neutron?
2. Two separate masses are floating around in space. They are each positively charged. Will the masses repel or attract each other?
3. Name some materials that are good conductors of electricity. Name some materials that are good insulators of electricity. How many of the materials that you listed are common in your home?
4. Name some devices in your home that you think use magnetic forces to perform their function.
5. Define what the inputs, processes, and outputs are for the following.
   a. Home heating system
   b. Laptop computer
   c. Cell phone
   d. MP3 player
   e. Microwave oven
   f. Garage door opening system
6. Which of the following could be binary numbers?
   a. 1011
   b. 1111
   c. 1234
   d. 10112
   e. 00000
   f. 12345
7. Convert the binary number (10011) to decimal.
8. Add the binary numbers (101) and (011).
9. Build a circuit that acts like a night light; when it is light out, a display LED is off but when it becomes dark, the same LED turns on. In this circuit, place a resistor in series with the display LED that limits the current to the LED to half of its maximum rated current.
10. Like in the previous exercise, build a circuit that acts like a night light, but place a resistor in series with the display LED that allows the current to the LED to attain its full maximum rated current. Using the results of the two night light circuits (half max current and full max current), can you determine if Ohm's law is obeyed? (Hint: Does an LED [a diode in general] conduct current at low voltages?)

EXTRA MILE

ENGINEERING DESIGN ANALYSIS CHALLENGE

1. Design and make (from LEGO or Fischertechnik) a toy vehicle powered by a small DC motor that, when turned on, will travel forward for about two seconds, stop for about two seconds, and move forward for about two seconds. It should do this until it is turned off. Use a transistor and/or relay to switch the motor.
2. Design and make a toy that will flash lights when a ball is tossed to a target.
3. Design and make a drawbridge from foam-board materials. Using a relay circuit, design the bridge to open upon pushing a button and automatically stop when the bridge is fully open. Upon pushing a button, make the bridge close and automatically stop when the bridge is fully closed.
4. A flashing warning light is needed to attract the attention of a pilot when a plane's fuel is starting to get low. Design and make a simple astable circuit to flash an LED when a switch is closed. Using a second 555-timer chip or one 556 chip, redesign the circuit to produce a flashing pulse (on for one second and off for one second) for a total of one minute. What are some ideas of how to throw a switch inside a tank of fluid to indicate that the fluid is getting low (initiating the sequence of events described earlier)?
5. A store would like to keep track of the number of people entering during the day. It has separate entrance and exit doors. Design and make a circuit that will sense when each person enters and display the number of total people who have entered the store in seven-segment display format. Your circuit should count a total of nine people. Redesign the circuit to count up to ninety-nine.

CHAPTER 14
Designing Pneumatic Systems

Menu

Before You Begin
Think about these questions as you study the concepts in this chapter:

 How do liquids and gases differ under pressure and how can we take advantage of these characteristics to do work?

 How do hydraulic and pneumatics systems differ?

 How do pneumatics systems create mechanical advantage and develop enormous forces?

 What components are used in pneumatic systems?

 How are fluidic systems designed and how are forces, area, volume, and distance of travel calculated?

INTRODUCTION

Although mechanical and electronic systems are visible in many familiar consumer products, pneumatic and hydraulic systems are less well known. Pneumatic refers to the use of compressed air, which is air that has been raised to a pressure above the normal atmosphere. Hydraulic systems also use a liquid instead of a gas.

From a science point of view, both gases and liquids can be considered to be fluids. Therefore, the terms *fluidics* and *fluid power* are used to refer to both pneumatic and hydraulic systems. These systems are useful for applications in which linear, reciprocating, or rotary motion is needed. Electrical or mechanical devices can also produce these motions but usually at greater expense. Individual electrical components can produce linear motion, but requirements of more than a few inches of travel are impractical or costly. With pneumatic and hydraulic cylinders, linear travel up to 10 feet is easily accomplished. High forces are easily obtained with these systems as well.

For producing rotary motion, pneumatic and hydraulic motors are more compact and can provide higher speeds than electric motors. In applications in which high heat, flammable fumes, dust, and/or grit may be present, pneumatic systems are safer than electric motors. For example, pneumatic tools are common in the auto body repair industry, where dust and grit would cause rapid wear in electric motors. Figure 14-1 shows some of these tools.

Figure 14-1: A variety of pneumatic tools used in the automotive body and repair industry.

CHARACTERISTICS OF FLUIDS

By definition, a **liquid** is a fluid with a volume that, for practical purposes, does not change. In other words, a sample of liquid will always maintain the same volume no matter how much you try to either compress or expand it. A **fluid** also has the characteristic of being able to flow and to take the shape of its container. A gas is a fluid but with a very important difference: The volume of a sample of gas will change based on how much you try to compress or expand it. A gas has no fixed volume; its volume will depend on applied pressure and temperature.

According to **Pascal's law**, when a **pressure** or a force is applied to a confined liquid, that force is transmitted to all parts of the liquid in all directions. As a result, the force is transmitted to the inside walls of the container that holds it. This is called a hydraulic system. Let's say that you have a balloon filled with water (see Figure 14-2). If you press your finger on one part of the balloon, the balloon will change shape as the force you applied is transmitted throughout the liquid water. If you were able to enclose the water balloon completely with your hands and squeeze, you would not be able to change the volume of the liquid (the size of the balloon).

> **Pascal's law:**
> States that when a pressure or force is applied to a confined liquid, that force is transmitted to all parts of the liquid in all directions at the same rate.

In addition to being essentially uncompressible, a confined liquid can be used to gain *mechanical advantage*; that is, it can multiply force. For example, in Figure 14-3, a simple hydraulic jack is pictured (although without the safety and other features of a practical device). A 25-pound force applied to the piston on the left can lift the 250-pound load on the right. You don't get something for nothing, so what is being sacrificed in this system? If you said distance of travel, you would be right. If you pushed down one inch on the piston on the left, the piston on the right with the weight would only raise one-tenth of one inch. Can you explain why?

As was discussed briefly before, gases are very different than liquids. When a force is applied to a gas in a container, the force tends to decrease the distance between the molecules, compressing the gas into a smaller volume (but higher internal pressure). **Newton's third law**, which describes how each action is accompanied by an equal but opposite reaction, tells us that there must also be a force pushing the molecules away from each other. Therefore, when a force is applied to a gas, the gas exerts its own force in

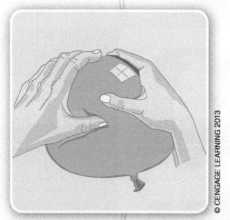

Figure 14-2: *Squeezing a water balloon will not decrease its volume. Liquids cannot be compressed.*

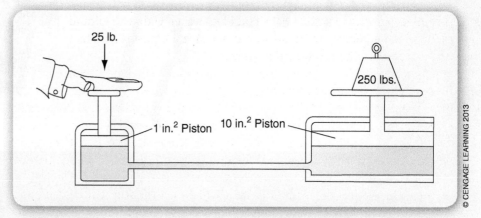

Figure 14-3: **Fluids can transmit and multiply force.**

response. Unlike the water balloon, a balloon filled with air will decrease in volume if you can get your hands completely around it. But the smaller you manage to get the volume of the balloon, the harder you will need to squeeze it to maintain that volume.

> **hydraulic system:**
> Uses a liquid to transfer force from one point to another.

PNEUMATICS VERSUS HYDRAULICS

There are some applications in which one specific system is better suited. A hydraulic system allows a precise linear movement, and that movement may be stopped at any point. For example, the operator of a backhoe needs control over each axis of movement of the hoe. The blade must be able to move to a location, stop, and move again. By opening valves that allow a certain volume of liquid to enter a cylinder, the piston will move a certain distance.

In a simple hydraulic system, such as a car jack, a person moves the handle on a pump, which forces hydraulic oil into a cylinder (see Figure 14-4). The pump can develop high pressure, but the volume of liquid it can pump is very low. Many strokes are needed to force enough liquid into the cylinder to raise the jack a few inches. Most hydraulic systems use motor- or engine-driven high-pressure pumps to force the liquid into a cylinder to extend a piston. When the piston must be withdrawn, the liquid in the cylinder must go somewhere, so return hoses or lines are used to route the fluid back to a reservoir. Liquid is recycled in hydraulic systems.

There are many kinds of hydraulic fluids. Some are petroleum-based and some are not. Newer fluids include those that are based on vegetable oils and are biodegradable. They all have several requirements, including low viscosity within the operating temperature range of the equipment. Hydraulic equipment operating in low temperatures often encounters problems with fluids becoming resistant to flow, which puts a lot of stress on the system components. The opposite problem occurs when operating in high-temperature environments. The fluid can become too thin and not work as effectively.

Pneumatic systems, in contrast, have valves that allow a gas under pressure, usually air, to enter a cylinder. The pressure of the gas pushes the piston, but the pressure builds up rapidly and the piston is typically forced from its starting location to the end of its travel. Unlike the liquid system, in which a volume of liquid would displace the piston a certain distance, gas does not have a specific volume.

Although liquid-filled systems transmit force more efficiently than gas-filled systems, there are advantages to the gas-based pneumatic system over the liquid-based hydraulic system for many applications. Pneumatic systems are much cleaner and require less piping than hydraulic systems, because they do not require liquid return lines and no leak cleanup is necessary. Pneumatic systems are widely used in factory automation, braking systems on large trucks, small air-driven sanders, wrenches and other small tools, and many other applications. Because of these reasons, this text will deal primarily with pneumatic systems. However, many of the same principles and applications will apply to hydraulic systems (see Figure 14-5).

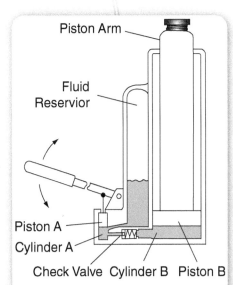

Pushing down on handle moves piston A down which closes off passage to reservoir hydraulic fluid. Fluid is forced past check valve and enters cylinder B where it displaces piston B a small distance upward. Lifting the handle raises piston A until it uncovers hole in the cylinder wall which is the passage to reservoir. Hydraulic fluid fills cylinder A by gravity. When handle is pushed down again, more fluid is forced into cylinder B displacing piston B upward again. The check valve prevents the fluid in cylinder B from pushing back into cylinder A.

Figure 14-4: Diagram showing the operation of a hydraulic jack.

> **viscosity:**
> The resistance of a fluid to flow. Among other factors, temperature can cause a change in a fluid's viscosity.

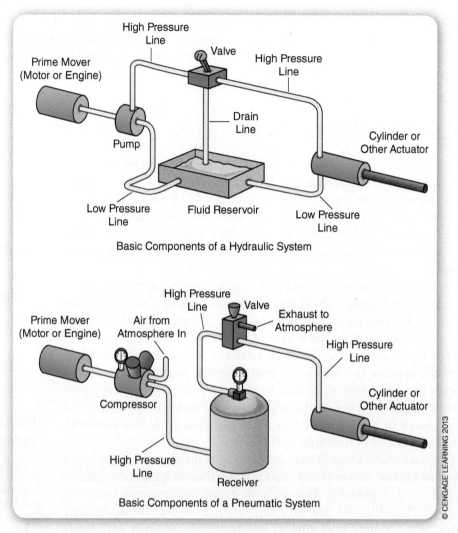

Figure 14-5: Comparison of hydraulic and pneumatic systems.

PRINCIPLES OF PNEUMATICS

The normal air pressure at sea level is 14.7 pounds per square inch (psi). The unit of psi is only one method of describing air pressure, however. Sea level air pressure in other widely used units is 29.92 inches of mercury, 101.3 kilopascals, 1.013 bars, and 1,013 millibars. The pressure of the atmosphere at sea level is a result of the weight of the huge number of molecules of air above that point. The atmosphere extends about 62 miles (100 kilometers) above the earth's surface.

Pressure differences in the atmosphere are also pneumatics. The uneven heating of the earth's atmosphere causes these pressure differences, which we see as changes in barometric pressure. Moving molecules of air, in the form of wind, try to get from high-pressure areas to low-pressure areas. Humans have used the pneumatics of the atmosphere to power sails and windmills for thousands of years (see Figure 14-6).

When you inflate a bicycle tire, a basketball, or a balloon, you are compressing the air. At normal air pressure, the molecules of air are a certain, consistent distance apart. When you force more air into a container, such as a basketball, you are pushing more molecules into that space so that the distance between the molecules decreases. The result of this compression is that the molecules of air exert a higher force against

Figure 14-6: Humans have harnessed the power of pneumatics for thousands of years.

the walls of the container. This force can do work. (From a mathematical point of view, pressure is simply force per unit area, so the internal forces go up as air is added, resulting in an increased pressure and increased force.)

> ### New York City's Pneumatic Subway
> In 1870, New York City's first subway opened. The demonstration line was only 312 feet long by 8 feet in diameter and was powered by pneumatic pressure created by a giant fan. The brainchild of Alfred Ely Beach, the system was constructed in secret below the streets of the city until it was opened to the public. It operated for only three years. The fan, located at one end of the tunnel, created enough pressure to push the passenger car to the other end of the line. The fan was then reversed, creating a partial vacuum in the tunnel, which moved the car back to its starting station. In 1912, the construction of the current New York City subway system destroyed what was left of the pneumatic subway.

In pneumatics, pressure is created by means of a compressor. A compressor requires a motor or an engine to operate. Two common types are the **reciprocating-piston compressor** and the **sliding-vane compressor**, although other types of compressors are in use. A piston compressor draws air from the atmosphere into a cylinder where it is squeezed into a small space by a moving piston (see Figure 14-7). Sliding-vane compressors have movable seals or vanes on an eccentrically mounted rotor (see Figure 14-8). This type of compressor is usually quieter in operation than

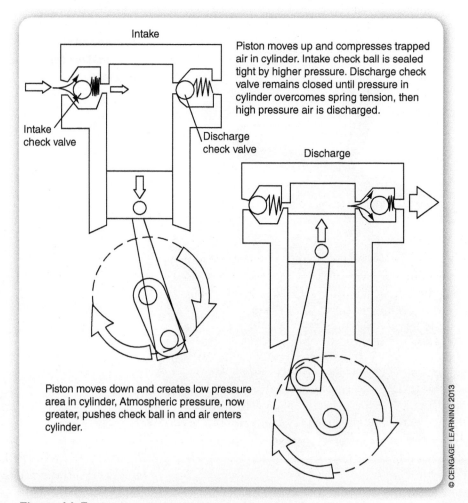

Figure 14-7: A reciprocating-piston compressor.

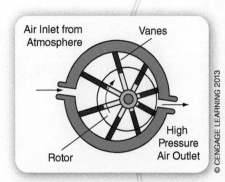

Figure 14-8: A sliding-vane compressor squeezes air into a smaller volume as the vanes rotate.

the reciprocating-piston compressor. From the compressor output, the compressed air is fed into a storage tank, called the receiver. The use of a tank ensures that an adequate supply of compressed air is available.

All pneumatic systems require pressure regulators to limit and maintain the air pressure at a fixed level. Moisture from the atmosphere also enters a pneumatic system along with the air, so a device called a dryer is often used to remove moisture from the pneumatic line. In addition, pneumatic systems sometimes have lubricators that provide a fine mist of oil droplets in the air line to lubricate the moving parts of pneumatic components. Rotating motors, such as those found in impact wrenches and other air tools found in automobile repair or body shops, need this lubrication. However, a lubricator in the line would damage other pneumatic systems, such as paint-spraying or sand-blasting equipment.

PNEUMATIC-SYSTEM COMPONENTS

A container with one movable wall is called a pneumatic cylinder. The cylinder is made up of the cylinder body, piston, piston rod, seals, and fittings (see Figure 14-9). As air pressure is applied to the cylinder body, the force of the tightly squeezed air molecules pushes outward, equally in all directions. Because the cylinder body itself is not designed to move, the force exerted causes the piston to slide out of

Chapter 14: Designing Pneumatic Systems   393

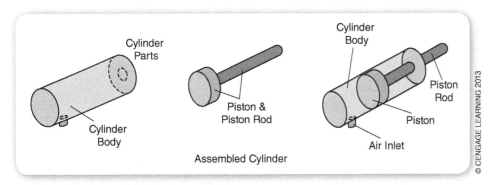

Figure 14-9: A pneumatic cylinder.

the cylinder body (see Figure 14-10). The seals are used to prevent air loss between the piston and the cylinder walls and often to keep dirt on the piston rod from entering the cylinder body. The fittings are the connectors for the compressed air lines or hoses.

Cylinders and Control Valves

Single-Acting Cylinders. Cylinders that are constructed so that air pressure is applied to only one side of the piston are called single-acting cylinders. Single-acting cylinders have only one port or opening that can receive or discharge gas. The syringes in Figure 14-11 are single-acting cylinders.

Figure 14-12 shows a CAD drawing of a single-acting cylinder. The view is transparent so you can see the interior containing the piston. Air entering the cylinder on the left end pushes the piston forward. A return spring is sometimes installed on the opposite side of the piston to return the piston to its original position when the air pressure is removed. In most cylinders, the spring is internal and cannot be seen.

Commercial cylinders are typically made from brass, steel, aluminum, or other noncorrosive material so that rust or corrosion will not damage the smooth interior or the cylinder. A plastic or rubber seal is used to form a tight barrier to prevent fluid from escaping around the piston. If a good seal is not made, the escaping fluid will reduce the force that can be exerted by the piston.

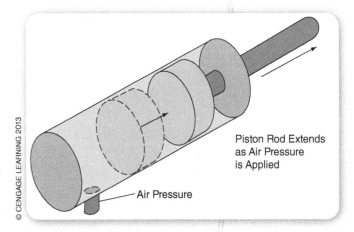

Figure 14-10: When air is forced into the cylinder body, the piston is pushed to the far end of the chamber.

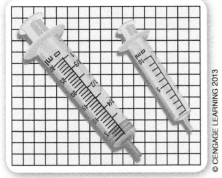

Figure 14-11: A syringe is a simple pneumatic or hydraulic cylinder.

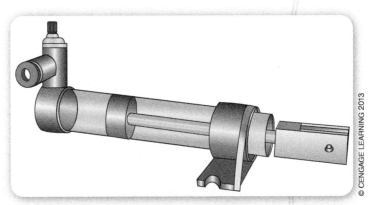

Figure 14-12: A CAD drawing of a single-acting cylinder showing the piston inside.

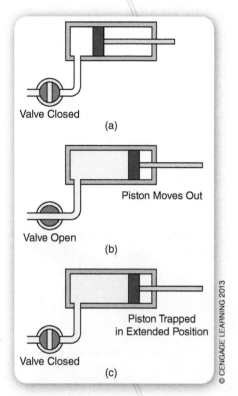

Figure 14-13: A simple on/off valve will not operate a pneumatic cylinder.

Three-Port Valve. To control the action of a single-acting cylinder, a special valve is required. If a simple flow valve were used, the air pressure applied to the piston in the cylinder when the valve was opened (on) would remain in the cylinder even when the valve was closed (off), as shown in Figure 14-13. To correct this problem, a **three-port valve** was created that provides an exhaust path for the trapped compressed air in the cylinder when the valve is off. It has one air inlet port, one cylinder port, and one exhaust port.

The three-port valve has two positions. In the "on" position, the air inlet port is connected to the cylinder port, allowing compressed air to travel through the valve and enter the cylinder. In the "off" position, the cylinder port is connected to the exhaust port, allowing air pressure trapped in the cylinder to escape through the valve into the atmosphere (see Figure 14-14). These valves come in a variety of types, including push-button, toggle, plunger, and roller-trip (see Figure 14-15).

Double-Acting Cylinders. Cylinders constructed so that air pressure can be applied to either side of the piston are called double-acting cylinders. In Figure 14-16, you can see two ports, one at the front and one at the back of the cylinder. In this way, pneumatic pressure is used both to extend and to retract the piston, and no return spring is needed. It is possible to control a double-acting cylinder with two three-port valves—one controlling the extension of the piston and one controlling the retraction. Remember, the compressed air must be allowed to escape from one side of the cylinder as air pressure pushes the piston from the other side.

Five-Port Valve. Many applications require that the double-acting cylinder be operated by one valve. The five-port valve is a combination of two three-port valves in one package (see Figure 14-17). There are two cylinder ports, two exhaust ports, and one air inlet port (so we don't call it a six-port valve). The five-port valve, like the three-port valve, has two positions. In the "on" position, the air inlet port is connected to one cylinder port. At the same time, the other cylinder port is connected to one exhaust port. In the "off" position, the air inlet port is connected to the second cylinder port. Also, while in the "off" position, the first cylinder port is connected to

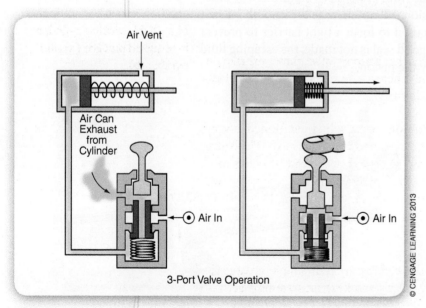

Figure 14-14: A three-port valve is used to control a single-acting pneumatic cylinder.

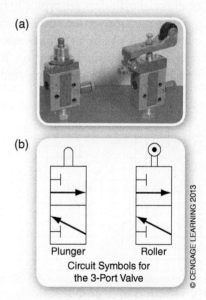

Figure 14-15: Common three-port valves (a) and their symbols (b).

Chapter 14: Designing Pneumatic Systems 395

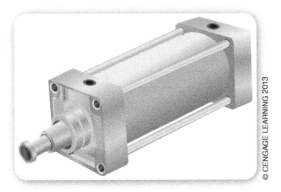

Figure 14-16: A double-acting cylinder used in commercial systems. Note the two air ports on the top of each end of the cylinder.

Figure 14-17: A five-port valve is necessary to control a double-acting cylinder picture (a) and symbol (b).

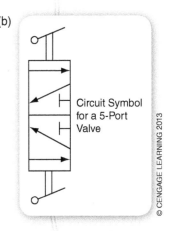

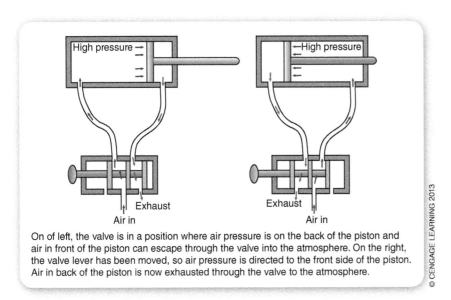

Figure 14-18: Air flow through a five-port valve to a double-acting cylinder.

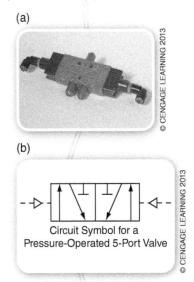

Figure 14-19: A pressure-operated five-port valve is used when remote control of a double-acting cylinder is required picture (a) and symbol (b).

an exhaust port (see Figure 14-18). In this way, the five-port valve allows exhaust on the side of the piston that is not under pressure. There are a variety of five-port valves, including push-button, toggle, plunger, and roller-trip.

Pressure-Operated Five-Port Valve. In addition to the valves mentioned, a very important valve in pneumatics is the pressure-operated five-port valve. This valve is similar to the five-port valve just discussed, except that, instead of being controlled manually or mechanically, it is controlled by air pressure. Normally, two three-port valves or one five-port valve control(s) the air pressure on the pressure-operated five-port valve (see Figure 14-19).

Other Components

The components discussed thus far allow the development of simple pneumatic systems. Other components are required, however, to make these systems practical.

Flow Regulator. The flow regulator allows control of the speed of piston travel in a cylinder. It does this by restricting the flow of air in one direction only (see

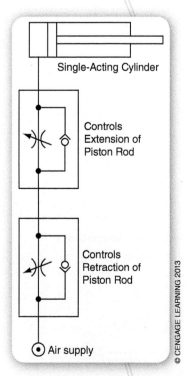

Figure 14-21: **Using two flow regulators back to back to control the extension and retraction rate of the piston rod on a single-acting cylinder.**

Figure 14-23: **Solenoid-operated pneumatic valve controls a pneumatic circuit with an electrical signal.**

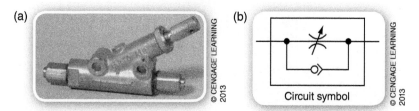

Figure 14-20: **Adjustable flow regulator with bypass picture (a) and symbol (b).**

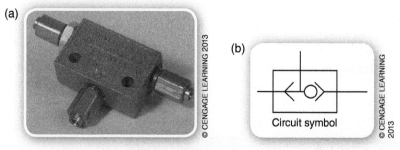

Figure 14-22: **Shuttle valve allows two valves to control one air line picture (a) and symbol (b).**

Figure 14-20). Air flowing in the opposite direction pushes a ball or a disk inside the valve that allows the free flow of air around the restriction. Flow regulators can be placed in a circuit so that either the compressed air inlet flow or the exhaust flow can be regulated. To control the motion of a double-acting cylinder, two flow regulators must be used. By adjusting the two flow regulators, the speed of extension and retraction of the piston can be controlled.

Because flow regulators only control the flow of air in one direction, one flow regulator can only control the piston speed in one direction in a single-acting cylinder. By using two flow regulators back to back, both the extension and the retraction piston speed of a single-acting cylinder can be controlled (see Figure 14-21).

Shuttle Valve. A shuttle valve allows a single-acting cylinder to be controlled from two locations. It is a "T" connector with three ports, two air inlets, and one air outlet. It has a small valve inside that shifts to close off one inlet port if the other inlet port has pressure applied (see Figure 14-22).

Solenoid Valve. Controlling pneumatic systems by means of electronic circuits is possible with a solenoid valve (see Figure 14-23). The solenoid was described as an actuation device in Chapter 13. When electric current is connected to the solenoid coil, a plunger moves to open a valve. Both three-port and five-port solenoid valves are available.

CALCULATING FORCES IN FLUIDIC SYSTEMS

Forces in pneumatic and hydraulic systems are usually very high. For example, the forces present in the hydraulic braking system of a car reach nearly 1,000 pounds psi. Generally, the reason a pneumatic or hydraulic system is used in the first place is because high forces are required.

There are situations where you need to know forces or length of piston travel to design a system. For example, the calculations required to design a simple system to rotate a model robotic arm are shown in the following example.

Example: Calculating Force in a Hydraulic System

Situation: You are designing a simple hydraulic system to rotate the base of a robotic arm. When you push on the piston in cylinder A, piston B extends and rotates the circular base, Figure 14-25. [Figure 14-24 gives symbols for common pneumatic parts.]

Problem: Find the distance piston B will travel. You need to know the distance that piston B will travel if piston A moves 1 1/4 inches. You also need to know the force that piston B can exert if a force of 3 pounds is applied to piston A.

You Know:
Diameter of piston A = 1 1/4 inches
Length of travel of piston A = 1 1/8 inches
Diameter of piston B = 3/4 inch

You Need to Find:
Volume of liquid that the travel of piston A will displace
How far piston B will travel if that amount of liquid enters cylinder B (see Figure 10-25)

Solution:
Find volume displaced by travel of piston A:
Diameter = 1.25 inches so
radius = 0.625 inch
Area of piston A = πR^2
$= 3.14(0.625 \text{ in})^2$
$= 3.14(0.39 \text{ in}^2)$
$= 1.22 \text{ in}^2$
Volume = (Area)(Length)
$= (1.22 \text{ in}^2)(1.125 \text{ inches})$
Volume displaced = 1.37 cubic inches = 1.37 in^3

Find area of piston B:
Diameter 0.75 inch so radius = 0.375 inch

Symbols and basic circuits

Single-acting cylinder

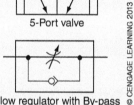
Double-acting cylinder

3-Port valve

5-Port valve

Flow regulator with By-pass

Figure 14-24: **Schematic symbols for basic pneumatic components.**

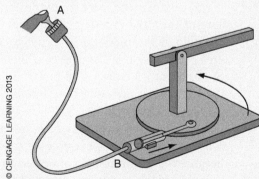

Figure 14-25: **Physical design for a rotating arm.**

(continued)

Example: Calculating Force in a Hydraulic System (continued)

$$\text{Area of piston B} = \pi R^2$$
$$= 3.14(0.375 \text{ in})^2$$
$$= 3.14(0.14 \text{ in}^2)$$
$$\text{Area of piston B} = 0.44 \text{ in}^2$$

Find piston travel of B:

$$\text{Length of travel} = \frac{\text{Volume}}{\text{Area}}$$
$$= \frac{1.37 \text{ in}^3}{0.44 \text{ in}^2}$$

Length of Travel of Piston B: = 3.11 inches

Finding Forces: How much force does piston B exert if the input force on piston A is 3 pounds?

Find the pressure that exists within piston A:
Pressure = (input force) / (area of piston A)
= 3 pounds / 1.22 square inches
Pressure = 2.46 psi

[Key: Pressure is the same throughout the liquid so the pressure at piston B is the same at piston A.]

Find force that piston B exerts:
Force = (pressure)(area of piston)
= (2.46 psi)(0.44 square inches)
Force = 1.08 pounds

Figure 14-26: Log splitter with a horizontal hydraulic cylinder. The piston is capable of extending 24 inches and exerting 34 tons of pressure.

In hydraulic systems, liquid displaces the piston in a cylinder, causing the piston to move. The volume of liquid must be known to calculate that distance. For example, a log splitter pictured in Figure 14-26 has a 5-inch-diameter piston that can extend 24 inches. The amount of hydraulic fluid required to extend the piston the full distance is 471.24 cubic inches, or just over 2 gallons. This does not include the additional fluid that must fill the hoses, valves, and pump.

To calculate the force a cylinder can exert, you must know the pressure of the fluid entering the cylinder and the area of the piston. For example, if the pressure is 10 newtons per square centimeter and the area of the piston is 5 square centimeters, so the piston can exert a force of (10 n/cm^2)(5 cm^2), or 50 newtons. To create the 34 tons of force for the log splitter previously mentioned, a fluid pressure of almost 3,500 psi is required.

Double-acting cylinders have less piston area on one side than the other, so they cannot exert equal pressure in both directions. This is because the piston rod is connected to the piston on one side and takes up some of the piston area. When the force is calculated for movement of the piston back into the cylinder (negative direction), the figure for piston area is less, so the resulting force is less.

Example: Finding the Area of a Piston

The area of the piston is found by using the formula πR^2. For example:
A piston with a diameter of 6 cm has a radius of 3 cm.
Area of a circle = πR^2 = 3.1416(3 cm)^2 = 3.1416(9 cm^2) = 28.31 cm^2.
The area of the piston, then, is 28.31 cm^2.

Your Turn

1. A pneumatic cylinder with a piston diameter of 0.45 inch has an air line connected with a pressure of 44 pounds. About how much force can the piston exert?
2. If the air pressure in the pneumatic system previously mentioned is increased to 95 psi, what is the force that the piston can exert?
3. A simple hydraulic braking system has two cylinders, an input cylinder with a piston diameter of 0.4 inch and an output cylinder with a diameter of 1.25 inches. If the input piston is pushed 1.0 inch, how far will the output piston travel?

BASIC PNEUMATIC CIRCUITS

Using the schematic symbols introduced earlier in this chapter, here are a few basic pneumatic circuits. Circuit one (see Figure 14-27) shows a single-acting cylinder connected to a three-port valve. On the left, the system is shown in the off position. The exhaust port is connected to the cylinder port through the three-port valve, so there is no air pressure on the cylinder piston. On the right, the system is shown in the on position, where the air supply is connected to the cylinder through the three-port valve. The piston is shown extended.

schematic:
A diagram of a system that uses graphic symbols to represent pneumatic, hydraulic, or electrical components.

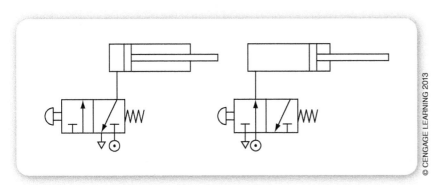

Figure 14-27: Single-acting cylinder controlled by a three-port valve, shown in both the inactivated (left) and activated (right) positions.

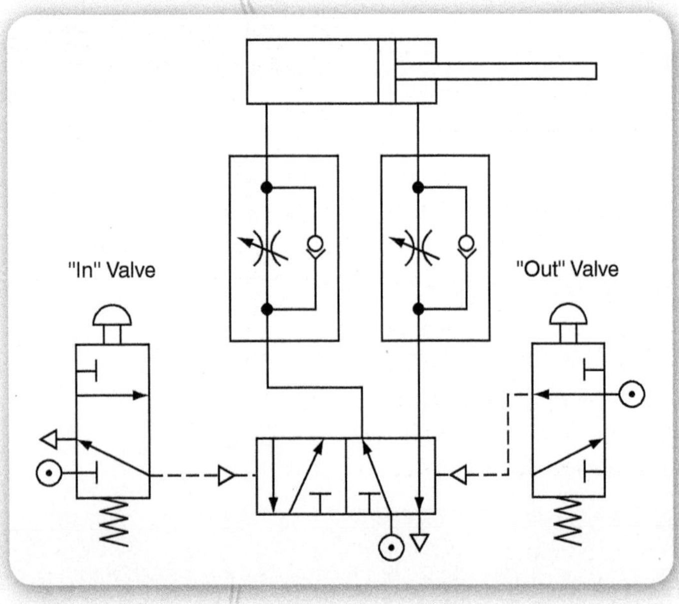

In Figure 14-28, a more complex system is represented. A double-acting cylinder is shown connected to a pressure-operated five-port valve through two flow regulators. Two three-port valves control the five-port valve, one to control the extension of the piston (out), and the other to control the retraction of the piston (in). Note that the air pressure to operate the cylinder goes through the five-port valve. The air connections to the two three-port valves are only used to operate the five-port valve.

A more interesting variation on the previously mentioned system is shown in Figure 14-29. Here, a double-acting cylinder is actuated by a push-button three-port valve. When the piston extends, it contacts another three-port valve and depresses it, causing the piston to retract. The two three-port valves are used to operate the pressure-operated five-port valve. In operation, when the push button is pressed, the piston extends and automatically retracts. A fully automatic system could be created if the three-port valve that extends the piston were placed in such a way that the piston would depress it when it retracted. In such a system, the piston would constantly extend and retract until the air pressure was removed. The flow regulator would control the speed of the extension.

Figure 14-28: A double-acting cylinder controlled by a pressure-operated five-port valve. Three-port valves control the pressure-operated valve, and flow regulators control the speed of the piston on both the extend and retract movements.

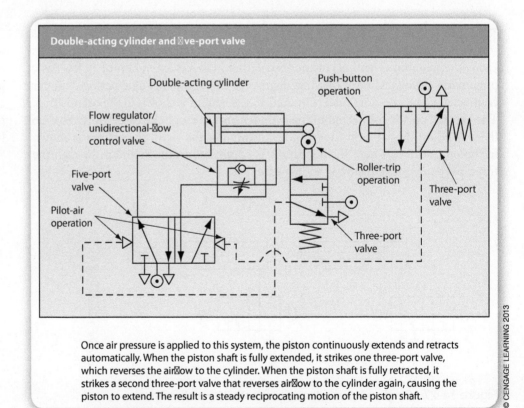

Once air pressure is applied to this system, the piston continuously extends and retracts automatically. When the piston shaft is fully extended, it strikes one three-port valve, which reverses the airflow to the cylinder. When the piston shaft is fully retracted, it strikes a second three-port valve that reverses airflow to the cylinder again, causing the piston to extend. The result is a steady reciprocating motion of the piston shaft.

Figure 14-29: Simple automated pneumatic system.

SAFETY IN FLUIDIC SYSTEMS

Pneumatic and hydraulic systems can develop extremely powerful forces. Compressed air itself can be dangerous. Here are a few simple rules that will help make work with pneumatic and hydraulic systems safe:

▶ Never blow compressed air at yourself or anyone else. Under high pressures, air can enter the bloodstream through the skin and cause serious injury or death. Eye injury can easily result from high-pressure air blown in the face.

▶ Components that operate on compressed air or hydraulics can also be hazardous. A moving piston rod will not stop if your finger or your hand is in the way. Crushing pressures are common in the moving parts of these systems, and even low air line pressures can translate into large forces. Think about the log splitter and you will gain some respect for the forces that can be generated in these systems.

▶ Always make connections and remove connections to pneumatic components with the air pressure line disconnected. Working with hydraulic systems, in addition to the hazards of pressure, can be very messy. Be certain that the circuit is completed and that no hoses or lines are left disconnected when pressure is applied.

▶ Hoses strung across the floor or between tables invite tripping and falling accidents. Make certain all air lines are out of the way.

Safety Note
Working with pneumatic and hydraulic systems presents many hazards. Please review the list of safety precautions before working with system components.

OFF-ROAD EXPLORATION
For further information about hydraulic and pneumatics principles and components, go to the website *www.hydraulicspneumatics.com* and select "Fluid Power Basics."

SUMMARY

Pneumatic and hydraulic systems are both classified as fluid power systems. To do work, pneumatic systems use air, whereas hydraulic systems use liquids. Both systems have advantages and disadvantages.

Pneumatic systems require a source of compressed air. Through special valves, the compressed air is directed to actuating components such as cylinders, which provide linear motion and can exert a great deal of force. Other applications use pneumatic motors, such as in air wrenches, air sanders, and many industrial applications. Other pneumatic components, such as pressure regulators and flow regulators, control the amount of air pressure to the system and to specific components.

Similarly, hydraulic systems use cylinders to create linear motion and hydraulic motors to create rotary motion. Because hydraulic systems are more complicated and messy, they are most commonly used in industrial applications, or where pneumatic systems are not up to the task.

BRING IT HOME

OBSERVATION/ANALYSIS/SYNTHESIS

1. Explain the differences between pneumatic and hydraulic systems, and provide examples of the application of these systems.
2. What are the advantages of hydraulic systems over pneumatic systems? What are the advantages of pneumatic systems over hydraulic systems?
3. Design and make a hydraulic-operated remote gripper for a robotic arm using acrylic, wood, or other material and two syringes filled with water. The gripper should be able to pick up a "D" size battery when attached to the end of a 4-foot-long pole.
4. In a hydraulic system, an input force of 60 pounds is applied to a 2-inch-diameter piston and the output piston diameter is 4 1/4 inches. What is the pressure in the system? Also, what force will the output piston exert?
5. Design a simple hydraulic system with two cylinders so that, when the piston of a 5/8-inch-diameter cylinder is pushed in 1 1/2 inches, the piston in the other cylinder moves 4 inches. You will need to determine the size of the second cylinder.
6. Using standard pneumatic component symbols, do the following:
   a. Design and draw a simple pneumatic clamp system with one double-acting cylinder, one pressure-operated five-port valve, and separate valves to apply and remove clamping pressure.
   b. Design and draw a simple pneumatic door lock system that would extend and retract a metal bolt into a hole in the edge of a door. A person must be able to operate the system on either side of the door.
7. Show mathematically how the log splitter in Figure 14-26 could exert 34 tons of pressure, when the piston diameter is 5 inches.

EXTRA MILE

ENGINEERING DESIGN ANALYSIS CHALLENGE

Using a kit or commercial pneumatic components, simulate the following system:

- Design and make a system that will clamp a piece of material and drill a hole. The system should use a single-acting cylinder to clamp and a double-acting cylinder to hold the drill head. Remember to regulate the speed of the drill bit entering and exiting the material. Can you redesign the system to automatically clamp and drill in one operation?

CHAPTER 15
Human Factors in Design and Engineering

Menu

Before You Begin
Think about these questions as you study the concepts in this chapter:

1. What is human factors (ergonomics)?
2. Why is it necessary to understand human factors to be a successful designer/engineer?
3. How is anthropometric data used during the design process?
4. How are principles of human physical characteristics, behavior, and abilities applied during the design process?
5. How can universal design principles help everyone?
6. What steps are used to do human factors design?
7. How is a design evaluated for safety, comfort, and effective human use?

INTRODUCTION

Good design must always take into consideration human users. Did you ever try to ride a small tricycle as a teenager (see Figure 15-1) or use a product that was too big or too small or uncomfortable? The degree to which the things in our environments are comfortable and safe is a measure of good design and the application of principles of human factors.

We expect to live and work in environments and to use products that "fit" us as human users, but that was not always the case. During the early Industrial Revolution, humans were expected to fit into their working environment. By the early twentieth century, some designers and engineers were beginning to learn more about fitting design to people. One of the important leaders was Henry Dreyfuss. His important work in this area has earned him recognition as the father of human factors in the United States. Dreyfuss began his career in stage design but later apprenticed under Norman Bel Geddes where he learned industrial design. His award-winning telephone designed for Bell Laboratories was produced from 1937 to 1950. He designed the first modern refrigerator for GE by concealing the motor and compressor. He also designed Hoover vacuum cleaners and Mercury diesel locomotives for the New York Central Railroad. His leadership in the field of industrial design led to the establishment of the Society of Industrial Designers in 1944, later to be renamed the Industrial Designer Society of America (IDSA). His work led to the book *Designing for People* (1955), which contained anthropometric charts for fictional characters Joe and Josephine, and the standard-setting book *Measure of Man* (1960).

Figure 15-1: Products must be designed for the intended user.

Human factors is also called ergonomics, from the Greek *ergo* for work and *nomics* for natural law. Stated another way, human factors is the application of the knowledge about human physical characteristics, behavior, and abilities to the design of products, systems, and environments for safe and effective use. In fact, human factors uses knowledge from psychology, anthropology, physiology, biology, and engineering. Human factors principles are used by engineers and industrial designers to create better tools and products, by architects and engineers to create better environments in which people live and work, and by professional societies and governmental agencies to establish safety standards.

The example given in Figure 15-1, where the tricycle is too small, relates to the importance of product size fitting the human users. Most people understand the importance of product size and may consider purchasing something that is advertised as "ergonomically" designed. But what does ergonomically designed mean and are there other areas of human factors to consider when doing technological design? In this chapter, in addition to learning about physical measurements, such as size and weight, you will learn how to consider characteristics such as human behavior and abilities in the design of tool, systems, and environments for safe and effective human use.

All people at every stage of their lives interface with technology. Technology can help you be healthier, help you study and play, and make life easier and more enjoyable. Our technological world also has risk associated with the things we

human factors (HF):
The application of the knowledge about human physical characteristics, behavior, and abilities to the design of products, systems, and environments for safe and effective use.

OFF-ROAD EXPLORATION

For additional information about human factors, see the following:
About Ergonomics:
www.ergonomics.about.com
Center for Ergonomics at the University of Michigan:
http://sitemaker.umich.edu/center-for-ergonomics/home
Ergo Web: www.ergoweb.com

design and make. For example, each year in the United States, approximately 3,000 children die from falls or drowning, with nearly one-half occurring in or near their home. Children are also injured using toys or by other products such as a child touching a hot stove. Over 250,000 children ages 14 and under are treated for bicycle-related injuries alone. In the workplace, over 5 million injuries and 6,000 deaths result annually. Common workplace injuries include repetitive stress injuries (RSIs) and cumulative trauma disorders (CTDs) such as carpal tunnel syndrome and tendonitis (tennis elbow). The Occupational Safety and Health Administration (OSHA) is responsible for enforcing safety and health in the workplace and to assure that American employers are providing for the well-being of working men and women. To improve working conditions, OSHA sets and enforces standards and requires training and a continual process of improvement in the workplace (see www.osha.gov for additional information).

HUMAN SCALE

Because almost all problems or opportunities in technology and engineering involve human users in some way, designers will need detailed information about human physical characteristics. Anthropometry is the branch of science that deals with human measurement by describing human size, shape, and other physical characteristics. Anthropometric data are available for designers in statistical form. Data are based on the probability, or likelihood, of a human characteristic happening at a given frequency. Suppose you were to observe all students entering your school. It would be much more probable that you would see substantially more (a higher frequency) average-height students entering your school than either very short or very tall students. If you made a graph of the number of students with certain heights that you observed, the curve would most likely look like a Gaussian distribution or normal bell-shaped distribution curve (see Figure 15-2). This is because the frequency distribution of the people entering the school is likely to represent a population reflecting a normal distribution of high school students. This would not be true if you made a distribution curve of students of, for example, all male high school basketball players, which would still be a bell-shaped curve but likely with a higher average height. Engineers, industrial designers, and architects use anthropometric data to design products and environments for the safe and productive use of all people.

anthropometry:
The branch of science that deals with human measurement by describing human size, shape, and other physical characteristics. Anthropometric data are available for designers in statistical form.

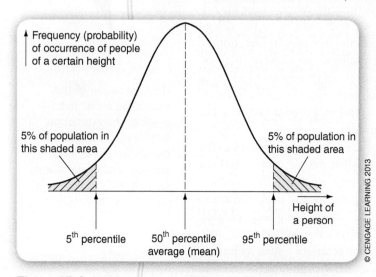

Figure 15-2: Data representing the height of a population of all high school students would take on a bell-shaped form, called a Gaussian curve.

The Myth of the Average Person

Most people believe that things are designed for the average person. This isn't always true. Some products are designed for adult men or women while others are designed for children (see Figure 15-3). In addition, some products are designed for a single person or very limited group of people. For example, Dr. Mae Jemison has a space suit that was made specifically for her (see Figure 15-4). When something is designed for one person, we call it a custom product.

Figure 15-3: Golf clubs designed for adults and children.

Figure 15-4: Dr. Mae Jemison's space suit was custom-designed for her.

Adult Male and Female Weight* in Pounds and Kilograms by Age, Sex, and Selected Percentiles†																
		18 to 79 (Total)		18 to 24 Years		25 to 34 Years		35 to 44 Years		45 to 54 Years		55 to 64 Years		65 to 74 Years		75 to 79 Years
		lb	kg	lb	kg	lb	kg	lb	kg	lb	kg	lb	kg	lb	kg	lb	kg
99	MEN	241	109.3	231	104.8	248	112.5	244	110.7	241	109.3	230	104.3	225	102.0	212	96.2
	WOMEN	236	107.0	218	98.9	239	108.4	238	108.0	240	108.9	244	110.7	214	97.1	205	93.0
95	MEN	212	96.2	214	97.1	223	101.2	219	99.3	219	99.3	213	96.6	207	93.9	198	89.8
	WOMEN	199	90.3	170	77.1	191	86.6	204	92.5	205	93.0	211	95.7	196	88.9	193	87.5
90	MEN	205	93.0	193	87.5	208	94.3	207	93.9	209	94.8	203	92.1	198	89.8	191	86.6
	WOMEN	182	82.6	157	71.2	173	78.5	184	83.5	190	86.2	195	88.5	183	83.0	178	80.7
80	MEN	190	86.2	180	81.6	195	88.5	193	87.5	194	88.0	190	86.2	183	83.0	170	77.1
	WOMEN	164	74.4	145	65.8	152	68.9	165	74.8	171	77.6	176	79.8	169	76.7	162	73.5
70	MEN	181	82.1	171	77.6	185	83.9	184	83.5	185	83.9	180	81.6	172	78.0	161	73.0
	WOMEN	152	68.9	137	62.1	143	64.9	153	69.4	158	71.7	165	74.8	160	72.6	155	70.3
60	MEN	173	78.5	164	74.4	177	80.3	177	80.3	178	80.7	172	78.0	166	75.3	150	68.0
	WOMEN	144	65.3	131	59.4	136	61.7	144	65.3	149	67.6	154	69.9	151	68.5	147	66.7
50	MEN	166	75.3	157	71.2	169	76.7	171	77.6	171	77.6	165	74.8	161	73.0	146	66.2
	WOMEN	137	62.1	126	57.2	130	59.0	137	62.1	143	64.9	146	66.2	145	65.8	137	62.1
40	MEN	159	72.1	151	68.5	162	73.5	164	74.4	163	73.9	158	71.7	153	69.4	141	64.0
	WOMEN	131	59.4	122	55.3	125	56.7	131	59.4	137	62.1	140	63.5	138	62.6	127	57.6
30	MEN	152	68.9	145	65.8	154	69.9	158	71.7	156	70.8	151	68.5	146	66.2	137	62.1
	WOMEN	125	56.7	117	53.1	120	54.4	125	56.7	130	59.0	134	60.8	132	59.9	119	54.0
20	MEN	144	65.3	140	63.5	146	66.2	151	68.5	149	67.6	143	64.9	138	62.6	132	59.9
	WOMEN	118	53.5	111	50.3	114	51.7	119	54.0	122	55.3	129	58.5	125	56.7	113	51.3
10	MEN	134	60.8	131	59.4	136	61.7	141	64.0	139	63.0	131	59.4	126	57.2	120	54.4
	WOMEN	111	50.3	104	47.2	107	48.5	113	51.3	113	51.3	120	54.4	114	51.7	105	47.6
5	MEN	126	57.2	124	56.2	129	58.5	134	60.8	131	59.4	123	55.8	117	53.1	107	48.5
	WOMEN	104	47.2	99	44.9	102	46.3	109	49.4	106	48.1	112	50.8	106	48.1	95	43.1
1	MEN	112	50.8	115	52.2	114	51.7	121	54.9	116	52.6	112	50.8	99	44.9	99	44.9
	WOMEN	93	42.2	91	41.3	92	41.7	100	45.4	95	43.1	95	43.1	92	41.7	74	33.6

*All measurements were made with the examinee stripped to the waist and without shoes, but wearing paper slippers and a lightweight, knee-length examining gown. Men's trouser pockets were emptied.
†Measurement below which the indicated percent of people in the given age group fall.

Figure 15-5: Weight of men and women by age.

Custom products can include the relatively rare professional bicycle and space suit or much more commonly used products such as a tailored shirt or dress suit, or custom-made furniture. All custom products are more expensive than mass-produced products. **Mass-produced products** are designed for a broad range of consumers and as such are less expensive to produce and sell. A house designed for a very tall professional basketball player may have custom doorways with more than a seven-foot-high clearance, while a countertop for a person in a wheelchair will need to be set lower than the standard 36 inches height, and clearance will need to be provided under the countertop.

Should a doorway in a standard house, which is normally six feet, eight inches high, be sized instead at five feet, five inches for the average height of all adult men and women? Obviously not! The standard doorway is designed to allow nearly all adults to pass through without bending over. If you go into a very old home in this country or a foreign country, you may see doorways that are less than 6 feet 8 inches. Can you speculate why these doors were made shorter than today's standard height?

> **OFF-ROAD EXPLORATION**
> To see an example of an innovative handicapped-accessible kitchen design, go to http://www.handicapped-accessible-kitchens.com

Picking the Right Numbers

Most students understand basic descriptive statistics. When test grades are returned, you hope to be above the average, or mean, score for the test. Another concept used in statistics to describe data is **percentiles** (see Figure 15-5). This figure provides data organized to show the frequency distribution by weight for all adults (ages 18 to 79) in the United States. As can be observed, the average or 50th percentile weight is 75.3 kg (166 lbs) for men and 62.1 kg (137 lbs) for women. Each percentile describes a point on the frequency distribution. For example, the 5th percentile indicates that 5 percent of the population is below this point and 95 percent of the population is above this point. The 5th percentile weight of this population is 57.2 kg (126 lbs) for men and 47.2 kg (104 lbs) for women.

Conversely, the 95th percentile indicates that 5 percent of the population is above this point and 95 percent of the population is below this point. The 95th percentile weight for this population is 96.2 kg (212 lbs) for men and 90.3 (199 lbs) for women. Looking at the Gaussian curve, you will observe that the highest frequency of people is around the 50th percentile. It is important to observe that fewer people fall at the two extreme ends of the curve.

The chart in Figure 15-6 shows the differences in heights in inches and centimeters between the 1st and 99th percentiles of men and women in several age categories between 18 and 79 years of age. As can be observed, the average or 50th percentile height is 173.5 cm (68.3 in) for men and 159.8 cm (62.9 in) for women. Each percentile describes a point on the frequency distribution. For example, the 5th percentile for women is 149.9 cm (59 in) and 95 percent for men is 184.9 cm (72.8 in). Note that the doorway height of 6 feet, 8 inches discussed earlier will accommodate over 99 percent of the population.

Adult Male and Female Stature* in Inches and Centimeters by Age, Sex and Selected Percentiles†

%ile	Sex	18 to 79 (Total) in / cm	18 to 24 Years in / cm	25 to 34 Years in / cm	35 to 44 Years in / cm	45 to 54 Years in / cm	55 to 64 Years in / cm	65 to 74 Years in / cm	75 to 79 Years in / cm
99	MEN	74.6 189.5	74.8 190.0	76.0 193.0	74.1 188.2	74.0 188.0	73.5 186.7	72.0 182.9	72.6 184.4
99	WOMEN	68.8 174.8	69.3 176.0	69.0 175.3	69.0 175.3	68.7 174.5	68.7 174.5	67.0 170.2	68.2 173.2
95	MEN	72.8 184.9	73.1 185.7	73.8 187.5	72.5 184.2	72.7 184.7	72.2 183.4	70.9 180.1	70.5 179.1
95	WOMEN	67.1 170.4	67.9 172.5	67.3 170.9	67.2 170.7	67.2 170.7	66.6 169.2	65.5 166.4	64.9 164.8
90	MEN	71.8 182.4	72.4 183.9	72.7 184.7	71.7 182.1	71.7 182.1	71.0 180.3	70.2 178.3	69.5 176.5
90	WOMEN	66.4 168.7	66.8 169.7	66.6 169.2	66.6 169.2	66.1 167.9	65.6 166.6	64.7 164.3	64.5 163.8
80	MEN	70.6 179.3	70.9 180.1	71.4 181.4	70.7 179.6	70.5 179.1	69.8 177.3	68.9 175.0	68.1 173.0
80	WOMEN	65.1 165.4	65.9 167.4	65.7 166.9	65.5 166.4	64.8 164.6	64.3 163.3	63.7 161.8	63.6 161.5
70	MEN	69.7 177.0	70.1 178.1	70.5 179.1	70.0 177.8	69.5 176.5	68.8 174.8	68.3 173.5	67.0 170.2
70	WOMEN	64.4 163.6	65.0 165.1	64.9 164.8	64.7 164.3	64.1 162.8	63.6 161.5	62.8 159.5	62.8 159.5
60	MEN	68.8 174.8	69.3 176.0	69.8 177.3	69.2 175.8	68.8 174.8	68.3 173.5	67.5 171.5	66.6 169.2
60	WOMEN	63.7 161.8	64.5 163.8	64.4 163.6	64.1 162.8	63.4 161.0	62.9 159.8	62.1 157.7	62.3 158.2
50	MEN	68.3 173.5	68.6 174.2	69.0 175.3	68.6 174.2	68.3 173.5	67.6 171.7	66.8 169.7	66.2 168.1
50	WOMEN	62.9 159.8	63.9 162.3	63.7 161.8	63.4 161.0	62.8 159.5	62.3 158.2	61.6 156.5	61.8 157.0
40	MEN	67.6 171.7	67.9 172.5	68.4 173.7	68.1 173.0	67.7 172.0	66.8 169.7	66.2 168.1	65.0 165.1
40	WOMEN	62.4 158.5	63.0 160.0	62.9 159.8	62.8 159.5	62.3 158.2	61.8 157.0	61.1 155.2	61.3 155.7
30	MEN	66.8 169.7	67.1 170.4	67.7 172.0	67.3 170.9	66.9 169.9	66.0 167.6	65.5 166.4	64.2 163.1
30	WOMEN	61.8 157.0	62.3 158.2	62.4 158.5	62.2 158.0	61.7 156.7	61.3 155.7	60.2 152.9	60.1 152.7
20	MEN	66.0 167.6	66.5 168.9	66.8 169.7	66.4 168.7	66.1 167.9	64.7 164.3	64.8 164.6	63.3 160.8
20	WOMEN	61.1 155.2	61.6 156.5	61.8 157.0	61.4 156.0	60.9 154.7	60.6 153.9	59.5 151.1	59.0 149.9
10	MEN	64.5 163.8	65.4 166.1	65.5 166.4	65.2 165.6	64.8 164.6	63.7 161.8	64.1 162.8	62.0 157.5
10	WOMEN	59.8 151.9	60.7 154.2	60.6 153.9	60.4 153.4	59.8 151.9	59.4 150.9	58.3 148.1	57.3 145.5
5	MEN	63.6 161.5	64.3 163.3	64.4 163.6	64.2 163.1	64.0 162.6	62.9 159.8	62.7 159.3	61.3 155.7
5	WOMEN	59.0 149.9	60.0 152.4	59.7 151.6	59.6 151.4	59.1 150.1	58.4 148.3	57.5 146.1	55.3 140.5
1	MEN	61.7 156.7	62.6 159.0	62.6 159.0	62.3 158.2	62.3 158.2	61.2 155.4	60.8 154.4	57.7 146.6
1	WOMEN	57.1 145.0	58.4 148.3	58.1 147.6	57.6 146.3	57.3 145.5	56.0 142.2	55.8 141.7	46.8 118.9

*Height, without shoes. See Table 1A for definition of stature.
†Measurement below which the indicated percent of people in the given age group fall.

COURTESY OF WATSON-GUPTILL PUBLICATIONS.

Figure 15-6: **Height of men and women by age.**

Choosing the right numbers becomes very important for appropriate design solutions. If the product only fits a few people, it may be easier to design because the constraints are more clearly defined but the potential market will be very small. A product designed for a larger population is much more difficult to design and produce but has a larger potential market. An adjustable chair illustrates how a well-planned design can lead to a product that fits a larger population (see Figure 15-7). The cost benefit of designing for a larger population, however, diminishes near the extremes of the normal distribution curve. For this reason, most designers and engineers plan to use measurements for the middle 90 percent (5th to 95th percentiles) of the population.

Figure 15-7: **An adjustable chair is a solution that fits a larger population.**

Not All Measures Are Equal

We have already learned that many measures of human beings are important. In the case of custom production, for someone such as a cyclist or an astronaut, anthropometric measurements would need to be made for only that individual. Because most products and environments are designed for mass production, the best or **optimal match** must be planned. Designers must first decide on the group of people for whom they are designing. This group is known as the **target population**. Is the design going to be used primarily by adults or children? If designing for children, what are the children's ages (children's heights vary dramatically with age)? Can the product be made adjustable? Is the design going to be used primarily by men or women, or both? Will it be used by the elderly or physically challenged? Typically, the measurement for the 5th to 95th percentiles will be used, but not always.

The amount of difference in the population is also important to know. **Standard deviation (SD)** is a measure of the degree of variation from the mean, or the "width," of the bell curve. Stated another way, SD describes how far the population is stretched out along the base of the normal distribution curve. A measure of 1 SD above and 1 SD below the mean (± 1 SD) includes 68 percent of the population, while ± 2 SD includes 95 percent of the population. The 5th to 95th percentiles are nearly two standard deviations on either side of the mean. These differences occur in a population for a variety of reasons. The greatest differences in a population are related to age and gender.

target population: A group of individuals for which a design is being made or targeted. Target populations are usually described in terms of gender and age, but may also include other characteristics such as income, left-handedness, or physical disability.

Reach and Clearance

Reach is a measurement that is selected from data representing the smallest member of the population and is known as a one-way measurement (see Figure 15-8). Reach, as the name implies, is a measurement associated with a person's ability to reach something such as an automobile gas pedal or house light switch. When you were a small child, you could not reach the overhead light switch when entering a room. The proper measurement for reach is based on the smallest member of the population and is expressed as a maximum dimension. It is logical that if the smallest person can reach the object in question, such as a control switch on an appliance, then all other (larger) members of the population will also be able to reach the same control switch.

Clearance is a measurement that is selected from the largest member of the target population and is also a one-way measurement. Clearance, as its name implies, is a measurement associated with a person's ability to fit between or under some object such as a doorway or table. Would Shaquille O'Neal (Shaq) have to duck under your front doorway to enter your home? The clearance dimension is based on the largest member of the population and is used to specify the head-room,

Figure 15-8: This is an illustration of reach and clearance. In each case, the smaller figure is based on the 5th percentile body size data, and the larger figure on the 95th percentile data.

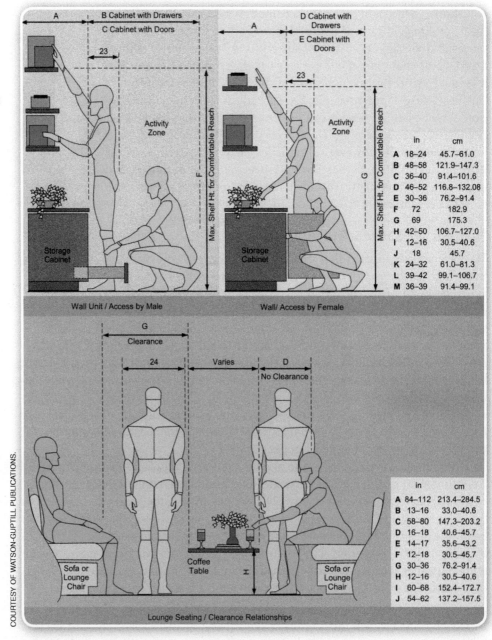

hip-room, or shoulder-room needed in a given situation, such as the seat of a chair. Unlike reach, which accommodates the smallest member of the population, the clearance dimension must accommodate the largest member of the population. For this reason, clearance is specified as a minimum dimension. It is important that the knee clearance under a table be enough (minimum) for the largest member of the population. If the largest person fits the clearance, then the other (smaller) members of the population will also fit.

HUMAN BEHAVIOR

People are very interesting to observe. Have you ever driven with your family to a major event and noticed when you arrived that the parking lot had a "Lot Full" sign? Did someone in your family say, "I think that there is a parking spot in that lot!"? Did you ever purchase a new product and not read the directions and safety notices? The way we normally react to certain situations reflects the human factors (HF) principle of human behavior. When interacting with our human-designed world, it is very important that we are aware of the dynamics of our environment; this is sometimes referred to as situational awareness. You already know to be careful when working at a hot stove or turning on the hot water faucet to wash your hands. Is the iron hot that is sitting on the ironing board? You know to wear safety glasses when working in a shop and not to touch a sharp blade!

Some of you may be driving a car or will be soon. Driving a car requires careful attention to situational awareness because you will need to know if anything is behind you when backing out of a driveway or parking spot. You will need to know if someone is entering an intersection before you pull out from a stop sign. But what if you are talking on your cell phone or adjusting your car radio—you may not be paying attention and an accident can result. The way we react to technological situations is an example of human behavior that is important to the design of new products.

Compatibility

Every time we use a tool or work in an environment, we make assumptions about how things work. Compatibility describes the relationship between how a person expects something to work and the way it was designed to work. In an ideal world, the way something was designed to work and how the person uses it would be the same. Unfortunately, this is not true and the fact that human expectations of their environment vary can lead to serious consequences. In a machine laboratory, we should be aware that a power saw is present in the room (another example of situational awareness), but if we are planning to turn the saw on, we need to know which direction the blade is designed to turn. What can happen if you feed a piece of wood into a machine like a table saw in the wrong direction?

In our home we should be aware of how the burner controls on our stove work. Compatibility determines how we decide which control turns on which burner (see Figure 15-9). Because there are a number of logical possibilities, most modern stoves use a graphic to indicate which burner is being controlled by which knob. Without this graphic aid, a person could be seriously burned by a gas or electric burner being accidentally lit. We also need to know which faucet controls the hot water, and which direction on the faucet is open and which is closed. Again, as with the stove control, serious consequences can result if the wrong assumptions of how the faucet was designed and installed are made.

We make compatibility decisions every day. Something as simple as turning the wrong way to go to our next classroom or to finding the elevator in an office building is of little consequence if our assumption is wrong. Other decisions, such as turning the steering wheel the wrong way while backing your car out of the garage or a parking space, have potentially more serious consequences. What if you are working at a nuclear power plant and you need to add more cooling water to the reactor? What if you turned the valve the wrong direction? Our conditioned responses or assumptions about how something should work relate to the human factors issue of compatibility. Good design dictates that the potential for misunderstandings based on compatibility principles between the user and designer be minimized.

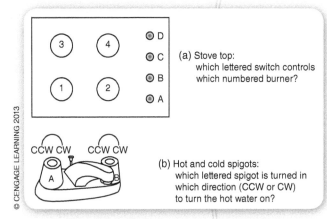

Figure 15-9: Stove and spigot illustration.

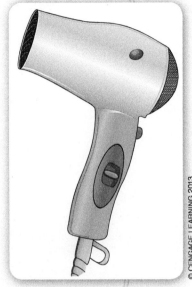

Figure 15-10: A hair dryer handle and controls must be designed to fit human users.

POSTURE AND MOVEMENT

Human biomechanical capabilities are established by an individual's skeletal structure, muscles and connecting tissue, and nervous systems, giving the typical person a wide variety of capabilities. In considering posture, we need to know various measurements when a person is sitting, bending, standing, reaching, gripping, and kneeling, to name a few. When sitting, the body is better supported but able to generate less force. When standing, the body is able to exert greater force but is under more stress and becomes fatigued more easily. When working, it is often better to alternate between sitting and standing. If this is not possible, then periodic breaks from standing or sitting are recommended. Posture and movement are important when you are designing chairs, storage cabinets, control panels, tools, and all other products and environments in which the body position and ability to move matters.

Capability (Abilities and Limitations)

The design for each product or environment requires different anthropometric measurements. Although we will need to consider the height of the target population to design a counter height, what human factors principles are important in designing a new hair dryer? One group of important measures will relate to human capability, including the abilities and limitations of the target population. While designing a hair dryer, anthropometric measurements for hand size, range of hand motion, gripping ability, and the best location for the on/off and temperature controls would be needed (see Figure 15-10).

Range of Motion

Humans are capable of a wide range of motion. Moving hands, arms, legs, or any other body part is measured in degrees of movement and varies in individuals based on bone structure, muscle strength, and joint flexibility (see Figure 15-11).

Body movement occurs around joints that hinge and pivot. As a joint moves, it passes from a neutral position to a maximum closing point, called joint flexion. The joint can also move from a neutral position to a maximum open point, called joint extension.

Figure 15-11: Range of motion of college males in the 5th, 50th, and 95th percentiles.

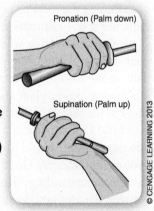

Figure 15-12: The forearm rotates to pronation (top) or supination (bottom) hand position.

The body can also rotate around a longitudinal axis. For example, hold your forearm in front of you and rotate from a palm-up position, known as **supination**, to a palm-down position, known as **pronation** (see forearm illustration in Figure 15-12).

Generally speaking, the body is most comfortable in a neutral position where stress and tension are at a minimum. The body is also able to exert the most force and is the strongest in the neutral position. Designing products, systems, and environments that minimize stress and tension allow people to work and play more efficiently and more comfortably, reduce injury, and improve general health.

The Hand

The human hand is often the direct link to our natural and technological world. We need our hands to get dressed in the morning, eat meals, open doors, answer the phone, and control the television. We use our hands to write or type messages and turn the pages of a book. Our hand is a complex structure of bones, arteries, nerves, ligaments, and tendons. Our hand is controlled by muscles in our forearm connected by tendons that run through the carpal tunnel in our wrist to our fingers. Hand movement occurs in two 90-degree planes. Moving the hand toward the palm is called **palmar flexion** and away from the palm is called **dorsiflexion**. Moving the hand toward the little finger and ulna forearm bone is called **ulnar deviation**. Moving the hand toward the thumb and radius forearm bone is called **radial deviation** (see Figure 15-13).

Cumulative (or repetitive) trauma disorders (CTDs) can result when repetition or misuse of a body part occurs. Many CTD injuries involve the hand, wrist, forearm, and back. Cumulative trauma disorders occur when muscles and joints are stressed, tendons become inflamed, nerves pinched, and blood flow is restricted. Common injuries include trauma-related weakening such as tears and breaks (stress fractures) or inflammation, causing carpal tunnel syndrome, tendonitis, or lower back pain. Good tool and product design and proper training to use tools and products can reduce CTDs, especially related to the hand. Designers should consider the following list of design principles when designing something to be controlled by the hand. Consider the capability of the target population to

1. Use a neutral hand position;
2. Avoid tissue compression;
3. Protect the palm of the hand;
4. Avoid repetitive finger action;
5. Maximize grip strength;
6. Plan for safe use.

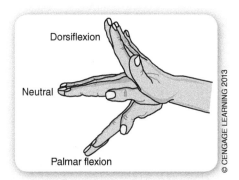
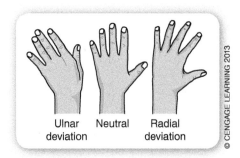

Figure 15-13: **Motion of the hand around two planes.**

Your Turn

What is the standard suitcase weight limit for airline passengers? Is that limit fair for female airline employees who must handle passenger luggage?

OFF-ROAD EXPLORATION

To learn more about neurological disorders, go to the National Institute of Neurological Disorders and Stroke (NINDS) website: www.ninds.nih.gov.

To learn more about preventing workplace illnesses and injuries, go to National Institute of Occupational Safety and Health (NIOSH) website: http://www.cdc.gov/niosh/

Lifting

Lifting, like all body movement, is controlled by biomechanical principles. Many people get hurt because they try to lift something that is too heavy or they lift incorrectly. Lifting incorrectly or lifting too much weight is not smart and can lead to lifelong back pain. The National Institute of Occupational Safety and Health (NIOSH) has set a maximum safe lifting weight at 23 kg (50 lbs) for a healthy adult male and 17 kg (37 lbs) for adult females.

When lifting, the stress on a joint is the result of the moment ((moment) = (force)(distance)). The ability to lift something safely depends on body position and movement issues. For example, most people have been told to lift with their legs and keep the weight near their bodies. This is good advice and is true to physical laws. When lifting, it is best to be standing and to be in a neutral position. It is stressful to be lifting while bending forward, or twisting. You should not use a jerking motion while lifting and you should avoid prolonged and repetitive lifting motions. The distance the load is lifted vertically and horizontally and the frequency of lifts reduces maximum safe lifting load. The Occupational Safety and Health Administration has established a multiplier for each lifting condition. Multipliers start at 1.0 for ideal lifting conditions and become <1.0 as the lifting conditions become less ideal. For example, if a load is lifted every 30 seconds for less than one hour, the frequency (F) multiplier is set at 0.91, but if the load is lifted every 30 seconds for eight hours, then the multiplier becomes 0.65. Complete information about NIOSH lifting equations is available at http://www.cdc.gov/niosh/docs/94-110/.

What is the safe lifting weight for an adult female if the load is lifted (VM) 30 cm three times per minute (FM) for less than one hour? Information to solve a problem will need to be drawn from a good human factors reference such as the OSHA website located at http://www.osha.gov/dts/osta/otm/otm_vii/lifting_analysis_worksheet.pdf.

Table 2 (VM) multiplier for 30 cm is 0.87

Table 5 (FM) multiplier for three times per minute is 0.88

(maximum weight for an adult female)(VM)(FM) = safe weight

(17 kg)(0.87)(0.88) = 13.0 kg

universal design: The design of a product, system, or environment that reduces barriers and assists not just the special needs population but all individuals.

Lifting is greatly aided by the ability to grip the object, such as having handles or grab bars that allow you to avoid twisting while holding a load. If you must position the object in another location, it is best to keep the object in front of your torso and turn your entire body to place the object in the desired location.

If you want to push or pull an object, the maximum force recommended by NIOSH is 20 kg (44 lbs). When moving an object, it is important that you maintain clearance so as not to get your foot caught, especially when pulling the object. When pushing an object, it is important that you have a clear vision of the area in front of the object and be careful when pushing or pulling objects up or down grades.

UNIVERSAL DESIGN

Governments and businesses are attempting to use designs that reduce barriers and assist individuals with disabilities. When the designs reduce barriers and assist not just the special needs population but all individuals, the concept is known as **universal design**. Individuals in wheelchairs would have a great deal of difficulty getting across a street if they had to first get over standard curbs. Most communities have replaced the standard curb at each corner with a ramp (see Figure 15-14). Because this change not only helps a person in a wheelchair, but also people pushing a baby stroller or a child on a scooter, or a teenager on a bicycle, we refer to the design as *universal*—a design solution that benefits everyone.

Today, all entrances to commercial buildings must be accessible without stairs, and elevators must be provided for all multiple storied buildings. Bathrooms must contain a lower urinal and an enclosed stall capable of wheelchair entrance. Door handles and controls such as elevator buttons must be reachable from a wheelchair. These requirements not only assist special populations and make life somewhat more comfortable for all ages and abilities; they are proving to be a good business practice. The design and marketing of universal design products is having a strong market impact.

Figure 15-14: *A curb cut is an example of universal design because it benefits not only someone in a wheelchair but all people crossing the street.*

The following principles of universal design should be used to design and evaluate potential new products:

1. Equitable: the design is useful to a wide range of people with different abilities.
2. Flexible: the design is adaptable for a wide range of people with different abilities.
3. Simple and intuitive: the design is easy to use regardless of the person's experience, language, knowledge, or ability.
4. Perceptible: the design can be understood regardless of the person's sensory ability.
5. Tolerance: the design minimizes the risk of accidental or unintended use.
6. Low physical effort: the design can be used with a minimum of fatigue.
7. Size and space: the design allows for people with different abilities to reach, grasp, or manipulate an object to use the product.

OFF-ROAD EXPLORATION

Center for Universal Design: *www.design.ncsu.edu*

Assistive or Adaptive Technology

Closely related to universal design is the principle of **assistive or adaptive technology**. Assistive technology refers to a product that enables people with disabilities to accomplish daily living tasks. Products may assist hearing- or sight-impaired people

assistive or adaptive technology:

Products, devices, or equipment, whether acquired commercially, modified, or customized, that are used to maintain, increase, or improve the functional capabilities of individuals with disabilities (Assistive Technology Act of 1998).

Figure 15-15: A three-wheel mobility scooter assists a person with difficulty walking.

with reading a book, or physically challenged people with getting around their home or to work (see Figure 15-15). Assistive technologies also include needs or difficulties associated with education and recreation such as a lift to help a wheelchair-bound person into a swimming pool.

Assistive technology products help people achieve greater independence and enhance their quality of life. Assistive-technology devices can improve physical or mental functioning or help overcome a disorder or impairment. Some devices prevent some conditions from getting worse, or are used to strengthen a physical or mental weakness. Some can improve a person's capacity to learn (see Figure 15-16). Some devices such as prostheses are used to replace missing limbs.

Assistive or adaptive products include programmable, on-screen, or voice-activated keyboards. Closed-circuit television systems can be used to magnify a page for the visually impaired. Optical character recognition systems can convert a printed page to a computer file. Telephones can have enhanced sound. Flashing lights are used to alert hearing-impaired individuals to someone at their front door or some emergency.

Elderly and Physically Challenged

More data are becoming available for designing products for the elderly and physically challenged (see Figure 15-17). Today, the elderly represent over one in eight Americans, which is over 33 million people, and that number will increase substantially over the next 10 to 20 years. Why is this? In addition to the elderly, a significant number of people are classified as physically challenged. Worldwide, these numbers are considerably larger. The elderly and physically challenged have greater difficulty coping with their physical environment, primarily due to diminished reach and mobility

OFF-ROAD EXPLORATION

Learn more about adaptive and assistive technology at *www.adaptiveenvironments.org*.

Figure 15-16: The MagniSight is a closed-circuit television (CCTV) product for the visually impaired.

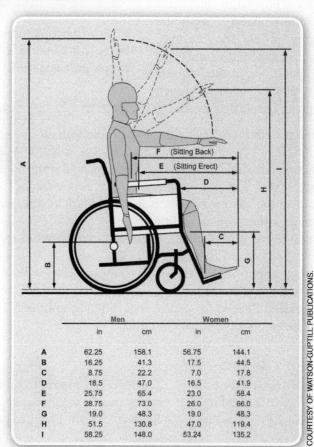

Figure 15-17: Anthropometric measurements for chair-bound people.

	Men		Women	
	in	cm	in	cm
A	62.25	158.1	56.75	144.1
B	16.25	41.3	17.5	44.5
C	8.75	22.2	7.0	17.8
D	18.5	47.0	16.5	41.9
E	25.75	65.4	23.0	58.4
F	28.75	73.0	26.0	66.0
G	19.0	48.3	19.0	48.3
H	51.5	130.8	47.0	119.4
I	58.25	148.0	53.24	135.2

Case Study

Toy Guide for Differently Abled Children

In the early 1990s, Toys "R" Us, Inc. founder and CEO Charles Lazarus had received so many letters from parents of children with disabilities voicing frustration over the lack of toys for their children that he decided to do something about the problem.

The company is one of the world's leading retailers of toys and children's apparel, with 1,600 stores in 27 countries. Starting as a baby furniture business in Washington, DC in 1948, Mr. Lazarus expanded his business to include toys for children of all ages. In 1957, he opened the first toy supermarket, and the idea was so successful that the company went public in 1978 as Toys "R" Us. Today the business also includes Kids "R" Us and Babies "R" Us stores.

Moved by the many letters from parents, Charles Lazarus asked Tom DeLuca to have his product development staff look into the problem. Tom noted that, "We realized that this was a sensitive issue, and we knew we didn't have any internal experts on disabilities. We wanted to respond in the right way, without embarrassing ourselves or our customers." DeLuca contacted various organizations for help. Diana Nielander of the National Lekotek Center, a nonprofit organization that promotes play for children with disabilities, was among the first to respond.

After several months of testing their most popular toys, Mr. DeLuca realized that many of their toys could be used by kids with disabilities. With this information, he proposed to Toys "R" Us management that they consider asking the Lekotek Center to help the company develop a guide for parents to select toys appropriate to their own children. Tom noted, "We thought at first about whether to begin carrying assistive technologies in our stores, or creating a 'special' section for toys for kids with disabilities. But we knew we didn't want to do that. Parents want their child to be treated like every other child."

Figure 15-18: *The Toys "R" Us "Toy Guide for Differently Abled Kids" helps parents select toys that are appropriate for their own children.*

In 1993, Toys "R" Us published the first "Toy Guide for Differently Abled Kids" (see Figure 15-18). The guide describes features of each toy based on the assessment of trained play experts. These experts observe children with disabilities playing with each toy and prepare comments about the skill level required in the following nine human factors areas:

1. Gross motor skills
2. Fine motor skills
3. Creativity
4. Auditory abilities
5. Language abilities
6. Self-esteem
7. Social skills
8. Tactile abilities
9. Thinking abilities

In addition to producing the toy guide, Mr. DeLuca believes that the company is helping to change the toy market. He noted that, in 1999, Mattel released a Cabbage Patch Kids Playtime Friend with mobility limitations, and Fisher Price in 2000 developed Aiden Assist, a rescue character in a wheelchair. The market for toys suitable for children with a handicap is as much as $2 billion a year and is growing faster than the general toy market. According to the American Academy of Pediatrics, over six million children have some form of disability. The company considers the project a success and believes that it has made a difference in the lives of children with disabilities and their parents.

Your Turn

Do you know someone who has a physical disability? Inquire what gives him or her the most difficulty in everyday life and work on a design to assist in that chore. It could be as simple as something to help someone open a door or a jar.

capabilities. Fortunately, in the United States and other developed countries, new construction for public buildings must comply with handicap access regulations. In addition, because of the general affluence of our society, many products are available to help extend the human capability of the physically challenged and elderly. These products include motorized wheelchairs, Braille readers, prosthetic devices, and hand-operated vehicles. Nevertheless, many needs and opportunities still exist for solving problems for the elderly and physically challenged populations.

DESIGNING WITH ANTHROPOMETRIC DATA

The following steps are necessary when considering human factors in your design:

1. Identify the target population.
2. Find appropriate human factors data about the target population's physical characteristics, behavior, and abilities.
3. Apply that data to all possible design solutions.
4. Test a solution using members of the target population.

Most designs will need to use one of the basic measurements illustrated in Figure 15-19 during the design process. The members of the design team must agree to the target population from which a correct set of measurements can be

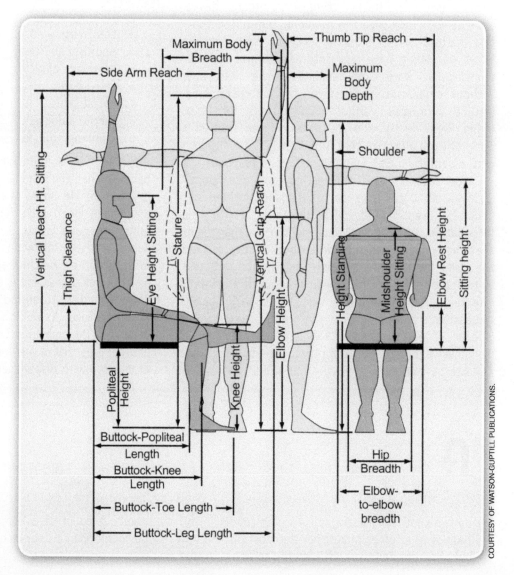

Figure 15-19:
Anthropometric body measurements that are most useful to designers of interior space.

selected. For example, stature would be considered when designing the correct clearance for a work environment with an overhead obstruction. Eye height would be important to determine the placement of an instrument panel or a privacy screen height in an office complex. Each product or environment must be designed to fit the human user.

Designing a Chair

An example of using anthropometric data can be illustrated in designing a chair. Let us assume that the chair will be a general-purpose chair to be used by any adult, primarily in the North American market. For this population, the 5th to 95th percentiles will be used, making the chair reasonably comfortable for 90 percent of the population. All elements of the chair must be planned. The seat height is determined by the **popliteal height** measurement "A" in Figure 15-20. The popliteal height is the measure from the floor to the back part of the leg behind the knee joint while seated. As can be seen from the chart, the popliteal height for the identified population is between 14 inches (5th percentile, females) and 19.3 inches (95th percentile, males). Most studies indicate that people like to sit one

popliteal height: The measure from the floor to the back part of the leg behind the knee joint while seated.

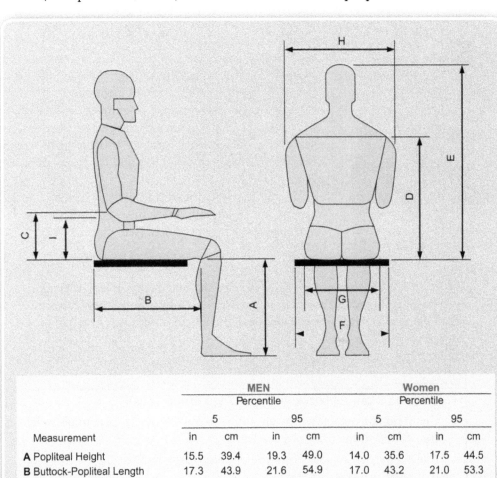

Figure 15-20: Key anthropometric dimensions required for a chair design.

	MEN				Women			
	Percentile				Percentile			
	5		95		5		95	
Measurement	in	cm	in	cm	in	cm	in	cm
A Popliteal Height	15.5	39.4	19.3	49.0	14.0	35.6	17.5	44.5
B Buttock-Popliteal Length	17.3	43.9	21.6	54.9	17.0	43.2	21.0	53.3
C Elbow Rest Height	7.4	18.8	11.6	29.5	7.1	18.0	11.0	27.9
D Shoulder Height	21.0	53.3	25.0	63.5	18.0	45.7	25.0	63.5
E Sitting Height Normal	31.6	80.3	36.6	93.0	29.6	75.2	34.7	88.1
F Elbow-to-Elbow Breadth	13.7	34.8	19.9	50.5	12.3	31.2	19.3	49.0
G Hip Breadth	12.2	31.0	15.9	40.4	12.3	31.2	17.1	43.4
H Shoulder Breadth	17.0	43.2	19.0	48.3	13.0	33.0	19.0	48.3
I Lumbar Height	See note.							

inch to two inches higher than their popliteal height. For this reason, the standard height for most chair seats is between 17 and 18 inches. However, if the chair seat we are designing were set at 18 inches, not all people in our population would find the chair comfortable. If the popliteal height is too high, then additional pressure is put on the underside of the knee joint, cutting off lower-leg circulation. Poor circulation leads to leg cramping and swelling in the feet after long periods of sitting. On the other hand, if the seat is too low, the legs must extend forward or backward, reducing good posture. In addition to the popliteal height information, adjustable seat mechanisms for up to four inches are readily available at a reasonable cost. Given this information, it would be appropriate to design a chair seat with a seat height that is adjustable between 14 1/2 inches and 18 1/2 inches.

The remainder of the chair design would take into consideration other measures identified in Figure 15-19. Each additional element of the chair, such as the seat width and seat depth, tilt of the seat, angle and size of the back, lumbar support, and location and size of arm rests adds additional design complexity and production cost.

Figure 15-21: **Effect of chair design on posture and comfort.**

One of the problems of seat design, especially for female workers, is the table height of the workstation. Think about the workstation, which is comprised of not only the chair but also the table holding a keyboard or other work-related tools. A petite female must raise her chair well above her popliteal height to reach the top of her desk or workstation set at 29 inches high.

The design team must consider more than the physical size of the chair. In an airplane, the spacing between seats is important. Have you ever noticed the difference between the seats in first class and coach, as shown in Figure 15-21?

In addition, the design team must consider choosing appropriate materials, mechanisms, aesthetics (style), and manufacturing processes (see Figure 15-22).

Figure 15-22: **Standards are developed by the Business and Institutional Furniture Manufacturer's Association (BIFMA) Engineering Committee and recommended to the American National Standards Institute (ANSI) for adoption every five years.**

ANSI/BIFMA Commercial-Grade Chair Specifications Summary:
• Chair back can support a load of 113 kg (250 pounds).
• Chair remains stable when an adult load of 79 kg (173 pound) weight is transferred to the front or back legs.
• Chair leg maintains its structural integrity when a 52 kg (115 pound) weight is applied to a side.
• Chair seat can withstand a static test of 136 kg (300 pounds) dropped from 152 mm (6 inches) above the seat.
• Chair is considered durable if a weight of 57 kg (125 pounds) dropped 100,000 repetitions from 51 mm (2 inches) above the seat does not change its structural integrity.

Careers in the Designed World

L. STEPHEN SCHMIDT, P.E.: CONSTRUCTION MANAGER, AND CONSULTANT FOR WATER RESOURCE MANAGEMENT

Building on a Vision

Steve Schmidt was always asking questions as a youth: "What does the pattern of roads on a map tell about the landforms and land use? Where does our tap water come from and how does it get to my faucet? Why is Missouri hilly and southern Illinois so flat?"

For Steve, it was the process of asking questions that helped him understand how something worked, and how ideas could be fabricated as real, physical designs. Steve's inquisitive nature led him to study engineering in college, and then to explore a variety of job experiences in engineering.

On the Job

Currently, Steve is the construction project manager for a $50 million dollar factory in McPherson, Kansas. The factory is large enough to cover 10 acres under one roof. Steve is the link between the factory owners, who know what they want the facility to do, the designers who created the plans and specifications for the building, and the contractors who are building it. "My goal is to see that the owner's money is spent wisely and the building meets the needs of the workers and all is built to the specifications needed to be technically correct," says Steve.

Steve's work in engineering has covered a wide range of projects. For one project he designed the emergency construction of a diversion canal to carry the water of Walnut Creek runoff from the Rocky Flats Nuclear Weapons Plant around the Great Western Reservoir in Colorado. This canal was designed as the construction occurred, with completion of a diversion structure, stilling basin, 9,100 feet of earth-lined canal, and 1,200 feet of a concrete-lined chute in two weeks from a cold start. The work for this diversion canal was precipitated by an FBI raid on the Rocky Flats Nuclear Weapons Plant, and it involved coordination with numerous state and federal agencies. As Steve says, "It was quite an exciting project but also very rewarding knowing this canal was protecting the water for a whole community."

Inspirations

Steve was inspired by great engineering achievements like the Brooklyn Bridge, Golden Gate Bridge, Hoover Dam, Panama Canal, and the Transcontinental Railroad. He says he gravitated to civil engineering because of his love of trains and fascination with bridges. This foundation has opened engineering opportunities for him in soil mechanics, water resource engineering, and now project management.

Education

Steve studied at Oklahoma State University, earning undergraduate and master's degrees in civil engineering with an emphasis in soil mechanics and foundation engineering.

Advice for Students

"First, keep your curiosity active," says Steve. Always ask how things work and why.

"It is also important to pay attention to detail as well as the big picture. Disasters are caused by overlooking very simple and basic things such as the space shuttle *Challenger* disaster and the Hyatt Regency walkway collapse. No matter how small the task you are given, give it your full attention and best effort. If you can prove you can handle the little things, you will earn a part in the big things."

Most importantly, Steve urges students to bring respect to the team environment. "When working in design teams there are no unimportant people; you can learn something important from just about anyone."

EVALUATING DESIGN FOR HUMAN FACTORS

Step 9 of the engineering design process (described in Chapter 2) calls for the testing and evaluating of the design proposal using design specifications. Although some of the design specifications will be concerned with mechanical, electronic, or other operational capabilities, most designs must also be evaluated against human factors principles. In the last section, you learned how to design using anthropometric data. The following section shows how a new product or an existing product can be evaluated for the safe and productive use by an intended audience.

Computer Workstation

The computer is a very important tool used by millions of people daily. Although the common view may be that the computer is a very safe tool to use, business and the U.S. Department of Labor, Occupational Safety and Health Administration are concerned about the growing number of health-related issues. Of increasing concern are the number of teenagers and younger children who are developing computer-related injuries such as eye strain, carpal tunnel syndrome, and neck stiffness.

All computer users should consider human factors principles including posture, component placement, and environmental issues in designing their computer workstation both at home and at work (see Figure 15-23).

Establish a Neutral Body Position. Have you ever had an arm or leg "fall asleep"? If so, the problem was most likely due to a lack of blood flow caused by some improper body alignment. The body is a complex musculoskeletal structure. If the muscles, tendons, and skeletal system are not properly positioned and aligned,

Figure 15-23: *General posture and component placement for a safe work environment.*

stress and strain result. Serious and prolonged stress and strain can result in musculoskeletal disorder (MSD). Health-related problems can be avoided by achieving and maintaining a neutral body position. For a computer workstation, a neutral body position is achieved by:

- Keeping hands, wrists, and forearms straight and parallel to the floor.
- Keeping your head level or bent slightly forward and balanced.
- Keeping your shoulders relaxed and your upper arms hanging relaxed beside your torso.
- Keeping your elbows close to your body and bent between 90 and 120 degrees.
- Keeping your feet flat on the floor (or footrest).
- Using a chair that keeps the back fully supported and positioned vertically or tilted slightly back.
- Keeping your thighs and hips supported and knees aligned with your hips.

A well-designed chair is essential for maintaining a neutral body position. A computer chair with rollers should have a natural spine curvature shape, a five-leg base for stability, and be adjustable for different body sizes and shapes. Arm rests are optional, but if provided, they should be soft and allow arms to be in a neutral position. A footrest should be provided for anyone who cannot have their feet flat on the floor when the chair height is placed in the proper position.

As you know from personal experience, prolonged work at a computer even with good posture will cause some stress and strain. If you plan to work at the computer for an extended period of time, make small adjustments in your posture, stretch frequently, and stand up and walk around for a few minutes periodically—go get a glass of water!

Establishing Component Placement. Computer components usually consist of a monitor, keyboard, and mouse. Placing these components in the proper position and proper environment reduces awkward body positions and strain due to glare.

- Position the keyboard and mouse so that your arms, wrists, and fingers are as relaxed as possible.
- Keep the monitor between 50 and 100 cm (20 to 40 in) and directly in front of the user.
- Position the top of the monitor at or slightly below eye level.
- Position the monitor perpendicular to any window or strong light source.
- Keep the keyboard directly in front of you.
- Elbows should be close to your body and forearms parallel to the floor.
- Wrists should be straight (neutral position); use a wrist rest to aid in keeping the wrist straight.
- Use keyboard shortcuts to reduce repetitive motions on the keyboard or mouse.
- Use light pressure to reduce tension when typing.

If the computer workstation does not allow for the proper positioning of monitor, keyboard, or mouse, changes should be made in the workstation rather than

Figure 15-24: **Ergonomically designed keyboard.**

using the components in a stressful position. Many types of monitors and keyboards are available. Flat-panel displays will take up less space on the workstation. Do not place the monitor on top of the computer (CPU) if it raises the top of the monitor above eye level. Arrange the workstation to avoid glare from overhead or desk lights and windows. A glare screen can be added to the monitor to reduce eye strain caused by ambient light glare. There should be adequate air circulation but avoid direct contact with air vents.

The standard straight keyboard requires the user to have their wrists bent slightly to the little finger side (ulnar deviation). Avoid using keyboard feet as they may cause the wrist to be bent up (dorsiflexion). Wrist rests on the keyboard or mouse should be used to reduce ulnar deviation and dorsiflexion wrist deviations. An ergonomically designed keyboard will allow your wrists to be in the most relaxed position (see Figure 15-24).

Rest your arms and especially your wrists periodically. Stretch your fingers and move your wrists to relax muscles. Rest your eyes periodically by looking away from the screen and focusing on a distant object for a short period of time. If you wear glasses, especially bifocals, be sure to not tilt your head back to achieve screen focus. Adjust the height of the screen, if necessary.

With an increasing number of people using laptops, PDAs, and game controllers, additional guidelines should be considered. As with the desktop computer, the placement and use of all the components is important. Although the mobility of the laptop and the small size of the PDA have desirable advantages, these conveniences can lead to incorrect use and physical harm. When using a laptop, follow these additional guidelines:

1. Select a laptop with a large screen for better viewing.
2. Select a workplace that is comfortable, and use good posture.
3. Place the computer at elbow height; do not bend over to work on the computer.
4. Position the top of the screen two to three inches below seated eye level and tilt the screen upward.
5. Position your screen to avoid glare from windows and overhead lights.
6. Consider using a wireless keyboard and mouse for improved ergonomics.

When using a PDA or game controller, follow these additional guidelines:

1. Limit your time to 10 to 15 minute sessions.
2. Stretch often by holding your palm upward (supination) and open your thumb and fingers, gently massaging the hand.
3. Use an external keyboard and stylus if possible.
4. Support your arms.
5. If your hand is sore, stop and use a cold pack and seek medical attention if the pain persists.

SUMMARY

Human factors is the application of the knowledge about human physical characteristics, behavior, and abilities to the design of products, systems, and environments for safe and effective use. Henry Dreyfuss, the father of American human factors, wrote many books on the subject including his first, *Designing for People* (1955), and the groundbreaking *Measure of Man* (1960). He helped found and served as the first president of the Industrial Designers Society of America.

Because almost all problems or opportunities in technology and engineering involve the human user in some way, designers need to be able to get information about human physical characteristics. Anthropometry is the branch of science that deals with human measurement by describing human size, shape, and other physical characteristics. Anthropometric data are available for designers in statistical form.

Designers may design a custom product for an individual or more commonly a product to be mass-produced. When designing for a broad target population, it is important to pick the right numbers from the anthropometric data. Typically, designs are made for the 5th to 95th percentiles of the target population. Data will vary considerably based on age and gender differences. In attempting to achieve an optimal match or best fit, some one-way measurements must be considered. Reach is a maximum measurement considering the smallest member of the target population while clearance is a minimum measurement for the largest member of the target population.

Many aspects of human behavior are important in designing products, systems, and environments. Compatibility describes the relationship between how a person expects something to work and the way it was designed to work. In an ideal world, the way something was designed to work and how the person uses it would be the same. Unfortunately, this is not true and the fact that human expectations of their environment vary can lead to serious consequences.

Human biomechanical capabilities are established by an individual's skeletal structure, muscles and connecting tissue, and nervous systems, giving the typical person a wide variety of capabilities. In considering posture, we need to know various measurements when a person is sitting, bending, standing, reaching, gripping, and kneeling, to name a few. Humans are capable of a wide range of motion. Moving hands, arms, legs, or any other body part is measured in degrees of movement and varies in individuals based on bone structure, muscle strength, and joint flexibility.

The human hand is often the direct link to our natural and technological world. Cumulative (or repetitive) trauma disorders (CTDs) can occur when repetition or misuse of a body part occurs. Many CTD injuries involve the hand, wrist, forearm, and back. Good tool and product design and proper training to use tools and products can reduce CTDs. Care should be taken when lifting or pushing/pulling loads. The maximum safe lifting weight under ideal conditions is 23 kg (50 lbs) for adult men and 17 kg (37 lbs) for adult women.

Government and businesses are attempting to reduce barriers and assist individuals with disabilities and the elderly by using universal design solutions that benefit not only the special needs population but all individuals. Assistive or adaptive technology enables people with disabilities to accomplish daily living tasks.

SUMMARY

(continued)

Designing with anthropometric data begins with identifying the target population and then finding appropriate human factors information on that population. The data is used to guide the design process, and members of the population test any solutions to see if they fit. A chair design illustrated how human factors data can be applied to a design solution.

The computer is a very important tool used by millions of people daily. Of increasing concern are the number of teenagers and younger children who are developing computer-related injuries such as eye strain, carpal tunnel syndrome, and neck stiffness. Human factors data were used to evaluate the proper way to design and use a computer workstation.

BRING IT HOME

OBSERVATION/ANALYSIS/SYNTHESIS

1. Make a list of products that you use that are either too large or too small to fit your hand comfortably.
2. Ask someone that is either elderly or physically challenged for a list of tasks he or she finds difficult to accomplish. Ask why the tasks are difficult.
3. With the permission of your teacher, look up one of the human factors or ergonomic websites referenced in this chapter and prepare a list of interesting facts found on the site.
4. See if the book by Henry Dreyfuss, *Designing for People* (1955, 1967, or 2003) can be found in your school library or community library. Review the book to see his interesting sketches and designs based on human factors principles.
5. Collect anthropometric data for height and weight for each member of your technology/pre-engineering class. Prepare an anthropometric table for the data; calculate the mean, and the 5th and 95th percentiles. (See Chapter 16, Math/Science Applications for formulas.)
6. Using the data available from one of the figures in this chapter, make a two-dimensional anthropometric model for high school-age students. The model should be made of stiff cardboard or other appropriate material with movable joints. The model can be used to test the human scale of products or environments.
7. Determine the standard deviation for your class population. How do you think your class compares with the population of all students in your school? (See Chapter 16, Math/Science Applications for formulas.)
8. Go to the OSHA website at http://www.osha.gov/SLTC/etools/computerworkstations/checklist.html. Print and complete the checklist for your personal computer workstation in school or at home. Indicate areas where you need to make improvement in your workstation to bring it into compliance with OSHA recommendations.

EXTRA MILE

ENGINEERING DESIGN ANALYSIS CHALLENGE

Using data from the "Designing a Chair" section of the chapter, evaluate a chair from school or home by comparing the recommended sizes and your personal measurements for at least three human factors.

Was the chair a good fit for you and, if not, what needs to be changed? Using PowerPoint software, make your presentation to the class or a group in your school or community.

CHAPTER 16
Math and Science Applications

Menu

Before You Begin
Think about these questions as you study the concepts in this chapter:

1. Why must we account for variability in designs?

2. How are absolute and relative graphs constructed and why are they useful?

3. What are four main uses for spreadsheet software?

4. Which scientist in the seventeenth century revolutionized math and science?

5. What are some important examples of static structures human beings use, and what is the sum of all the forces in a static structure?

6. What is a rate of change and how is it useful in studying dynamics?

7. What are some examples of both natural and human-made oscillatory behavior?

8. What is one of the most widely used mathematical functions in science and engineering?

INTRODUCTION

In Chapter 1, we discussed the continuum into which designers fall, where the placement within the continuum depended on the level of use of math and science. Good math and science skills are typically needed in engineering disciplines, but math and science skills are also very useful across all design disciplines.

This chapter will review some of the math and science skills that are most useful in high school technology education and pre-engineering programs. This chapter cannot cover all the math or science concepts that engineers must learn—there is simply too much material! Consult other textbooks dedicated to math, science, or engineering principles for more complete information on these subjects. You can also research these topics and sharpen your skills through some very good websites.

MEASUREMENT: THE REAL WORLD OF VARIABILITY

Nothing in the real world can be measured or fabricated with infinite accuracy, so variability must be considered. Variability is a measure of the extent to which a dimension or parameter is expected to vary. **Variation** is a critical concept for designers to understand, and was discussed in detail in Chapter 8, Technical Drawing, along with measurement.

Why do engineers or technologists need to consider variability? As the following example shows, engineers consider variability to make sure parts fit and fasten together correctly. The correct use of variability is also vital to ensure that designs are reliable and can be manufactured. Consider the case of a simple box (blue) that needs to fit through a simple hole (see Figure 16-1). If the box and hole have the same specification (Figure 16-1a) then the box will not always fit through the hole, For example, a 2.1 m wide block will not fit through a 1.9 m wide hole. This situation is corrected in Figure 16-1b by altering the variability (tolerance) specifications of the block to 1.7 ± 0.1 m (a 1.8 m wide block will always fit through a 1.9 m wide hole). This "box-through-a-hole" situation is a simple example but could easily apply to male and female connectors on electrical devices or a large number of real-world devices.

> **variation:**
> Also known as variability; a measure of the extent to which a dimension or parameter is expected to vary in magnitude.

OFF-ROAD EXPLORATION

For interesting discussions on the art, science, and math of origami, as well as photographs of some origami art, see *http://www.langorigami.com/*.

Your Turn

List items where the acceptable variability is (1) very small or (2) very large. Estimate the specifications and compare them to similar specifications on other objects. Try to think of examples of variability for important specifications that are not simply units of length.

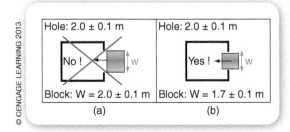

Figure 16-1: A "box-through-a-hole" example illustrates the need to accurately define variability to achieve successful designs. Unless the variances in both the hole and box are correctly accounted for, the box will not fit through the hole, resulting in failure (a). Figure (b) shows variances that will allow the box to always fit through the hole.

Point of Interest
Math and the art of origami

The design continuum includes graphic artists on one end who typically use less math or science, compared to engineers at the other end of the continuum whose livelihood depends substantially on their skill level in math and science. Artistic principles such as aesthetics, color, and scaling are utilized much more on the graphic artist end of the design continuum but sometimes need to be considered by engineers also. An interesting question arises: Are there examples of art that math and science can influence? Indeed, there are! There is even art that can influence math, science, and engineering.

Robert Lang is a prolific laser scientist and an accomplished origami master. Origami is an ancient Japanese art that involves folding paper. The term *origami* in Japanese actually means "folding paper." Origami artists use a single sheet of paper and fold the paper to form an object. The key constraint is that no cutting is allowed—only folding.

Historically, it was always a challenge to fold realistic objects with numerous appendages, let alone have the appendages appear in the right position and be the right size. That is until Robert Lang and some fellow mathematically minded origami artists discovered several new methods for describing more complicated folding techniques. Among other things, these new folding design techniques enabled designers to create more realistic structures with multiple appendages. One of the new mathematical methods is the "circle-river packing" layout technique. An example of this is the crease-pattern diagram shown in Figure 16-2. This crease pattern results in a crab with a realistic asymmetric

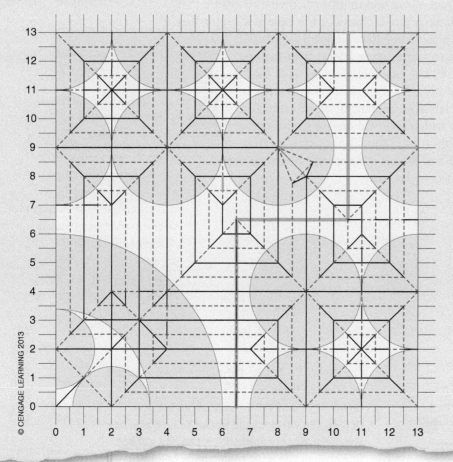

Figure 16-2: Mathematics has recently been applied to the art of origami, resulting in the ability to fold, among other things, multi-appendage structures that are very realistic. A new technique called "circle-river packing" produces the necessary crease pattern. Can you determine which of the animals shown in Figure 16-2 is represented by this crease pattern?

Point of Interest *(continued)*

claw design. With the aid of computer programs, origami artists are now able to fold more detailed and realistic forms.

These advancements in the art, and mathematics, of origami also resulted in improvements in engineering. The same techniques developed to advance origami have been applied in space applications (large, foldable telescopes), the automotive industry (optimization of folded airbags), robotics (manipulators), and medicine (unfoldable blood vessel stents).

Figure 16-3: Images of realistic animals folded using origami.

Variability in Wood
Measure twice, cut once

Wood is a common material and serves as an excellent example of accounting for size variability in design. We typically harvest wood from the trunk of a tree. Wood is comprised of millions of fibers running parallel to one another vertically up and down the trunk of the tree (see Figure 16-4). An accurate analogy would be a set of parallel straws tightly packed next to one another. As one might expect, depending on the amount of moisture in the environment, the "diameter" of these wooden fibers (the "straws") can either be wide and fat (if the fibers have absorbed moisture from the environment) or thin and skinny (if the moisture has evaporated out of the fibers). Therefore, the width across the fibers can vary greatly with moisture content, which changes seasonally. In this transverse direction (across the fibers, or *grain*) wood can

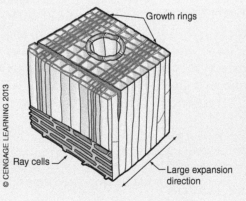

Figure 16-4: Wood is made up of straw-shaped fibers stacked parallel to one another. The variability in the size of wood is especially tricky in that, depending on moisture content in the air, wood changes in size substantially in one direction but very little in the other.

(continued)

Variability in Wood (continued)

shrink or expand by as much as 2 to 10 percent. For example, a board that is nominally one foot wide could expand to a length of 1.02 to 1.1 feet, an increase in width of 0.02 to 0.1 foot, which is 0.2 to 1.2 inches. This is a very substantial movement and must be accounted for. In contrast, the length of the fibers changes very little with moisture content. In this longitudinal direction (parallel to the fibers), wood only shrinks or expands by approximately 0.1 to 0.3 percent, resulting in almost negligible size changes. Therefore, due to changes in moisture content, wood changes in size in a very nonsymmetric fashion with substantial changes across the grain.

In the winter, air typically contains very little moisture. However, it can be quite humid in the summer to the point where the relative humidity of air can exceed 80 percent. To achieve a high-quality design made of wood, the resulting nonsymmetrical expansion or contraction must be accounted for. Cherry wood is a fairly stable wood that contracts little with moisture. For this reason, cherry wood was routinely used for printing blocks. However, even furniture made out of cherry wood must be designed with structures that do not allow the expansion and contraction to, literally, rip the wood apart over even just a few seasons.

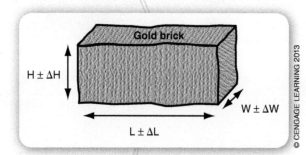

Figure 16-5: Example of a solid, three-dimensional gold brick with known variability in each dimension.

Let's put the concept of variability, or error, into practice. Suppose in the year 2006, a museum curator of marine archeological sites gives you one of several roughly identical ancient solid gold bricks that were uncovered in an archeological dig. The curator is thinking of selling a few bricks to pay for more research work, so she wants to know the volume of the brick along with an estimate of the brick's monetary value. The volume of a rectangular brick is straightforward. The volume, V, is given by $V = (L)(W)(H)$, where L is length, H is height, and W is width (see Figure 16-5). The dimensions of the brick are nominally 21 cm by 11.5 cm by 6 cm, but each dimension is only accurate to about ±0.5 mm (0.05 cm), mainly due to the surface roughness. The maximum volume of the brick is given by the case where each dimension is at its largest value, while the minimum volume is given when each dimension is at its smallest value.

Therefore, the minimum, nominal, and maximum volumes are given as follows:

$$V_{min} = (20.95 \text{ cm})(11.45 \text{ cm})(5.95 \text{ cm}) = 1{,}427.3 \text{ cm}^3 \qquad \text{Eq. 16-1a}$$

$$V_{nom} = (21 \text{ cm})(11.5 \text{ cm})(6 \text{ cm}) = 1{,}449.0 \text{ cm}^3 \qquad \text{Eq. 16-1b}$$

$$V_{max} = (21.05 \text{ cm})(11.55 \text{ cm})(6.05 \text{ cm}) = 1{,}470.9 \text{ cm}^3 \qquad \text{Eq. 16-1c}$$

The difference between V_{max} and V_{nom}, as well the difference between V_{nom} and V_{min}, is approximately 20 cm³. Therefore, the volume of the brick can only be known to the "tens" place. Therefore, knowledge of the volume of the gold brick can be summarized by stating that $V_{brick} = 1{,}450 \pm 20$ cm³. [Notes: (1) Written in this form, both the nominal value and the variability are each quoted to the tens place; (2) each of the two zeros in the volume $1{,}450 \pm 20$ cm³ are simply placeholders, which is made clear by using scientific notation, $(145 \pm 2) \times 10^1$ cm³, where now each number is written to the ones placeholder.] This calculation indicates that the volume of the brick is nominally 1,450 cm³, but there is a total possible variation of ~40 cm³ (±20 cm³). It does not make sense to quote the volume of these bricks more accurately than ±20 cm³. Therefore, this answers the curator's first question about volume.

Units Analysis

OFF-ROAD EXPLORATION
See the following for interesting math games and activities:
http://cte.jhu.edu/techacademy/web/2000/heal/mathsites.htm.

We now need to calculate the monetary value, which gives us the opportunity to review units analysis. Units analysis is simply working with units (letter symbols) exactly as one would work with numbers. Units analysis is one of the more powerful processes used in engineering and design, and in "everyday life." When calculating items, it is extremely important to carry along the units with each number to enable a units analysis to occur, ensuring correct calculations. This will become more clear as we work through this example.

Gold is typically purchased by weight measured in troy ounces. In 2011, the price of gold was approximately $1,700 per troy ounce (Toz). Given that we know the volume of the gold bricks, we now need to know the density of gold. The **density** of a material is its mass divided by its volume. In the metric system, some possible units of density are grams per cubic centimeter (g/cm³) or kilograms per cubic meter (kg/m³). Gold has a high density and is 19.3 g/cm³. What we do not know is how to convert troy ounces to grams. Multiple sources verify that one troy ounce equals 31.1 grams. The calculation of the monetary worth of a gold brick can now be completed and is summarized in Equation 16-2, with the calculation in two steps. Equation 16-2a uses the fact that density equals mass divided by volume, or equivalently mass equals volume times density $M = (V)(D)$, and Equation 16-2b calculates the monetary value.

density:
A material's mass divided by its volume, nominally a constant. A few examples of density are air (1.2 kg/m³), water (1,000 kg/m³), and gold (19.3 gm/cm³).

Mass = (1450 ± 20 cm³)(19.3 g/cm³) = 28,000 ± 400 g   Eq. 16-2a

Value = (28,000 ± 400 g)(1 Toz/31.1 g)(1700 $US/Toz)   Eq. 16-2b
      = 1,530,000 ± 20,000 $US

In Equation 16-2a, there is a cm³ on both the top and bottom (numerator and denominator), thus canceling each other, leaving only the units of grams (gm), the mass of a brick. In the second equation, Equation 16-2b, two sets of units cancel. The units of grams (g) and troy ounces (Toz); each appear in both the numerator and denominator, thus canceling, leaving simply a unit of $US, which is the desired result. Units analysis like this is a common and very important technique for verifying correct calculations. For example, if the units obtained (via units analysis) are not the desired units, then the calculation is very likely incorrect and needs to be reworked.

In summary, looking at each brick as simply a source of gold, the value of each brick is $1,530,000 with a variability of $20,000. Knowledge of this variability is very useful to the curator. For example, $20,000 could feed and house many archeology graduate students for a whole summer. This could be a very useful resource to complete the archeological research.

Your Turn

Find an object in your home or classroom that is attached with multiple screws, perhaps even several kinds of screws. Estimate how accurately the holes for the screws must be specified and why.

Point of Interest
The Codes of Hammurabi

Hammurabi was a Babylonian king living around 1700 B.C. and was the creator of the first written laws. We refer to these laws as the Code of Hammurabi. They consist of over 280 detailed laws written on 12 tablets. In Hammurabi's era, engineers had to get their designs correct or forfeit their freedom, or possibly even their lives! The list that follows paraphrases a few of these laws:

1. If a builder builds a house, and constructs it well, the owner will pay two shekels for each surface of the house.
2. If, however, he does not succeed, and the house falls in, killing the owner, the builder will be killed.
3. If the son of the owner dies, the son of the builder shall be killed.
4. If one does not take good enough care of a dam, and the dam breaks, he shall be sold for money, which will replace the corn ruined due to the overflooding of the crops.

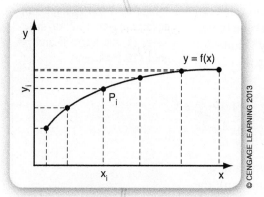

Figure 16-6: Example of a Cartesian coordinate system, where the placement of points is determined by rectangles. This type of system can also be referred to as a rectilinear system.

Graphing: Absolute, Relative, and Polar. A unique representation of a point in one dimension (somewhere along a line) requires one number. For example, the number 123.5 represents a unique and clearly defined point on the number line exactly halfway between the numbers 123 and 124. Similarly, two numbers are required to represent uniquely a point in a two-dimensional space somewhere in the x-y plane. Continuing on this train of thought, three numbers are required to uniquely define a point in a three-dimensional space.

The most common coordinate system is the Cartesian system, and is named after René Descartes, a very famous mathematician and philosopher who lived in the early 1600s. The Cartesian coordinate system uses a rectilinear (rectangular) system to plot ordered sets of numbers, such as pairs (x, y) in two dimensions and triples (x, y, z) in three dimensions. Figure 16-6 illustrates the two-dimensional (2-D) Cartesian coordinate system, showing several points of an arbitrary function $y = f(x)$. Each point on the graph, P_i, is determined by a rectangle, which is defined by a particular position x_i on the x-axis and a particular position y_i on the y-axis.

A Cartesian coordinate system is very useful for the numerous important phenomena and structures in the world that are rectangular in form. Some examples of important rectangular shapes are buildings, bridges, furniture, gravity (items fall perpendicular to the ground), dams, weapons, and various tools and machines.

Absolute versus Relative Graphs (Using Cartesian Plotting)

You can plot a graph as either an absolute or a relative plot. In an absolute plot, the real (absolute) numerical values are used in the plot, while only relative values are used in a relative plot. For example, Figure 16-7a shows the price of a gallon of gasoline during the first eight months of 2007. The prices began to increase sharply in February, rising from approximately $2.20 per gallon to over $3.20 per gallon. This is an example of an absolute plot because the graph shows the actual values of price.

Your Turn

Go onto the Internet, perhaps to financial portions of Yahoo! or Google, and construct an absolute plot of the stock price of several stocks that interest you. The stocks could be companies you are familiar with as a consumer or perhaps a company that employs one of your parents. Next, try to find two companies that are in the same industry and compare their stocks using a relative plot. Try to find stocks that have changed substantially over time. Did the y-axes appear to be made up of unequally spaced grids? These are likely "semi-log" plots. What are semi-log plots and why are they used?

Figure 16-7b shows the same information as a relative plot, where the y values are shown relative to the lowest price in this period ($2.20 per gallon in February). The relative data in this plot was obtained by dividing the absolute data by the data point to be compared to $2.20 per gallon, in this case. The comparisons offered in a relative plot can be useful to emphasize certain desired results. For instance, this relative graph makes relative features such as the percentage of change in the plot immediately accessible to readers, allowing readers to process the information quickly. In this example, it is obvious that the price per gallon in the month of June (month 6) increased by about 50 percent (1.5 versus 1.0) relative to the month of February (month 2). Relative plots are often used to plot investments. For example, if you purchased 100 shares of both Ford Motor Company and Honda in the year 2003, the net worth of each of these investments, plotted relative to the value in 2003, is shown in Figure 16-8. From this graph, it is immediately apparent that the investment in Honda would have more than doubled in value, compared to Ford, where the investment would have lost about 20 percent. Financial websites supported by companies like Yahoo! and Google each support both absolute

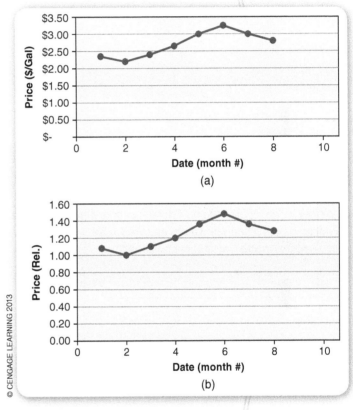

Figure 16-7: Gasoline prices plotted as (a) an absolute plot and (b) a relative plot. The data in plot (b) is relative to the lowest price of $2.20 (month 2).

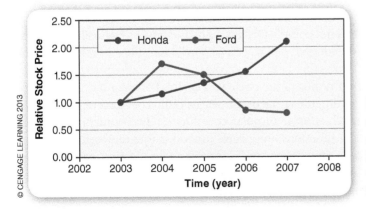

Figure 16-8: Relative plots are used often in displaying stock information. Shown here is the stock price of two automotive companies, Ford and Honda.

and relative plotting features. When you request the history of the stock price of a company on these websites, an absolute plot is computed and displayed. However, if you request to compare the history of one stock to another, a relative plot is computed and displayed.

Polar (Non-Cartesian) Graphing

OFF-ROAD EXPLORATION

For a review of trigonometry, see http://en.wikipedia.org/wiki/Trigonometry.

Many phenomena are not rectangular in nature, or certainly cannot be easily described in terms of rectangles. Circles, for example, are very important in both nature and human-designed objects. One of the most important human "inventions" was a circle, the wheel. A *polar* coordinate system is typically the most useful for circular-like phenomena. Just like in the Cartesian system, a unique point in the two dimensions of a polar plot is represented by two numbers called r and θ (Greek letter *theta*). For the ordered pair (r, θ), the number r refers to the *radius* (the distance between the origin and the point). The number θ is the magnitude of the angle between the line formed by the point and the origin (that is, the radius line) and the positive x-axis. Figure 16-9 shows examples of points defined by both rectangular and polar coordinates. Figure 16-9 also illustrates the formulas used to convert between the two coordinate systems. Note that only basic knowledge of the Pythagorean theorem and trigonometry are required to fully derive and understand polar coordinates.

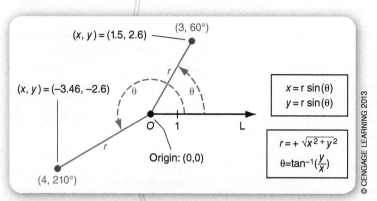

Figure 16-9: Examples of points expressed in polar (colored, ordered pairs) and Cartesian coordinates (colored, ordered pairs). Also shown are equations for translating between polar and Cartesian coordinates.

An investigation of circles demonstrates the substantial advantages of the polar coordinate system. The equation for a circle centered at the origin in Cartesian coordinates is $x^2 + y^2 = R^2$, where R represents the specific radius of the circle and x and y are the Cartesian coordinates. The Cartesian form for the equation of a circle does not explicitly give y as a function of x, making the equation $x^2 + y^2 = R^2$ appear simpler than it really is. For example, solving for y yields the following two equations:

$$y_1 = +\sqrt{R^2 - x^2} \qquad \text{Eq. 16-3a}$$

$$y_2 = -\sqrt{R^2 - x^2} \qquad \text{Eq. 16-3b}$$

Therefore, to represent y as an explicit function of x now requires two equations! It is indeed getting more complicated. Both equations are required though: One equation (the positive root) describes only the top half of the circle, and the second equation (the negative root) describes only the bottom half of the circle. If having two equations was not complicated enough, one now has to deal with square roots of squares. In stark comparison, in polar coordinates, the equation of a circle is extremely simple: r = R, where r is one of the variables in the polar system and R is simply a number, the radius of the circle you desire. For example, the equation r = 10 meters is a circle centered at the origin with a radius of 10 meters. Equations of circles are indeed much simpler in polar coordinates. The same holds true for many other "circular-ish" functions, important functions that represent both natural and man-made shapes and processes. Ellipses, spirals, electron orbital shapes, and radiation patterns for cell phone signals are more easily expressed using polar coordinates (or the three-dimensional equivalent to polar coordinates).

Your Turn

Imagine you are the captain of a military submarine and that a fellow submarine is in distress and stuck on the bottom of the sea, but you do not know exactly where. You only know that they are located somewhere in a 20-mile by 20-mile square region. As part of a rescue process, you want to send sound signals from your submarine through the water to the other submarine. However, you need to keep in mind that you must function as stealthily as possible to avoid being detected by enemy vessels. Which type of radiation pattern in Figure 16-10 would you choose to use and why? Would it be beneficial to be able to use more than one type of radiation pattern in your search?

As a real-world example of polar coordinates in action, Figure 16-10 shows "pickup" patterns for a variety of microphones, each of which are much easier to describe mathematically using polar coordinates. There are three important types of microphones shown: (1) omnidirectional, (2) bidirectional, and (3) shotgun. Each of these microphones is designed for a very specific use. To understand these figures, one has to think in polar coordinates, not Cartesian. For each point on each curve in Figure 16-10, the radius (the distance from the origin) represents how easily sound is picked up at that particular angle θ: The larger the radius at a point,

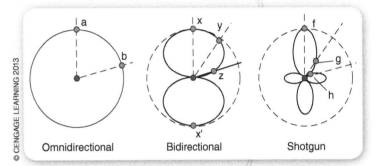

Figure 16-10: *Polar coordinates have many uses. Shown here are pickup patterns, in polar coordinates, of three commonly used microphones.*

the better the sound coming from that angle is picked up. For the omnidirectional microphone, with a circular radiation pattern, sound from all angles is picked up equally well. The radius is the same for all pickup angles: Points a and b are equidistant from the origin. Businesses typically use this type of microphone in conference rooms so that each person's voice at a table is equally heard. The bidirectional microphone has a two-lobed pickup pattern (at points x and x′ that are 180 degrees apart from each other) so a bidirectional microphone picks up voices equally well only in these two directions. Bidirectional microphones are used primarily in a "cohost" format, where two people are seated opposite each other (180° apart). With a bidirectional microphone, the voices of a cohost and a guest are picked up quite well but voices at the sides, like the voices of the crew (point z), are not picked up well, minimizing any noise from the crew being transmitted to the audience. This gives the crew substantial freedom to communicate about important items without disturbing the show. Another interesting microphone is the shotgun microphone. A shotgun microphone picks up sound from primarily one angle (point f), which is very useful for spying, cell phones, and certain theater applications (like a solo in a musical where only a particular singer's voice is desired).

Spreadsheets and Structured Programming

Computer spreadsheets are powerful tools, and engineering professionals must be adept at using them. Microsoft's Excel program is now the standard for many tasks that engineers and technologists perform. Excel is highly capable, has broad usage, and is relatively low in cost. Other widely used tools are MathCAD and Mathematica—even your calculator can perform powerful tasks.

In this chapter, we use Microsoft Excel (2004 version) but all features used in this chapter are also available in newer versions. The most common uses of spreadsheet programs fall into the following four categories:

1. Organization of data/information
2. Calculations
3. Graphing
    a. Pie/bar charts (one-dimensional)
    b. Plotting (two-dimensional)
    c. Curve fitting
4. Statistics

Following, we discuss each of these areas in more detail using several examples.

Organization of Data/Information. It is valuable to organize data so that it is easily retrievable. Figure 16-11 shows an example of an ongoing summary of maintenance work on the cars of a household. Organizing records in this way makes information more accessible and readable than stuffing receipts in the back of the glove compartment. A scientific example of data organization would be an archeologist recording the location of all items found in a dig site. Spreadsheets easily accomplish this. In this application of organizing data, the primary purpose of the spreadsheet is not to calculate but to simply record. However, because the information is in a spreadsheet, a calculation or graph could also be easily completed if desired.

In both engineering and management, decision matrices are often recorded using spreadsheet software. Figure 16-12a shows an example of a decision matrix.

TRUCK			Oil &	Tire		
Date	Mile.	Brks	Filter	Rot'n	Insp.	Comments
11-Jan-02	35,092	no	yes	no	no	STS: 4 Tires;P245/70R16;@$117
4-May-02	40,344	yes	yes	yes	no	STS: $87 Pads; $77 Lbr; Oil $34
14-Jul-02	43,276	no	yes	yes	yes	STS: ~$115; Oil $30; Insp $24

VAN						
12-Jan-02	95,806	no	no	yes	no	STS: note high spots on rotors
15-Jun-02	100,560	no	no	no	no	STS Auto: 2 new Front tires
14-Jul-02	104,200	yes	no	no	yes	STS Auto: ~$355;brakes&rotors
14-Aug-02	108,000	no	no	no	no	STS Auto: ~oil chng
22-Dec-02	116,741	no	yes	no	no	Trans. Flush,Valve cvr gaskets, air filter, pwr steer. Flush, New spark plugs, Minor tune-up inspect window [$878.57 !]

Figure 16-11: Summary of maintenance work on the cars of a household. Organization of this type of information can be more accessible and readable than those wrinkled and dirty receipts stuffed in the back of the glove box.

Figure 16-12: Examples of decision matrices constructed in a spreadsheet used to help determine which vendor to buy a product from. Shown are decision matrices with (a) textual information and (b) numerical information.

(a)

Vendor	Devel. Cost	Vol. Cost	On-time Delivery	Cust. Support	Reliability	Document.	Quality
XYZ	OK	G	B	OK	G	G	VG
TNET	O	VG	VG	VG	OK	OK	G
ABC	Bad	VG	VG	OK	OK	OK	G
OCS	OK	G	OK	OK	O	O	O

Scale: Bad/ OK/ Good/ Very Good/ Outstanding

(b)

Vendor								Ave.
XYZ	3	3	0	3	5	5	8	**3.9**
TNET	10	8	8	8	3	3	5	**6.4**
ABC	0	8	8	3	3	3	5	**4.3**
OCS	3	5	5	3	10	10	10	**6.6**
Scale:	0	3	5	8	10			
	Bad	OK	Good	VG	O			

Your Turn

Use a spreadsheet software to create the following in successive cells: (1) a column and a row that contain the odd numbers between 23 and 47, (2) a column that contains the first 10 months of the year, (3) a column that contains the last eight months of the year, and (4) a column that contains the numbers between 1 and 13.2 with a difference of 0.2.

This matrix helps to determine the best vendor to use. Choosing which vendor to purchase materials, processes, and parts is often one of the most important decisions an engineer or technologist can make. Figure 16-12b shows a quantitative decision matrix with numerical ratings. The spreadsheet software calculated a quantitative comparison. With which vendor listed in Figure 16-12 do you think the company will choose to work?

Calculations. The calculation capabilities of spreadsheet software programs such as Excel displace calculators and, in some cases, other more expensive software packages. Spreadsheet software comes with many built-in functions, a list of which is easily obtained via the "Help" function.

Before working through a few example calculations, let's look at an overview of a few spreadsheet basics: (1) automatically enumerating cells, (2) iterative calculations and formula entry, and (3) freezing cells.

Automatic enumeration. Spreadsheet users often need to create cells that have a distinct repetitive relationship. This is true for both numbers as well as text. For example, say you want to plot the functions $y = x$, $y = 5x$, $y = x^2$, and $y = 2^x$. One of the first problems encountered is how to create the input x-values efficiently. Manually typing in the numbers by hand would be tedious and time-consuming. Fortunately, a sequence of numbers or text is easy to generate by simply "dragging" cells. For example, to create hundreds or even thousands of input x-values in only a few seconds, you only need to type the first two values in successive rows of the same column, and then highlight both cells and drag down (or up, or left or right) on the small black square in the lower right corner of the highlighted cells. Excel will automatically load the new cells (the cells you are dragging over) with numbers, or text, that have the same pattern as the two initial cells you highlighted. For example, if the difference between the numbers in the first two (highlighted) cells was 1, as in column A in Figure 16-13, then the new cells will be loaded with numbers incremented by 1 until you stop dragging. You can perform similar automatic enumerations with non-unity numerical differences, dates, and even text. Figure 16-13 shows several examples of automatic enumeration in columns. The same procedure also works by rows ("left" and "right").

Iterative calculations and formula entry. Excel performs calculations when a formula is entered into a cell. To enter a formula, select a cell and then enter the equal sign (=). Excel always recognizes a first character of an equals sign as the beginning of an equation. In Figure 16-14, the "x-values" have

	A	B	C	D
1	Difference	Difference		
2	(unity)	(non-unity)	Date	Text
3	−3	2.56	Jan	Year1
4	−2	2.68	Feb	Year2
5	−1	2.8	Mar	Year3
6	0	2.92	Apr	Year4
7	1	3.04	May	Year5
8	2	3.16	Jun	Year6
9	3	3.28	Jul	Year7

Figure 16-13: Examples of automatic enumerating, by column. Columns A and B show examples of numeric enumeration (difference of 1 and 0.12), while columns C and D show examples of date and text enumeration, respectively.

	A	B	C	D
1	x	5x	x^2	2^x
		=5*A2	=A2*A2	=2^A2
2	0	0	0	1
3	1	5	1	2
4	2	10	4	4
5	3	15	9	8
6	4	20	16	16
7	5	25	25	32
8	6	30	36	64

Figure 16-14: Examples of calculating, on a column basis. Three functions are calculated: $y = 5x$, $y = x^2$, and $y = 2^x$.

been loaded into column A using automatic enumeration. To calculate the function $y = 5x$ in column B, select cell B2 and type in "=5*A2" followed by <return>. Excel will understand this action to mean that cell B2 will take on the value of A2 times 5. In this case, 5*A2 will be the number 0. To load up the other functional values for the function $y = 5x$ values, highlight the first calculated cell in row B (in this case B2) and drag down on the small black square in the lower right corner of the highlighted cell, staying in column B. This process results in the selected cells in column B having the value of five times the number in the same row in column A. The same process was used to calculate the functions $y = x^2$ and $y = 2^x$ in columns C and D, respectively. (The yellow and blue cells in Figure 16-14 were inserted here for instructional purposes only and do not appear in the Excel display. However, this information does appear elsewhere. Can you find where?)

Freezing cells. As previously described, Excel can calculate iteratively the entry in one cell depending on an entry in a corresponding cell. For example, in Figure 16-14, columns B, C, and D were all calculated using the corresponding entries in column A (where the x-values are stored). However, this column-by-column format is not always desired. For example, sometimes you may want to calculate with a constant but not want a whole column dedicated to holding a single number, the constant. Neither may you want to hide the constant within the formula window, where it is only visible when a cell is selected. Excel provides the ability to "freeze"

Point of Interest
Cost calculator

A family is in the market for a car, but one parent is a teacher and travels a fairly long distance to work about 40 weeks out of the year, so gas mileage is very important. An Excel calculator is helpful for determining which car to buy (see Figure 16-15). A larger, potentially safer car has a fuel efficiency of about 18 mpg. However, a smaller, more efficient car (gas, diesel, or hybrid) can achieve a fuel efficiency up to 45 mpg. The decision of which car to buy may depend largely on cost savings for the higher fuel efficiency car. The worksheet shown in Figure 16-15 summarizes the savings and uses a format that clearly defines inputs and outputs for the problem. Both the layout and color of the cells help create a very easy-to-use format. Four cases are summarized at the bottom of the worksheet (green cells and red text).

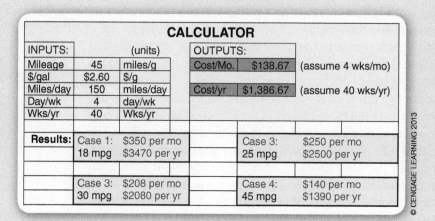

Figure 16-15: An Excel calculator for comparing gas expenses for cars with varying gas mileages. The format defined in this calculator uses clearly defined regions for inputs and outputs that can be very helpful.

a cell, so that Excel will not perform iteratively with the constant as it did when we constructed the example for Figure 16-14. An example with freezing a cell should make this clearer.

Shown in Figure 16-16 is an Excel sheet that calculates how far an object falls in a vacuum due to gravity. The object is given an initial **velocity**, $V_{initial}$. The equation that governs the fall is $D = 0.5gt^2 + (V_{initial})t$, where D is distance, g is the acceleration of gravity (a constant), and t is time. For these calculations, we require g and $V_{initial}$ to be constants that can be easily changed to see how any change makes a difference. These values of g and $V_{initial}$ can then be considered to be the "inputs" in our investigation. For example, we may want to increase $V_{initial}$ to see how it affects distance, or perhaps even change g to see how the distance traveled might change if the object were on the moon, as opposed to earth. (Due to its smaller mass, the moon has a value of g that is a factor of 0.0123 smaller than that on earth.) The formula typed into cell B4 would appear as follows: "= 0.5 * D1 * A4 ^ 2 + D3 * A4." Note that there is a dollar sign ("$") typed before the column letter and the row number for cells D1 and D3, the cells that hold the numerical values for g and $V_{initial}$. The placement of the symbol $ before a column or row entry freezes that column or row. In this case, cells D1 and D3 are "frozen." What this means is that the row that holds the calculated distances, D, can be filled in by dragging as we performed previously in the example for Figure 16-14, but in this case, we are varying time (as the x-variable) in the A column. However, the g and $V_{initial}$ values are always taken from cells D1 and D3 and not from some iterating cells as they were in the example provided in Figure 16-14. Figure 16-16 shows two cases: Case (a) is for the

velocity:
The rate of change of position with time. Some typical units of velocity are meters/second (m/s), centimeters per second (cm/s), feet per second (ft/s), and miles per hour (mi/hr or mph).

	A	B	C	D
1	Gravity	(Earth)	g=	9.8
2				m/s^2
3	Time (s)	Distance (m)	V-initial=	3.0
4	0	0.0		m/s
5	1	7.9		
6	2	25.6		
7	3	53.1		
8	4	90.4		

(a)

	A	B	C	D
1	Gravity	(Moon)	g=	0.121
2				m/s^2
3	Time (s)	Distance (m)	V-initial=	3.0
4	0	0.0		m/s
5	1	3.1		
6	2	6.2		
7	3	9.5		
8	4	13.0		

(b)

Figure 16-16: Example of freezing a cell in a calculation. In this case, cells D1 and D3 are frozen in the calculation, while all other cells are calculated iteratively. The use of freezing a cell, or cells, for a calculation can lead to a very desirable "input/output" format. These calculations show the distance an object drops on Earth (a) or the moon (b), given the same initial velocity (assuming no air friction).

Point of Interest
Break-even point

High-efficiency and low-cost wind-powered generators are now available for private use. The idea of being able to generate a good fraction of your own power is increasingly attractive to many homeowners, especially with lower-cost generators on the market. However, does it make financial sense to install one?

Figure 16-17 shows the results of some preliminary calculations. This figure shows an estimate of the monthly, yearly, and cumulative yearly savings achieved with a wind-powered generator for an inland site with moderate average wind speeds. Column B is the input data, the measured average wind speed for each of the 12 months of the year. The month column was created using automated enumeration. Column C, the generated power in kWH, was calculated using a formula that translates average wind speed into generated power per month. (This formula is actually obtained in the curve-fitting section of this chapter. The conversion from average wind speed to generated power turns out to be a quadratic function.) Column D is the initial output—the cost savings per month. The savings are calculated by knowing the average cost per kWH, which is about $0.08 per kWH (see Figure 16-20). The yellow columns F and H are the final outputs and are calculated as follows: The savings in the first year is $371.80. The savings for successive years are estimated by knowing that the cost per kWH has increased on the average about 4 percent per year. Therefore, column F generates the next year's savings by multiplying the current year's savings by 1.04. Column H adds each of the previous year's savings to achieve a total cumulative measure of savings. For example, the total savings in the third year is $371.80 + $386.70 + $402.20 = $1,160.70. (Column H is generated by taking cell H5 = H4 + F5 and repeating the cell calculations by selecting and dragging down on cell H5.) Given that the least expensive installation of the wind generator costs about $5,500, it would take about 12 years to break even. This is a substantial amount of time, so a wind generator installation in an area with only moderate average wind speeds (average speed = 12.4 mph) may not make economic sense. However, using the same calculator for a higher average wind location, a coastal site with an average wind speed of 17.4 mph, gave a break-even time of 7 1/2 years, which is a very reasonable payoff time. Independent of economics, it certainly is more ecologically sound to use a renewable resource.

	A	B	C	D	E	F	G	H
1	Inland: Mid-Atlantic							
2		Ave. Speed	Generated Pwr (kWH)			Yrly. Savings		
3	Month			Savings			Yr	Cum. Sum.
4	Jan	10.5	295.8	$23.7		$371.8	Yr1	$371.8
5	Feb	12	375.9	$30.1		$386.7	Yr2	$758.6
6	Mar	10.5	295.8	$23.7		$402.2	Yr3	$1,160.7
7	Apr	12	375.9	$30.1		$418.3	Yr4	$1,579.0
8	May	14	474.7	$38.0		$435.0	Yr5	$2,014.0
9	Jun	16	564.5	$45.2		$452.4	Yr6	$2,466.4
10	Jul	15	520.8	$41.7		$470.5	Yr7	$2,936.9
11	Aug	7.5	120.4	$9.6		$489.3	Yr8	$3,426.3
12	Sep	14	474.7	$38.0		$508.9	Yr9	$3,935.1
13	Oct	11	323.1	$25.8		$529.2	Yr10	$4,464.4
14	Nov	18	645.3	$51.6		$550.4	Yr11	$5,014.8
15	Dec	8.5	181.2	$14.5		$572.4	Yr12	$5,587.3
16	Ave:	12.4	4648	$371.8	Sums			
17			15.0%	percent of total power used				

Figure 16-17: A more complicated Excel calculator used for assessing the financial viability of a low-cost wind generator. Cost savings are estimated by month, for a year, using average monthly wind speeds. Higher average wind speeds create higher power levels, which save costs by not having to buy power from a public utility.

earth, which has an acceleration of gravity of 9.8 m/s², and case (b) is for the moon, which has an acceleration of gravity of 0.121 m/s². You will notice that a ball will fall at a substantially slower rate on the moon than on earth—13 meters versus 90.4 meters in 4 seconds.

Graphing. Graphical representation of information is very important. Not only is it important for a technologist to graph data but it is also important in everyday life. It would be very difficult to find a daily newspaper that does not routinely use some type of graphical representations of information. The following are types of graphical information:

▶ Pie, bar, and column charts: Information in the media is routinely represented using pie, bar, and column charts. These types of charts effectively represent one-dimensional information, but can also represent two- and three-dimensional data. Figures 16-18, 16-19, and 16-20 show examples of pie, bar, and column charts, respectively, summarizing utility costs for a private residence.

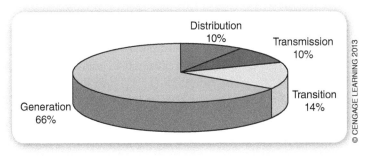

Figure 16-18: Components of a utility bill displayed as a pie chart. A utility bill is composed of several fees: (a) generation, (b) transition, (c) transmission, and (d) distribution. Generation costs are by far the largest contributor.

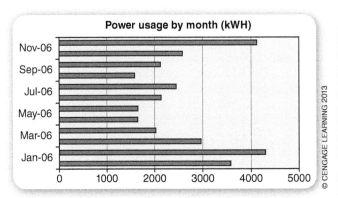

Figure 16-19: Bar chart displaying electric power usage in kilowatt-hours (kWH) by month for a state on the East Coast. The cold winter months of December, January, and February dominate the power usage for this residence. Power usage depends on many factors, including local climates, building size, heating/cooling technology, and cost for power.

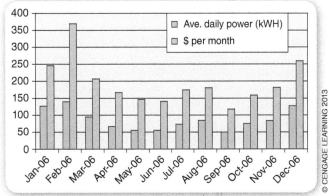

Figure 16-20: A column graph showing, by month, both average daily power usage in kilowatt-hours (kWH) and monthly cost. From this plot, you can estimate that the average cost per kWH is about $0.08.

Your Turn

Using data in the utility bills from your home, construct pie, bar, and column charts like those shown in Figures 16-18, 16-19, and 16-20. Is the makeup of the charges (generation, transmission, distribution, and transition) about the same as in Figure 16-21? How does your cost (per kWH) compare to that shown in Figure 16-20? Comment on your results.

▶ Two-dimensional graphing: The ability to plot data quickly and effectively is very important for many occupations, not just engineers, and is therefore a central capability of spreadsheet software. Graphing can be effectively self-taught using routines integrated with the spreadsheet software. Shown in Figure 16-21 (in blue) is a graph of a fourth-order polynomial: $y = 12x^4 - 950x^2 - 36x + 12$. (The order of a polynomial is simply the highest exponent—in this case, the exponent in the term $12x^4$.)

▶ The ease of graphing in Excel can be particularly useful in demonstrating some basic concepts in math. For example, in a matter of seconds Excel can calculate the slope, point by point. Figure 16-21 shows this type of calculation. The slope between two points, $P_1(x_1, y_1)$ and $P_2(x_2, y_2)$, is given by the change in y-values divided by the change in x-values:

$$\text{Slope} = \frac{(y_2 - y_1)}{(x_2 - x_1)} = \frac{\Delta y}{\Delta x} \qquad \text{Eq. 16-4}$$

▶ The slopes shown in column C are generated by entering the equation in Equation 16-4 (also shown in Figure 16-21) and dragging down, as demonstrated previously. The name Y' (y prime) was chosen for the slope of Y. The slope process is repeated in column D, which shows the slope of the slope of Y. The name chosen for the slope of the slope is Y" (y double prime). The pattern for the naming of the slopes is perhaps now apparent: A prime mark for each time a slope is taken. We have just performed calculus! Calculus is all

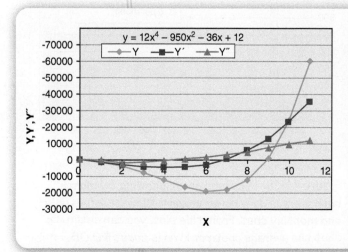

	A	B	C	D
1	X	Y	Y'	Y"
2	0	12	"C3=(B3-B2)/(A3-A2)"	
3	1	-962	-974	"D4=(C4-C3)/(A4-A3)"
4	2	-3668	-2706	-1732
5	3	-7674	-4006	-1300
6	4	-12260	-4586	-580
7	5	-16418	-4158	428
8	6	-18852	-2434	1724
9	7	-17978	874	3308
10	8	-11924	6054	5180
11	9	1470	13394	7340
12	10	24652	23182	9788
13	11	60358	35706	12524

Figure 16-21: An example of plotting a more complicated function: $y = 12x^4 - 950x^2 - 36x + 12$. The slope at points on almost any function can be estimated very quickly using Excel. The slope of the function is named Y' (y prime) and is itself another function. The slope of the slope is named Y" (y double prime) and is also a function.

about lines and slopes. The slope calculations we just completed with Excel accurately simulate the derivative process in calculus. The derivative process is simply a slope-generation process. Slopes are very important in math and science. For example, perhaps a function P gives the position of an object as a function of time. The object might be an electron traveling around an atom or a spacecraft traveling somewhere between the moon and earth. Imagine further that you have the equipment to measure accurately the object's position P (in meters) with time (seconds). Further, assume you use Excel and actually plot your measured positions versus time in a graph, just as we plotted y(x) in Figure 16-21. The slope of the function P will give the velocity of the object as a function of time. This makes sense because the slope is the change in y-value (meters) divided by the change in time (seconds), which gives units of meters per second (m/s) for the slope, which are the units of velocity. Similarly with Excel, if you were to calculate the slope of the slope, your units would be meters per second divided by seconds, which is meters per second squared (m/s^2), which are the units of acceleration. In just a few seconds, using only position and time information, we have used calculus, via Excel, to calculate the velocity and acceleration, two very important attributes of motion.

▶ Curve-fitting: Especially in experimental situations, one often has plotted the experimentally measured points on a graph but you want to know the functional form of the data. In Excel, once a graph is plotted, it is easy to perform a curve-fit to the data to get the function that best fits the data. Curve-fitting can be used to verify that the data is following the expected mathematics, or simply give a mathematical form to the data where the form is not known.

Use the following procedure to complete a curve-fit with Excel:

1. Right-click on any of the plotted data points on a graph to bring up a menu.
2. Select the Add Trendline option from the menu.
3. The next menu will bring up six different function-type options to fit the data: (a) Linear, (b) Logarithmic, (c) Polynomial, (d) Power, (e) Exponential, and (f) Moving Average. Choose one function type and any additional functional information (like the order of the polynomial).
4. In the Options menu, choose to Display the equation and/or the R^2 quality factor of the fit on the graph.

Figure 16-22 shows an example of curve-fitting. Here, the efficiency of a wind-powered generator is fit with a second-order polynomial. The mathematical function obtained here was used in Figure 16-17 to calculate the yearly savings by installing a high-efficiency, low-cost wind-driven generator.

Statistics. The modern world is inundated with statistical data, some valid and some invalid. Therefore, it pays off to understand the basics of statistics. Excel can quickly provide several useful statistical calculations. Here, we review two widely used examples of statistical calculations that Excel easily performs: (1) descriptive statistics and (2) histograms.

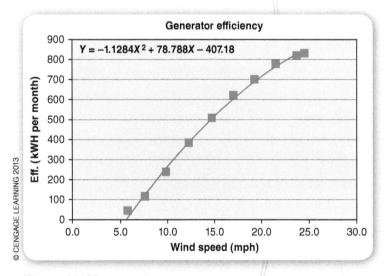

Figure 16-22: A wind generator generates more power for higher wind speeds, but the dependence is not linear. Excel can be used to plot the raw power efficiency data (obtained from the vendor) and then to perform a curve-fit. These data are best fit with a quadratic function.

446 Part II: Resources for Engineering Design

Statistical calculations with Excel are accessed through the Data Analysis option under the Tools menu. If the Data Analysis package is not visible, you must install it by going to "Add-Ins . . ." in the Tools menu.

Descriptive statistics. Descriptive statistics is an option within the Data Analysis option. With this option, you can calculate, among other things, mean (average), median, mode, standard deviation, maximum, minimum, range, sum, and the total count of the input numbers. Figure 16-23 shows an example of the input and output of the Descriptive Statistics package. To operate the Descriptive Statistics option, choose this option within the Data Analysis menu and fill in the required data into the menu. For example, to enter the input numbers (numbers that are already recorded in an Excel sheet), simply put the curser in the "input" portion of the pop-up menu and then use the curser to drag across the input data within the Excel cells (the green cells in Figure 16-23). Then, in a similar fashion, select the cell where you want the output printed. In the case of the wind speed, the cell to the right of the 6.4 mph cell was chosen for the output data (yellow cells in Figure 16-23). The output data consists of two columns of data: the first column for the title of the data output and the second column for the actual statistical data. In the example shown, the average wind speed over the 31 days of March turned out to be 12.4 mph.

Histograms. Creating histograms in Excel is also a very useful capability. For example, the wind speed data in Figure 16-23 indicated an average wind speed of 12.4 mph, but was this average a result of each of the 31 days having an average speed of 12.4 mph? Or rather, was this average a result of having a few days with gale-force winds with the

> **mean:**
> Also called *average*; the sum of the numbers divided by the number of numbers in the set. For example, the mean of the first four odd numbers is $(1 + 3 + 5 + 7)/4 = 16/4 = 4$.

Wind Speed (mph)			
6.4			
14.2		Column1	
13.6			
11.2	Mean	12.39119804	
11.8	Standard Error	0.822303344	
10.9	Median	12.12713936	
9.4	Mode	10.91442543	
10.0	Standard Deviation	4.578391256	
18.8	Sample Variance	20.9616665	
12.7	Kurtosis	-0.082252525	
22.7	Skewness	0.338069707	
16.4	Range	18.79706601	
13.9	Minimum	3.941320293	
12.7	Maximum	22.73838631	
10.6	Sum	384.1271394	
12.4	Count	31	
8.8			
20.6			
20.0			
3.9			
17.6			
10.9			
9.4			
16.7			
9.7			
5.2			
12.1			
12.1			
15.2			
5.2			
8.8			

Figure 16-23: Shown here are the measured average daily wind speeds in the 31-day month of March. Excel's Descriptive Statistics package is used to quickly calculate values like averages (means), ranges, and standard deviations.

Chapter 16: Math and Science Applications   447

Your Turn

Use spreadsheet software to calculate the descriptive statistics of your home's monthly utility bills for the last year. Compare the numbers you obtain to the descriptive statistics you obtain by just using the three months of June, July, and August. Look especially at the average, or mean, and the standard deviation.

remaining days having very little wind? The answers to these questions are important. First, the generator certainly has a maximum wind that it can handle. Second, if the average wind speed is largely dependent on a few storms, which may or may not come with each passing season, a wind generator may be too risky.

As the name implies, a histogram is a kind of "history" of data. Figure 16-24 shows a histogram for the wind speed data of Figure 16-24. Excel easily generates histogram data using the Histogram option within the Data Analysis option under Tools. The Histogram submenu appears. Enter the input data (data that is already entered into an Excel sheet), which in this case would be the green cells in Figure 16-24, into the histogram menu. As before, this is done by placing the curser in the "input" portion of the pop-up menu and then using the curser to drag across the input data within the Excel cells. Another piece of data is required for a histogram: the bin definition. The bin definition determines the size of bins into which the data (wind speeds) will be separated. The way to think of this is literally as bins, or buckets. Pretend that you write each recorded wind speed on a piece of paper and you throw each piece of paper, depending on its magnitude, into a bucket where each bucket, or bin, will hold wind speeds within a specific range of values. After you place all the

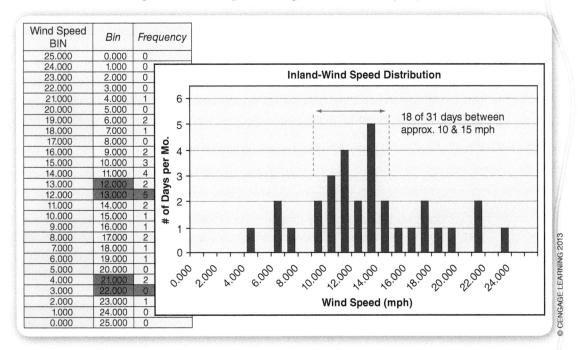

Figure 16-24: Excel's Descriptive Statistics package can also be used to quickly plot a histogram (history) of a set of data. This histogram shows graphically how many days had wind speeds of certain magnitudes. For example, there are five days with wind speeds between 12 and 13 mph, and zero days with wind speeds between 21 and 22 mph (see purple cells).

pieces of paper in the appropriate buckets, one counts how many pieces of paper are in each bucket. The histogram is a plot of how many pieces of paper are in each bucket. Excel calls this count "frequency." For example, for the histogram in Figure 16-24, there would be five pieces of paper in the most populated the bucket, or bin.

This bin definition in Excel defines how many buckets there are and what range of wind speeds go into each bucket. It is typical to have buckets of the same width (range of values accepted). Figure 16-24 shows in green the bin definitions that were entered and subsequently used for the wind speed histogram; each bin has a width of 1. (The actual input wind speeds are those shown in Figure 16-23, and are not repeated here.) Similarly as before, the output is printed, starting in a user-defined cell position. The output consists of two columns (yellow): the bin edges and the number of wind speeds in each bin ("frequency"). Once you write this output into the Excel cells, you can obtain a graph of the output using a column chart.

This histogram certainly indicates that there were no gales, and that, in fact, 18 days had wind speeds between approximately 10 and 15 mph. There were 4 days with mild wind (8 mph or less). Given the efficiency of the generator, there would be little to no power generated on these low-wind days. Similarly, there were 4 days with high wind (20 mph or more). These windy days would generate a substantial amount of power.

Newton's Three Laws of Motion

As mentioned in previous chapters, Sir Isaac Newton was a famous physicist. Newton was born in 1605, which interestingly is the same year as Galileo's death. Newton successfully described two of the most important works in science and math: (1) Newton's laws of motion and (2) calculus. (Calculus was independently described at about the same time by the Polish mathematician Gottfried Wilhelm Leibnitz, and today we use much of the notation defined by Leibnitz.) Newton made very significant contributions in math and science; our technology-rich society is a direct result of his accomplishments.

Newton and others from the era recorded much data on naturally occurring phenomena, particularly the motion of the planets. However, no one other than Newton explained successfully the governing science behind these motions. Newton's three laws of motion describe the motion of almost all matter (certainly up until relativity and quantum mechanics were required to accurately describe the worlds of the very large and the very small, respectively). Newton's laws of motion so dominate human technology that they are respectfully referred to as Newtonian physics. The following is a summary of Newton's three laws of motion:

1. *"Every body persists in its state of rest or uniform motion (constant speed) in a straight line unless it is compelled to change that state by forces impressed on it."*

Unlike most scientists in his day, Newton actually believed that Galileo was correct in stating that a body that was moving at a constant velocity had no forces acting on it. Most people thought that if something was moving, then something must be acting on it. This did not make sense to Newton or to Galileo. Galileo thought, for example, if a block resting on a surface were given a push, then it would surely come to rest (due to friction). However, if one were to put a very smooth block on a very smooth surface, then, after the push, the block would continue to move at a constant velocity for a long time, as long as no additional forces acted upon the block. Newton's first law states this (correct) belief explicitly. A body is either at rest (zero velocity) or at a constant velocity unless, and only unless, a force acts upon the body.

When you roll (impart force) a bowling ball, the ball will continue rolling forever at a constant velocity unless a force acts upon it. Of course, if the bowling lane were long enough, the bowling ball would certainly stop eventually, but only because frictional forces between the floor and the ball are acting on the ball to slow it down. With the realization of space flight, this law perhaps is much easier to understand because we routinely see footage of small objects being ejected from a space shuttle or space station that look like they could move on forever. This is, of course, readily apparent in space because there is extremely little friction.

2. *"The acceleration of an object is equal to the force imparted on an object divided by its mass."*

This is the famous "F = ma" law (or equivalently, a = F/m). It states in clear mathematical terms that when a force acts on a body, the body undergoes a very specific reaction: It accelerates in the direction of the force with the magnitude of the acceleration determined by the force divided by the object's mass.

3. *For every action there is an equal and opposite reaction. Or, as Newton stated, "To every action there is always opposed an equal reaction; or, mutual actions of two bodies upon each other are always equal, and directed to contrary parts."*

This law is often the hardest to understand. If one applies a force and the other object pushes back with the same force, then nothing should ever move because the forces would cancel each other. This is clearly not the case because objects do move, so what is the third law really stating?

Two forces exist when you exert a force on a body. Take the case of throwing a basketball directly down to the ground. The basketball pushes on the ground (earth) just as hard as the earth pushes back on the basketball—they each apply an equal force on each other. However, the reaction of each of the objects to the forces, per Newton's second law, is very different. The action, or reaction, force results in an acceleration that is inversely proportional to the bodies, mass: a = F/m. The basketball moves away very quickly because its mass is extremely small, leaving the quotient (F/m_{ball}) being relatively large. In reaction to the collision, the earth also moves, but in the opposite direction with a much smaller acceleration because the mass of the earth is extremely large (F/m_{earth} is a very small number). The acceleration of the earth due to the collision is so small that it can be completely ignored.

The ball-earth system previously discussed is an example of an elastic system because there is no permanent deformation of the two bodies. We also experience daily systems that have inelastic reactions, reactions that result in permanent deformation. For example, replace the basketball in the previous example with an equal mass of wet clay. When thrown to the ground, the force that the earth imparts onto the clay mass results in a permanent deformation of the clay—the force imparted by the earth moves clay molecules, resulting in an internal deformation of the clay. This deformation of the mass uses appreciable energy. In the clay-earth system, the two bodies join as one because the clay attaches itself to the earth and they continue in motion together as one combined body. This is very different than the ball-earth system where two bodies do not attach to each other and the two bodies move off in opposite directions. There are also cases in between elastic and inelastic reactions. Consider the case where an opened door is kicked, causing the door to shut violently. The door reacts quite quickly to the kick by closing quickly (rotating on its low-friction hinges), slamming shut. The person's foot, however, has not fared so well. Some of the cells in the foot have undergone a substantial deformation, causing severe damage (and pain). These cells have been permanently deformed and have died, only to be replaced over time as the body heals. The foot, however, did bounce back somewhat from the collision with the door (the door pushed back) so this collision was neither completely elastic nor completely inelastic but rather some of both.

> **acceleration:**
> The rate of change of velocity with time. Some typical units of acceleration are meters per second per second (m/s^2), feet per second per second (ft/s^2), miles per hour per hour (mi/hr^2), and miles per hour per second (mi/hr-s).

Acceleration has the units of distance per unit time per unit time. Therefore, a possible unit of acceleration is miles per hour per second (mi/hr-s). Why might this measure be more useful to us than mi/hr² or mi/s²?

Statics and Vectors

Statics is the study of structures that do not move. For example, the structure of a bridge or a house does not, to first order, move. Therefore, these structures are static, or unmoving. Buildings, bridges, furniture, cars, and planes are all structures that have major portions that do not move substantially and are important examples of static physics. When items in a structure are not moving, the sum of all the forces must equal zero. If the sum of forces were not zero, then, per Newton's second law, there would be movement (an acceleration).

Static structures, as well as many other subjects in physics, are studied with vectors. Put in simple mathematical terms, a vector is an arrow of a particular length pointing in a particular direction. These two aspects of a vector, length and direction, are exactly the important aspects of force. Forces have direction and strength (length of the vector), so vectors are very useful in describing forces. A vector can also represent a multitude of physical quantities: velocity, force, acceleration, stress, strain, or even polarization (of light).

Figure 16-25 shows examples of four vectors. Each vector is defined by a single point in the Cartesian coordinate system. For example, vector A is defined by the point (3, 4) while vector B is defined by the point (−4, 3). Vectors A and B are exactly the same length (magnitude) but point in very different directions. Vectors C and D are defined by the points (2, −2) and (4, −4). Vectors C and D are pointing in the same direction but have different magnitudes (lengths). Vector D is twice as big as vector C. For a vector, the length corresponds to strength while the angle of the vector corresponds to the angle of the action one is modeling.

Just like numbers, vectors can be added and subtracted as well as multiplied and divided. Here, we only describe how to add and subtract vectors. As shown in Figure 16-26, two vectors are added by graphically connecting the vectors tail to head, with the resultant vector (V_R) being the vector between the origin and the head of the last vector. This same procedure also works for more than two vectors. Figure 16-27 shows an example of

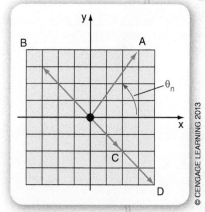

Figure 16-25: Diagram showing four different vectors. A vector is defined by its endpoint and consists of a line with a specific direction (arrow) and length. The length represents the strength (magnitude) of a quantity, like force or velocity. The direction (angle) of the vector represents the direction of the quantity.

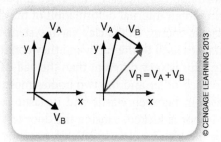

Figure 16-26: Two vectors are added by simply stacking them tail to head as shown here. The resultant vector, V_R, is the sum of the two vectors.

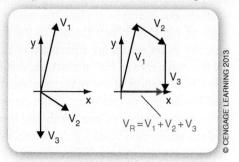

Figure 16-27: Three or more vectors are added by simply stacking them tail to head. The resultant vector in this case is a vector that is exactly horizontal. The initial three vectors are equivalent to the resultant vector, V_R.

the addition of three vectors. The vectors chosen in this example form a special case: The net resultant vector is pointed strictly in the positive x direction. What this means is that when the three forces (V_1, V_2, and V_3) all act on the same body, it is exactly like there was only one force (V_R) acting in the positive x direction.

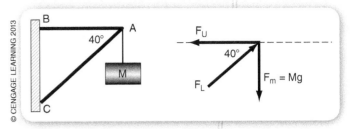

Figure 16-28 shows an example of a static system. This system consists of a weight hanging from a brace. Brace structures are commonly used. Plant hangers and shelves both represent brace structures. In the system, a mass M is hanging from the end of the brace

Figure 16-28: A static brace design and its free-body diagram. This model of a brace accurately represents a variety of real-world items like shelving and plant hangers.

and the two members of the brace form an internal angle of 40 degrees. All of the forces in this system are acting through the end point of the brace, point A, and because the system is static, the sum of all the forces equals zero. The corresponding diagram that shows only the forces, called the *free-body* diagram, is also shown in Figure 16-28. The three forces are labeled F_M, F_U, and F_L, and correspond to the forces of the mass, the upper leg, and the lower leg of the brace. Subscripts can be very helpful in clearly defining, and remembering, which variable corresponds to which real-world object or activity. In this case, the subscripts clearly identify the three important parts of the problem: the mass, the upper leg, and the lower leg.

Let's assume that the mass M is 1,000 kg, and we want to solve the system for how the force of the mass is distributed between the two legs of the brace. This is important because both the material used to make the brace as well as the fasteners used to connect the brace to the wall must be able to withstand these resultant forces. The first step in solving a static system is usually breaking down each force into its horizontal and vertical components. Then, separately for the horizontal and vertical components, sum all the forces and set the sum equal to zero. Remember, there is no motion in either the x or y directions, so the sum of all the forces for each of these directions must be zero. This should become clear as we work through this example. Referring to Figure 16-28, force F_L angles up and to the right, so it has both a horizontal and vertical component. The vertical component is shown in blue and has a magnitude of $F_L\sin(40°)$. The horizontal component of F_L is shown in red and has a magnitude of $F_L\cos(40°)$. Note that the vertical component of a vector added to the horizontal component simply results in the initial vector. Therefore, it is equivalent to using the two component vectors in place of the initial vector. Force F_M is special in that it only points down so it only has a vertical component (pointing down) and has no horizontal component. Similarly, force F_U has only a horizontal component and no vertical component. Mathematically, you need to distinguish if a force is a push or a pull. A sign convention must be established (which way is positive?) and is shown in Figure 16-29. The resulting equations for the horizontal and vertical components are as follows:

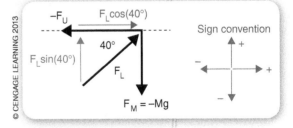

Horizontal: $F_L\cos(40°) - F_U = 0$ Eq. 16-5a

Vertical: $F_L\sin(40°) - Mg = 0$ Eq. 16-5b

Figure 16-29: A breakdown of the initial forces (F_U and F_L) of a brace in terms of their equivalent horizontal (red) and vertical components (blue).

Figure 16-30: A summary diagram showing the actual force values of the brace members. In this system, the upper beam is under tension while the lower beam is under compression.

With M = 1,000 kg, the downward force of the mass is given by the equation $F_M = Ma = Mg$. Newton's second law states that $F = Ma$, but the acceleration of gravity on earth is g, the acceleration of gravity, which is a constant and equals 9.8 m/s² (or 32.2 ft/s²). Therefore, $F_M = (1,000 \text{ kg})(9.8 \text{ m/s}^2) = 9,800$ N, where N is the abbreviation for a newton, the unit of force (which equals 1 kg − m/s²). In addition, because the trigonometric terms are known, we are left with two equations and two unknowns, which can be solved. In this case, the terms F_L and F_U can be obtained quickly because the equation for the vertical forces, Equation 16-5b, directly yields a value for F_L because it is the only unknown in Equation 16-5b. For this system, it turns out that $F_L = 15,246$ N and $F_U = 11,679$ N. Figure 16-30 shows a summary free-body diagram for this system. This diagram shows all the forces and indicates that the lower beam is under compression (being compressed) while the upper beam is under tension (being pulled apart).

Dynamics

The term *dynamics* refers to motion. Something that changes with time is dynamic. Dynamics in physics refers to the description of objects in motion. The objects could be Newton's famous planets, cars, space shuttles, bullets, arrows, soccer balls, or missiles coming at your F-16 Tomcat at Mach-2 speed. The successful description of objects in motion is obviously critically important.

Unlike statics, a substantive discussion of dynamics requires detailed knowledge and understanding of rates of change, which is essentially calculus. We promised that we would not delve deeply into calculus, and we will not. However, we will discuss calculus briefly in the following few sections.

Acceleration. Newton's second law explicitly mentions acceleration. But what is it? Acceleration is how quickly velocity is increased (or decreased) over time. A

Your Turn

Figure 16-31 shows a weight (force) of 1,000 pounds pushing down on a 3-member truss. This 3-member truss is a repeated element in bridges. The 1,000-pound force is representative of car traffic on the bridge. Can you solve for the forces F_{AB} and F_{AC} that are required to counteract the 1,000-pound downward force?

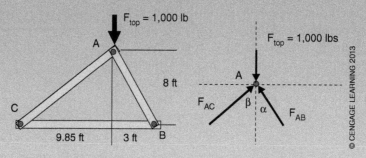

Figure 16-31: A 1,000-pound weight (force) pushing down on a 3-member truss.

Lamborghini or Ferrari sports car accelerates fast (0 to 60 mph in only 3.2 seconds) while a standard car will take two to four times longer to achieve the same speed. So, for example, the acceleration of the sports car is (60 − 0 mph)/3.2 s, which equals 18.75 miles per hour per second or 0.0052 miles per second per second (0.0052 mi/s^2).

How fast something changes with time is a rate of change with respect to time. Therefore, if acceleration is the rate of change of velocity, what is velocity? Velocity is itself a rate of change. Velocity is a rate of change of position with respect to time. Common units of velocity are meters per second (m/s), miles per hour (mph), and feet per second (ft/s). These units are straightforward to understand because velocity is determined by a distance traveled divided by the time it takes to travel that distance. Taking a common example, if you travel 60 miles in a car and it takes 60 minutes (one hour), you have traveled an average speed of 60 mph (60 miles per one hour). Assume in this trip that you were on a highway traveling at a uniform speed of 60 mph, if you plot the distance traveled versus time, then you would obtain a line with a slope of exactly 60 mph. In such a plot, the slope of the graph of distance versus time gives the average speed (units of slope give miles divided by hours, or mph). Similarly, when the velocity of the car changes with time, like when you accelerate forward at an intersection after a red light changes to green, the car experiences acceleration. When the acceleration is constant, meaning that the speed increases the same amount for each equal unit of time, then the car experiences a *uniform* acceleration.

Calculus is simply a very detailed treatment of these slopes, except not on an average basis but on an instantaneous basis—or, how fast something is changing exactly at any particular time, not averaged over some relatively large period of time. At its root, calculus is a lot about lines and slopes. Calculus takes the concepts of lines and slopes of lines, and applies these to the general cases of any function. Because calculus deals directly with rates of change, you can see that calculus is an integral tool for studying dynamics.

Free-Falling Objects. Objects falling freely toward earth is a common dynamic situation. Neglecting air resistance, the movement of free-falling objects are governed by the following three equations. The starting point for these three equations is that F = ma = mg:

$$a = g \qquad \text{Eq. 16-6a}$$

$$V = gt + V_{initial} \qquad \text{Eq. 16-6b}$$

$$S = \frac{1}{2}gt^2 + V_{initial}t + S_{initial} \qquad \text{Eq. 16-6c}$$

In these equations, F is force, m is mass, a is acceleration, g is the acceleration of gravity on earth (9.8 m/s^2), V is velocity, t is time, and S is position. It is important to notice that in the equations in Equation 16-6, no mass appears. Indeed, the speed and position of an object falling, subject to earth's gravity (without air resistance), does not depend on mass. A heavy hammer falls at the same rate as a marble, a human being, or even a feather. Clearly in practice, if you were to drop a feather and hammer at the same time from the same height, the hammer would hit the ground first but only because the feather is subject to a relatively large frictional force (air). If you were to drop a hammer and a feather at the same time on the moon where there is no air, the two objects would hit the moon's surface at exactly the same time. (Search the Internet for "Apollo 15 video hammer feather" to see this actual experiment conducted on the moon.) Equation 16-6b indicates that the velocity of the falling object increases continuously at a linear rate with time. Equation 16-6c indicates the position of the object as described by a quadratic function of time.

When first encountering dynamics, air resistance is typically neglected. However, as pointed out with the case of a feather, air resistance actually has a dramatic effect on an object's motion. When an object is falling in air, or any other material like water or quicksand, the speed does not continuously increase like it does in a vacuum but rather it increases only to "terminal velocity" (that is, the velocity "terminates" at this value). The terminal velocity is given in Equation 16-7, where V_T is the terminal velocity, m is the mass, g is the acceleration of gravity, r is the fluid density (density of air in this case), A is the cross-sectional area of the object, and C_d is the drag coefficient. Notice that now mass is present:

$$V_T = \sqrt{\frac{2mg}{\rho A C_d}} \quad \text{Eq. 16-7}$$

The terminal velocity for a person (that is, a skydiver without the parachute open) is approximately 120 mph. Including air resistance, a falling skydiver's velocity increases rapidly but the velocity increase eventually terminates at about 120 mph, and never gets any faster. Terminal velocity also limits the speed of falling rain and hail, so, being the target of such objects, we certainly benefit from these slower speeds caused by air resistance.

Projectile Motion. The motion of projectiles, like arrows, bullets, and thrown rocks, is important to understand. Projectile motion was the problem that motivated many scientists in the fifteenth to seventeenth centuries, as well as the inventors of one of the first computers, the ENIAC, in about 1946. The x and y coordinates for a projectile thrust at an angle θ with respect to the horizontal are as follows:

$$x = V_{initial}[\cos(\theta)]t \quad \text{Eq. 16-8a}$$

$$y = \tfrac{1}{2}gt^2 + V_{initial}[\sin(\theta)]t + H_{initial} \quad \text{Eq. 16-8b}$$

In Equation 16-8, x is the horizontal position, y is the vertical position, $V_{initial}$ is the initial velocity, g is the acceleration of gravity (9.8 m/s² or 32.2 ft/s²), t is time, $H_{initial}$ is the initial horizontal position, and θ is the angle that the projectile makes with the horizontal (ground) at the time it leaves the machine producing the motion. Figure 16-32 shows an example of a projectile motion calculation. This Excel sheet uses an "input/output" format for the variables of time, acceleration of gravity (g), initial velocity ($V_{initial}$), and angle being the inputs, and the position of the object (x and y) being the outputs. The shape of projectile motion is parabolic. The maximum speeds occur when the projectile first launches and when it lands; in fact, these launch and landing speeds are identical. The minimum velocity (of zero) occurs at the peak of the parabola.

Rotational Motion. In addition to translational motion (movement up or down, left or right, or forward and back), some items, like wheels and balls, rotate. Rotational motion is also governed by Newtonian mechanics but the expressions need to be modified slightly. With translational motion, we deal with how position changes with time, leading us to the concepts of velocity and acceleration. Rotation motion is the movement in angle over time not position over time, so one replaces position with angle to arrive at the equations of motion for rotating objects.

Chapter 16: Math and Science Applications 455

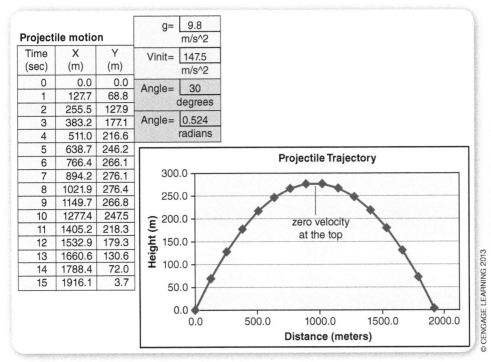

Figure 16-32: An Excel calculator for projectile motion. The path of projectile motion is always a parabola.

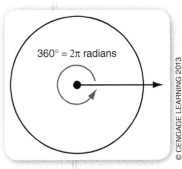

Figure 16-33: A circle consists of 360 degrees. That is, when a wheel (circle) is rotated by one revolution, it has rotated 360 degrees, which is the same as 2π radians.

Because the study of rotation depends largely on angle, let's first review some basics about angles. An **angle** is the amount a line needs to be rotated to bring the line into coincidence with another line. Angles are measured in degrees or radians, but a mathematician will usually prefer radians. There are 360 degrees, or equivalently 2π radians in a circle. Figure 16-33 depicts this, which shows a circle centered at the origin along with only the positive x-axis drawn. If the x-axis is rotated about the origin until it ends up where it began, the x-axis has rotated 360 degrees, or equivalently 2π radians. Taking a real-world example, if a wheel were to rotate 810 degrees, the wheel would have rotated exactly two and a quarter rotations.

Translational motion centers on the concepts of acceleration, velocity, and position. Rotational motion, being about angular motion, centers on the concepts of angular acceleration, angular velocity, and angle. Typical units of these rotational motion concepts are as follows:

Angle: radians

Angular velocity: radians per second (rad/s)

Angular acceleration: radians per second per second (rad/s²)

angle:

The amount a line needs to be rotated to bring the line into coincidence with another line.

OFF-ROAD EXPLORATION

http://en.wikipedia.org/wiki/Rotational_motion

Your Turn

Why do we use multiples of 60 so often in some measurement systems, such as 360 degrees per rotation, 60 arc minutes in a degree, 60 arc seconds in an arc minute, 60 minutes in an hour, and 60 seconds in a minute? (Hint: Search on "sexagesimal" on the Internet.)

Your Turn

A wheel of radius 15 cm is rotating at 33.3 rpm. What is its angular speed in radians per second? If the wheel is in contact with the ground (like a wheel on a car), how far would the wheel travel along the ground after one minute? (Hint: You'll need to use the equation for the circumference of a circle as a function of radius.)

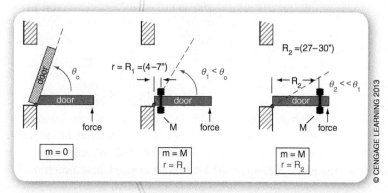

Figure 16-34: *Top views of a swinging door. A swinging door can be used to understand rotational forces, torques, by experimenting with position of mass and position of the applied forces.*

Rotational speed is measured in rotations per minute (rpm), so it is useful to be able to convert rotational speed (rpm) to angular velocity (rad/s). Equation 16-9 converts rotational speed, S_{Rot}, in rpm to angular velocity, v, in radians per minute.

$$\omega = 2\pi S_{Rot} \qquad \text{Eq. 16-9}$$

Taking a real-world example, a motor that is rotating at 1,000 rpm is rotating at 6,283.2 rad/min, or 104.7 rad/sec.

Torque and Moment of Inertia. With a good understanding of angle and how angles change with time, we are ready to summarize Newton's laws of motion applied to rotation. A good way to understand rotational motion is to conduct an experiment. Referring to Figure 16-34, find a standard door in your house or school. For best results, pick a door that has reasonably good hinges to minimize the effects of friction. If necessary, lubricate the hinges to lower the friction. The following section outlines the experiment.

SWINGING DOORS

1. Pick a door in your house or school that has reasonably good hinges to minimize the effects of friction.
2. Find two items that weigh between 2 and 10 pounds each and attach them to each other with a rope or heavy string. Leave about one foot of rope or string between them.
3. With the tied weights set aside, open the door about three-fourths of the way and apply a force near the outer edge of the door, so that the door approaches the closed position, as shown in Figure 16-34. Using a piece of masking tape, place a mark on the door where you apply the force. Apply the force with your hand open but do not grab the door; simply push on it. Apply enough force so that the door almost shuts but never shuts all the way. Practice by applying the same amount of force, resulting in the door reaching the same "almost closed" position. Quantify your results by measuring how far the door has rotated (in angle). When you are able to apply a reproducible force, go to the next step.
4. Have a partner hang the tied weights over the top of the door (so that the weights counterbalance each other) a distance of about four inches from the hinge axis.

5. Repeatedly apply the same force you practiced earlier and measure how far the door moves (in angle). Remember to apply your force in the same spot (same radius). Try this several times and calculate the average. Record your results.
6. Now move the tied weights to the outer portion of the door, as far as they can go without falling off.
7. Repeatedly apply the same force you practiced earlier and measure how far the door moves (in angle). Remember to apply your force in the same spot (same radius). Try this several times and calculate the average. Record your results.

These experiments deal with applying forces in a rotational system, and the results should give you a very good understanding of how Newton's laws of motion apply to rotational systems. In your experiment, you should have discovered that the door rotated substantially more with the weight closer to the hinge axis. With the weights hanging in the outer position, the door should have moved only slightly. The total mass is the same in your experiments: the mass of the door plus the mass of the two weights. The only difference in your experiment is *where* the mass is placed on the door. This, at first, seems like a contradiction because Newton's laws of action depend on mass (a = F/m). However, in your experiments, the reaction (the door going into motion) depends on where the mass is, not the actual mass. This is the first important difference in rotation motion: Mass must be replaced with the distribution of mass. More specifically, it is the square of the distance of the mass from the rotational axis that is important. This attribute of distributed mass is called the *moment of inertia*. The moment of inertia is represented by the letter I, and is the sum of mass times the square of the distance from the mass to the axis of rotation. The moment of inertia of a solid object is a constant, just like its mass. For simple structures, like dumbbells, the moment of inertia is easy to calculate, but, in general, calculus is needed to calculate the moment of inertia of most objects. Let's look at a dumbbell as an example (see Figure 16-35), where the mass of the bar is assumed to be negligible. Assume there is a mass M_1 at each end of the dumbbell and the dumbbell is going to be rotated about its center. Then the moment of inertia of the dumbbell, I_{DB}, is given by Equation 16-10.

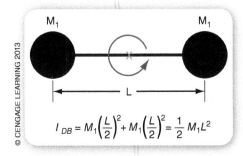

Figure 16-35: A dumbbell of length L with a mass of M_1 at each end. The moment of inertia of a dumbbell, rotating about its center, is $0.5\, M_1 L^2$.

$$I_{DB} = M_1\left(\frac{L}{2}\right)^2 + M_1\left(\frac{L}{2}\right)^2 = \frac{1}{2}M_1 L^2 \qquad \text{Eq. 16-10}$$

If each mass M_1 is 50 kg and the length of the dumbbell is 2 meters, then $I_{DB} = 100$ kg $-$ m². In summary, for rotational motion, the mass is replaced with moment of inertia.

There is a second difference with rotational motion. In rotational motion, it matters where the force is applied. For example, in the swinging door experiment, if you repeated the experiment exactly as you did before but instead of applying your push (force) to the end of the door, you applied it halfway toward the hinge axis, the reaction of the door would be less. Another common experience related to forces acting at a distance is when you open a very heavy door. It is much easier to open the door by pushing near the outer portion of the door as opposed to the inner portion of the door. For a very heavy door, it might even be impossible to open when you push it near the hinge. Therefore, not only does it matter *where* the mass is, but it also matters *where* the force is applied. This introduces the concept of **torque**, τ (the Greek letter "tau"). Torque is force times distance, where distance is the direct distance to the axis of rotation. Figure 16-36 shows a few examples of torque. For Newton's laws applied to rotation, force is replaced with torque.

torque:
The rotational equivalent to force. Torque is defined as force times distance [$\tau = (F)(d)$], where the distance is the distance between the acting force and point of rotation. The Greek letter tau (τ) is often used to represent torque. Typical units of torque are newton-meters (n-m), foot-pounds (ft-lbs), or inch-pounds (in-lbs).

Your Turn

Due to the forces being applied to the bar in Figure 16-36, will the bar rotate and, if so, in what direction?

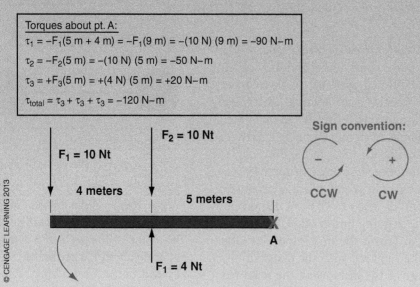

Figure 16-36: Diagram of a rod fixed at one end with a hinge. Different forces are applied at various distances from the point of rotation, resulting in different torques. Torque is force times distance so, for example, the same force applied at different distances results in different torques.

rotational velocity:

The rate of change of angle with time. Two typical units of rotational velocity are radians/second (rad/s) and degrees/second (deg/s). (Rotational speed can also be measured in rotations per unit time with some examples being rotations per minute [rpm] or rotations per second [rot/s].)

The equations for rotational motion are summarized in Figure 16-37, alongside the translational equivalents.

As with translational motion due to gravity, with constant accelerations, the following apply: (1) Rotational speed depends linearly with time and (2) position (angle) depends quadratically with time. As an example of rotational motion with a constant acceleration—in this case, a deceleration—Figure 16-38 shows calculations for a wheel that is initially rotating at 1,000 rpm (104.7 rad/sec) but due to friction, it slows down at a rate of 2 rad/s^2. As shown in Figure 16-38a, the rotational velocity changes linearly with time while the angle ("rotational position") changes quadratically with time. The wheel rotates a total of 2,740.5 radians or 236.2 rotations before it stops.

Figure 16-37: Table of important quantities for both translational and rotational motion.

	Translational	Rotational
Displacement:	x, distance (meters)	τ, angle (radians)
Velocity:	v (m/s)	ω (rad./s)
Acceleration:	a (m/s^2)	α (rad./s^2)
Mass:	m (kg)	I (Kg · m^2)
Force:	F (Newtons)	$\tau = F \cdot d$ (Nt · m)
2nd Law:	F = ma	$\tau = I \cdot \alpha$ (Kg · m^2/s^2)
Kinetic energy:	$\frac{1}{2} mv^2$	$\frac{1}{2} I\omega^2$

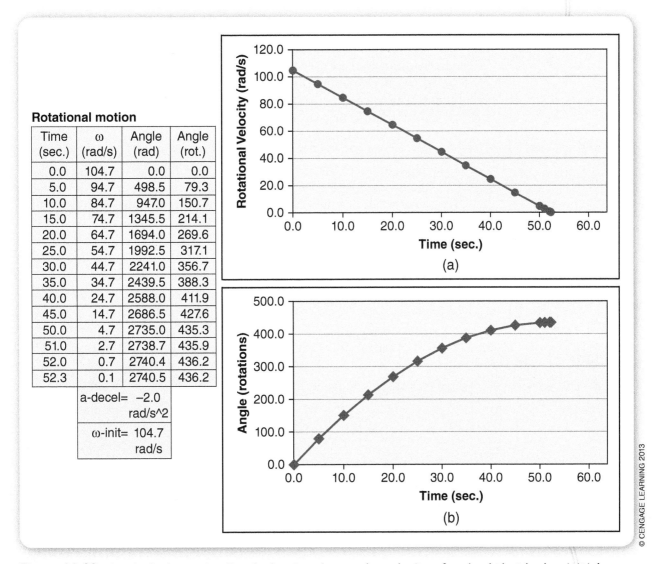

Figure 16-38: A calculation, using Excel, showing the angular velocity of a wheel that had an initial velocity of 104.7 rad/s (1,000 rpm) but is subject to friction (a deceleration of 2 rad/s²). The wheel's rotational speed decreases linearly, while the angle moved (how many rotations) is shaped like a parabola.

Your Turn

Plot the upper half of a circle of radius 1 (see Equation 16-3a) using Excel to create the data points and the graph. Use at least 16 points. This is a very special circle and is called the "unit circle." Place a particular point on the circle in the upper right quadrant. What do the coordinates of this point have to do with the sine or cosine of the angle defined by that point? (Hint: Stay in the first quadrant for your first few examples.)

Springs: Nature's Trigonometric Function

Engineers use trigonometric functions for a wide variety of applications. Trigonometric functions are used to describe oscillatory motion (like pendulums). Here we review the definition of the trigonometric functions, as well as some applications. The trigonometric functions are most easily defined by using a right triangle, a triangle that has one internal angle of 90 degrees ($\pi/2$ radians).

As shown in Figure 16-39, the sine, cosine, and tangent functions are simply ratios of the lengths of the appropriate sides: the sides either adjacent (A) to the angle, opposite (O) to the angle, or the side that is the hypotenuse (H) of the right triangle. The hypotenuse is always the longest side of a right triangle. Figure 16-39 also shows a plot of the sine and cosine functions.

OFF-ROAD EXPLORATION

For more information on springs, see http://en.wikipedia.org/wiki/Spring_%28device%29.

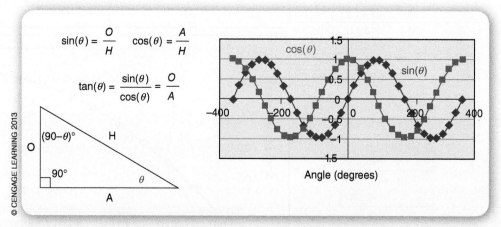

Figure 16-39: Definitions of the trigonometric functions along with a plot of the sine and cosine functions, showing their oscillatory shape.

Many important phenomena are oscillators. Some common everyday items that oscillate are pendulums (clock technology circa the year 1500), electronic crystal oscillators (clock technology circa the year 1900), light, sound, water waves, and lasers. All of these phenomena oscillate in one form or another. This oscillation can be described by sine waves. Because oscillators are so common and important, in this section we want to briefly describe the basic math and science behind oscillators. The common, and very ordinary, spring is a human-made device that best illustrates a natural sine wave oscillator.

Figure 16-40 shows a frictionless, horizontal spring system. The system is chosen to be horizontal so that the effects of gravity can be ignored (the surface is frictionless). The equilibrium point of a spring is the point, without the influence of any external forces, where the mass is naturally at rest. After playing with a spring for just a few minutes, it becomes apparent that the further one pulls a spring away from its equilibrium point, the harder the spring pulls back. More specifically, the amount of force with which the spring pulls back depends linearly with distance away from the equilibrium point. This is represented mathematically by the simple equation $F_s = -kx$, where F_s is the force that the spring exerts, k is a constant (the *spring constant*), and x is the displacement away from equilibrium (how far you have pulled the end of the equilibrium point).

The minus sign indicates that if you are pulling the spring in one direction, the spring pulls back in the opposite direction. Further, the more you pull the mass away from its equilibrium point (increasing x), the stronger the spring pulls back. This restoring force, the force that goes opposite to the direction the mass is traveling, is the key to oscillatory motion. (That simple little minus sign in the force equation, $F_s = -kx$, makes a big difference.) Newton's laws govern this system, so let's see what they tell us here. We know that $F = ma$ but we also know that $F = -kx$, so we have the following:

$$ma = -kx \qquad \text{Eq. 16-11}$$

Rearranging terms, we obtain the following equation:

$$a = -\left(\frac{k}{m}\right)x \qquad \text{Eq. 16-12}$$

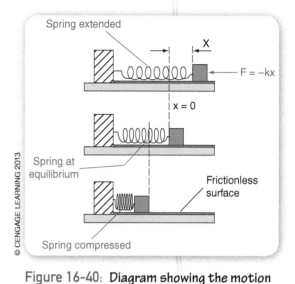

Figure 16-40: Diagram showing the motion of a mass on a spring. The spring always acts against the motion of the mass in a linear fashion. That is, the force applied by the spring to the mass, F_s, is given by $F_s = -kx$, where k is a constant (the "spring" constant) and x is the position of the spring.

This equation is relatively simple—but can we deduce something useful from it? Equation 16-12 gives acceleration, a, as a linear function of x (k and m are just constants). Keeping in mind that we are looking for the equations of motion (movement), perhaps Equation 16-12 is telling us something after all, because if the acceleration is large, the mass will be undergoing substantial changes in motion. However, can we get more specific? Can we find the actual position as a function of time, x(t)? We can do this but we have to go back to our brief discussion of calculus. Acceleration is the rate of change of velocity with respect to time, and velocity is the rate of change of position with respect to time. So, acceleration really is related to position. If you remember from a previous section (discussion of Figure 16-24), a rate of change only involves taking a slope. Slopes are easily calculated, especially with Excel. So let's rewrite Equation 16-12 in a way that tells us that acceleration is the slope of the slope of x (position) with respect to time. As we did in Figure 16-21, we use a prime (an apostrophe) for each time a slope is taken. Let's replace acceleration (a) in Equation 16-12 with the slope of the slope of position (x) with time (an x''):

$$a = x'' = -\left(\frac{k}{m}\right)x \qquad \text{Eq. 16-13}$$

Your Turn

Using Excel, show that the slope of the slope of a sine wave is minus a constant times the initial sine wave. Let's give the constant a name, K. For example, take the sine function, y = sin(3t), where the argument (3t) is presumed to be in radians, and use Excel to determine its slope and then the slope of the slope (as was done in Figure 16-24). Then plot the three functions [(1) sin(3t), (2) the slope of sin(3t), and (3) the slope of the slope of sin(3t)]. Also, what is the constant K? (Note that the independent variable is t, which stands for time.)

In this notation, the term x″ (*x double prime*) means you have to take the slope of x twice. This prime notation is exactly what mathematicians use (in calculus). So Equation 16-13 is asking us is to find a function x of time, x(t), such that when you take its slope twice, what is returned is –x back again times a constant (the k/m). Such a function is exactly a sine or cosine function. This solution makes sense: When a mass on a spring is pulled away from equilibrium and released, the resulting behavior is an oscillation, or a sine wave oscillation.

Exponential Functions

The exponential function is one of the most prevalent functions in science and engineering. Phenomena such as electricity, electronic circuits, light, radioactivity, technology growth, cell growth, chemical reaction rates, oscillatory motion, interest-bearing financial accounts, and an endless number of other phenomena are accurately described with exponential functions. Even our sense of hearing has inverse exponential (logarithmic) response.

OFF-ROAD EXPLORATION

For more details on exponential functions, see http://en.wikipedia.org/wiki/Exponential_growth.

An exponential function is of the form $y = B^x$, where B is called the base and is any positive real number (not equal to unity) and the independent variable x is in the exponent. Examples of some exponential functions are $y = (1.7)^x$, $y = (3.45)^x$, and $y = 10^x$. Exponential functions grow, or decline (for negative values of x), very fast. The larger the base, the faster the function will increase (or decrease). The function $y = 2^x$ was discussed in detail in Chapter 13, and approximates the advancement of semiconductor technology (Moore's law). Semiconductor microprocessor capability has grown at an unbelievable rate in the last 30 to 40 years. Figure 16-41 shows some examples of exponential functions. (An exponential function should not be confused with a power function, which is of the form x^B, which is sort of the "opposite" in that B is in the exponent and the x is the base. Power functions are themselves very useful but do not grow nearly as fast as exponential functions.)

A very special form of the exponential function is called the *natural exponential function* and is $y = e^x$, where the base is the number e (Euler's number), which

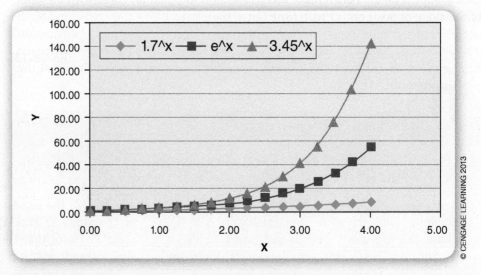

Figure 16-41: Plots of three exponential functions: $y = (1.7)^x$, $y = e^x$, and $y = (3.45)^x$. Exponential functions grow extremely fast.

is approximately 2.7183. The natural exponential function is special because the slope at any point on the function is exactly equal to the functional value of the function. For example, at x = 0, y = e^0 = 1 (anything to the 0th power equals unity), and the slope at the point (0, 1) on the graph is exactly 1. Another simple point on the curve y = e^x is (1, e) because anything to the 1st power is just itself. The slope at the point (1, e) on the curve y = e^x is exactly e (~2.7183).

Population growth. The study of population growth is a very important topic. The rate that animal and plant populations grow or decline is of obvious importance in nature. Population growth is also very important on the microscopic level. The growth rate of cells, both good cells like red and white blood cells as well as cancer cells, is governed by exponential-based functions. The reason why exponential functions govern population growth rate is similar to our earlier discussion on spring oscillators. The rate at which a population grows, its rate of change with respect to time, depends on how many individuals are in the population to begin with: The more individuals available to procreate, the more offspring can be produced. Therefore, the equation that governs the growth of populations is a function whose rate of change (slope) is proportional to itself. This is exactly the definition of an exponential to begin with. Common equations for population growth are given in Equations 16-14 and 16-15. In these equations, the independent variable is now time (t), not x. Equation 16-14 could describe the increasing population for embryonic cells. Equation 16-15 describes the decreasing population of a set of N_o radioactive atoms as the material decays. Each radioactive material has its own characteristic time constant t (t is proportional to the half-life of the material [t_{half} = 0.69t]).

$$C = C_o e^{kt} \qquad \text{Eq. 16-14}$$

$$N = N_o e^{-\frac{t}{\tau}} \qquad \text{Eq. 16-15}$$

A better mathematical model for population growth accounts for increased competition within the population as the population size gets larger. For example, studies have shown that as a rodent population increases (exponentially) processes begin that decrease the rate of growth. For example, death rates increase and birth rates decrease due to stress (overcrowding), disease, and infighting. The equation that describes this more accurate model is also given by an exponential-based function (Eq. 16-16), where C, D, and α are constants and t_o is the time that the growth process starts:

$$P(t) = \frac{C}{1 + De^{-\alpha(t-t_0)}} \qquad \text{Eq. 16-16}$$

A plot of this modified growth model is shown as curve I in Figure 16-45. This type of curve is called an *S-curve* or a *logistic curve*. It is evident that the population grows exponentially at first, but then the growth slows, resulting in the population saturating at a constant population level of 400 for Case-I.

In addition to accurately describing populations of living things, this S-curve model also accurately describes the spread of new technology, so-called *technology curves*. As a new technology (that is, product or process) emerges, say a new cell phone, video game, medicine, or high-density memory process the rate of adoption depends both on how

OFF-ROAD EXPLORATION

For more information on S-curves or logistics curves, see
http://en.wikipedia.org/wiki/Logistic_curve.

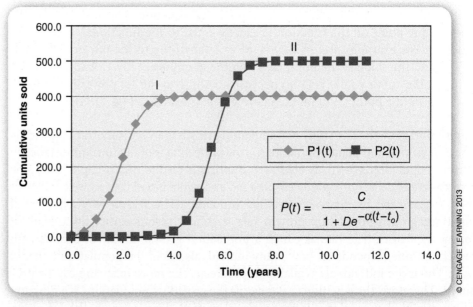

Figure 16-42: Exponential functions are also used to represent S-curves, also called logistic or technology curves. S-shaped curves accurately model growth rates of populations as well as technology maturity.

many consumers have adopted the technology ("good news travels fast") and the number of consumers who have not adopted the technology (as more consumers adopt the technology, fewer customers remain). Also shown in Figure 16-42 is a second technology curve (curve II) showing how a new technology will start to take over and displace the older technology. Luckily for this company, the second technology (red curve II) starts to enter the market just as the old technology stops selling any more. In high-technology industries, the economic impact of these exponential curves is very important. A company that depends largely on new technology development needs to find ways to bring to market technologies with steep growth curves, like in Figure 16-42. Additionally, they must also make sure that the timing is good so that the newer technology is ready by the time the older technology stops selling.

Probability/Statistics: Applications

As you read this paragraph, look at your surroundings and note the various types of objects nearby, or any processes that are occurring. There may be objects like chairs, lights, pencils, desks, clothing, plastic cup, and other books. In addition, there are likely several processes occurring like a clock ticking, an aquarium tank bubbling, a heart beating—or even an electron buzzing around its orbital inside a hydrogen atom inside your left index finger! When many people think of probability they typically think of flipping coins or games of chance like poker or "21." However, probability really has its hand in all of the objects and processes that surround you. Sometimes the probability at work is readily apparent, but other times the presence of probability at work is subtle. Let's try to understand why probability abounds in your environment by looking at a few examples.

Are you male or female? The chromosomes passed on to you by your parents determine your sex. In addition to all of the amazing biological processes involved in creating life, the outcome is described accurately by probability. There are two human gender types, male and female. In everyone's DNA, there are 23 chromosomes, two of which are the "gender" chromosomes. These two chromosomes can be one of two types, either X or Y. A male's two gender chromosomes are XY, whereas a female's two gender chromosomes are XX. Given that the offspring gets one of its two gender chromosomes from each of its parents, when a male and female mate, there are four possibilities for the offspring: XX, XY, XX, and XY (see Figure 16-43). The probability is 2 out of 4 for the offspring to be male, and 2 out of 4 to be female. Therefore, the probability is equal of being male or female, 50 percent each, which is approximately what actually happens. An interesting aspect of this process is that only the father determines gender because the female can pass on only an X chromosome. If the father passes on an X chromosome, the offspring will be female. If the father passes on a Y chromosome, the offspring will be male. Another interesting conclusion of this process is that nothing substantive to life can be a part of the Y chromosome because females do not get the Y chromosome. Gender and other genetic processes are large sets of complicated chemical processes involving complex proteins. However, from the outside it can be described by probability. The probability (math) of genetics can get very complicated because there are 23 chromosomes and tens of thousands of genes. Similar analyses describe other common genetic processes such as blood type, eye color, and several genetic-based diseases like cystic fibrosis.

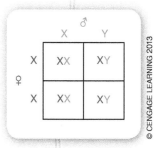

Figure 16-43: Diagram showing how probability determines gender. One gender chromosome is supplied by each of your parents for a total of two chromosomes that determine gender. A male is an XY while a female is an XX.

Let's return to the inanimate objects you noted around you. These items are made up of materials like metal, plastic, wood, or fabric. When these objects were fabricated, there were variances in all aspects of the manufacturing process. As discussed at the beginning of this chapter, nothing can be made with complete accuracy, and further, many of the same items cannot be made identical to one another. Therefore, in looking at the objects around you, engineers and technologists put a lot of thought into how the critical dimensions of these objects need to be specified to achieve high enough yields and still satisfy the customer. Let's be more specific. A chair typically has machined holes into which screws or other pieces of wood fit. The manufacturer can specify the hole placement and size very accurately, so that the probability of the hole being offset (an error) is very small. However, such precision will cost more. Therefore, there is a never-ending battle between the level of accuracy, or tolerance, of a specification and cost. It is with a strong understanding of probability that a manufacturing or design organization makes such decisions and sets goals. To be ISO (International Organization for Standardization) certified, an organization typically implements several procedures to monitor the statistics (probability) of their processes. This involves using sophisticated statistics and probability software packages.

Your Turn

List some industries and products that use an unusually large amount of probability and statistics and state why.

Manufacturing processes, as well as human factors like those discussed in Chapter 15, are often governed by a probability function that is called a normal, Gaussian, or bell distribution. As discussed in Chapter 15, a Gaussian distribution often accurately describes the variability of a quantity. An example would be the wind speed data that we analyzed using the "Histogram" feature in Excel (see Figure 16-24). The data in Figure 16-24 told us that the probability of experiencing a wind speed between 10 and 15 mph was good (18 out of 31 days) while the probability of experiencing 25 to 30 mph wind speed was essentially zero (0 out of 31 days has wind speeds in this range). The mathematical form of a Gaussian distribution is given in Equation 16-17, where σ is the standard deviation and μ is the mean. The standard deviation is directly related to the width of the curve while the mean is the point on the x-axis where the curve is centered. Figure 16-44a gives several different plots of Gaussian distributions that represent the height distribution of adult males living in a certain city. Assume that these four sets of data were measured by four different research groups.

$$y = \frac{1}{\sigma\sqrt{2\pi}} e^{-\frac{(x-\mu)^2}{2\sigma^2}}$$ Eq. 16-17

In Figure 16-44a, the blue, green, and red curves are all centered about the same point on the x-axis (6 feet), so these three curves all represent data that have the same mean (average) of 6 feet. However, the widths of the blue, green, and red data vary substantially. The red data is a very narrow distribution, which shows that all of the men are close in height, varying by only approximately ±0.1 feet. In contrast, the blue curve shows that the heights vary much more, by ±0.4 feet. The purple curve indicates a very different mean of approximately 5.8 feet, 0.2 feet less than the other three sets of data. An engineer looking at these data might find the purple data suspicious because the other three curves all give the same mean. Why would the mean height of males in the same region have a lower height?

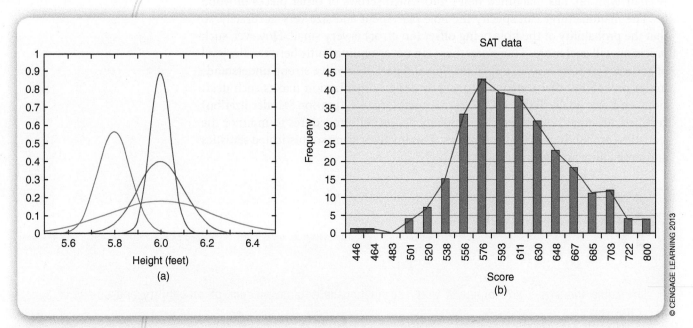

Figure 16-44: Plotted are (a) several examples of Gaussian curves and (b) real SAT test data, which also simulate a Gaussian distribution. Gaussian distributions are also called *normal distributions*.

Perhaps that company made an error, or perhaps they only took data of older males or younger males, which would have smaller heights. Figure 16-44b shows a plot of some actual data, the average incoming Scholastic Aptitude Test (SAT) score (quantitative test) for 285 four-year colleges and universities. These SAT data appear a little asymmetric, with a tilt toward higher scores, but might be modeled reasonably well with a Gaussian distribution.

> **OFF-ROAD EXPLORATION**
>
> For more information about the discussions in this chapter, visit the following websites:
> *http://nssdc.gsfc.nasa.gov/planetary/lunar/apollo_15_feather_drop.html*
> *http://www.langorigami.com/*
> *http://en.wikipedia.org/wiki/Trigonometry*
> *http://en.wikipedia.org/wiki/Vector_(spatial)*
> *http://en.wikipedia.org/wiki/Rotational_motion*
> *http://en.wikipedia.org/wiki/Spring_%28device%29*
> *http://en.wikipedia.org/wiki/Exponential_growth*
> *http://en.wikipedia.org/wiki/Logistic_curve*
> *http://cte.jhu.edu/techacademy/web/2000/heal/mathsites.htm*

SUMMARY

Humans use a variety of math and science principles daily. Quantitative analysis of our world and experiences is beneficial to us. Many prominent scientists and technologists thrust our society forward by providing enlightened methods of looking at our world. In the past, however, higher education was typically only accessible by a small fraction of the population. For example, just a few hundred years ago, only a few individuals could comprehend Newton's physics and calculus. However, in modern times calculus and Newtonian physics have been common classes for millions of high school students. Today, nearly everyone has access to advanced educational opportunities. However, still a surprising number of people have limited quantitative skills (math and science). This can be cumbersome in our world dominated by high technology and statistical data. Studies have shown that taking a math class past Algebra II correlates with future success, so attention to quantitative and logical thinking is worth the investment. This chapter provided you with a brief overview of the math and science skills common in high school pre-engineering and technology courses.

This chapter also emphasized quantitative analysis for certain situations using computers. Spreadsheet software in particular has become so powerful and accessible that it is hard to imagine not using it in math, science, and technology classes, as well as in our everyday lives. We worked many real-life examples using Excel software that included recording car maintenance information, calculating gas savings to help decide which car to purchase, analyzing the financial viability of a wind-powered generator, and calculating both static and dynamic systems and several other science-based situations. We concluded this chapter with discussions on probability and statistics using biological examples. The twentieth and twenty-first centuries brought substantial advances in the biological sciences, which do not appear to be slowing down. Improved medical diagnostic tools and medicines are examples of how humankind will substantially benefit from continued quantitative investigations in biological sciences.

BRING IT HOME

OBSERVATION/ANALYSIS/SYNTHESIS

1. Why must designers consider variability in a design?
2. You have two measuring tapes, one 100 feet long and the other 15 feet long. They are otherwise identical. Which measuring tape will give you a more accurate measure of a span of about 65 feet and why?
3. Briefly describe Newton's three laws of motion, giving real-world examples of each in action.
4. What is the term that describes a system where the sum of all the forces equals zero?
5. How does the velocity of a free-falling object (in a vacuum) change with time? How does the position of a free-falling object (in a vacuum) change with time? How does the velocity of a free-falling object (in air) change with time?
6. Convert a wheel rotating at 670 rpm to a rotational velocity in radians per second. If the wheel were used on a cart and had a radius of 2 decimeters, how far would the wheel travel after 12 seconds?
7. Which function grows faster: $y = x^4$ or $y = e^x$?
8. In a manufacturing plant, the two features M and N are measured on every part. If the standard deviation of dimension M is 1.9 mm and the standard deviation of dimension N is 2.4 mm, which dimension is known more accurately? How much more accurately?

EXTRA MILE

ENGINEERING DESIGN ANALYSIS CHALLENGE

1. Make a simple projectile launcher using a rubber band for the "gun." Use cardboard or foam board and pins to construct the mechanical holder for the rubber band. Devise a method for reproducing the same launch conditions (pull the rubber band back the same length every time). Pick a projectile that will work well with your design and with a mass that can be altered without substantially changing the launch conditions or the aerodynamics during flight. Test your design by doing the following: Take multiple "identical" shots and measure the distance that the projectile travels. Perform a statistical analysis on your data (using Excel). Have someone else use your launcher. Is the distribution of their shots about the same as yours? How can you improve your launcher? Put in a projectile of different mass. Does the mass affect the distance traveled? Can you devise a way to map out, experimentally, the parabolic path followed by your projectile?
2. Obtain an assortment of springs. Pick three very different springs and attach one end of each of the springs to a fixed surface so that they hang down. Put an identical mass on each spring, pull it down a little (against gravity), and let it go. Measure the period (how long it takes the mass to return to the same position) of oscillation for each of the springs. (To obtain a more accurate measurement of period, it might be easier to measure the time of 5 to 10 oscillations and then divide by the number of oscillations.) Are the periods the same or different for the different springs? Is there a correlation between the strength of a spring and the period of oscillation? Replace the mass with one that is different by a factor of 2 to 4, and repeat. Does the oscillation period change with mass?
3. Materials can be cut with band saws, table saws, milling machines, and wire- or laser-EDM machines. Through personal experience, and a little searching on the Internet, estimate the accuracy of these types of cuts.

CHAPTER 17
Design Styles

Menu

Before You Begin
Think about these questions as you study the concepts in this chapter:

 Why is the visual appearance or style of an object an important part of the design process?

 How do different styles get their names?

 Who gives style to structures and products?

 How can different styles be identified and classified?

 Who are some of the great architects and industrial designers of the last two centuries?

INTRODUCTION

Earlier in this text we talked about design as an iterative decision-making process that results in a plan to produce a new product. It was noted that the product must function as expected and must be of value to the consumer who will purchase it and the producer who will make it. Obviously the solution should minimize negative trade-offs and risks to users and the environment. To get consumers to purchase a product, designers also realize the product must look good and have visual appeal.

In this chapter, we will introduce design appearance. Appearance is subjective and is studied as a branch of philosophy called aesthetics. Aesthetics deals with the response to our visual sense in terms of making judgments about the things we see. These judgments usually result in expressions of like or dislike and are related to cultural, economic, political, and moral values. All products have visual appearance.

Appearance has shape or form, which is influenced by the object's size or scale, proportion, and symmetry. Appearance has texture, which is influenced by the surface smoothness, patterns, and material.

Appearance also has color, which is influenced by hue, tone, and contrast.

When the appearance of an object relates to a particular time period or is associated with the work of a particular person, the term style is often used. For example, the Victorian style is associated with design elements made popular during the reign of Queen Victoria (1837–1901).

Style is also used to describe design elements used in the design of structures and products. Architectural style is usually associated with the work of architects who give shape or form to residential or commercial buildings. Product style is usually associated with the work of industrial designers who give shape or form to a product such as a cell phone or automobile. Some designers such as Michael Graves (see Case Study) are both architects and industrial designers. Engineers use the design process to create functional components used in products and structures but rarely deal with aesthetics. Many talented professionals must work together to create a product that is functional, attractive, and valuable to both the producer and consumers or society.

Sometimes a style such as Victorian is associated with both products and architectural structures (see Figure 17-1).

(a)

(b)

Figure 17-1: Victorian style used in the design of (a) products (phone and table) and (b) architectural structure (house).

INTRODUCTION

Your Turn

Select a product and describe its design appearance using terms related to shape, texture, and color.

As noted in Chapter 1, the Industrial Revolution (IR) moved production from craft to industry. As products began to be produced in industry, the person creating the style or form of the object no longer made the object.

By the middle of the 19th century, mass markets for new industrial products were increasingly important. To reach a growing population of consumers, new printing technologies were used to create illustrated mail order catalogs. Quickly expanding transportation technologies, specifically the railroad industry, were used to transport these products throughout the United States. For new mail order businesses such as Montgomery Ward and Sears Roebuck to compete, the appearance or style of the product being illustrated was becoming an important marketing factor (see Figure 17-2).

Figure 17-2: The 1895 Montgomery Ward Catalogue advertised a large amount and variety of goods including tools, books, the new improved Singer Sewing Machine, side saddles, straight-edged razors, high-button shoes, and hunting rifles—some 25,000 items in all. This catalogue is an excellent illustrated record of the industrial products available in the late 19th Century

aesthetics:
Having to do with appearance; a branch of philosophy that deals with human response to visual stimuli leading to judgments about the things we see. These judgments usually result in expressions of like or dislike and are related to cultural, economic, political, and moral values.

ARCHITECTURAL DESIGN

Architects design residential and commercial structures. These designs require a team of design professionals. The architect is the head of the team and is responsible for working with the client to establish the scope of the project. Architects are concerned with the intended use of the project and must meet all code requirements. Codes, which are a form of standards, are established and enforced by local municipalities. Some of the issues facing architects in the 21st century are designing new urban communities that are:

- Sustainable/environmentally friendly
- Transformed for the increasing number of females in the workplace
- Support a multicultural population
- Safe

> **architect:** A professional who designs residential and commercial structures. Architects head a design team responsible for working with the client to establish the scope and style of the project. Architects are concerned with the intended use of the project and must meet all code requirements.

Most architects belong to the American Institute of Architects (AIA). The AIA was established in 1857, as the professional organization for architects in the United States. This organization has programs designed to introduce students to the profession and its rich history. You can learn how architects improve the life of a community by visiting AIA's website at http://www.aia.org.

Frank Lloyd Wright

Frank Lloyd Wright (1867–1959), best known for his "Prairie style," was one of the most prominent and productive architects of the 20th century (see Figure 17-3). He credits his mother for instilling in him an interest in architecture. As a child he spent playtime using building blocks and other educational material developed by the originator of the kindergarten, Friedrich Froebel. The blocks contained various geometric shapes that he credits as having influenced his later architectural designs. Wright began his professional career in 1890, with the famed Chicago architects Adler and Sullivan who believed in the design philosophy "form follows function," which Wright later refined to "form and function are one." He started his own firm in Chicago, which he later moved to his home that he designed in the Chicago suburb of Oak Park (see Figure 17-4). His Prairie style houses were designed in the early 1900s, and can be found in many cities around the United States.

Figure 17-3: Frank Lloyd Wright.

Wright created two architectural schools, Taliesin East in Wisconsin and Taliesin West in Scottsdale, Arizona, where he trained future architects in his design philosophy. Over his 70 years as a practicing architect, he designed over five hundred residential and commercial buildings. His design genius was sought after by the rich and famous, leading to work such as the Guggenheim Museum in New York City. His influence on architectural design is unequaled, and he is appropriately recognized as one of America's most famous architects.

Figure 17-4: Frank Lloyd Wright's home and studio in Oak Park, Illinois.

Figure 17-5: Fallingwater is one of Wright's best known designs. Completed in 1936, the house and particularly the cantilevered porches have recently undergone extensive renovation to preserve this landmark structure.

Fallingwater is considered one of Wright's architectural masterpieces of the 20th century. Designed in 1935 for Edgar J. Kaufmann Sr., a successful Pittsburgh businessman, Fallingwater presented some very unique design and engineering challenges. Originally, Kaufmann expected Wright to design a home overlooking the waterfall in the western Pennsylvania Allegheny Mountains, but Wright surprised him by designing a home that was actually partly over the waterfall (see Figure 17-5). The stunning design created some interesting engineering challenges. The home included extensive cantilevered balconies. Wright used a structural system of reinforced concrete. This system was relatively new for construction, and the balconies eventually began to sag.

In an attempt to save this architectural landmark, the **cantilevered** balconies were temporarily supported by a steel structure in the river. The original design was extensively studied, and the cantilevers were found to be inadequately reinforced. In 2002, the floors over the cantilevers were carefully removed and post tensioning was added (steel cables were stretched between both ends of the beams). The increased support for the cantilevers has allowed Fallingwater to be restored to its original beauty.

OFF-ROAD EXPLORATION

Learn more about the design and preservation of Fallingwater at www.fallingwater.org.

The design of the cantilevered balconies at Fallingwater requires the application of both science and math principles. The supporting structure was designed to withstand all dead (the weight of the building) and all live loads (the weight of furniture and people on the balconies). Because the Kaufmanns used Fallingwater to entertain, calculating the live load for the smaller (15 ft × 15 ft) first floor cantilevered terrace balcony would need to take into consideration how many people could fit in that space. What is the average weight of an adult? What could happen to the live load if the Kaufmanns had a band playing and people were moving to the music, creating a dynamic load? For more information, see http://www.paconserve.org/43/fallingwater.

Architectural History

Egyptian architects were the first to design great structures, including the pyramids (see Figure 17-6). One of the first great architects was an Egyptian chancellor, first in line under the pharaoh. The architect Imhotep (circa 2600 B.C.), also known as a doctor, poet, and astrologer, was responsible for building Egypt's first pyramids. Even in ancient times, the size of the structure indicated the importance of the individual who occupied it; therefore, the burial tombs of the pharaohs were larger than other pyramids. The Great Pyramid is the only original Seven Wonders of the Ancient World still remaining.

The Egyptian architects, known for their meticulous planning, established rules of proportion for building elements such as columns. Columns and other architectural features first appeared in Egypt (see Figure 17-7). Columns are especially important to commercial building, including those built during the Federal period in the United States. Today, we continue to recognize the importance of proportion and scale in building design.

Figure 17-6: Great Sphinx with pyramid in background.

classical style: Latin for *elite*, it represents the highest order of architecture established by the Roman architect Palladio. The term is used today to describe architectural design that is based on original Greek and Roman design elements.

Greek Architecture Greeks considered architecture to be the greatest form of the arts. Greek architecture built on the principles developed by the Egyptians. Greek architecture reached its zenith with the building of the Parthenon in Athens (448–432 B.C.) (see Figure 17-8). This high order of architecture became known as *classical*, which is Latin for *elite*. The term **classical style** is used today to describe architectural design that is based on original Greek design elements such as the pediment gables and elaborate entablatures shown on the Parthenon.

Figure 17-7: Early Egyptian column and open papyrus column capital.

Figure 17-8: Parthenon in Athens, Greece, represents classic architecture. The Parthenon is placed on the Acropolis, which gives it visual prominence. The Doric columns support an elaborate entablature between the column capital and roof. Note the pleasing overall building proportions.

Figure 17-9: The Pont du Gard aqueduct (circa 19 B.C.) is one of France's top tourist attractions and an example of superior Roman engineering understanding.

Roman Architecture Roman architects transformed Greek architecture with superior engineering, including new construction techniques such as the arch and new materials like concrete (see Figure 17-9). The Romans built elaborate civic structures such as the Forum, amphitheaters and stadiums for sporting events, roads and large aqueducts, triumphal arches to honor their military achievements, and mausoleums to honor the dead.

One famous Roman architect was Andrea Palladio (1508–1580). Palladio studied classical architecture through writings and measurements. Of his many designs, he is best known for his country houses or "villas." These structures were elevated with stairs leading to each side, and had a central dome and four porticoes with Ionic order columns and pediments, and elaborate entablatures. Palladio used a classical proportional system, (see Figure 17-10). Proportion, which is the relationship in size of one thing to another, is very important in design. For example, the columns in the villa would not look in proportion if the building were much smaller or much larger.

Figure 17-10: A Palladian-Villas in Veneto, Italy.

Figure 17-11: An architectural application of the classic Palladian window style.

Many people have studied Palladio's work. His classical design elements have influenced architectural style for over five hundred years. Today, many homes have Palladian windows as a major architectural element in the design (see Figure 17-11).

Thomas Jefferson used many elements of classical design in the building he designed for the University of Virginia and his Monticello plantation. The White House, designed by James Hoban, also reflects Jefferson's study of Palladian principles (see Figure 17-12).

Figure 17-12: **The White House, 1700 Pennsylvania Avenue, Washington, DC.**

What Gives a Building Style?

We usually think of a building as having four walls containing interior rooms, exterior and interior doors, windows, and a roof for protection. Although these elements are necessary to create a structure, they are also used by architects to give a structure its distinctive look or style. When evaluating a structure for style or when we are interested in designing a new structure of a given style, we need to consider architectural form, texture, and color. The following elements help describe the anatomy of a house, which is very much influenced by technological development (engineering and industrial production):

- Floor plan (number and placement of rooms and hallways)
- Elevations (number of floors and overall shape and proportion)
- Shape of roof (shape and placement of any chimneys)
- Shape of the eaves (these are the elements between the roof and the exterior walls)
- Shape of porches and entranceways
- Shape and placement of windows and doors
- Exterior walls (materials, patterns, and finishes)

Floor Plan Most indigenous housing consisted of one room with four walls made of natural materials, a door opening, and a roof. As technological capability increased, window openings were added. To add protection, skins were used to cover window and door openings. As people became more prosperous, additional rooms and hallways were added, and chimneys were needed if fireplaces were used for heating. New building materials, such as dimensional lumber that included manufactured 2 × 4s, 2 × 6s, and so on, became more available to designers and builders. Dimensional lumber is measured in inches in its nominal size (a 2 × 4 is actually 1⅝ inches by 3½ inches). New building techniques, such as balloon framing commonly used in today's homes, allowed the design of larger homes with multiple floors. Balloon framing that uses dimensional lumber replaced post and beam construction used in barns and Colonial era homes. Today, floor plans can be very complex and typically include garages, which became common in the 20th century (see Figure 17-13).

478 Part II: Resources for Engineering Design

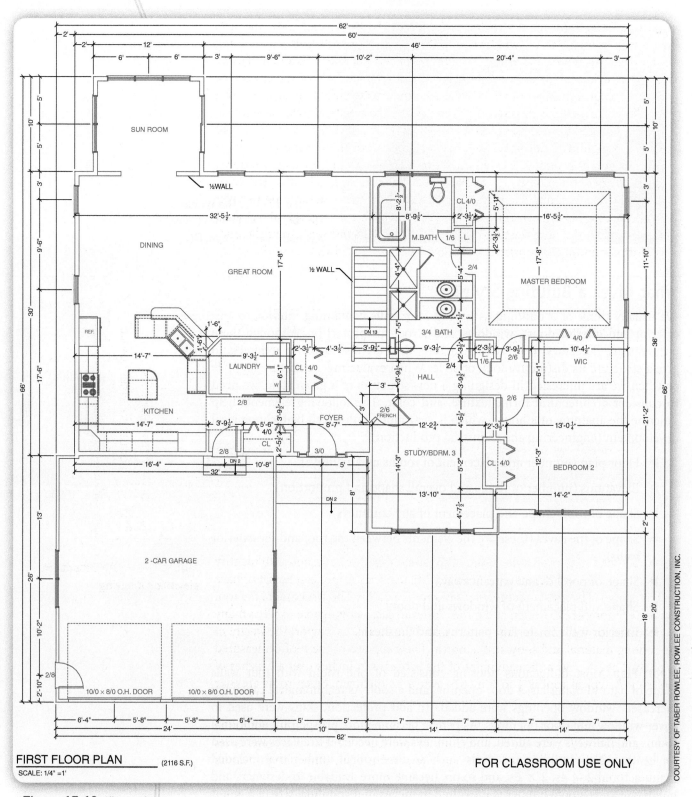

Figure 17-13 Floor plan.

Elevations Elevations are drawn for each side of the house and show the size, style, and placement of all windows and doors, exterior wall finishes, molding around windows, doors, and entablature (the area between the walls and roof), any porches and columns (posts), and some information about the roof and chimneys.

When an equal number of windows and chimneys are placed on each side of a central door, the design is considered symmetrical. Many classical styles have strong symmetry. In addition to the placement of windows, elevations will show wall finishes and moldings. Entrances can be just a plain door or an elaborate portico or covered porch. Molding is often added to a front entrance or over windows. A pediment is a low-pitched triangular gable structure over a front entrance or molding above a window. The elevation will also show the material used to cover the wall. In colonial America, materials readily available in a geographical area, such as wood in the northeast and adobe brick in the southwest, were the ones used most commonly to construct buildings (see Figure 17-14).

Figure 17-14 Architectural elevation drawing.

Shape of Roof The roof style is one of the easiest features of a house to identify and is one of the most dominating architectural features. A roof can be flat or have a pitch (slope). The most common style of roof is a gable. The structure of the roof can be very complex to design and build. For example, a cross-gable roof has many complex angles at each ridge and valley, and must be able to support the weight of the roofing material and snow in the north. The steepness of the roof is measured as pitch. The pitch is a measurement of the roof rise in inches over 12 inches of roof run. A typical roof is 4 to 6 inches over a 12-inch run. Steep roofs are 6 or more inches for each 12 inches of roof run (see Figure 17-15).

Shape of the Eaves The eave of the building is the connection of the roof to the walls. In most structures the roof extends beyond the exterior wall. Some eaves extend over the exterior of the wall by only a few inches while other designs feature large and elaborate eaves. In classical structures this area is called the entablature. The entablature consists of three major horizontal elements supported by columns. The lowest element is the architrave, which serves as the beam running from column to column. The central element or frieze is plain or highly ornamented. The frieze of the Building Museum in Washington, DC, depicts a series of Civil War relief images. The top element or cornice usually consists of molding just below the roof edge. In classical architecture, the entablature is supported by Doric, Ionic, or Corinthian order columns (see Figure 17-16).

480 Part II: Resources for Engineering Design

Figure 17-15: *Common roof styles.*

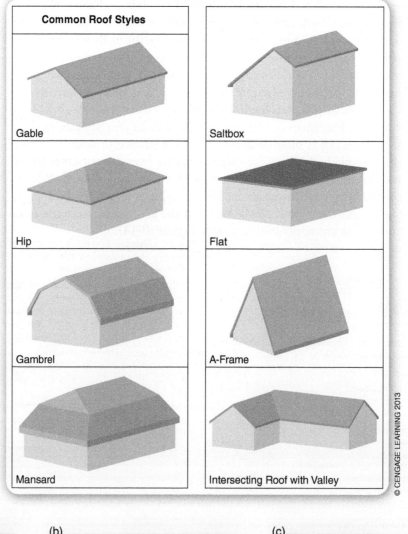

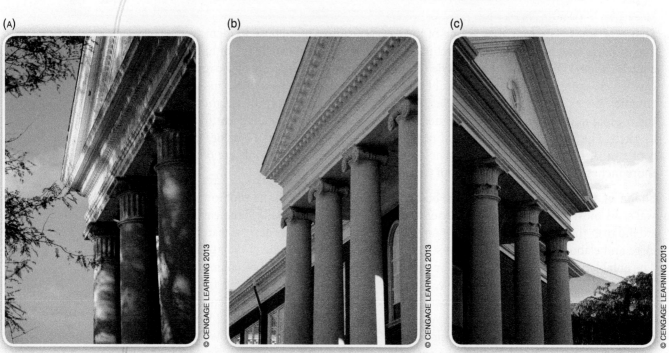

Figure 17-16: *Classic entablature with (a) Doric, (b) Ionic, and (c) Corinthian columns.*

Window and Door Styles Windows are architectural elements designed to fit into exterior wall openings. A variety of materials are used to construct the frame of a window, which holds a glass pane. Windows can be fixed or open for ventilation (see Figure 17-17).

(a)

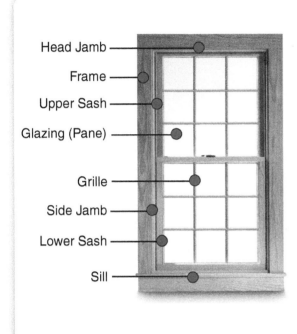

Head Jamb – The window frame members which compose the top, sides and bottom (sill) of a unit.

Frame – Outside member of window unit which encloses the sash, composed of side jambs, head jamb and sill.

Upper Sash – The framework holding a glass in the window unit. Composed of stiles (sides) and rails (top and bottom).

Glazing (Pane) – The glass panes or lights in the sash of a window. Also the act of installing lights of glass in a window sash.

Grille – Ornamental or simulated muntins and bars which don't actually divide the lights of glass. Generally made of plastic or wood, they fit on the inside of the sash against the glass surface for easy removal. They can also be secured to the exterior of the sash.

Side Jamb – The window frame members which compose the top, sides and bottom (sill) of a unit.

Lower Sash – The framework holding a glass in the window unit. Composed of stiles (sides) and rails (top and bottom).

Sill – Horizontal member that forms the bottom of a window frame.

(b)

Insulating Glass (IG) – A combination of two or more panes of glass with a hermetically sealed air space between them. This spaces may or may not be filled with an inert gas. IG with a special low emissions coating to restrict the flow of radiant heat is called Low-E insulating glass.

Cladding – A material secured to exterior or interior faces creating a more durable, low-maintenance surface.

Extension Jamb Groove – Flat wood parts which are fastened to the inside edges of the window jamb to extend it in width and adapt to a thicker wall. The inside edge of extension jambs should be flush with the finished wall surface; the inside casing is then secured to it.

Weatherstripping – Metal, plastic or felt strips designed to seal between a window sash and frame or stops to prevent air and water leakage.

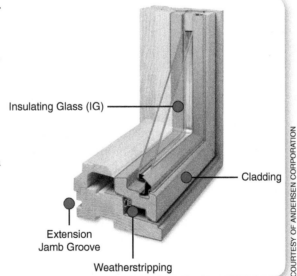

Figure 17-17: *Anatomy of a Window.*

> Glass is formed when certain types of rock are heated to high temperatures. Early human-made glass beads dating as early as 3500 B.C. have been found in Egypt. The Romans were the first to use glass for architectural purposes. They made a form of clear glass using manganese oxide. The glass was cast into sheets and used in windows that appeared in the elite structures in Western Europe and the Mediterranean region around 100 A.D. Glassmaking became an important industry during the Industrial Revolution. New discoveries and production methods improved the optical and thermal properties of glass. Glass for windows became available in larger sizes and quantities and at reduced prices. Plate glass was formed by pouring molten glass on a large table. The glass was cooled and polished on both sides. New technological developments in the early 20th century allowed for continuous glass sheets to be produced. Windows and doors are rated for energy efficiency using a U-factor, which measures thermal transmission or heat loss of the window. U-factor values typically range from 0.25 to 1.25 and are measured in $(Btu/h)(ft^2)(°F)$. The lower the U-factor, the better the window insulates. Additionally, today low-emittance (Low-E) glass is also used. Low-E glass is coated to suppress the flow of radiant heat transfer from the warm pane to the cooler pane. These coatings save energy by keeping radiant heat inside the house in the winter and outside the house in the summer. Glass for doors is also made shatterproof to protect people from being cut by broken glass.

Windows are manufactured in a variety of standard styles, materials, and sizes. Fixed windows are used in both residential and commercial construction and are usually specified for very large openings or where ventilation is not needed. In residential structures, most windows are designed to open. Windows may be designed to slide up and down, side to side, or tilt (hinge) outward or inward.

The oldest operating window style is the **double-hung window**. This window style has an upper and lower sash that slides in the frame. Original double-hung windows contained many smaller panes of glass held in the frame by vertical and horizontal bars called **muntins**. Today, muntins are often simulated by a grill that is inserted into large window frames to give the appearance of a 6 over 6 or 8 over 8 divided windows. Double-hung windows use a spring or weight (older style) to hold the sash in place when open. Sliding windows are similar to double-hung windows, but move sideways.

Casement windows operate by a mechanism holding the sash on the top and bottom. When the mechanism is cranked, the window opens typically outward on a pair of hinges. Awning windows are similar to casements but open on hinges placed on the top of the sash. **Jalousie** and **Hopper** are variations of the casement design.

Individual fixed or operative windows can be combined to form large and dramatic window units. Examples of multiple window units include the bay and bow window often found in residential designs. The vertical or horizontal joint between individual window units is called a mullion. The **Palladian window** combines three or more fixed or double-hung windows with an arched window over the middle unit. Seamless bent-glass corner windows or full-wall windows are newer products that offer unique unobstructed views. Glass blocks are used to cover openings in interior and exterior wall openings. Glass blocks are manufactured in a variety of patterns for varying degrees of privacy (see Figure 17-18).

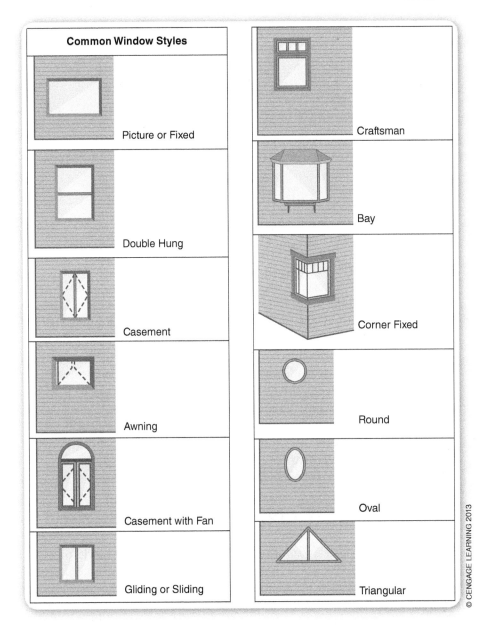

Figure 17-18: *Common window styles.*

Exterior Walls Historically, indigenous materials such as logs or stone were used to construct exterior walls. In addition to these naturally occurring materials, one of the earliest examples of material design was the making of brick. Materials such as wood, stone, or brick can be used to make a complete structural wall or be used to cover the exterior of a structural wall.

Dating back over 10,000 years in the Middle East, people mixed mud or clay with water and other materials such as straw and sand. The mixture was placed in a form or mold and sun-dried. The Romans were the first to employ kilns (a type of furnace) to dry and harden the brick. Roman brick structures including aqueducts can still be seen throughout Italy. Bricks were used extensively in England during the Industrial Revolution. After the Chicago fire of 1871, new skyscraper designs could not be built with brick or masonry. The Monadnock Building in Chicago, which opened in 1896, remains one of the tallest masonry buildings at 17 stories. Because of the weight of masonry construction, the ground floor of the building needed to be six feet thick to support the weight of the upper floors. Today bricks are manufactured in large quantities and in many colors and sizes. Most bricks

Your Turn

Working with your teacher, identify an appropriate house to collect the following data: Sketch and dimension a floor plan, sketch the front and side elevations including the roof, estimate the roof pitch (from the ground), identify the type of windows and exterior wall material, and include any special features such as elaborate eaves, columns, or porches.

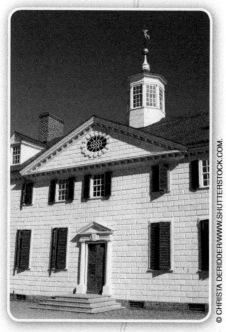

Figure 17-19: Washington's Mount Vernon with wood exterior that was made to look like stone.

have a red hue, but higher kiln temperatures result in darker purple and brown colors. The common brick in the United States is 8 inches long by 4 inches wide by 2¼ inches thick.

Exterior wall material is usually fairly obvious, but sometimes a material such as wood is made to look like another material, such as stone. A well-known example is Washington's Mount Vernon (see Figure 17-19).

Often less expensive materials are substituted to save money and give the appearance of a more elegant house. For example, in the Georgian style, wood siding was made to look like stone, especially on corners where quoins were employed. Today, stucco material can be made to look like brick or stone. Vinyl siding is often made to look like traditional clapboard or cedar shingles. A careful examination of the exterior wall will usually reveal the real material. Architects will use various materials and finishes to add color and texture to the exterior walls of the structure.

Today, a wide variety of exterior wall materials are used.

ARCHITECTURAL STYLES

Some confusion can occur when discussing architectural style of buildings. Although all structures have windows, doors, and other architectural elements, most structures are not designed with architectural style as the primary concern. Structures built by individuals or local craftspeople to provide basic shelter with indigenous materials are called "folk houses." In contrast, "classical architectural styles" are associated with a desire to make a fashion statement and reflect a high standard of living.

The Historic American Buildings Survey (HABS), administered by the U.S. Department of the Interior, has documented drawings, photographs, and historical information about 37,000 structures and sites in the United States and its territories. The American Institute of Architects cooperated with the Library of Congress and the National Park Service in forming HABS, see www.loc.gov/pictures/collection/hh/. The site shows excellent examples of periods and styles from the 17th through the 20th centuries of residential, commercial, public, monumental, religious, military, and industrial structures.

Major architectural styles are typically named for historical periods, influential people, or art movements. The Colonial style (1600–1800) was named for the colonial period in American history. The Victorian style (1860–1900) was named for Queen Victoria, a popular monarch who reigned over the British Empire from 1837 to 1901. The Arts and Crafts style (1880–1910) was named for an art movement of the late 19th century. Excellent examples of original architectural styles are maintained by individuals or historic preservation societies.

When a new structure is designed using most of the classic elements associated with a particular design, the term "neo" or "revival" is used. These neo-styles differ primarily from the original in the construction methods and materials used to build the house. House construction was greatly impacted by the increased availability of construction materials that could be moved more easily and more cheaply by the growing railroad industry during the second half of the 19th century. Curators will look for examples of handmade materials associated with the original period versus machine-made elements associated with modern construction to determine if the building is an original or revival. The term *eclectic* is used to describe an architectural style that incorporates a mix of design elements from various recognized styles.

Colonial Style (1600–1800)

Early American homes reflected the style and building practices of people's native countries. This cultural influence can be seen in Colonial style homes where English, Dutch, French, and Spanish style elements were used. Because most colonists were of English heritage, the English Colonial style was most prevalent. Early Colonial homes were simple, one room deep structures with an end gable roof and small eaves. Chimneys were often a dominant feature of the houses. Windows were relatively small with diamond shaped panes of glass in fixed casements or small, operative, double-hung style. These houses reflected the culture, economy, and materials available in the geographic region of the time.

In New England, clapboard sheathing on a wood frame was typical. In Pennsylvania, field or quarried stone was often used. Brick, which was more expensive, was used in Philadelphia and Virginia. Dutch Colonial homes often had a flared eave (arc) and parapet wall that extended above the roof and a divided (top and bottom) front door. French Colonial homes were characterized by paired French doors and windows, shutters, and a hipped roof. The French Quarter in New Orleans has many examples of the French Colonial style. The Spanish Colonial influence is seen in homes with thick masonry walls, low or flat pitched roofs, few window openings, and rooms opening to a central courtyard. Many of these can be seen in historic districts in the American Southwest.

Figure 17-20: Colonial house in historic Williamsburg, Virginia.

Early Colonial homes often consisted of one large room with a loft for sleeping or two large rooms with a central fireplace. Classic Colonial homes similar to what is seen today in the Colonial Revival style consists of a well-developed style of at least four rooms placed around a central hallway and stairs and multiple fireplaces and chimneys. Windows are double-hung 9 over 9 windowpanes with shutters. Gable roof with dormer windows are also common (see Figure 17-20).

Georgian Style (1700–1780)

The Georgian style reflected the increasing wealth in the American colonies. The improved economic conditions allowed designers to move beyond essentials, including an increasing interest in art, science, and technology. This

Figure 17-21: Historic Georgian house.

Figure 17-22: *Federal style house.*

style reflected the classical work of Palladio and was influenced by English architects working during the reign of the first three British King Georges. The style has an elaborate, raised, central pediment entrance with pilasters (flattened columns) and double-hung 12 over 12 windowpanes separated by muntins with elaborate pediment molding. Palladium windows above the door and rooftop balustrades were common but are not shown in Figure 17-21. Wood, brick, or stucco masonry was used on exterior walls with quoins on the front corner. Georgian structures are symmetrical with identical window and chimney elements, a hipped roof with an elaborate dentil (toothlike) or eave molding (see Figure 17-21).

Federal Style (1780–1840)

The Federal style reflects the cultural independence of the new United States from England. The style is also called "Adamesque" for the influential Scottish-born architect Robert Adam who worked during the American Revolution period in Scotland. While having some characteristics of the Georgian style, the Federal style created by architects Charles Bulfinch in Boston and Benjamin Henry Latrobe in Philadelphia had refined proportions and was more vertical than Georgian buildings (see Figure 17-22). Refinements included more moldings that were lighter and more elegant. The style used oval or elliptical central stairs with domed or arched ceilings. Porticoes were often oval with triple sash windows starting at the floor level. The style used graduated windows with the windows getting progressively smaller from first floor to the top floor. Front doors had fanlights and recessed arches.

Figure 17-23: *Greek Revival house.*

Greek Revival Style (1825–1860)

Greek and Gothic Revival and Italianate are part of the Romantic period in architectural style. Greek Revival was made popular by interest in Greek democratic ideals. It has a dominant appearance and is considered a high classical style (see Figure 17-23). The strength of the visual appeal includes full-length pilasters and a multistory portico. The entrance is a dominant feature, often containing a gable pediment with Ionic columns. The front door is surrounded by narrow sidelights and/or transom lights above the door incorporated with elaborate molding. Double-hung windows with six-pane glazing were common but molding surrounds were less elaborate than what was used around the front door. This style usually has a low-pitch gable or hipped roof with a wide frieze panel as part of the entablature. Stucco masonry walls were common but wood and brick were also used in various geographical regions.

Victorian Style (1860–1900)

The Victorian period was named for Queen Victoria, the popular British queen who reigned from 1837 until 1901. Her long reign was characterized by political and social reform and a strengthening of English politics and expansion of the British Empire. Her youthful energy and enthusiasm was very popular in England and was generally reflected in the personal tastes of the middle class.

In the United States, the Victorian era was embraced by the exploding population, which was experiencing increasing wealth from the industrial economy. The hard-working American middle class was looking for new homes that reflected their newfound success and enthusiasm. Improved building techniques including the flexibility of balloon framing, which allowed designers to include towers and turrets, bays and overhangs, and irregular floor plans. New utilities included indoor plumbing and gaslights.

Improvements in tools, such as the scroll saw, allowed for more complex architectural molding detail. Millwork factories created spindles, molding, columns, brackets, and other architectural details at reasonable costs for designers and builders.

Queen Anne and Second Empire are two of the more popular Victorian period house styles. Other more regional Victorian styles include Stick, Richardsonian, and Shingle. The Queen Anne style is asymmetrical with a steep-pitch front gable or front-to-back hip roof. The style has a large one-story wrap-around porch with decorative spindle between each supporting post, gingerbread millwork at the eaves, and a corner tower. Distinctive towers are common in all Victorian styles. A central tower and mansard roof are most common in the Second Empire style. Windows and doors have large single panes of glass and simple surrounds. Exterior walls are typically covered with shingles and can be painted bright Victorian colors. Many wall inserts and cutaways under windows add to the complexity of the house design (see Figure 17-24).

Figure 17-24: Victorian style house.

Other new industries began producing new building materials, including pressed brick and plate glass. Paint companies created the vibrant colors for the Victorian styles and improving railroad transportation systems allowed for construction materials to be distributed throughout the expanding territories at affordable costs. Newly published house plans and pattern books inspired a new generation of middle-class homeowners (see Figure 17-25).

Figure 17-25: The Victorian "Painted Ladies" in San Francisco's Alamo Square.

Twentieth-Century Styles

Architectural design in the 20th century reflected both new styles and a nostalgic return of 19th century styles. The century began with excitement for a growing industrial economy. Consumers were encouraged to purchase new products from "chain" stores or national mail-order catalogs, buy on "credit," go to a movie, or even consider buying a car from one of the new automobile companies such as the Ford Motor Company.

The Industrial Age, while providing many new products, brought with it nostalgia for a simpler lifestyle. Most people were not Luddites in the sense that they wanted to destroy the new industries mentioned in Chapter 1, but they were increasingly interested in handcrafted products and pleasing art forms such as Japanese art. In architecture, these interests were reflected through the Prairie and Craftsman styles.

Prairie Style (1900–1920)

The Prairie style is credited to a creative group of Chicago architects, of which Frank Lloyd Wright is considered the master. The style, developed between 1900 and 1917, was uniquely American and was intended to complement the surrounding landscape. Wright used the term *organic* to describe how a building should fit into rather than stand out from the environment, a design principle he established. His designs were characterized by an open floor plan with low profiles, horizontal lines, large overhangs and terraces, uninterrupted walls of windows, clerestory and stained glass windows, and natural building materials, including extensive use of stone and unpainted wood (see Figure 17-26).

Figure 17-26: The Frederick C. Robie House in Chicago is widely considered Frank Lloyd Wright's finest example of the Prairie style. It was built in 1909.

Craftsman Style (1905–1930)

The Craftsman style was based on the design principles of the Arts and Crafts movement. The Arts and Crafts movement was a rejection of the elaborate designs and industrialization associated with the Victorian period. Beginning in England with William Morris, the movement focused on designing and producing products in a craft setting. Production relied on the creativity of individual craftspeople who produced simple yet beautiful handcrafted items for the middle class. Unfortunately, the wages of the craftspeople made the products expensive so they were only available to the wealthy. In the United States, the style was inspired by the use of beautiful natural wood grains, simple yet elegant graphic patterns, ceramic tile, and a strong emphasis on rich earth tones. While individual craft was incorporated, the use of machines in an industrial setting allowed the products to be affordable to the growing American middle class. Some important individuals associated with the movements in the United States were Gustav Stickley (furniture manufacturer and publisher of *The Craftsman* magazine), Elbert Hubbard (founder of the Roycroft Community), Louis Comfort Tiffany (a stained glass artist, potter, and jeweler), and Dard Hunter (ceramic artist and graphic designer).

The Craftsman style house originated in southern California and was inspired by the work of Pasadena architects Charles and Henry Greene. Originally, the brothers designed simple Craftsman bungalows. A bungalow is a small house that can be built in any style, but the names Craftsman and Bungalow are often interchanged. The style was characterized by exposed roof rafters at the eaves, low-pitched roofs, front porches with heavy columns, extensive use of wood trim, decorative ceramic tile, and earth tone colors. The style was largely spread by popular magazines' published plans and pattern books and was relatively inexpensive to build (see Figure 17-27). Some companies offered complete packages of pre-cut lumber and assembly plans.

Although most Craftsman houses were relatively small, some notable exceptions exist (see Figure 17-28). New Craftsman-inspired homes are larger than the

Figure 17-27: Craftsman style home and illustration of a mail-order house plan.

Figure 17-28: *Greene and Greene "Gamble House" in Pasadena California*

originals, but the impact of natural materials, earth tone colors, and extensive use of wood throughout the home create a calming, comforting, and informal environment for owners, just like the intent of the originals.

Art Deco (1920–1940)

In Paris, France, in 1925, an exhibition was held called the "International Exposition for Decorative Arts and Modern Industry." At this exposition, a number of well-known designers unveiled their latest work. Even though all designers have their own style, the influence of new materials and industrial techniques were clear. For instance, a new material called plastic was introduced around this time, as were neon lights.

The Art Deco style is characterized by strong stepped forms, sweeping curves, and geometric shapes using machine-made materials such as glass, aluminum, stainless steel, lacquer, and inlaid woods. Art Deco is a style that was seen as elegant, functional, and modern, with the design principles applied to buildings, aircraft, trains, cars and ocean liners, radios, clocks, and furniture (see Figure 17-29).

Figure 17-29: *Radio City Music Hall and the Chrysler Building in New York City and Norman Bel Geddes's "Patriot" Radio (1939).*

International Style (1925–present)

International style in architecture emerged in Europe and the United States during the 1920s. Identifying features included flat roofs, unornamented wall surfaces, and metal windows set flush with stucco or glass walls. Designers and architects, such as Ludwig Mies van der Rohe (Seagram Building), Philip Johnson (Farnsworth House), Le Corbusier (United Nations Headquarters), and Walter Gropius (Bauhaus School) believed that buildings should be functional. Commercial buildings were designed with an emphasis on using modern materials such as exposed structural steel and plate glass, forming curtain walls to reduce building mass. Pure geometrical forms were used to create simple solutions with an emphasis on clutter-free open space. Le Corbusier used the phrase "a house is a machine for living" (see Figure 17-30).

Many designers of the modern period were also minimalists. **Minimalism** is a term used to describe product and architectural style where the design is reduced to only its necessary elements. Minimalism was influenced by the Bauhaus and De Stijl movements and the work of individuals such as architect Ludwig Mies van der Rohe and designer Richard Buckminster Fuller of the early 20th century. Minimalism is built on the principle of "less is more," which influences the aesthetics of the design but also has environmental impacts.

One of the most interesting minimalist designers was Richard Buckminster "Bucky" Fuller who is known for the geodesic dome. Fuller was an American engineer, author, designer, inventor, and well-known futurist, publishing over 30 books and popularizing terms such as *Spaceship Earth* and *synergetics*. One of his more interesting designs was the Dymaxion car (1933), which could transport 11 people, had a top recorded speed of 90 miles per hour and a fuel efficiency of over 30 miles per gallon. As an environmental activist, he believed that good design was the best way to manage human and material resources. His "less is more" design philosophy led to mass-produced products using the simplest and most sustainable materials and means possible. His work focused on seeking long-term, technology-driven solutions to world problems in architecture and transportation (see Figure 17-31). Despite his early innovations, it was not until the international success of his 1967 Montreal Expo geodesic domes that he gained recognition for all his work. There are more than 500,000 geodesic domes worldwide, including the Spaceship Earth at Epcot in Disney World.

Figure 17-30: The United Nations Headquarters on the East River is an easily recognizable modern-style skyscraper in Manhattan, New York. Designed by Le Corbusier and others, the 39-story building completed in 1953 is made of reinforced concrete with a glass curtain wall. It was designed to accommodate 3,400 employees.

Modern Style (1945–present)

Most Americans live in a modern style house. These homes are part of a post-WWII eclectic movement that borrowed design details from earlier 20th century Arts and Crafts movement or earlier classic styles. New ranch, split-level, and contemporary homes filled the new American suburban landscape (see Figure 17-32).

Revival styles such as Neo-Georgian, Spanish Colonial, Dutch Colonial, and English Tudor were also very popular during this time period. Some early revival styles were built between WWI and WWII (see Figure 17-33).

By far, the most popular style in America was Colonial Revival, with variations built extensively throughout the 20th century (see Figure 17-34).

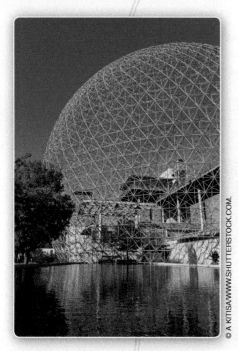

Figure 17-31: Buckminster Fuller's Montreal Biosphere (1967).

Chapter 17: Design Styles 491

Figure 17-32: Modern style ranch house.

Figure 17-33: English Tudor style house.

Figure 17-34: The American Colonial Revival house.

POSTMODERN ARCHITECTS

Postmodernism is a design movement that began in the late 20th century as a reaction against the creative limitations of modernism and scientific objectivity. Postmodern designers look for multiple solutions with cultural and historical contexts. Design solutions are based on experience rather than abstract principles and are focused on the needs of individuals. Postmodern designers are free to add surface ornament with historical or cultural decorative form and context.

Each year since 1979, the Hyatt Foundation has awarded the Pritzker Architecture Prize to recognize one of the world's premier architects. Architects from the United States, Mexico, United Kingdom, Ireland, Austria, Germany, Japan, Italy, Portugal, France, Spain, Norway, the Netherlands, Switzerland, Australia, Denmark, and Brazil have received the $100,000 prize. One of the most famous awardees is Ieoh Ming Pei, a Chinese American architect. Pei is considered a postmodernism architect. One of Pei's most recognized works is his 1989 Louvre Pyramid in Paris, France (see Figure 17-35).

postmodernism: A design movement beginning in the late 20th century reacting against the creative limitations of modernism and scientific objectivity.

Figure 17-35: Pei's Louvre Pyramid, which he designed with Peter Rice.

Case Study

Michael Graves, Postmodern Architect, Industrial Designer, and Professor

If you know the name Michael Graves, it is probably because of his product design work for Target stores. But product design for the mass market is only a more recent endeavor of Graves, who is a Princeton professor with an international reputation and extensive record of accomplishments. He is recognized as one of America's most accomplished contemporary architects and an industrial designer (see Figure 17-36). He is the principal owner of two design firms, Michael Graves & Associates (MGA), which provides planning, architecture, and interior design services, and Michael Graves Design Group (MGDG), which specializes in product design, graphics, and branding. His firms have done design work for over 350 buildings worldwide and brought to market over 2,000 products for clients such as Target, Alessi, Stryker, and Disney. Michael Graves and the firms have received over 200 awards for design excellence (see Figure 17-37).

Figure 17-36: *Michael Graves, FAIA.*

In architecture, his postmodern style incorporates changes in scale to accentuate understated components of his buildings. Geometric shapes are common in his designs, and at times, different shapes are included in one building complex. Graves typically uses a small palette of colors such as light "French" blue, dark blue, oranges, yellows, reds, greens, and grays. Materials include metals, woods, stone, brick, plaster, fabric, and marble.

The Graves-Target partnership started under interesting circumstances. In 1996, Target donated $100,000 to preserve the historic Virginia farm believed to be the site of the cherry tree a young George Washington could not lie about. In addition to saving the site, Target agreed to fund the scaffolding needed for the renovation of the Washington Monument. Given that scaffolding is generally not attractive, Target turned to Michael Graves for help.

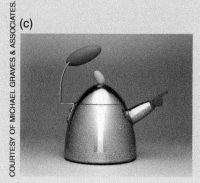

Figure 17-37: *(a) Walt Disney World Hotel, Lake Buena Vista, Florida; (b) United States Embassy Compound, Seoul, South Korea; and (c) the Teakettle for Target are some examples of the over 350 architectural and 2,000 industrial design products designed by Michael Graves & Associates and Michael Graves Design Group.*

Case Study

(continued)

Before his work with Target, Graves did mostly architectural work and some high-end product design. The well recognized Coach's whistle kettle was part of the first run of products that also included kitchen gadgets, countertop appliances, garden furniture, clocks, frames, and candleholders (see Figure 17-38). After five years, his design firm had designed over 800 products for Target's home and garden departments. Initially, Graves was not sure how this new venture would be taken but nearly everyone loved the new product line and the affordable price. Quality is important to Graves, and as he noted, it is just as difficult to design an affordable product as an expensive one, sometimes even more difficult.

In 2003, Michael Graves became paralyzed as a result of a bacterial infection. As he lay near death in the hospital, he recalls thinking that he didn't want to die in this ugly place. Being confined to a wheelchair has given the designer a new appreciation for accessibility, functionality, and style in the design of health care products. His Active Living Collection under MGDC works to improve products for people with all forms of illnesses and handicaps (see Figure 17-39).

The general public views accessibility issues covered in the Americans with Disabilities Act as relatively unimportant. To physically challenged people, accessibility is essential for everyday life. Graves notes that when he incorporates universal design principles in any design, the design benefits not only the physically challenged but also all human users (see Chapter 15 Human Factors in Design and Engineering). In 2006, he completed the St. Coletta School in Washington, DC, which serves 260 children with various disabilities. He notes that many of his design solutions will be of benefit for everyone (see Figure 17-40).

Michael Graves wants to make good design available to the widest possible audience. Now that audience includes individuals with all forms of physical challenges. He considers himself to be not limited to the postmodern style but to "figurative form" that allows both the traditional and familiar language of form and color to be part of his work. His work is

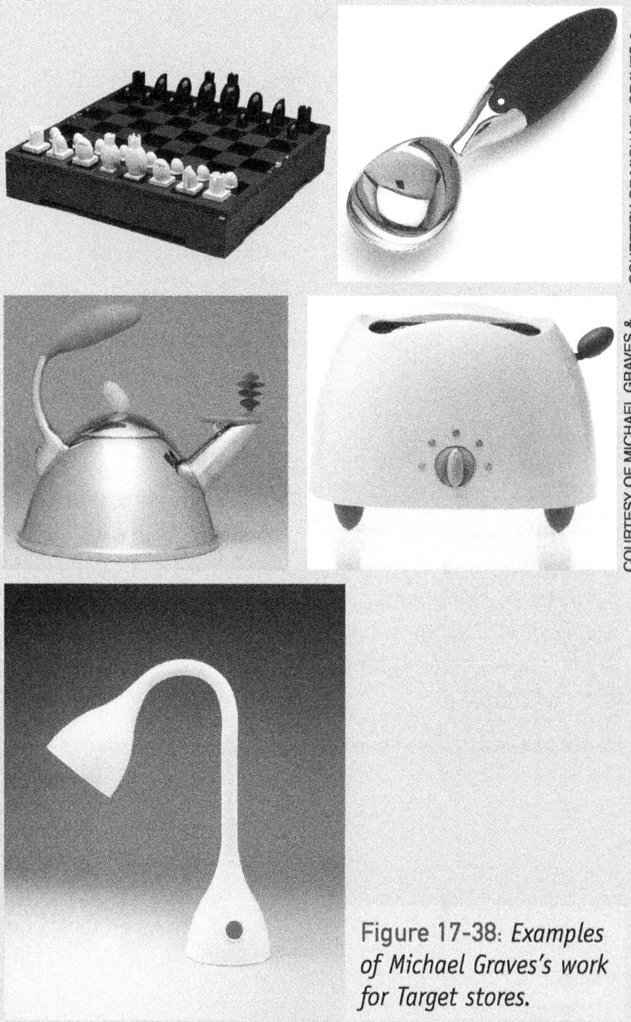

COURTESY OF MICHAEL GRAVES & ASSOCIATES.

Figure 17-38: *Examples of Michael Graves's work for Target stores.*

COURTESY OF MICHAEL GRAVES & ASSOCIATES.

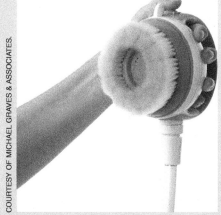

Figure 17-39: *The Michael Graves Active Living Collection is designed to respond to users' health care needs.*

(continued)

Case Study (continued)

Figure 17-40: St. Coletta of Greater Washington, DC, is a special education charter school that serves children with severe and multiple disabilities. The nearly 100,000 square-foot building is organized as a series of two-story schoolhouses and includes a full-court gymnasium and a community room.

witty, playful, comforting, and always fits human users. He has received over 200 awards, including the 2001 Gold Medal of the American Institute of Architects, the highest award that it bestows upon individual architects. He also received the 1999 National Medal of Arts from President Bill Clinton. In 2010, he received the prestigious Topaz Medallion, a joint award of the AIA and the Association of Collegiate Schools of Architecture for his illustrious 39 year teaching career at Princeton University.

Graves would love to walk again, but he is using his experiences as an opportunity to learn and design new products that will make life more comfortable and enjoyable for all.

Some designers have done both architectural and industrial design. Most designers, however, specialize in either architecture or industrial design. Frank Lloyd Wright, while being recognized for his architectural creativity, often designed furniture and even clothing for the people for which he designed houses. Unfortunately, the furniture and clothing were not always comfortable. One noted architect who made a successful transition to industrial design is Michael Graves. His achievements are presented in a case study in this chapter.

Your Turn

Look up one of the Pritzker Architecture Prize winners and learn more about his or her award-winning design. Why do you think the person received the award?

industrial designer: A professional who designs products and systems. Industrial designers work to create designs that optimize the function, value, and appearance of products and systems for the mutual benefit of the consumer and the manufacturer.

INDUSTRIAL DESIGN

Industrial designers create designs for products and systems. Like architects, they create or use styles and they work with other design professionals, especially engineers. For example, your next cell phone will have been designed by industrial designers who will have given it form and contributed to the function, while an electrical engineer would have developed the improved circuit to enhance voice clarity and reception.

Chapter 17: Design Styles   495

Figure 17-41: Apple iPod nano Armband and BMW.

The Industrial Designers Society of America (IDSA) states that industrial design (ID) is the professional service of creating and developing concepts and specifications that optimize the function, value, and appearance of products and systems for the mutual benefit of both user and manufacturer.

Industrial designers work to improve product function by applying human factors and other principles. They select or design materials that are appropriate for the product. These materials will contribute to the product appeal. They improve appearance by creating an effective product form or style, and by adding texture and color. By improving function and appearance, consumers will want to purchase the product and manufacturers will have to produce more products to meet consumer demand. If industrial designers do their work well, then the value of the product will be increased. Although we don't think of products as works of art, the iPod or BMW certainly are beautiful to many people (see Figure 17-41).

Industrial design began to take shape in the 1930s through the work of Raymond Loewy, Henry Dreyfuss, Walter Teague, and other people creating the new products of the early 20th century. ID was born in the postwar economic depression. Businesses wanted to increase consumer interest in their products, and industrial designers proved to be very successful in creating product forms that consumers liked.

Raymond Loewy, the Father of Industrial Design

Raymond Loewy (1893–1986) is considered the "father of industrial design." Born in France, he began designing products for American businesses in 1929, and successfully promoted his New York based design firm. He began his career as a fashion illustrator but soon turned his creative talents to product design. Over his career as an ID, he consulted for more than 200 companies, creating products from postage stamps to spacecraft. He believed that, "Between two products equal in price, function and quality, the one with the most attractive exterior will win" (see Figure 17-42).

Figure 17-42: Raymond Loewy, Time Magazine "Man of the Year", 1949

RAYMOND LOEWY™ IS A TRADEMARK OF LOEWY DESIGN LLC. WWW.RAYMONDLOEWY.COM

His designs reflected his famous "MAYA principle—Most Advanced Yet Acceptable." His redesign of the Sears "Coldspot" refrigerator created a smooth modern look. The appearance was possible because Loewy utilized a unibody structure that incorporated the exterior enameled metal and the structural frame into one piece. He also added efficient shelf storage in the door, ice cube trays, and other advanced features. The design was so successful that sales jumped from 60,000 to 275,000 units in just two years.

Much of his work was in the field of transportation. He began consulting with the powerful Pennsylvania Railroad in 1936, creating new modern rolling stock. For many of his designs, he employed "streamlining" principles.

Loewy also designed cars for the Studebaker Motor Company and the Greyhound Bus Company (see Figure 17-43).

Figure 17-43: Studebaker Avanti.

Figure 17-44: Loewy's design for the Coca-Cola and major corporate logos.

Although you may not recognize the name Studebaker as a car company, you will recognize Loewy's work for the Coca-Cola Company and the logos designed for other major corporations (see Figure 17-44).

His autobiography, *Never Leave Well Enough Alone*, was published in 1949, the same year he appeared on the cover of *Time* magazine.

Norman Bel Geddes and Streamlining

Streamlining was the first design style created through scientific research in a wind tunnel. The style, however, gained popularity through the designs of Norman Bel Geddes, in his book *Horizons*, published in 1932 (see Figure 17-45). During the middle part of the 20th century, many products reflected streamlining principles, especially in the automobile, train, ship, and airplane industries.

Figure 17-45: Norman Bel Geddes's model of a "streamlining" (teardrop-shaped) automobile design.

Henry Dreyfuss and Human Factors

Henry Dreyfuss (1904–1972) is considered the "father of human factors" in the United States (see Figure 17-46). Recognized as one of the pioneers of American industrial design, he founded his own design company in the late 1920s, and helped to create the Society of Industrial Designers in 1944 (later to become IDSA). Dreyfuss served as the society's first vice president. He worked for many of the largest American corporations including Bell Laboratories, John Deere, Hoover, New York Central Railroad, Honeywell, and Polaroid Land Company.

He is well known for his design of the "combined handset" desktop phone in 1937. The phone was developed in collaboration with Bell Laboratories and was produced in black phenolic plastic. The success of the phone is evident in the fact that it remained in production for over 20 years (see Figure 17-47).

In 1937, Dreyfuss began working for the John Deere Company. He redesigned the Model A tractor seat for safety and comfort. Wanting to make the tractor distinctive and recognizable from considerable distances on country roads, he created the distinctive John Deere yellow logo on the universally known John Deere "green" (see Figure 17-48).

In 1934, Dreyfuss was commissioned by the Hoover Company to design an upright vacuum cleaner. Prior to his work, most vacuum cleaners did not shroud the motor. Dreyfuss's stunning design for the Hoover Model 150 upright vacuum cleaner (completed in 1936) won him a yearly $25,000 retainer fee, which is nearly $400,000 in today's dollars (see Figure 17-49).

Dreyfuss appeared on the cover of *Forbes* magazine in 1951. His most important contribution to industrial design was not a product design but his introduction of human

COURTESY OF HENRY DREYFUSS ASSOCIATES, INC.

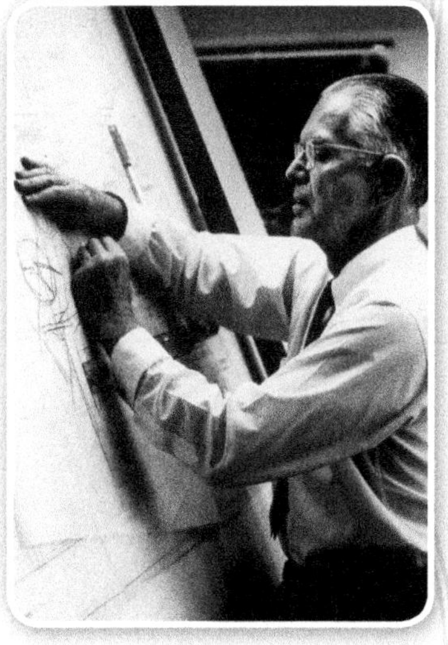

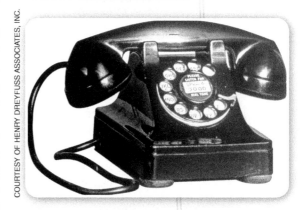

Figure 17-46: **Henry Dreyfuss, father of human factors in the United States.**

COURTESY OF HENRY DREYFUSS ASSOCIATES, INC.

Figure 17-47: **#302 Combined Desk Telephone made for Western Electric.**

COURTESY OF HENRY DREYFUSS ASSOCIATES, INC.

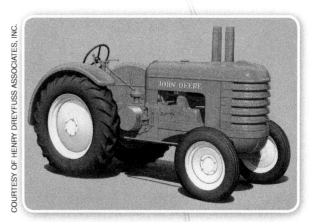

Figure 17-48: **Dreyfuss' 1937 John Deere Model A tractor.**

COURTESY OF HENRY DREYFUSS ASSOCIATES, INC.

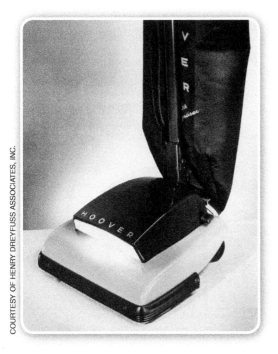

Figure 17-49: **Dreyfuss' Hoover Model 150 Upright Vacuum Cleaner.**

Your Turn

Research the work of Norman Bel Geddes, Harley Earl, Walter Teague, and Peter Behrens, and list the contributions they made as industrial designers for major U.S. corporations.

factors principles to design problems. The 1955 autobiography *Designing for People* included the first publication of "Joe" and "Josephine" anthropometric charts. He believed that designs adapted to people would be the most efficient. *Measure of a Man* was published in 1960, and included data compiled from military records. In 1967, he created Henry Dreyfuss Associates, which carries on his work in human factors.

Louis Comfort Tiffany and Art Nouveau

The Art Nouveau design style is based on natural shapes and is used for product design, including jewelry and architecture. In this style, patterns and designs are often based on curving, intertwining plant and animal shapes, such as climbing vines and flowers, leaves, birds and insects, and women with flowing hair. The stained glass windows and lamps of Louis Comfort Tiffany are good examples of the Art Nouveau style (see Figure 17-50). Begun in the early 1880s, Art Nouveau lasted until the beginning of the 1900s.

Art Nouveau was quite popular in many parts of Europe. Many cities, like Prague (Czech Republic), Paris (France), and Barcelona (Spain), have examples of Art Nouveau architecture. In Paris, for example, a number of the Metro underground subway stations were designed in the Art Nouveau style (see Figure 17-51).

Figure 17-50: A Tiffany stained glass window.

Figure 17-51: Art Nouveau style used for the Metro entrance in Paris, France.

William Morris and Arts and Crafts

What became known as the Arts and Crafts movement began in England in the 1880s, as a reaction to the newly manufactured products being produced in the new factories of the time. Some artists and craftsmen felt strongly that the quality and style of the mass-produced consumer products of the time, including home furniture, was poor. They also believed that the new factories were dehumanizing labor and wanted to pose an alternative lifestyle to the assembly-line work of machine operators. The style is based on simple lines and high-quality craftsmanship using traditional hand production techniques. Only natural materials were used, including wood, ceramic tile, and handcrafted textiles. Products were designed using earth tone colors.

William Morris (1834–1896) was one of the founders of the British Arts and Crafts movement. He is known for his wallpaper and patterned fabric designs. In the United States, Gustav Stickley was one of the leaders of the Arts and Crafts movement. Stickley started creating "mission-style" hand-made furniture of oak in Syracuse, New York, in 1904 (see Figure 17-52).

Today, the Arts and Crafts style continues to be popular in both architectural and furniture (product) design.

Figure 17-52: Gustav Stickley's bookcase in the Arts and Crafts style (circa 1900).

Walter Gropius and Bauhaus

The architect Walter Gropius founded a school of design in Weimar, Germany, in 1919 called the Bauhaus. Right after World War I, Germany needed much rebuilding, and the Bauhaus was instituted to help with that effort. Philosophically, the Bauhaus architects rejected the decoration that prevailed in architecture of the time and began a new, modern design style.

One of the basic aims of this school of design was the design of products suitable for machine and mass production. The phrase "form follows function" has come to be associated with the Bauhaus, because the designs that emerged were very functional and reflected the materials from which they were made. Materials were not disguised but instead became a statement about the product itself. Many of the product designs from the Bauhaus look as if they were produced today (see Figure 17-53).

Postmodern Industrial Designers

Most industrial designers are postmodern. At times, "retro" or "neo" designs draw from classic designs of the past. Contemporary designers are influenced by the advances in computer, materials, and manufacturing capabilities. By the early 1980s, miniaturization in electronics radically changed

Figure 17-53: Classic Bauhaus furniture design from the 1930s.

design. The designers of the first half of the century, such as Dreyfuss, Loewy, Teague, and Bel Geddes, worked to put functional objects in attractive packages. Today, smaller parts, and new materials and production processes in plastics allow for shorter production runs and more flexibility in design. One of the most successful postmodern designers—who is using all of the new advances to create new successful products—is Jonathan Ive of Apple Computers. His design work for Apple is profiled in the following case study.

Case Study

Jonathan Ive

Jonathan Ive was named Senior Vice President of Industrial Design at Apple Inc., by CEO Steve Jobs in 1997. At a time when many companies were outsourcing their design work, Apple was building an exceptional internal design team. Working in close collaboration with Jobs, Ive and his design team have created some of the world's most successful electronic products, starting with the revolutionary iMac G3 in 1998, and continuing with the more recent iPod (2001), iPhone (2007), and the iPad (2010). The iPad was awarded *TIME* magazine's "Invention of the Year" award. Steve Jobs called the iPad revolutionary and magical.

Born near London, England, in 1967, Ive is one of today's most influential industrial designers. As a youth, he was interested in made objects. He would take apart everything he could get his hands on. As his interest developed, he became more aware of product form, materials, and how things are made. He quickly learned that the form and color of the product defined people's perception of the object. During these early years, he started to appreciate the importance of the historical and cultural context of product design. He never fully developed his drawing skills and later wished that he had worked more on them during his younger days. He studied industrial design at Newcastle Polytechnic (now Northumbria University). In 1985, he became a partner at Tangerine, a London-based design consultancy. While at Tangerine, his creativity was recognized by one of its clients, Apple Computer, Inc.

In 1992, Jonathan joined Apple as a full-time member of its California design team. He noted that when he first looked at an Apple product, he was impressed by the design. Beginning with the Apple I introduced in 1976 and the Macintosh introduced in 1984, Apple products have been recognized for user-friendly design innovations, including component integration in an attractive form. Ive continued that tradition when, in 1998, he led the redesign of the iMac, which helped revive the company with two million units sold in the first year it was introduced. The innovative design introduced color such as blueberry, grape, tangerine, lime, and strawberry for the first time into a computer industry that was characterized by boring grey and beige plastic boxes (see Figure 17-55).

Figure 17-54a: *Jonathan Ive, Apple Vice President and 2004 Designer of the Year.*

Figure 17-54b: *Steve Jobs, Apple President.*

Figure 17-55: *Jonathan Ive lead the design of the iMac in 1998.*

Case Study

(continued)

The iMac was considered Apple's computer for the new millennium. The design integrated the cathode ray tube (CRT) with the central processing unit (CPU) and simplified connecting peripheral devices including a newly designed universal serial bus (USB) keyboard and mouse. Designed for the broad consumer market and with the Internet in mind, the G3 had a 233 MHz processor and a 15-inch 1024 × 768 resolution screen. Although the iMac was the fastest and easiest-to-use low-end computer on the market, it quickly became recognized for its iconic visual appeal.

Jonathan Ive is a leader who nurtures a close-knit team. He created a large workspace with an incredible sound system for the exclusive use of his team. The team uses this space to foster creativity and lateral thinking. Their attention to detail and innovation led to the development of some of the most innovative new products in the industry.

Ive and Steve Jobs had a similar business philosophy and shared an obsession for detail. They collaborated on all-important decisions and they worked extremely well together.

Jonathan Ive doesn't overlook any product details. The design team at Apple wants to make everything intuitive and accessible. In the Power Mac G4, Ive created a single-piece plastic core. In designing the new case, clutter was removed and air for ventilation entered holes in the bottom without the use of a fan, making the computer very quiet. Components such as the circuit board and even the CD ejecting vertically make it easy to remove components for servicing. Advances in polymer materials have allowed the Apple group to focus on functional goals because most shapes are possible. By developing new processing techniques such as twin-shot plastic, the team allowed for the iPod to be produced fully sealed without the need for fasteners or battery doors. New advances in adhesives and methods for joining metal also support design innovation.

Ive focuses on simplicity, elegance, and innovation in his design work. About his early work, he notes: "There's an applied style of being minimal and simple, and then there's real simplicity . . . the G5 looks simple, because it really is." An important influence in his work is nature, organic shapes, and color. He believes that these natural shapes make the products more friendly and understandable. In a recent interview, Ive noted that designing and producing successful products is really hard and complicated. He constantly asks the question, "Why is it like that?" when looking at a prototype or product. He believes that design should "get out of the way"—noting that when it is right, you should feel that "of course, this is the way it should be." Most people who have used Apple products feel exactly that way—the product is easy to use, as it should be!

In 2003, the same year Apple launched the PowerBook, Jonathan Ive was recognized by London's prestigious Design Museum with their first "Designer of the Year" prize. His design of the iPod changed the way the world listens to music, and the iPhone has changed the way we use our phones. His design of the iPad may change the way people read books, browse the Web, listen to music, look at photographs, watch videos, read and send emails, play games, and much more. As Apple's Senior Vice President of industrial design, this private and unassuming man is having a huge impact on contemporary product design (see Figure 17-56). He pays serious attention to the obvious stuff while constantly looking for new tools, materials, and production processes to continue to innovate new products for his company. Jonathan noted that many companies today are simply looking to be different in the marketplace rather than working to make their products better though investing in product innovation. You can see Mr. Ive's early works at New York City's Museum of Modern Art and the Georges Pompidou Center in Paris.

Figure 17-56: *Examples of new products from Apple.*

SUMMARY

Design is both process, such as what an engineer does to solve a technical problem, or appearance (aesthetics), such as style associated with a historical period, or the work of a particular person. All products have visual appearance qualities such as shape, symmetry, naturalness, scale, proportion, smoothness, shininess, texture, pattern, or color. Style then is used to describe design elements used in the design of structures and products. Architectural style is usually associated with the work of architects who give form to residential or commercial buildings. Product style is usually associated with the work of industrial designers who give form to a cell phone or automobile. Some designers are successful in both architecture and industrial design.

Architects design residential and commercial structures. These designs require a team of design professionals to complete. The architect is the head of the team and is responsible for working with the client and setting the style of the project. Architects are concerned with the intended use of the project and must meet all code requirements. Codes, which are a form of standards, are established and enforced by local municipalities. Most architects belong to the American Institute of Architects (AIA). Frank Lloyd Wright is one of America's most famous architects.

Architectural design started with the Egyptians, who established principles of proportion and created columns and other structural elements. The Greeks refined architectural elements, especially the development of columns designs. The Romans who possessed superior engineering knowledge built large amphitheaters, roads, bridges, and water systems. Most notably, the Renaissance architect Andrea Palladio wrote extensively about classic design principles that are still used today. Architectural style can be seen in the shape and proportion of the structure itself, the roof, doors and windows, eaves (entablature), and building materials used. Styles reviewed include Colonial (1500–1800), Georgian (1790–1780), Federal (1780–1840), Greek Revival (1825–1860), Victorian (1860–1900), Prairie (1900–1920), Craftsman (1905–1930), Art Deco (1920–1940), International (1925–present), and Modern (1945–present). Today, most designs reflect new applications of classical design elements or an eclectic mixture of elements.

Industrial designers create designs for products and systems. Most industrial designers belong to the Industrial Designers Society of America (IDSA). Early industrial designers included Raymond Loewy who practiced his MAYA principle while designing for the Pennsylvania Railroad, Greyhound Bus, and Coca-Cola companies. Norman Bel Geddes popularized streamlining, the aerodynamically efficient design style created through scientific research. Henry Dreyfuss, the father of human factors, created designs that fit human users. Most product design reflects postmodern style principles, but some products are created as variations of neo-classic styles such as Victorian, Art Nouveau, Arts and Crafts, or Art Deco. The work of Michael Graves and Jonathan Ive represent important postmodern design that is influencing contemporary architecture and product design.

OFF-ROAD EXPLORATION

American Institute of Architects (AIA) http://www.aia.org/

Dating your house http://www.bricksandbrass.co.uk

Restoration of Fallingwater http://paconserve.org/43/fallingwater

Drawings and photographs of important American architectural structures are available as ABS and HAER documentation http://www.loc.gov/pictures/collection/hh/

Information about Henry Dreyfuss http://www.hda.net

Information about Raymond Loewy http://www.raymondloewy.com

Information about the Craftsman style http://www.americanbungalow.com

For more information on British design and architecture http:// designmuseum.org/designinbritain

Information about Buckminster Fuller http://www.bfi.org

Interview with Ive at http://www.designmuseum.org/design/jonathan-ive

Information on Michael Graves http://www.michaelgraves.com

BRING IT HOME

OBSERVATION/ANALYSIS/SYNTHESIS

1. Select an important architectural structure in your community. Take a digital photograph and identify architectural elements including roof type, window and door style, overall shape, and materials used. Using material from the chapter and the architectural elements identified, place the structure in one of the historical style periods.
2. Select an architect and do a 10-minute PowerPoint presentation on the person's most significant buildings.
3. Select an architectural style period and do a 10-minute PowerPoint presentation on the identifying characteristics of the style period.
4. Select an industrial designer identified in this chapter and do a 10-minute PowerPoint presentation on the person's most significant designs.

EXTRA MILE

ENGINEERING DESIGN ANALYSIS CHALLENGE

- Select a similar product or structure from around 1850, 1900, 1950, and 2000. Develop a list that describes the major materials used in the design. Compare and describe how size and weight of the product or structure have changed over time. Identify the style used in the design. Discuss how technological developments and increased math and science understanding are reflected in the differences you found.

CHAPTER 18
Graphics and Presentation

Menu

 Before You Begin
Think about these questions as you study the concepts in this chapter:

 Why are graphics important to successful product design and effective professional presentations?

 What graphic design principles do I need to apply to create more effective designs?

 What do I need to know about type (typography) to make better graphic designs?

④ What do I need to know about photographs and illustrations to create more attractive designs?

⑤ How are graphics used in product and package prototyping?

 How is creating a design for the Web different than creating a design for printing?

 What do I need to know to make better PowerPoint presentations?

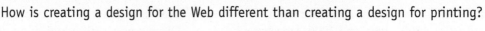

INTRODUCTION

Every day, hundreds of newly printed products enter the market. A printed advertisement or color package must be designed and produced and is also considered a product. For the purposes of this chapter, the word *product* will be used to refer to things produced from graphic designs.

Skilled tradespeople print most graphic designs using advanced computer software, technologically complex full-color printing presses, and specialized paper for Web distribution. However, no matter how well the product was made, it must be well designed to be successful (see Figure 18-1). Good graphic design requires hard work, careful planning, appropriate application of graphic design principles, and considerable revisions. Graphic designs must be viewed in two ways: (1) what they say, and (2) how they look. Good designing is more than a matter of aesthetics. Unlike fine art, which must only satisfy the artist's needs and desires, a graphic design must satisfy the needs of the client who buys it and the audience for whom it is intended.

Graphic designers create visual designs. Training to become a graphic designer usually involves a four-year college education, but some designers have developed the necessary talent to create successful designs without any formal training.

Today, most designers work with high-end design software such as Adobe InDesign or CorelDRAW Graphics Suite for page layout, Adobe Photoshop for image editing, Dreamweaver for Web design, or PowerPoint for professional presentations (see Figure 18-2). Many other software programs are available free or at low cost on the Web. To design effectively, designers use knowledge of the product, the design process, and the design principles. This knowledge—along with design talent—is used to create effective designs.

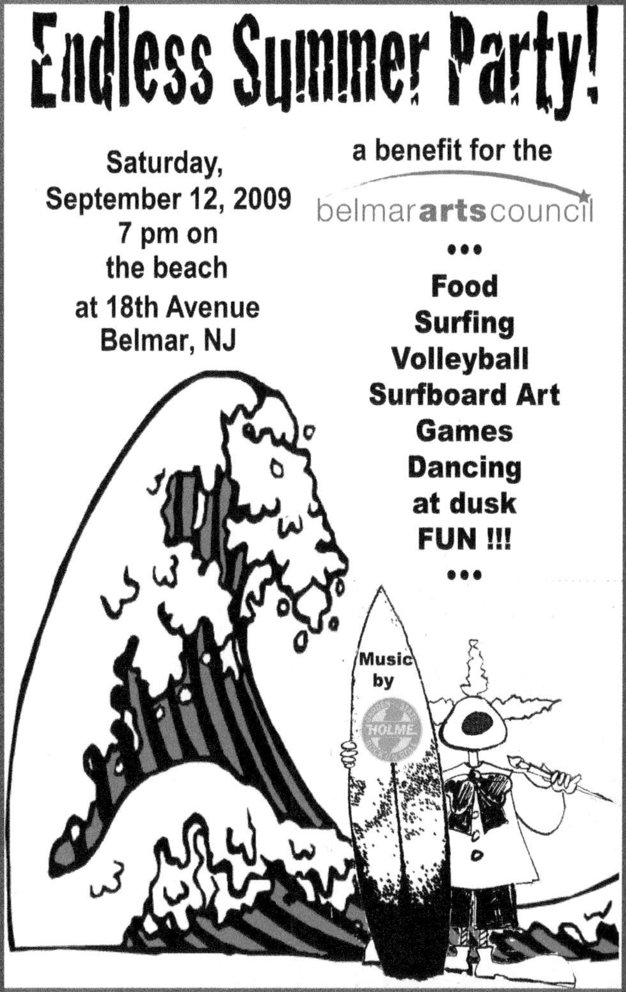

Figure 18-1: *This advertisement uses graphic design effectively to promote a product. In this chapter, we will refer to the graphic design itself as a product.*

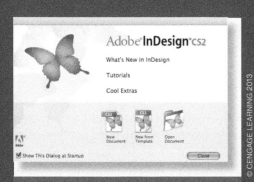

Figure 18-2: *Many graphic designers use Adobe InDesign software for typesetting and layout.*

product: In commerce, the term refers to all goods and services that are bought or sold. Products are designed to satisfy human wants and needs. In retail, a product is usually referred to as merchandise. In manufacturing, a product may be the raw materials purchased as commodities such as sheet metal or cabinet-grade plywood to make the finished good.

graphic design: The process of planning for some form of visual communication using type, photographs or video, and illustrations arranged in an aesthetic layout for a printed page or digital display for the Web or multimedia presentation.

audience:
An identified group of people whose characteristics are of interest to a designer for product development purposes. Characteristics may include age, gender, income, interests, and other descriptive items that relate to the specific product.

OFF-ROAD EXPLORATION
Directory of schools for studying graphic design www.graphicdesignschools.com/
High-end graphic design examples artworksdesign.com/

THE GRAPHIC DESIGN PROCESS

The graphic design process is iterative like the twelve-step process described in Chapter 2, but has its own unique characteristics like all design processes. The design process begins with a clearly defined problem (Step 1: Define the Problem). Most graphic designs are done for a client who has a need to communicate to an intended audience. The graphic designer must fully understand the client need and the nature of the intended audience. The graphic designer begins by placing all available information about the problem into some logical order. The designer will use brainstorming (Step 2), do research (Step 3), and identify all problem constraints (Step 4). Constraints will always involve budget items such as whether or not the job is to be printed in full color and how much time is needed for completion. Once the designer collects all the information for the design, he or she will begin to explore possibilities, sometimes using thumbnail sketches (Step 5). Although thumbnails are typically done as hand-drawn sketches, most designers do as much work as possible on the computer. The designer plays with the design by moving elements around, trying different elements, or adding color (Step 6). When the designer is satisfied with the solution, he or she makes a final design proposal (Step 7). The designer makes a comprehensive drawing (Step 8, a graphic prototype), and it is shown as part of the design presentation to the client. The client must give final approval for the design to be produced in large quantities.

Usually the client wants some changes or revisions in the design (Step 9). The designer will make the necessary changes and show the revised design to the client (Step 10). When final approval is given, the graphic designer will send the revised design to a company for either reproduction by printing or uploading on a Web server (Step 11). The form of production will depend on the intended use. For example, designs for posters, brochures, or packages will be prepared for a conventional printing process while a design for a cell phone logo may become part of a manufacturing or postmanufacturing process. A design for a website will need to be created in code and transferred to the Internet service provider (ISP), while a design for a presentation may be created in Microsoft PowerPoint, Keynote, or another presentation program.

Collecting Information for the Design

All graphic products convey ideas through the visual sense. This communication process involves a sender and receiver. In most cases, the designer bridges the gap between the sender (client) and the receiver (audience). Each client has a unique need (problem) and is looking for a product (solution) to communicate effectively to a particular audience. To meet this need, the designer works with the client to identify the elements, such as written copy, photographs, and illustrations, with which the designer is to work in producing the design. The client, for example, might be the Motorola Corporation, who is looking for a new logo for its new cell phone, or General Mills, who is planning to redesign its packaging for a new breakfast cereal promotion (see Figure 18-3).

The audience is the group of people with whom the client wants to communicate. The makeup of the audience is very important to the designer. Specifics such as age, gender, education, ethnic background, interests, values, language, and economic status are important characteristics that help the designer to better understand the audience.

(a)  (b)

Figure 18-3: Branding a new product or promoting an existing product involves the use of graphic design.

Figure 18-4: The graphics on this snowboard deck are designed to appeal to specific consumers.

Figure 18-4 shows how Burton Snowboards uses graphic design to appeal to their intended audience. The product and package design represent the designer's concern based on the makeup of the audience. The client chose a high-energy design to appeal to the target market and to "brand" the board as a Burton product. Because the main purpose of the graphic product is to communicate a message between a sender (client) and receiver (audience), the designer must understand the makeup of the audience if the design is to be effective.

Creating Possible Solutions

Designers begin the creating process once they receive all of the information about the audience and the product. Designers of two-dimensional (2-D) images may also use a form of reverse engineering (see Chapter 6) in the design process. Graphic designers are always sketching new visual design techniques by looking at and analyzing new graphic products. If a designer is considering a new logo or new package design, he or she will do a Web search or visit stores to look for interesting uses of type, photographs/illustrations, and color. The creative process involves the development of many initial ideas through to the final client-approved visual solution.

The best way to explore many possible designs quickly is to develop thumbnail sketches. **Thumbnail sketches** are preliminary visuals of a possible idea for the design (see Figure 18-5). They may be drawn on paper or a computer. Most thumbnail sketches are not full-size and have little detail. They are intended to explore possible alternative designs quickly. In the thumbnail sketches, designers often use symbols to indicate various design elements, as shown in Figure 18-6.

508 Part II: Resources for Engineering Design

Figure 18-5: Thumbnail sketches.

Figure 18-6: Thumbnail sketch symbols for display, text, photographs, and illustration images.

© CENGAGE LEARNING 2013

(a)

(b)

(c)

(d)

Figure 18-7: Rough layout.

When the designer has exhausted multiple possible design ideas, he or she then evaluates the thumbnails. The evaluation involves reviewing the problem and design brief statements. The designer selects the best design from this evaluation. To better visualize the finished product, the designer can prepare a rough layout of the best thumbnail. Sometimes the designer will use the rough layout to explore a combination of ideas from the thumbnails. Most designers will begin to introduce color at this point in the design process. A rough layout is advantageous for a couple of reasons: It is a quick, full-size drawing that resembles the finished product and it is created in a fraction of the time (see Figure 18-7). Body type, which is the type making up the paragraphs, may still be shown as lines, but display type, such as heading and titles, is drawn to size with some concern for style. Photographs and illustrations are sketches of the visual information for presentation. Rough layouts allow designers to work out any visual problems and finalize space allocations for individual design elements. The client may be shown the rough layouts, especially if there are different directions that the final design could take.

The last stage of creating the design involves the preparation of the final prototype, called a *comprehensive* or *comp*. The comprehensive is a visual model that is either drawn as art using markers or acrylic paints or created on a computer. The comprehensive must be a representation of how the final product will appear in final production. It is prepared to full size, has considerable detail, and shows all colors. Because it is so easy to make adjustments in image size and image position on a computer, the designer may choose to give the client a number of options for the final product (see Figure 18-8). The client reviews the comprehensive for final approval to proceed with production. The comprehensive also accompanies the product through production so that production personnel can see how the final product should appear.

510 Part II: Resources for Engineering Design

Figure 18-8: *Comprehensive samples.*

Producing a Graphic Image

In principle, the design process is complete when the client approves the designer's comprehensive. Because most graphic products will contain text, photographs, and illustrations, original images will need to be created to commercial standards. Most graphic images are created in a digital format, and the quality of the image is specified as a level of resolution. For example, the lowest level of resolution specified for Web work and some dot matrix printers is 72 dots per inch (DPI), while high-resolution images and laser printers are capable of 2,400 DPI or higher. Each method of production has different requirements. Offset lithography is a commonly used method of production for general commercial work such as brochures, posters, magazines, or books. When extremely long runs are required, such as cereal boxes or other packaging products, gravure is often the best production method. When unusual images or materials are required, especially for placing an image such as a logo on a skateboard, screen printing may be the only appropriate method to use. For example, snow skis are printed by a flexible screen printing system capable of printing the completely formed ski.

Although we can still learn a lot about the reproduction of images for product design and presentation, designers typically consider the following questions for a graphics project:

What Specifications Are Needed When Placing an Order? Printers price the jobs by the quantity (number of finished products), format (size and colors needed), paper, and finishing. For example, 10,000 8.5 by 11 inch four-color brochures printed on both sides of 80# coated paper and tri-folded would cost approximately $1,200. If the job had additional production work or was a rush job, the cost would be higher.

What Software Should Be Used to Create the Designs? Most businesses can work with standard software formats such as those output by Microsoft, Quark, Adobe, Corel, and Macromedia products. Usually the major concern is that the image file received has sufficient resolution to produce a quality image. To achieve a high-quality product, it is usually best to work with professional-quality software.

What Image Resolution Is Needed for the Job? Image resolution is measured in DPI or PPI (see Figure 18-9). Dots per inch (DPI) is a measure of printing resolution. As an image on a page, DPI represents the number of individual dots of ink or toner produced within one linear inch (2.54 cm). Another measurement that is used to determine image resolution is pixels per inch (PPI), a measurement of the resolution of the image on a computer screen, image scanner, or digital camera. It is generally assumed that a higher DPI/PPI means a sharper or higher-resolution image. Unfortunately, this may not always be true. For example, if a 3- by 4-inch photograph was scanned at 300 DPI, but then printed at 6 by 8 inches, the printed image would only be 150 DPI. To be accurate, DPI and PPI must be taken in the context of the recorded and output image sizes. As a rule of thumb, most people cannot differentiate (with visual acuity) resolution beyond 300 PPI. Therefore, high-quality photographs require at least 300 PPI when printed.

Product	Application	Resolution
Commercial Printing	Full-Color	300dpi or higher
Photograph	Standard Print	
Website	Web	72dpi
Video	S-VHS	400 x 480ppi
Television	High Definition (HD)	1900 x 1080ppi
Monitor	CAD, graphic design and video games 19"	1600 x 1200ppi (UXGA – Ultra-eXtended) 1280 x 1024ppi (SXGA – Super eXtended)
Canon S1 3.2 Mega pixel	General Photography	2048 x 1536ppi
Nikon D2K 12.2 Mega pixel	Professional	4288 x 2848ppi
HP Laser Printer	Industrial/Commercial Color Printer	4800 x 1200dpi

Figure 18-9: This chart shows the dpi and ppi for commonly used graphic products.

Electromagnetic radiation between 400 and 700 nm (nanometers) is known as a visible light because it is the part of the electromagnetic spectrum that the human eye can see (see Figure 18-10). Visible light is comprised of the primary colors of red, green, and blue (RGB). RGB combined creates white light. RGB reflects the **additive color principle**. When electromagnet radiation is absorbed by pigments in objects, the **subtractive color principle** is employed. As each of the subtractive colors—cyan, yellow, magenta, and black (CYMK)—are combined, the light reflected is darker. When all the subtractive colors are present, no light is reflected from the object and the visual sensation is black.

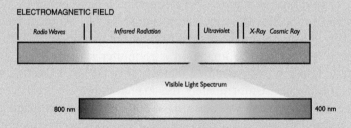

Figure 18-10: Electromagnetic spectrum showing the relationship of light to other forms of energy.

REPRINTED WITH PERMISSION FROM POPPY EVANS AND MARK A. THOMAS, *EXPLORING THE ELEMENTS OF DESIGN*, SECOND ED. COPYRIGHT © 2008 CENGAGE DELMAR LEARNING.

Chapter 18: Graphics and Presentation 513

What Is the Difference Between the RGB CMYK Color? RGB stands for the colors red, green, and blue. RGB are the primary colors of the additive color principle and they are the primary components of white light (see Figure 18-11). Another aspect of color that is important to design is the subtractive color principle. Subtractive color consists of the primary colors of cyan (process blue), magenta (process red), yellow (process yellow), and black. In graphics, both the additive and the subtractive color principles are applied in processing color images.

The additive color principle describes the process of combining the primary additive colors RGB to create new lighter colors (light energy is added). As can be seen in Figure 18-11, a new lighter color yellow results by combining red and green. When RGB are all combined, the resulting color is white. Additive color principles are applied when stage designers create new lighting patterns for a play or engineers design new color monitors, cameras, or scanners.

The subtractive color principle describes the process of combining layers of pigment that results in the absorption of light. When CMYK pigments are all combined, the result is the absorption of all light or the visual sensation of black (the absence of light). Again referring to Figure 18-11, by combining pigments of cyan and yellow, the new darker color green results. Subtractive color principles are applied when artists use color pencils or paints, when a finish is applied to a new car, or when ink is applied to a substrate during one of many printing processes. The inks used in printing CMYK are four-color process printing or full-color printing. When designing for four-color printing, it is best to convert your color files to CMYK (see Figure 18-12).

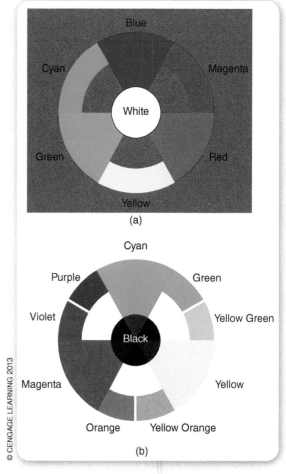

Figure 18-11: **These color wheels demonstrate the primary additive and subtractive color principles: (a) the additive color principle, and (b) the subtractive color principle.**

Figure 18-12: **This enlargement of a CMYK image shows how tiny dots of cyan, magenta, yellow, and black combine to create a range of colors.**

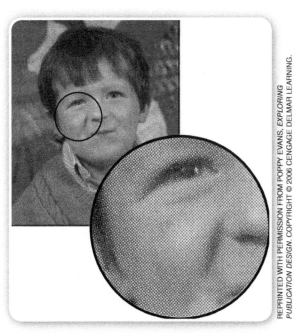

Figure 18-13: This is a page from a PANTONE® swatch book used to specify ink colors in the CMYK process. Each shade of blue on this page requires a different ink formula. Designers use the number on the left below each color swatch to specify that ink color.

Will the Color on the Finished Job Match the Original on the Computer? When viewing color, there is always a difference between what is seen in the original, what is seen on a monitor, and what is seen on the printed page. Color matching in the digital world is a universal problem not only for the printing industry but also paint, plastics, textile, and other types of manufacturing industries, where color consistency is important. HTML color codes are used in identifying color for Web content. In print-based design, Pantone created color systems to assist graphic designers in writing color specifications and in visualizing how the final product should appear (see Figure 18-13).

Additional Considerations. All graphic designers must consider the needs of the client and required production techniques. It is also important to plan the distribution and use of the product. For example, if the product is to be mailed, mail regulations must be taken into consideration (see http://pe.usps.gov/). It makes no sense to design a postcard slightly larger than the standard 4.25 by 6 inches. The rate to send a standard postcard is substantially lower than that of an oversized piece. If the client is sending 100,000 postcards, the difference to send the oversized piece may be an additional $15,000 in postal fees! Designers must also understand industry standards. The printing company must receive all materials in an appropriate form and resolution. All specifications for the job must be accurate. The printing company will send a proof to the designer or client for approval. The designer or client is responsible for proofreading and otherwise checking for mistakes. Once approved, any nonproduction errors found in the final job are not the responsibility of the printing company. The designer or client also should mention any deadlines in the specifications. Finally, the customer will receive and be billed for an exact quantity of products delivered by standards that can vary ± 10 percent from the order.

What Are the Legal Considerations? Designers must take into consideration a number of legal concerns in the design and reproduction of a printed piece involving original material or someone's intellectual property. Most of the restrictions for appropriate use are defined by copyright laws. Copyright laws protect original works as intellectual property of authorship, including literary works, musical works, dramatic works, pantomimes and choreographic works, pictorial, graphic and sculptural works, motion pictures, and sound recordings from being copied without permission. The law does allow for original work to be copied under limited conditions. For example, work may be copied for use in teaching, research, news reporting, or criticism. However, credit must be given to the creator or author. This practice is known as the fair-use doctrine. For most students, the issue with using another person's original material usually involves questions of plagiarism when writing a report. Plagiarism can also be an issue in a new graphic design. Remember: When designing graphics for a product or poster or when including visual images in a presentation, it is illegal to use copyrighted material without permission.

The Copyright Act of 1976 grants protection to original works for the lifetime of the creator or author, plus 50 years after the death of the creator or author. The law grants exclusive right to reproduce the original work for sale, rent, lease, or loan. A copyright is obtained by registering the original work with the copyright office and paying all fees. A copyright notice must appear on all copyrighted publications. This notice consists of either the word *copyright*, the abbreviation *copr.*, or the symbol ©, in addition to the name of the copyright owner and the year of copyright. International copyrights are possible in many but not all countries. New technological developments, such as the photocopier and the computer, have unfortunately made breaking copyright laws easier.

Figure 18-14: Electronic clip art.

intellectual property: An original work usually associated with innovation but also covering authorship, including literary works, musical works, dramatic works, pantomimes, and choreographic works, pictorial, graphic and sculptural works, motion pictures, and sound recordings. U.S. patent, trademark, and copyright law protects intellectual property.

Illustrations such as trademarks or other uniquely distinguishable symbols may not be used without permission. All original art is also protected. One exception to this rule is clip art. **Clip art** is a form of original copyrighted art that is designed and sold to be copied. When someone purchases clip art, he or she is granted permission to duplicate the art for any reason except to be sold again as clip art. Figure 18-14 shows examples of clip art.

Photographs can also be copyrighted. With photographs containing images of people, the photographer must receive permission to record the likeness of all individuals shown in the photograph. In the case of minors, parents or guardians must grant permission. In most instances, you should have written permission for all art and photographs that will be used in a publication.

CREATING EFFECTIVE GRAPHIC DESIGNS USING DESIGN PRINCIPLES

Although understanding the nature of the product and the design process is important, the ability to create effective designs involves understanding the visual relationship between the elements and the space where they will be displayed, such as on a page or a product. The selection, creation, placement, and visual qualities of these elements, such as color and tone, are controlled by graphic design principles. The principles can involve many visual subtleties, and new designers can achieve very effective designs by focusing on the following five design principles:

- ▶ Unity
- ▶ Balance

- Rhythm
- Emphasis
- Proportion and Scale

Graphic designers must also design with a variety of visual elements, such as written copy, illustrations, and photographs, and a certain amount of space. Once a designer identifies the specific elements, thousands of possible designs exist to fill that space. Many people assume only a few creative people can design. Some believe that good design is controlled by subjective feelings without rules for making judgments. Neither of these two beliefs is correct. Good design is a logical as well as an aesthetic function. It is based on easily understood principles. Knowledge of these principles will make any design task easier and more successful.

Unity

We begin with the principle of unity because all the design elements and the page itself must come together as a unified whole for the design to be effective. One important element of unity is the grouping of elements, called proximity. Consider the problem of designing a new business card for a client who works for Designed World Learning. You would need to collect all the appropriate information: name of the individual, title, address, and contact information. Grouping elements helps readers understand the design and improves the communication qualities. Consider the design in Figure 18-15, where the designer did not group the typographic elements. The design is not very effective because it is difficult to determine which elements go together, or which elements are more important to the readers.

Grouping improves proximity of elements and the visual or aesthetic appeal of the design. Figure 18-16 shows that the designer applied the principle of proximity and grouped the related information.

Another important element of unity is alignment of elements. The use of alignment is common on both the printed page and Web page. The relationship of the elements to the page and the elements to one another are important consideration in designing. Page margins and the alignment of the element with the page strengthen the design. Margins are the white space between the edge of the page and the element. A page's margin sizes can vary to provide different degrees of effectiveness. Even single-page book margins vary in size. Contemporary products have margins that range in size from no margin (known as an *image bleed*), to extensive space between elements and the page edge. Although internal white space between elements is usually kept to a minimum, larger margins are an acceptable form of white space.

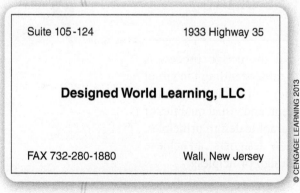

Figure 18-15: Related elements are not grouped, resulting in a poorly designed business card.

Figure 18-16: Grouping related elements to achieve visual unity through the principle of proximity.

The page edge and page proportion provide a boundary for the design. As each element is placed on the page, it is usually very effective to relate the elements to the page edge. Can you see how the grouped elements of the business card in Figure 18-17 relate to the page edge? The blue band with reversed type provides emphasis and the name is placed near the optical center of the page. The name and address are two groups placed on the left side of the page. Typographically, this layout is called *neat left* or *aligned left*. The address block is placed so as to create equal margins on the bottom and left of the page.

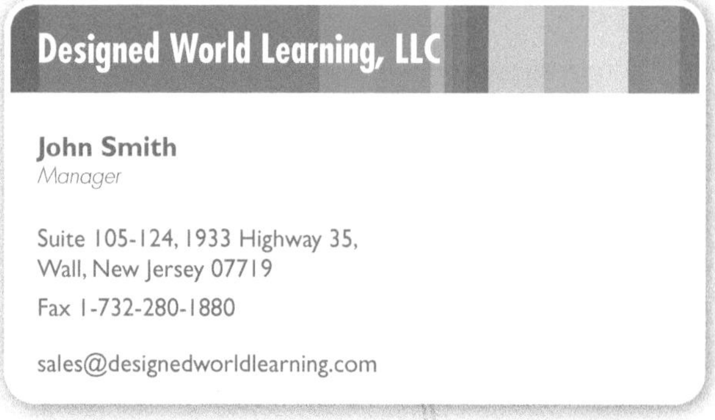

Figure 18-17: **The elements on this card have a strong relationship with the overall page shape.**

Your Turn

How does a graphic designer apply principles of unity, including proximity and alignment, to a design? Consider the information provided by the Belmar Arts Council to the design team for the Endless Summer Party poster project shown in Figure 18-18. Take a moment to organize the information using proximity. What did you take into consideration to organize the information? Although many solutions are possible, the poster in Figure 18-1 shows the solution created by the Designed World Learning team.

Endless Summer Party!

A benefit for the Belmar Arts Council

Food, Surfing, Volleyball, Surfboard Art, Games,

Fun Dancing (at dusk)

Saturday, September 12, 2009

5-10 pm on the beach at 18th Avenue, Belmar, NJ

Music by Holme

Figure 18-18: **Advertising copy is grouped and organized in proximity to the product features it describes.**

Balance

Visual and physical equilibrium is described as balance. Balance may be due to equity in area, mass, or attention within a defined area and may be symmetrical or asymmetrical. Symmetry means that balance has been achieved by elements being placed around a center point that could be on a vertical or horizontal line (see Figure 18-19a). In graphic design, balance describes the relationship between the elements and the page. Formal balance is symmetrical; that is, all elements are equally placed around the centerline of the page. Informal balance is asymmetrical, meaning there are design elements of varying sizes, shapes, weights, colors, and tones placed in various positions on the page (see Figure 18-19b). Because the rules for formal balance are easily understood, most beginning designers tend to select formal balance in their design work.

(a)

(b)

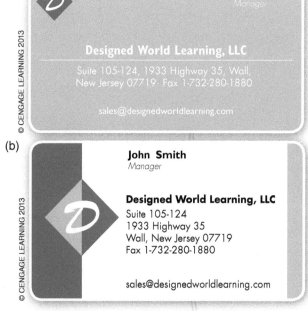

Figure 18-19: **(a) Formal balanced designs; and (b) Informal balanced designs.**

Jane Quincy Doe

4300 Sentry Drive
Anytown, IA 50703
(555) 255-5555

Objective

Lorem ipsum dolor sit amet, con secteteur adipsicing elit, sed diama nonnumy eusidnod tempor inci dunt ut labore.

Education

Joetown Technical School, August 1995-present. Vel illum dolor eu feugiat mulla facilsi at vero eos. Nam liber tempor cum soluta nobis.

Joetown High School. Graduated in top 10% of class. Special interests in automotive technologies, geography, geometry and related mathematics. Eligent option conque nibil impediet doming id quot maxim plecat facer possum omnis voluptas.

Experience

Work Study Automotive Lab Monitor, August 2005-present. Duties include eligent option conque nibil impediet doming id quot maxim plecat facer possum omnis voluptas.

Joey's Grill, December 2003-August 2005. Nam liber tempor cum soluta nobis.

Skills

Nam liber tempor cum soluta nobis. Lorem ipsum dolor sit amet.

Figure 18-20: ***Repetition strengthens the impact of this resume design.***

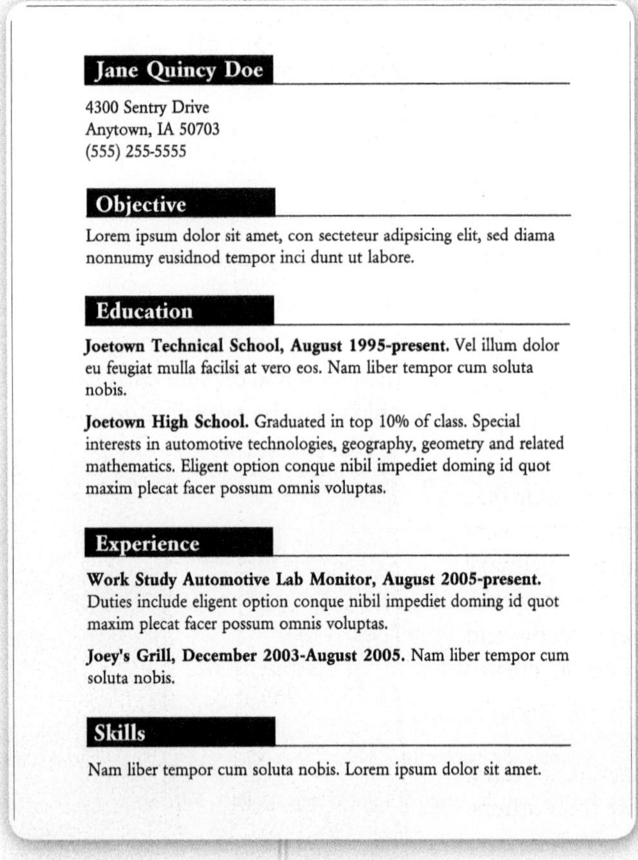

Rhythm

Rhythm uses repetition and other graphic techniques to hold the design together visually. Usually a design with good rhythm is easier to read because the more important elements have dominance in the design. **Repetition**, as the name implies, is the repeating of some graphic element throughout the entire design (see Figure 18-20). Often designers will repeat bold type for headings, or use bullets or numbers for a list of items. Designers use lines, also known as *rules*, to help set off sections of the design or guide the readers from element to element. Designers sometimes refer to this guiding through the design as *rhythm*.

To develop a standard format, designers use repetition when creating a multipage design. Page format involves column width, margins, major headings and subheadings, and text type styles. Designers may decide to place photographs in consistent positions or use a unique symbol to draw attention to important information.

Emphasis

Emphasis brings interest to the design by providing **contrast** between design elements. These contrasts can be in size (large and small), shape (rectangular and round), texture (smooth and rough), value (light and dark), or color (red and green or other complementary colors) (see Figure 18-21).

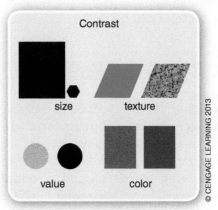

Figure 18-21: ***Contrast in size, texture, value, and color.***

Emphasis is often used in graphic design, such as print or Web media. The use of large bold headlines and smaller paragraph text next to each other creates a striking contrast. In photography, the lightest parts and the darkest parts of the photo provide contrasting visual elements. Two colors that are too close to each other in value or hue should not be used near each other. Similarly, two textures that are too close to each other should be avoided, because they may look more accidental than planned. Contrast must be bold to be effective (see Figure 18-22).

Color is extremely important to achieve design emphasis. The popularity of a color may be a trend that will change over time or may be timeless such as John Deere green. Color affects people in different ways. Notice that certain foods in the grocery store are packaged in similar colors. You will see sugar, for example, usually packaged in blue, associating the color with sweetness. Warm colors, including red, yellow, and orange, are associated with fire while cool colors, including blue, violet, and green, are associated with water. Dark colors cause the sensation of inactivity while light colors are stimulating. Vibrant colors like red can cause excitement by actually increasing pulse and respiration. Blue is considered calming, brown and green are considered earthy, and a color like purple denotes royalty. Because of its different qualities and effects, blue is the most popular color used in design. Understanding how color affects people will allow you to use color more effectively in creating your designs.

Hue, or the name of a color such as red, is one attribute of a color (see Figure 18-23). As we have already covered in this chapter, magenta, cyan, and yellow are the primary colors used in printing. When two primary colors are equally mixed, they create the secondary colors. For example, magenta and cyan make blue, cyan and yellow make green, and yellow and magenta make red. Equally mixing primaries and secondary colors creates intermediate colors. For example, the primary color yellow mixed with the secondary color green produces the intermediate color yellow-green.

Designers achieve emphasis by combining various primary, secondary, and intermediate colors. Some colors, however, contrast more than others. **Complementary colors** contrast the most, and consist of colors that are opposite each other on the color wheel. Yellow and blue, for example, are complementary colors that contrast greatly. The monochromatic color combinations contrast the least. These colors result from a mixture of white or black with any hue. One monochromatic color combination would be dark blue and light blue. Color creates contrast in any printed product and the colors selected determine the amount of contrast.

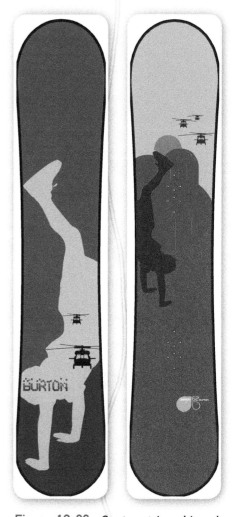

Figure 18-22: **Contrast is achieved in this snowboard design by using changes in color, size, and shape of the graphic elements.**

COURTESY OF BURTON SNOWBOARDS. GRAPHICS BY JAGER DI PAOLA KEMP DESIGN.

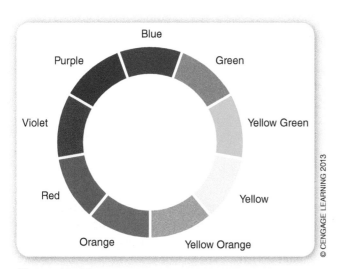

Figure 18-23: **This color wheel shows how primary colors mixed with secondary colors produce intermediate colors.**

Figure 18-24: Designers use tints to achieve tone variation.

Tone affects the saturation and brightness of the color by adding white to the color. Tone provides contrast and is similar to the concept of monochromatic color. However, tone only refers to the use of a single color value rather than a combination of color values. A process known as tinting achieves tone. Tinting allows for a percentage of the original area to print. For example, a 30 percent tint is a 30 percent dot surrounded by 70 percent paper. The nonprinted areas or paper color visually blend with the image and take on a lighter appearance (see Figure 18-24).

As you have seen, good design involves logic as well as the application of aesthetic principles. In creating effective designs, new graphic designers should consider ways to apply unity, balance, rhythm, emphasis, and proportion and scale principles to their design solutions.

Proportion and Scale

The relationship of one part of a design to another is known as proportion. Objects have overall length, width, and height, and features within an object have other dimensions as well. The relationship between these various dimensions is proportion.

Once the designer groups all elements of the design, he or she will place the elements on a page. A standard-size page in the United States and Canada is 8.5 by 11 inches, or letter size. Other countries use the A4-size page (210 mm by 297 mm). However, designers can use any page proportion desired. Page or book sizes are usually rectangular, with ratios of 2:3, 1:3, 3:5, and 5:8. Standard paper sizes, or those sizes sold by paper manufacturers, and standard printing press sizes relate closely to the traditional page proportion. The metric paper sizes specified by the International Organization for Standardization (ISO) are based on a rectangle with a ratio of 1:1.414. Standard page sizes are most commonly used and are most efficient to produce, for this reason. However, different page proportions such as a square or very long rectangle will help create visual interest—it will just cost more.

The optical center is the focal point of the page and should be considered when placing the most important element of the design. The optical center is visually above the physical center of the page. Divide a page into three equal parts vertically. The optical center is two-thirds up from the bottom of the page. Because the reader's eye is most comfortably attracted to this part of the page, most designers group or cluster important information on or around the optical center rather than the physical center of the page.

It was recognized long ago that a shape divided into equal parts is rather dull and lifeless. More interesting shapes are suggested when areas are divided into unequal

(a)
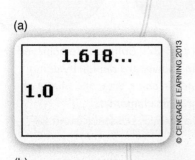

(b)
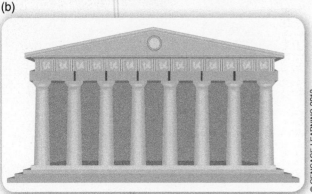

Figure 18-25: (a) The golden rectangle, sometimes called the golden section; (b) the Parthenon in Greece is a famous example of the golden section in architecture.

parts. The golden rectangle refers to a rectangle originated by the ancient Greeks whose proportions are based on the scale of the human body. The Greeks believed that this was the perfect proportion and so used it extensively in their art and architecture. It has a ratio of about 1:1.618 (see Figure 18-25). Although this is a good starting point, do not let yourself be limited to making all objects using the golden rectangle.

These and other common design concepts are applicable in other areas of design such as architecture and product design. Familiarity with the design elements and sensitivity to possible relationships make aesthetic designing a reliable process, accessible to anyone who is willing to pursue an idea.

USING TYPOGRAPHY EFFECTIVELY

Throughout modern history, the printed word has been the most effective form of communication. Human knowledge, for the most part, is stored as alphanumeric visual elements. Before Johannes Gutenberg invented movable type in about 1450, printed knowledge could only be stored in a handwritten or calligraphic form. As such, only a few people had access to man's accumulated knowledge, giving them control over the illiterate masses. Type for all printing was set by hand until 1886, when Ottmar Mergenthaler, an American mechanical engineer, invented machine composition. Mergenthaler's "Linotype" machine is considered one of the greatest inventions of the nineteenth century. The second major invention in composition occurred in 1984, when Paul Brainerd, president of Aldus Corporation, led the development of a special software program called PageMaker. This software combined with the Apple Macintosh computer and Apple LaserWriter printer created the first desktop publishing (DTP) system. The new system provided a WYSIWYG (pronounced "wis-e-wig") or "What You See Is What You Get" display of a full-page layout directly on the computer screen. The new DTP software allowed nearly anyone to create a basic typeset page.

OFF-ROAD EXPLORATION

See *How* magazine with creative ideas for design community:
http://www.howdesign.com/.

Composition is the process of assembling letters in varying formats to form words, sentences, paragraphs, and finally pages. Most authors, journalists, and copywriters today create their manuscript copy on a computer with specialized software to capture and display keystrokes. The primary advantage of word processing is the ease of editing and making corrections. For composition purposes, the major advantage is the capturing of keystrokes.

Modern graphic design software allows designers to compose type, photographs, and illustrations on a high-resolution color monitor (see Figure 18-26). Those images are sent via the Internet or by CD to a printer that outputs the image for production. Modern systems can output color-separated and fully imposed files (all elements properly positioned on a page format) to film or paper at more than 2,500 DPI.

Figure 18-26: **The pages of this text were composed using software to combine type, illustrations, and photographs on each page.**

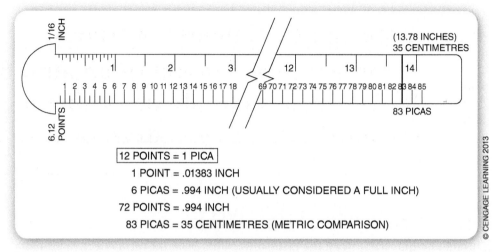

Figure 18-27: Printer's measurements (points and picas).

Understanding Type Specifications

Effective composition involves making good typographic decisions including specifying type size, line spacing, type style, line length, and formatting. Composition systems, such as desktop publishing, utilize typographic principles and are based on a typographic measurement system (see Figure 18-27).

Type Size. Type size is measured in points. A point (pt) is roughly .01383 of an inch but most designers think in terms of the number of points for the type size, such as 12 point type. The point measurement standard was adopted in 1886 by the United States Type Founders Association. It may appear awkward to us, but it did provide a practical system for dealing with small type sizes. In this system, 12 points are equal to one pica. The strength of the system is the ease of describing type sizes and other composition factors.

The numerical size of type is actually an expression of body size of the type (a block of an alloy of lead, antimony, and tin on which a typeface was cast) and not the visual element size. Furthermore, within one type size, the visual element size may vary. To understand these subtle differences among typefaces takes experience. To help those concerned with the visual element size, a system called *x-height* is used. In x-height, the lowercase x is measured and the height of each style is compared (see Figure 18-28).

Figure 18-28: The x-height.

REPRINTED WITH PERMISSION FROM POPPY EVANS AND MARK A. THOMAS, *EXPLORING THE ELEMENTS OF DESIGN*, SECOND EDITION. COPYRIGHT © 2008 CENGAGE DELMAR LEARNING.

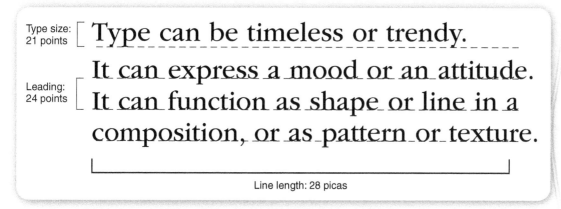

Figure 18-29: **The visual effect of varying line spacing.**
REPRINTED WITH PERMISSION FROM POPPY EVANS AND MARK A. THOMAS, *EXPLORING THE ELEMENTS OF DESIGN*, SECOND EDITION. COPYRIGHT © 2008 CENGAGE DELMAR LEARNING.

Line Spacing. Like type size, line spacing is measured in points. In the original setting of metal type, the space between lines was called *leading* (pronounced "ledding") and some designers still use the term today. Line spacing is the vertical distance between two lines and is measured from baseline to baseline (see Figure 18-29).

Type size and line spacing are expressed together. For example, 12/13 indicates that a 12 point typeface will be composed with 13 points of line spacing. When the type size and line spacing are equal, such as 12/12, it is referred to as *set solid*. On the average, one additional point of line spacing above a text face size is adequate. However, if the x-height is below normal, less line spacing may be used; or if above normal, more line spacing may be necessary. Normally, line spacing no greater than two points or 20 percent of the size of the type is appropriate. As line spacing is increased, so too is internal white space.

The spacing between the letters in words is called *letter spacing*. It can be set to loose, normal, or tight (see Figure 18-30). Letter spacing is automatically adjusted in standard digitized typesetting and, therefore, is usually of little concern to designers. However, sometimes a design calls for altered letter spacing such as a newspaper column that needs to be justified or a URL that needs to appear all on one line. Letter spacing is also used as a design element and can be manually adjusted in all publishing software—even Microsoft Word allows users to adjust letter spacing (Home>Font>Advanced>Spacing).

Another related concept is kerning. Kerning is the principle of adjusting letter spacing for visual purposes. Large display type often requires kerning to adjust for problem character combinations such as TY or LA where the letters will appear too far apart (see Figure 18-31).

Typeface or Font. With more than 1,500 different typefaces in use today, designers must carefully specify the proper typeface to use. A typeface or font is the name given to the actual shape of the letter. Differences between typefaces are often subtle. Typefaces fall into four major groups: serif, san-serif ("san" is Latin for "without"), cursive, and decorative. A serif is an additional visual detail added to the ends of some typeface elements. Cursive faces represent handwriting and decorative styles. These type faces, which are numerous, represent type styles that have limited or decorative usage and are typically not appropriately set in all caps (see Figure 18-32).

LETTER SPACING

LETTER SPACING

Figure 18-30: **Letter spacing.**
REPRINTED WITH PERMISSION FROM POPPY EVANS AND MARK A. THOMAS, *EXPLORING THE ELEMENTS OF DESIGN*, SECOND EDITION. COPYRIGHT © 2008 CENGAGE DELMAR LEARNING.

Figure 18-31: **Kerning.**

Type Group	Characteristics	Uses	Examples
Serif	Serif fonts have strokes, or feet, at the ends of the letters. They are considered traditional.	Body text, long passages, headlines, and display	Times New Roman Garamond Palatino
Sans-Serif	Sans-serif fonts are considered more contemporary.	Long passages, headlines, and display	Arial Helvetica Tahoma
Script	Script fonts resemble hand lettering. They can be serif or sans-serif, traditional or contemporary.	Display and decorative purposes	*Brush* Comic *Lucida Calligraphy*
Occasional	These typefaces include a broad range of style from traditional to contemporary.	Display and decorative purposes	Bauhaus Matura Aircut

Figure 18-32: **Different typeface groups have different uses.**

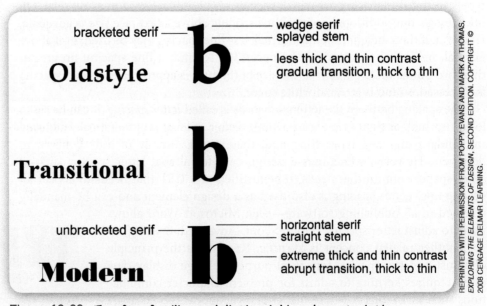

Figure 18-33: **Typeface families and distinguishing characteristics.**

Within each group are families of typefaces. A family consists of a basic typeface form with variations such as bold and italic (see Figure 18-33). This means that various parts of the typeface have similar characteristics. These similar characteristics will be in the shape of the serif, the contrast between thick and thin strokes, the position and direction of cross strokes, the length of ascending and descending stems, and the shape of the lowercase "g."

Line Length. Line length is the length of the column on which the type will be composed or set. It is measured in picas and points. Usually line measure is in full pica or pica and six-point (one-half pica) line lengths. Like the size of type, line length does not have to refer to the printed characters on the page, but rather to

Short lines of text are hard to read, because they create unnecessary hyphenation and awkward line breaks.

Long lines of text are hard to read because they cause the eye to track back to the beginning of the next line. Readers forced to read long lines of text often find themselves starting to read the line of text they have just read, instead of tracking down to the next line. Sometimes designers compensate for this factor by increasing the amount of leading between lines of text. Although adding space can help the eye to differentiate one line from the next, this option may be impractical when space needs to be saved. To save space and ensure reader-friendly text, set up a column width that allows for no more than fifty characters per line.

Above sample set in Garamond 8/10 at 25 picas.

Sample at left set in Garamond 18/21 at 10 picas.

Figure 18-34: **Determining appropriate line length.**

the maximum line length available to hold characters. Only in the case of justified composition, such as that found in most newspaper columns, are line length and printed characters equal in length. Proper line length ranges from one and one-half to two and one-half times the length of the lowercase alphabet (see Figure 18-34).

Line length is also determined by the number of words on the line. On the average, seven to nine sans serif or eight to ten serif words per line are acceptable. Line length is important because it affects the readability of the printed page.

Format. Type can be positioned anywhere horizontally on a specified line length from the left margin to the right margin. Format is the position of the type on the line. All interline spacing is based on a proportional system related to the size of the type. Word spacing is a visual factor and at times may need to be adjusted wider or narrower to make the spacing appear normal. In the case of justified copy, the space between each word is changed (usually increased) until the line is filled (see Figure 18-35).

Left-Justified	Right-Justified
Type can be left-justified, right-justified, full-justified, or centered.	Type can be left-justified, right-justified, full-justified, or centered.
Full-Justified	Centered
Type can be left-justified, right-justified, full-justified, or centered.	Type can be left-justified, right-justified, full-justified, or centered.

Figure 18-35: **Type justification can affect the readability of the copy.**

Pagination. Pagination is the process of formatting the captured keystrokes so that the typeset output will meet the original specifications. In addition, pagination often involves integrating illustrations (graphics) and photographs with the type. Pagination is the process of building composed material into books, magazines, newspapers, or whatever the final graphic form is to take.

The term *galley proof* comes from the fact that columns of composed metal type were placed on a galley, inked, and a proof was made for the proofreader. Today, it is more common to receive a page proof that can be reviewed online. As its name implies, a page proof shows the type in a page format. With the newest electronic prepress systems, type, illustrations, and photographs are included (sometimes in color) on the page proof.

A series of symbols have been developed to communicate changes from the proofreader. These symbols are called proofreaders' marks (see Figure 18-36). The marks are usually placed in the margin of the page and in a different color as an aid to the person making the corrections. Leaders or lines may be drawn from the mark to the area needing correction.

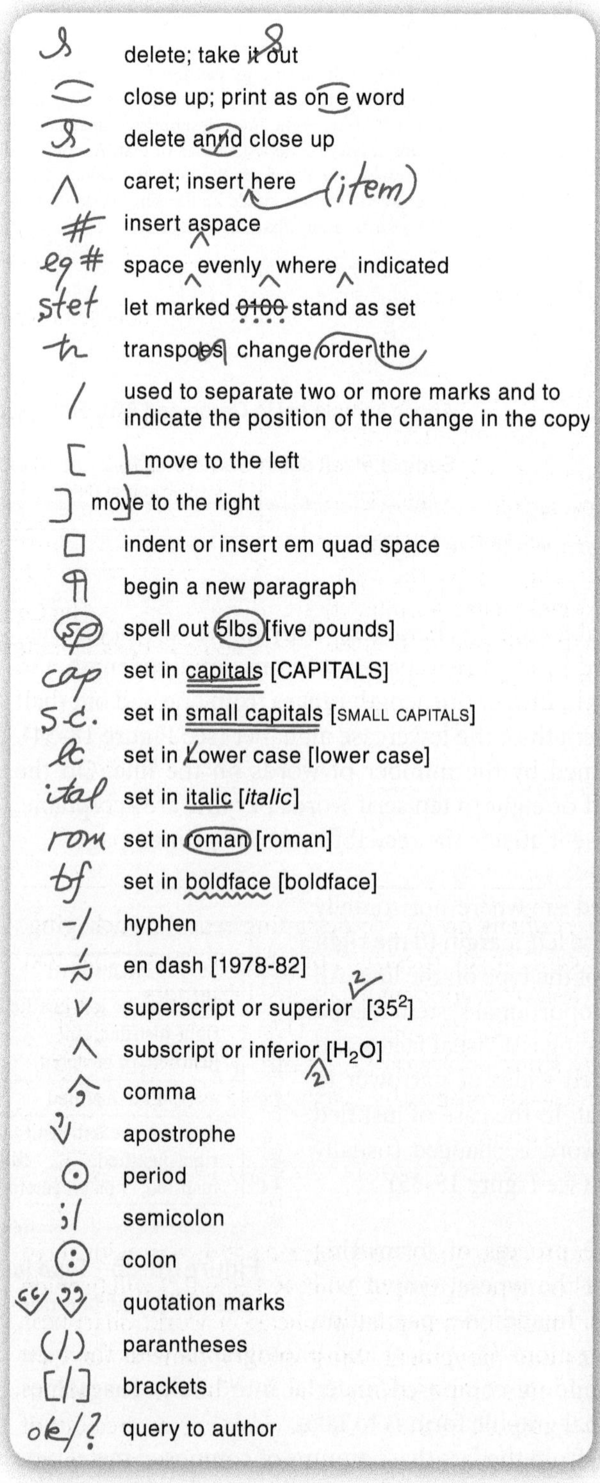

Figure 18-36:
Proofreaders' marks.

ℐ	delete; take it out
⌒	close up; print as on e word
ℐ̄	delete an/d close up
∧	caret; insert here (item)
#	insert a space
eq #	space evenly where indicated
stet	let marked 0100 stand as set
tr	transpoes change/order the
/	used to separate two or more marks and to indicate the position of the change in the copy
[	move to the left
]	move to the right
□	indent or insert em quad space
¶	begin a new paragraph
sp	spell out 5lbs [five pounds]
cap	set in capitals [CAPITALS]
s.c.	set in small capitals [SMALL CAPITALS]
lc	set in lower case [lower case]
ital	set in italic [*italic*]
rom	set in roman [roman]
bf	set in boldface [**boldface**]
-/	hyphen
⁄N⁄	en dash [1978-82]
∨	superscript or superior [35²]
∧	subscript or inferior [H₂O]
⸴	comma
⸲	apostrophe
⊙	period
;/	semicolon
⊙:	colon
❝ ❞	quotation marks
(/)	parantheses
[/]	brackets
ok/?	query to author

Today, it is more likely that designers will be asked to use an editing tool such as "Track Changes" in the Microsoft Word. When the Track Changes tool is being used, the recommended changes are identified by the reviewer. The Track Changes feature shows changes in content, grammar, and text formatting. This feature also allows for annotation for reviewers' comments or questions. When all reviews are returned, the copywriter must review all change recommendations and decide to accept or reject the recommendations.

Your Turn
Business Card Design

Designing a business card is relatively simple because the design constraints are well understood. The design starts with identifying all necessary text, grouping the elements, and then prioritizing them from most important to least important. Next, the designer identifies the graphics, such as a company logo or personal photograph. Some companies have corporate colors that are identified by their Pantone color system number. The design process results in a corporate template for the business card so that individual employees can plug in their personal information and receive a printed card for distribution (see Figure 18-37). How have design principles been employed in the design of the business card?

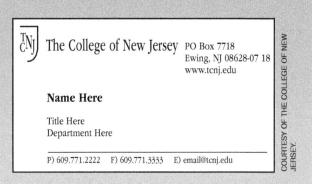

Figure 18-37: *The College of New Jersey Business card template.*

USING PHOTOGRAPHS

Photographs are an important part of any design. In designing reports, packaging, and presentations, photographs can convey messages or provide clarity that text alone cannot achieve. People like to look at good photographs. Designers will use both black-and-white and color photographs. Designers can take their own photographs, buy stock photographs, or hire a professional photographer. Taking your own photographs is acceptable as long as the quality and composition are adequate.

OFF-ROAD EXPLORATION
You'll find recommendations for taking great pictures from the Kodak Consumer Products Tips and Projects Center at *http://www.kodak.com/eknec/*.

Technical Qualities

Technical qualities include resolution, sharpness, and color. As shown previously in Figure 18-9, a good 3.2 megapixel camera capable of 2,048 × 1,536 PPI will provide adequate resolution and support most commercial publication work. Sharpness is a quality associated with focus and movement. Most quality cameras have an autofocus feature; however, this feature is not always reliable. The camera system can automatically focus on the wrong part of the scene, leaving the key area out of focus. Another more common reason for a loss of sharpness is camera movement. Many amateur photographers attempt to hold the camera by hand and consequently move the camera during shutter release. Therefore, photographers should use a tripod to steady the camera.

Finally, lighting influences color and contrast. Early morning outdoor lighting is best. High, direct sunlight on a clear day yields photographs with too much contrast; diffused lighting is more appealing. Indoor lighting and camera flash are usually inadequate for commercial-quality photographs, and professional photographers will often work with supplemental and specialized lighting. Figure 18-38 shows an example of the importance of lighting in a print advertisement.

Figure 18-38: **Lighting and resolution add to the detail and overall impact of this print ad.**

Composition

Composition refers to the selection of elements and the layout of those elements in the photograph. Composition is governed by aesthetic principles such as the rule of thirds. Although photo-editing software can crop out unwanted images, the general composition of the photograph cannot be altered. Unfortunately, if the composition of the picture is not appropriate for the design job, a new photograph will need to be taken—at considerable expense and loss of time. As with all aspects of design, good planning is necessary to get good pictures. The following suggestions will help in the planning and taking of photographs:

▶ Take many pictures. Use different camera angles and positions. Move in close to the subject. Get down to the subject's level when taking pictures of children or pets. Try both landscape and portrait formats.

▶ Check the background. Generally, a plain or out-of-focus background is best to draw attention to the subject.

▶ Divide the viewfinder into thirds both horizontally and vertically. Move the main subject from the middle of the picture by placing it on one of the thirds (see Figure 18-39).

▶ Be sure to focus on the subject and lock the focus before moving the subject from the middle of the viewfinder. Most camera systems use the center of the viewfinder to determine focal distance.

▶ Use early- or late-day lighting for scenic images.

▶ Know the direction of the sun and be sure unwanted shadows are not being captured. The diffused light of an overcast sky is better than direct sunlight.

▶ Be proactive by rearranging your subject, including props, and take your photographs from different viewpoints.

Figure 18-39: **The designer positioned the body of the biker at the intersection of the horizontal and vertical thirds.**

Figure 18-40: **This panoramic photograph was created in Photoshop by stitching together three individual photographs.**

Editing Photographs for Design and Production

When cropping alone will not yield usable photography, a photo-editing program, such as Adobe Photoshop, can be used. Photo-editing software is very useful in preparing photographs for publication. In addition to enhancing tonal ranges and color, these programs allow images to be extracted from their background and inserted as a component of an integrated design (see Figure 18-40).

USING ILLUSTRATIONS, SYMBOLS, AND LOGOS

The term *illustration* is commonly used to describe a wide range of graphic images. Designers may use charts and graphs to present categories of data, such as human factors measurements for the design population. Sometimes, designers use illustrations like artistic renderings in place of photographs to convey a message. Illustrations can be line drawings or tonal art rendered in black and white or color.

A specialized form of illustration is the symbol and logo. Symbols and logos are designed to be universally recognized and understood. These unique forms of illustrations are words or other images stripped down to their most basic form (see Figure 18-41).

Industrial designers in the early twentieth century introduced logos to improve product and corporate identity. The U.S. Patent and Trademark Office protects these trademark and service mark images (see Figure 18-42).

Figure 18-41: **Common symbols.**

Figure 18-42: **The "GE" logo has been widely recognized since the turn of the twentieth century.**

WEB DESIGN

The Internet has caused a revolution in the way people communicate and buy products and services. It is increasingly difficult to find a business, government agency, or private organization that does not have a presence on the Internet. Hobbyists rely on access to information and other individuals with similar interests through the Web. Writers and researchers rely on rapid access to millions of documents and search engines to provide them with current information. Even artists use the Web to show their work in the hope that it will be recognized and purchased. This has led to a business boom in designing and maintaining Internet sites.

Early Web design required knowledge of **hypertext markup language (HTML)**, a computer language developed for browsers to display graphic information on the Internet see figure 18-43. Soon after, software designers developed programs that allowed designers with little or no HTML knowledge to create Web pages in a WYSIWYG graphic environment. Today, programs like Adobe Dreamweaver and Microsoft FrontPage or WordPress are available for website development, allowing not only web page design but also the organization and management of complex websites that may include hundreds of separate web pages. These programs also integrate the mechanisms to transfer files from the designer's computer to the Internet server where the website resides. This function, called file transfer protocol (FTP), used to be a separate operation, and you can still obtain programs that will perform only this function.

OFF-ROAD EXPLORATION

Go to Google "web publishing platforms" for a more extensive list of web development sites or go to the web design gallery at *http://www.homestead.com/* or the showcase at *http://www.wordpress.org*.

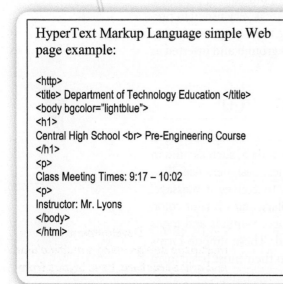
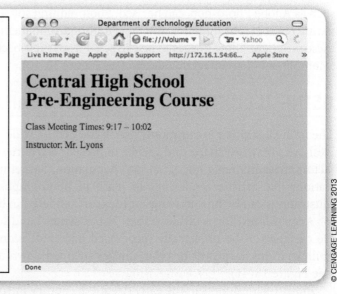

Figure 18-43: HTML code for a simple web page and the results seen in a Web browser.

Web Page Development

Much has been written about the development and evolution of the Internet and its rise to such a central place in our lives. In this text, we will provide you with information that will allow you to develop basic web pages, although you will need to learn considerably more to produce sophisticated websites.

Most web page development is done on a personal computer, with the files stored on that computer's hard drive. At some point, those saved files are transferred to the server of an **Internet service provider (ISP)** where they are accessible to anyone who knows the address. The address is called the **uniform resource locator (URL)**, and is identified by the prefix http:// that you have seen many times. The website designer creates and improves these files in one phase and transfers them to an ISP server in another phase. Web design software allows designers to perform both of these tasks.

Web design has much in common with other fields of graphic design and lends itself to the application of graphic design principles. It is more difficult, however, to apply some of those principles because of the variety of Web browsers in use, platforms and operating systems used, and the many screen resolution choices available. A design for the printed page does not change once it gets into the hands of the audience, but that is exactly what happens when visitors with different browsers and screen resolutions visit a website. Creating a web page that keeps these design elements consistent among all these variables is impossible, but the Web design professional must be able to make choices that will allow all website visitors to be able to use the site while maintaining an aesthetically pleasing look.

Brochure Design

Designing more complex items such as brochures involves a different set of constraints. Because there is more information to communicate, designers must develop a multipage format. All design principles still apply. A common format is to use multiple columns and a grid system (see Figure 18-44). The advantage of the grid system is to create a consistent look from page to page while still being able to vary the look. In the example shown, the name PT CRUISER is built on a grid that cuts across the columns and provides size contrast for the design. The photograph of the car has been cleaned up in a photo-editing program and integrated with the large type.

Figure 18-44: Development of multipage design using a multicolumn and grid structure. Page layout from @ issue.

Your Turn
Logo and Package Design

Design World Learning would like to create a new curriculum to promote environmental awareness. They are looking for a new logo and package design that conveys a universally understood message to support environmental awareness. Create at least six thumbnails for a new environmental awareness logo using uniquely created universal symbols.

MAKING PRESENTATIONS USING POWERPOINT

Many software programs such as Keynote by Apple or PowerPoint from Microsoft are available for preparing presentations. Design professionals make a variety of presentations and may provide status reports throughout the project. At the end of the design process, engineers and other design professionals will make a presentation that reports the final design solution to management or to the client. These reports may be made to a small or large group and they are always important to the success of the project. Good presentations result from hard work, careful planning, and well-designed supporting graphics. The presenter should also rehearse the presentation. Effective designers need to learn to be good presenters.

Planning a Presentation

Start by getting information about the presentation. The five Ws—who, what, when, where, and why—is a good place to begin.

- ▶ Who is going to make the presentation and who is going to attend? You want to know if you are making the presentation alone or if other members of the team will participate. Very importantly, a senior member of management may be planning to say something—you need to know this!
- ▶ What are the expectations for the presentation? Is it an informal meeting with business casual dress or a formal meeting where a suit is appropriate? How much time is being allocated for the presentation? Are other items on the meeting agenda?
- ▶ When is the presentation planned? You need adequate time to prepare and you need to have it on your calendar. Will you need to invite guests and provide information about travel and parking?
- ▶ Where will you make the presentation? You need to know how the room is organized. Is there a multimedia system to project your presentation or show other documentation? Will there be adequate seating and will everyone be able to see and hear you? Will you have a handheld microphone or clip-on Lavalier? How can you control lighting and temperature? As a presenter, your job is to make the audience as comfortable as possible.
- ▶ Why is the meeting taking place? This is probably the most important question as it will help the designer organize the presentation and include appropriate material. Is the meeting being planned for the design team or is the presentation needed to convince a client that the design solution should be supported?

With the five Ws answered, organizing the presentation can begin. You are the expert. The purpose for the meeting and the audience should suggest to you what you need to cover. Make notes and begin to develop an outline. What can you cover in the allotted time? What visuals or products should be shown? Do you need to make a short video? What are you going to hand out? Often presenters will write a summary of the important facts, sometimes called an *executive summary*. What will the executive summary look like graphically? Do you need to hand out a full report? Has it been printed? Should you make a PowerPoint presentation?

Making a PowerPoint presentation can be an effective aid in the presentation. You begin with your outline for the presentation and open the program. The first step is to select a new blank presentation or a new presentation from a design template. Sometimes a designer will create a unique template for the presentation or may use a corporate template (see Figure 18-45).

Templates come in a variety of styles called design templates, color schemes, and animation schemes. Each template allows the presenter to format text and other content such as photographs, charts, and video clips. Although the program templates are well designed, they may not work for your presentation. Designers can change the format by viewing the master slide. Any changes in the typography at this level will be available on all slides. Sometimes, the designer may want to make a change on just one slide. Making changes in the aesthetics of the slide presentation should be guided by the design principles covered in this chapter (review Figure 18-46). Presentation software typically allows for the integration of video clips or automatic jumps to websites. These programs can be easy to use and powerful but have potential negative consequences.

Figure 18-45: *An example of a custom template.*

When all the information for the presentation has been placed into the slide show, review your work by playing the slide show. View the show from the back of the room. Will your audience be able to read everything? Is there too much or too little information on the screen? Should the information, such as items on a list, be made visible one item at a time? If so, set a desired animation scheme. Play the slide show again. Are there other improvements that could be made in content, layout, or color scheme? When the presentation is done, you may want to make copies of the information for the audience. If the goal is to provide information, making copies with six slides per page is most efficient. If, on the other hand, you want the audience to participate by taking notes, making copies with three slides per page gives the audience room for writing.

> **OFF-ROAD EXPLORATION**
>
> See some commercial examples of graphic design for the Web and print at this graphic design and publishing center site at www.graphic-design.com/

Finally, practice the presentation. Slides can be advanced on a timer (usually not appropriate) or with the click of the mouse or arrow key on the keyboard. Nearly everyone is excited and nervous prior to a performance, even professional performers. What experienced presenters do to overcome jitters is focus on the performance and the audience and not their feelings. Novice presenters should know that jitters are natural and even to be enjoyed. The better people know their presentation and have practiced the material, the more they can focus on the audience and deliver an effective presentation:

- Be prepared, be early, be appropriately dressed, and check your setup.
- Have a well-designed slide presentation and be proud of it.
- Know the presentation very well and be comfortable with it.
- NEVER read the information on the screen!
- Have handouts prepared if appropriate.
- Greet your audience as they enter the room and focus on them during the presentation.
- Don't hide behind a desk; use a remote mouse if available.
- Be in charge; answer appropriate questions but stay on task.
- Finish on time.
- Mingle with the audience after the presentation.
- Enjoy a successful presentation!

Figure 18-46: A PowerPoint presentation using a designed template. The introductory slide uses the department logo; slides 2, 3, and 4 describe the department; and slide 5 introduces the design challenge. Note that slide 6 has an embedded video clip.

SUMMARY

Every day, hundreds of newly printed products enter the market. In most cases, these products have been carefully produced by a group of skilled tradespeople using advanced computer software, full-color printing presses, and specialized paper. Good graphic design requires hard work, careful planning, appropriate application of graphic design principles, and considerable revisions. Final design solutions must satisfy the needs of the client who buys it and the audience for whom it is intended.

The design process involves three unique activities. Collecting and organizing information is the initial part of the design process. It involves identifying and placing all available information about the product and the audience into some logical order. Creating involves the development of an idea through a completed visual model commonly called a comprehensive. A comprehensive is shown to the client who must approve the design prior to the design being produced in large quantities. Producing involves all the steps necessary to get the product ready for reproduction and the reproduction of the product itself. Reproduction, or making multiple copies of the graphic product, is a specialized form of manufacturing.

Although understanding the nature of the product and the design process is important, the ability to create effective designs involves understanding the visual relationship between the elements and the space where they will be displayed, such as on a page or a product. The selection, creation, placement, and visual qualities of these elements, such as color and tone, are controlled by graphic design principles. The principles can involve many visual subtleties, and new designers can achieve very effective designs by focusing on the rules associated with unity such as proximity and alignment, balance, rhythm, emphasis, and proportion and scale.

Effective composition involves making sound typographic decisions including specifying type size, line spacing, type style, line length, and formatting. Composition systems, such as desktop publishing, utilize typographic principles and are based on a typographic measurement system.

Photographs are an important part of any design. In reports, packaging, and presentations, photographs convey messages or provide clarity that cannot be achieved with text alone. People like to look at good photographs. Designers will use both black-and-white and color photographs. Designers can take their own photographs, buy stock photographs, or hire a professional photographer. Taking your own photographs is acceptable as long as the quality and composition are adequate.

Illustrations are a broad category of copy used in graphic presentations. Charts and graphs may be used to present categories of data. Sometimes, artistic renderings are used in place of photographs. These illustrations can be line drawings or tonal art rendered in black and white or color. A specialized form of illustration is the symbol and logo. Symbols and logos are designed to be universally recognized and understood. These unique forms of illustrations are words or other images that have been stripped down to their most basic form.

The Internet is widely used for communication, research, and to buy and sell products. hypertext markup language (HTML) is a computer language developed for browsers to display graphic information on the Internet. Software programs allow designers with little or no HTML knowledge to create web pages and manage complex websites. These programs also integrate file transfer protocol (FTP) mechanisms to transfer files from the designer's computer to the server of an Internet service provider (ISP) where they are accessible at an address

SUMMARY

(continued)

called the uniform resource locator (URL), identified by the prefix http://. Web design has much in common with other fields of graphic design and lends itself to the application of graphic design principles. The Web design professional must be able to make choices that will allow all website visitors to be able to use the site while maintaining an aesthetically pleasing look.

Design professionals are asked to make a variety of presentations. Early in a project, they may be asked to give a status report, and other status reports may follow. At the end of the design process, the designer will make a presentation that shows the final design solution to management or a client. These reports may be made to a small or relative large group and they are always important to the success of the project. Good presentation results from hard work, careful planning, well-designed graphics, and practice by the presenter. Effective designers need to learn to be good presenters.

BRING IT HOME

OBSERVATION/ANALYSIS/SYNTHESIS

1. Prepare a summary of your analysis of how the design in Figure 18-1 employs design principles.
2. Print a CMYK image on separate sheets of overhead material on a color copier (be sure the material is appropriate for the machine). Assemble the four sheets on a clear sheet of plastic by trimming and taping each color on a different edge. Finally, make a list of the effect of combining various CMYK colors by overlaying CM or MY or CY. What happens when K is added to an individual color or combination? What is the effect of all four colors?
3. Design a business card using design principles. The card can be printed on a color printer using perforated business card blanks.
4. Design a logo for a new cell phone to be used anywhere in the world.
5. Take a photograph of a product such as a cell phone and remove the background using a photo-editing program.

EXTRA MILE

ENGINEERING DESIGN ANALYSIS CHALLENGE

- Select one of the Greatest Engineering Achievements of the 20th Century and prepare a 10-minute PowerPoint presentation.
- Plan a presentation using the five Ws.
- Organize the presentation by taking notes and developing an outline.
- Collect visual images.
- Prepare the PowerPoint presentation. Select or design a template and modify the master slide if necessary.
- Review your work by playing the slide show and revise as needed.
- Make copies of the information for the audience if appropriate.
- Practice the presentation.
- Make the presentation.

Glossary

A

acceleration: The rate of change of velocity with time. Some typical units of acceleration are meters per second per second (m/s^2), feet per second per second (ft/s^2), miles per hour per hour (mi/hr^2), and miles per hour per second (mi/hr-s).

accuracy: The degree of conformity of a measured or calculated quantity to its actual value.

aesthetics: Having to do with appearance; a branch of philosophy that deals with human response to visual stimuli leading to judgments about the things we see. These judgments usually result in expressions of like or dislike and are related to cultural, economic, political, and moral values.

alternating current (AC): Electricity that is constantly switching the direction of flow. It is also often the case that the alternating current is sinusoidal in form; that is, it has a shape like a sine wave in time.

analysis: A detailed examination of the elements or structure of something.

angle: The amount a line needs to be rotated to bring the line into coincidence with another line.

anthropometry: The branch of science that deals with human measurement by describing human size, shape, and other physical characteristics. Anthropometric data are available for designers in statistical form.

architect: A professional who designs residential and commercial structures. Architects head a design team responsible for working with the client to establish the scope and style of the project. Architects are concerned with the intended use of the project and must meet all code requirements.

assembly line: A manufacturing process for the mass production of a product that involves adding interchangeable parts in sequence.

assessment: An evaluation technique that requires analyzing benefits and risks, understanding the trade-offs, and then determining the best action to take to ensure that the desired positive outcomes outweigh any negative outcomes.

assistive or adaptive technology: Products, devices, or equipment, whether acquired commercially, modified, or customized, that are used to maintain, increase, or improve the functional capabilities of individuals with disabilities (Assistive Technology Act of 1998).

audience: An identified group of people whose characteristics are of interest to a designer for product development purposes. Characteristics may include age, gender, income, interests, and other descriptive items that relate to the specific product.

B

bit: A bit refers to something that holds just a single piece of binary information. In other words, a bit is in effect a single placeholder binary number. A bit is either a 0 or a 1. The binary number 11011 is comprised of five bits, whereas the binary number 101 is comprised of three bits.

byte: A unit of computer storage containing eight bits.

C

cabinet oblique drawing: A form of oblique drawing in which the receding lines are drawn at half scale and usually at a 45-degree angle from horizontal.

CAD/CAM: The integration of computer-aided design with computer-aided manufacturing to improved efficiency.

classical style: Latin for "elite," it represents the highest order of architecture established by the Roman architect Palladio. The term is used today to describe architectural design that is based on original Greek and Roman design elements.

conductor: A material that allows the flow of charge (typically, but not always, electrons).

constraint: (1) A limit to a design process. Constraints may be such things as funding, space, materials, and human capabilities. (2) A limitation or restriction.

continuous improvement: A strategy used by industry to innovate or modernize existing products.

closed question: A question that has only one correct answer.

criteria: Principles or standards by which something may be judged or decided.

current: A measure of the flow rate of charge. Charge is measured in coulombs, and a typical flow rate is in coulombs per second or abbreviated C/s. Another name for C/s is an ampere, often abbreviated amp or the letter A.

D

density: A material's mass divided by its volume, nominally a constant. A few examples of density are air (1.2 kg/m^3), water (1,000 kg/m^3), and gold (19.3 gm/cm^3).

design: An iterative or repeating decision-making process that results in a plan to produce a new product.

design brief: A written plan that identifies a problem to be solved and its criteria and constraints. A design brief encourages thinking of all aspects of the problem before attempting a solution.

design process: A systematic and often iterative problem-solving strategy, with criteria and constraints, used to develop many possible solutions to solve a problem or satisfy human needs and wants and to narrow down the possible solutions.

design proposal: A written document, or set of documents, that clearly specifies how to fabricate a model, prototype, or final design. The design proposal should include documents that specify all (1) materials, (2) dimensions, and (3) processes used in the construction.

die: A term used for a variety of tools used in production to create a shape or 3-D form.

direct current (DC): Electricity that flows in only one direction and varies very little with time.

documentation: (1) The documents required for something, or that give evidence or proof of something; (2) drawings or printed information that contains instructions for assembling, installing, operating, and servicing.

durable and nondurable goods: A designation established by the U.S. Department of Commerce to describe the length of time a product is intended to be useful. Durable goods are those that are intended to last more than three years whereas nondurable goods are designed to be useful for less than three years.

E

element of design: A basic visual component or building block of designed objects (for example, line, shape, form, value, color, texture, or space).

engineering design analysis: The process of applying mathematical and scientific understanding to a proposed design solution to determine if there are any reasons why the design will not function as expected when it is prototyped or ultimately produced.

engineer's notebook: Also referred to as an *engineer's logbook*, a *design notebook*, or *designer's notebook*. Used as (1) a record of design ideas generated in the course of an engineer's employment that others may not claim as their own, and (2) an archival record of new ideas and engineering research achievements that can provide proof of an idea for patenting purposes.

entrepreneur: A person who takes initiative to establish a new enterprise or business and assumes risks in the hope of gaining financial and other successes.

ergonomics: The study of workplace equipment design or how to arrange and design devices, machines, or workspace so that people and things interact safely and most efficiently.

ethics: A branch of philosophy that considers how to apply concepts of right or wrong, good or evil, and taking responsibility for one's actions.

F

fatigue: In engineering, the fracture that occurs when a material is subjected to repeated or fluctuating stress that has maximum values less than the tensile strength of the material.

final design document: A complete set of design documentation prepared at the final stages of the design process. The final design documentation communicates clearly and completely what the design is and how well the design solves the problem. Final documentation should include all the necessary charts, graphs, calculations, CAD drawings, modeling, simulations, and text descriptions that, taken as a whole, represent the final design.

finite element analysis (FEA): A computerized numerical analysis technique used to solve mechanical engineering problems relating to stress analysis.

fixtures: Devices used to hold, clamp, align, or space during fabrication or a machining operation.

freehand: Done manually without the aid of instruments such as rulers.

G

graphic design: The process of planning for some form of visual communication using type, photographs or video, and illustrations arranged in an aesthetic layout for a printed page or digital display for the Web or multimedia presentation.

H

Hooke's law: The principle that stress applied to a material is proportional to the resulting strain (change in length). This is only true, however, within the elastic limit of the material.

human factors (HF): The application of the knowledge about human physical characteristics, behavior, and abilities to the design of products, systems, and environments for safe and effective use.

hydraulic system: Uses a liquid to transfer force from one point to another.

hypothesis: (1) An assumption based on limited evidence as a starting point for further investigation; and/or (2) a proposed explanation for an observation. A hypothesis is an educated guess, which forms a basis for investigation or analysis.

I

industrial designer: A professional who designs products and systems. Industrial designers work to create designs that optimize the function, value, and appearance of products and systems for the mutual benefit of the consumer and the manufacturer.

insulator: A material that does not allow the flow of charge (typically, but not always, electrons).

intellectual property: An original work usually associated with innovation but also covering authorship, including literary works, musical works, dramatic works, pantomimes, and choreographic works, pictorial, graphic and sculptural works, motion pictures, and sound recordings. U.S. patent, trademark, and copyright law protects intellectual property.

interchangeable parts: An important development during the Industrial Revolution where parts or components of a product were designed and made to fit in any product of the same type.

inverse square law: Any physical law stating that some physical quantity is inversely proportional to the square of the independent variable (often distance).

iteration: The act of repeating a set of procedures until a specified condition is met.

J

jig: Devices that help guide a tool for machining operation.

K

kinematics: A branch of engineering mechanics that studies motion without regard to the force or mass of the things being moved.

L

lateral thinking: Thinking that follows unconventional paths, sometimes called *low probability* thinking.

left brain: The left hemisphere of the cerebellum, where it is believed linear, verbal, analytical, and logical information processing are favored.

life-cycle cost: The total cost of a product that includes the amount spent on energy to run it. Also included may be the cost of recycling and/or the environmental impact cost to produce it.

linear perspective: In drawing, an approximate representation, on a flat surface, of an image as it is perceived by the eye. Typically, objects are drawn smaller as their distance from the observer increases, and items are somewhat distorted when viewed at an angle.

Luddites: An 18-century group from England who wanted to halt technological progress by smashing the new weaving machines being introduced into the textile industry. Today, Luddite is a name given to someone who is against technological progress.

M

machine: A device that is capable of transforming energy to accomplish a task. Machines can be mechanical such as a power tool or an automobile, or electrical such as a computer. The simple machines studied in science, including the lever, wheel and axle, pulley, inclined plane, wedge, and screw, only transform the direction or magnitude of a force. For mechanical engineering design purposes, students will need to understand the role of the lever and crank, wheel and gear, cam, screw, things that transmit tension or compression and things that provide intermittent motion to possible design solutions.

manufacturing process: The transformation of raw material into finished goods through one or more of the following: casting and molding, shaping and reshaping for forming, shearing, pulverizing, machining for material removal, or joining by transforming heat or chemical reaction to bond materials.

mean: Also called *average*; the sum of the numbers divided by the number of numbers in the set. For example, the mean of the first four odd numbers is $(1 + 3 + 5 + 7)/4 = 16/4 = 4$.

mechanism: An assembly of moving parts involved in or responsible for taking an input motion or force and transforming it to an output motion or force. Mechanisms are governed by physical principles and laws.

model: A detailed visual, mathematical, two-dimensional, or three-dimensional representation of an object or design. A model is typically smaller than the eventual, intended design. A model is often used to test ideas, make changes to a design, and to learn more about what would happen with a similar real object.

modeling: Creating a visual, mathematical, or three-dimensional representation in detail of an object or design, often smaller than the original. Modeling is often used to test ideas, make changes to a design, and to learn more about what would happen to a similar, real object.

O

one-point perspective: A method of realistic drawing in which the part of an object closest to the viewer is a planar face, and all the lines describing sides perpendicular to that face can be extended back to converge at one point, the vanishing point.

open-ended question: A question that has many possible correct answers.

P

parallel circuit: A circuit that has more than one path for the current to flow. Components connected in such a manner are often referred to as "in parallel".

parametric modeling: A CAD modeling method where each feature, such as a length of a side or radius of a fillet, uses a parameter to define the size and geometry of that feature and to create relationships between features. Because the software keeps a history of how the model was built, changing the parameter value updates all related features of the model at once when the model is regenerated.

Pascal's law: States that when a pressure or force is applied to a confined liquid, that force is transmitted to all parts of the liquid in all directions at the same rate.

patent: A form of legal protection granting exclusive rights to the inventor of a unique new product or process.

personal identification number (PIN): A number or alphanumeric that is used to gain entry to a secure site.

perspective: A form of pictorial drawing in which vanishing points are used to provide the depth and distortion that is seen with the human eye; perspective drawings can be drawn using one, two, and three vanishing points.

plan: (n) A detailed proposal for doing or achieving something; (v) the act of putting together a plan.

planned obsolescence: The conscious decision on the part of a designer to produce a product that will become obsolete in a defined time frame.

popliteal height: The measure from the floor to the back part of the leg behind the knee joint while seated.

postmodernism: A design movement beginning in the late 20th century reacting against the creative limitations of modernism and scientific objectivity.

precision: The degree to which several measurements or calculations show the same or similar results. *Repeatability* and *reproducibility* are also terms for precision.

primary source: Information that is original and has not been summarized or reported by someone other than the person or group responsible for the information.

problem solving: The process of understanding a problem, devising a plan, carrying out the plan, and evaluating the effectiveness of the plan to solve a problem, or meet a need or want.

product: In commerce, the term refers to all goods and services that are bought or sold. Products are designed to satisfy human wants and needs. In retail, a product is usually referred to as merchandise. In manufacturing, a product may be the raw materials purchased as commodities such as sheet metal or cabinet-grade plywood to make the finished good.

profit: In economics, profit is the difference between a company's total revenue, which includes all sources of income such as the sale of products and total costs (direct costs such as salaries and materials and indirect costs such as rent on facilities, equipment purchases, energy, and other costs that cannot be associated with any one project).

projection line: A horizontal or vertical line that can be used to locate entities in an adjacent view.

prototype: A full-scale working model of a design intended to have complete, or almost complete, form, fit, and function of the intended design. A prototype is used to test a design concept by making actual observations and making any necessary adjustments.

R

rapid prototyping: The process of creating a solid model in a computer-controlled machine using a CAD design file.

resistance: A measure of the resistance a material or component has to the flow of electricity (charge). The higher the electrical resistance, the more resistant the material or component is to the flow of charge (current). The unit of electrical resistance is volt-seconds per coulomb. These units of electrical resistance, being a bit complicated, are often referred to as an "ohm." One ohm is 1 volt-second per 1 coulomb. The symbol omega (Ω) is often used to represent an ohm. For example, a resistance of 12 ohms is the same as writing 12 Ω.

reverse engineering: A strategy used to find answers to questions about an existing product that are later used in the design of another product.

right brain: The right hemisphere of the cerebellum where it is believed simultaneous, holistic, spatial, and relational information processing are favored.

rotational velocity: The rate of change of angle with time. Two typical units of rotational velocity are radians/second (rad/s) and degrees/second (deg/s). (Rotational speed can also be measured in rotations per unit time with some examples being rotations per minute [rpm] or rotations per second [rot/s].)

S

safety factor: A measure of how much the product is overbuilt and the ratio of the ultimate stress (breaking point) and the working stress (maximum expected load).

schematic: A diagram of a system that uses graphic symbols to represent pneumatic, hydraulic, or electrical components.

science: A descriptive discipline that makes observations through experimental investigation to explain physical and natural phenomena. Science seeks to answer the question "Why?"

secondary source: Information that has been previously published usually by someone else.

sequential: Forming or following a logical order or sequence.

series circuit: A circuit that has only one path for the current to flow. Components connected in such a manner are often referred to as "in series."

sketching: Creating a rough drawing representing the main features of an object or scene, often as a preliminary study.

standard industry materials: Raw materials that have been processed into standard size, shape, or composition to be used in production of products, such as dimensional lumber, plywood, paper, cloth, and so on.

standardization: A process of establishing a technical consensus agreement that provides a common set of expectations for quality or compatibility of a material or product without creating an unfair competitive advantage in the marketplace.

structure: A structure is a body that supports a load and resists external forces without changing its shape, except for that due to the elasticity of the material(s) used in construction.

sustainable design: Also known as "green design", the process of producing products, systems, or environments that are environmentally friendly because they reduce the use of nonrenewable resources and minimize any negative impact on the environment.

synergy: Results when the unit or team becomes stronger than the sum of the individual members.

T

target population: A group of individuals for which a design is being made or targeted. Target populations are usually described in terms of gender and age, but may also include other characteristics such as income, left-handedness, or physical disability.

team: A group with a common purpose that achieves a specific goal using each individual's skills and mutual cooperation to produce the end product.

teardown: The process of taking apart a product to better understand it.

technical drawings: Drawings needed to actually produce the component or system.

technology: The human process of applying knowledge through innovation to satisfy our needs and wants by extending our capabilities and modifying our natural environment.

tolerance: The total permissible variation in a size or location dimension.

torque: The rotational equivalent to force. Torque is defined as force times distance [$\tau = (F)(d)$], where the distance is the distance between the acting force and point of rotation. The Greek letter tau (τ) is often used to represent torque. Typical units of torque are newton-meters (N-m), foot-pounds (ft-lbs), or inch-pounds (in-lbs).

two-point perspective: Two-point perspective is a realistic way of drawing objects in three dimensions using a horizon line, a key edge, and two vanishing points.

U

universal design: The design of a product, system, or environment that reduces barriers and assists not just the special needs population but all individuals.

V

variation: Also known as *variability*; a measure of the extent to which a dimension or parameter is expected to vary in magnitude.

viscosity: The resistance of a fluid to flow. Among other factors, temperature can cause a change in a fluid's viscosity.

visual brainstorming: A method of ideation in which drawing (in contrast to verbalizing) is used to generate large numbers of ideas. First, an existing object is drawn, then variations on that object are drawn, and then variations on one of those ideas are drawn, and so forth.

voltage: A measure of energy that can be imparted to a charge. The higher the electrical voltage, the more energy a charge will have. The unit of electrical voltage is volts, which is often abbreviated with a V. For example, a 12-volt battery is equally referred to as a 12 V battery.

# Index

Page numbers followed by f indicate a figure
Page numbers followed by t indicate a table

A

Absolute Beginner's Guide to Building Robots (Branwyn), 282
Absolute graphs *vs.* relative graphs, 434–436, 435f
ABS plastic. *See* Acrylonitrile-butadiene-styrene (ABS) plastic
AC. *See* Alternating current (AC)
Acceleration, 327, 449, 452–453, 461
Accountability, as virtual team challenge, 69
Accuracy, 212, 212f, 218, 465
Acid rain, 22, 22f
Acrylonitrile butadiene styrene (ABS) plastic, 173, 283
Active listening
   defined, 81
   techniques for, 82
Actuators, 366, 370, 371–374, 372f–373f
Adam, Robert, 486
Adding fraction, 209
Additive color principle, 512, 512f, 513, 513f
Adhesives, 279
Adobe brick, 291, 479
Advertising, 190, 259f, 517f
Aeolipile, 264f
Aerospace engineer, 7
Aesthetic(s)
   defined, 97, 471, 472, 505
   in design work, 252–253, 253f
   principles of, 5, 6, 520, 528
Agrarian revolution, 264, 264f
Agricultural Age, 8, 8f, 10, 10f
AIA. *See* American Institute of Architects (AIA)
AIPA. *See* American Inventors Protection Act (AIPA)
Air resistance, 356, 453, 454
Airtight seal, 167
Akashi-Kaikyo Bridge, 309, 310f
Aldus Corporation, 521
Aligned dimensioning, 230, 230f
Aligned left, 517
Alignment, as design principle, 217, 218f, 516
Alternating current (AC), 16, 17, 357, 357f
ALU. *See* Arithmetic logic unit (ALU)
American Civil War, 14

American Design Drafting Association, 207
American Institute of Architects (AIA), 473, 484
American Inventors Protection Act (AIPA), 192
American National Standards Institute (ANSI), 15, 15f, 216, 217, 217f, 420
American Psychological Association (APA), 198
American Society for Testing and Materials (ASTM), 15f
American Society of Mechanical Engineers (ASME), 12, 270, 324
American Standard Code for Information Interchange (ASCII code), 378
American system of production, 11
AMH. *See* Automated material handling (AMH)
Ampère, André-Marie, 355
Amplifier, operational, 376, 376f
Analogies, 89, 89f
Analogous colors, 125, 125f
Analog processor(s)
   defined, 374
   555-timer chips, 375–376, 375f
   operational amplifiers, 376, 376f
   transistors, 374–375, 374f
Analysis
   defined, 168
   engineering design, 298
Analytical thinking, 103
Angle, 208, 455, 455f
Annotated sketch, 148, 148f, 167, 167f
Annotation, 148, 149, 165, 216, 217f
ANSI. *See* American National Standards Institute (ANSI)
Anthropometric data, designing with, 418–420, 418f, 420f
Anthropometry, 406
Anytime™ Chair, 51, 284–285, 284f–285f
APA. *See* American Psychological Association (APA)
Appearance, 471, 495
   development, 97–99, 98f, 99f
Approach selection, in design process, 46–47

Arch bridge, 307, 309, 309f
Architects, 5, 5f, 108, 109, 471, 473
   Bauhaus, 499, 499f
   Chicago, 473, 473f, 488, 488f
   Egyptian, 475, 475
   Greek, 475, 475f
   Pasadena, 488, 488f, 489f
   postmodern, 491–494, 491f–494f
   Roman, 476, 476f
Architectural career information, 6
Architectural scale, 209–210, 209f, 210f
Architectural styles, 484–491
   Art Deco style, 489, 489f
   Colonial style, 485, 485f
   Craftsman style, 488–489, 488f, 489f
   Federal style, 486, 486f
   Georgian style, 485–486, 485f
   Greek Revival style, 486, 486f
   International style, 490, 490f
   modern style, 490, 491f
   Prairie style, 488, 488f
   twentieth-century styles, 487
   Victorian style, 486–487, 487f
Arithmetic logic unit (ALU), 381, 381f
Armstrong, Lance, 199
Armstrong, Neil, 17
Art Deco style, 489, 489f
Artifacts, 8
Art Nouveau design style, 498, 498f
Arts and Crafts movement, 488, 490, 499, 499f
ASCII code. *See* American Standard Code for Information Interchange (ASCII code)
ASME. *See* American Society of Mechanical Engineers (ASME)
Aspect ratio, 94
Assembling, 268
Assembling Etch A Sketch, 175, 175f
Assembly drawing, 216, 217f
Assembly line, 267, 267f
Assembly models, 220, 220f
Assessment, design process and, 46
Assistive/adaptive technology, 415–416, 416f
ASTM. *See* American Society for Testing and Materials (ASTM)

543

Atmospheric engines, 265, 266f
Atoms of elements, 350
Audience, defined, 505, 506
AutoCAD, 219, 219f
AutoCAD Version 1.0, 219
Autodesk Inventor, 171f, 220f, 239, 239f
Automated material handling (AMH), 280
Automatic enumeration, in spreadsheets, 439, 439f
Autonomous, defined, 35
Auxiliary view, 232, 232f
　drawings, 205
Average person myth, 199, 199f
Axonometric projection, 221, 221f

B

Background, adding, 144, 144f
Balance, principle of, 517, 517f, 516f
Barnick, Anita, 191
Barnick, Nicholas, 342
Baseline dimensioning, 233
"Base-2" system, 378
Bauhaus, 499, 499f
Beam bridge, 307–308, 307f, 308f
Beams, 305, 305f
Beam steam engine, Thomas Newcomen's, 265f
Bell crank, 333, 333f
Benchmark, 158
Bending, 269, 269f, 271, 271f, 302–303, 303f
Bidirectional microphone, 437
BIFMA. *See* Business and Institutional Furniture Manufacturer's Association (BIFMA)
Bilateral tolerances, 234, 234f
Binary numbers and arithmetic, 378–380, 380f
Binding of portfolio, 151
Biomedical engineers, 64
Bipolar transistor, 375, 383f
Bird's-eye view, 134, 134f
Bits, 382
Blackout, 18
Blanc, Honoré, 266
Blanking, 276
Blind contour drawing, 117–118, 117f, 118f
Bluehill® 2 software, 194f
Bluetooth microphone, 369, 369f
BMW, 258, 258f
Body position, safe work environment and, 422–423, 422f
Bonding, thermal, 279, 279f
Bookmark, 198
Boolean logic, 197
Bottle, plastic, 24
Bow's notation, 313–315, 313f–315f
Brainerd, Paul, 521
Brainstorming
　basic tenets of, 90
　defined, 38, 89
　in design process, 38–40, 39f
　documenting, 150
　session, 89–90, 89f
　visual, 147–148, 148f
Brand, 186
Branwyn, Gareth, 282
Brass stylus, 171f, 175f
Break-even point, 442, 442f
Breaking point, 301, 301f
Breaking the rules method, 90
Bridge structures, 305–310, 306f–310f
Britannica.com, 195
British Standards Institution (BSI), 15f
Brochure design, 531, 531f
Brooklyn Bridge, 306, 306f, 310
Brunel, I. K., 11
Brunel, Marc Isambard, 267
BSI. *See* British Standards Institution (BSI)
Buckminster, Richard, 294f, 490, 490f
Building, 477–491
　architectural style of, 484–491. *See also* Architectural styles
　eaves, 479, 480f
　elevations, 479, 479f
　exterior walls, 483–484, 484f
　floor plan, 477, 478f
　roof style, 479, 480f
　windows/doors, 481–482, 481f, 483f
Business and Institutional Furniture Manufacturer's Association (BIFMA), 285, 420f
Business card design, 527, 527f
Bytes, 382

C

C. Y. Lee and Partners, 291
Cabinet oblique drawing, 223
Cabinet projection, 223, 223f
CAD. *See* Computer-aided design (CAD); Computer-aided drafting (CAD)
CAD/CAM, 281, 281f
CAD software programs, future of, 239
CAD solid modeling programs, 235
CAD 3-D modeling software, 237
Calculations, spreadsheets and capabilities of, 439–443, 439f, 440f, 441f
Calipers, 168, 169, 169f
CAM. *See* Computer-aided manufacturing (CAM)
Camera phone technology, 158
Cantilever, 309, 309f
Cantilevered balconies, 474
Capability, 412
Capacitors, 364–365, 364f, 365f
Carbon resistor, 361
Career information
　for architects, 6
　for design profession, 4
　for engineering, 12
　for industrial design, 7
Careers in design
　Anita Barnick, 191
　Anne Fabrello-Streufert, 77
　Art Clark, 180
　Kristin Weary, 349
　L. Stephen Schmidt, 421
　Molly Hawthorne, 270
　Nicholas Barnick, 342
　Patty Ratchford, 258
　Randy Rausch, 9
　Stephen Douglas, 295
Carpal tunnel syndrome, 406
Cartesian coordinate system, 328, 434, 434f
Cartography, 118
Casement windows, 326, 326f, 482, 483f
Case study(ies)
　Anytime™ Chair, 284–285, 284f–285f
　designing new Ketchup bottle for Heinz Company, 186–187
　disassembling Etch A Sketch, 167, 167f
　functional analysis of Etch A Sketch, 170, 170f
　Jonathan Ive, 500–501, 500f–501f
　laser designer, 53
　manufacturing analysis of Etch A Sketch, 175, 175f
　material analysis of Etch A Sketch, 173, 173f
　Michael Graves, 492–494, 492f–494f
　Ohio Art Company and Etch A Sketch, 159, 159f, 164
　reverse engineering documentation, 176–178, 176f–178f
　sewing machine, 13, 13f
　space shuttle *Challenger*, 26
　structural analysis of Etch A Sketch, 171, 171f
　teams make sense for two printed circuit board designers, 67
　toy guide for differently abled children, 417
Casting, 269, 273, 273f
　drawing with, 206f
Caulk, John Jr., 219
Cavalier projection, 223, 223f
Cell phone, design of, 4–5, 5f
Center for Ergonomics at the University of Michigan, 406
Centerlines, 228–229, 229f
Center of gravity (CG), 97
Central processing unit (CPU), 381, 381f, 501
Centrifugal casting, 274, 275, 275f
Century of Innovation: Twenty Engineering Achievements That Transformed Our Lives, A (Constable and Somerville), 17
CG. *See* Center of gravity (CG)
Chain dimensioning, 233, 233f

Chair design, 409f, 419–420, 419f
Challenger, Space Shuttle, 26, 26f
Chamfer, 235, 236f
Charter, defined, 72
Checklist, in test results, 256, 257f
Chemical engineers, 64, 64f
Chemical fastening, 279–280, 279f
Chemical method, 277
Chicago architect, 473, 473f, 488, 488f
Chip removal, 276, 277f
Chroma, 124
CIM. *See* Computer-integrated manufacturing (CIM)
Circuit, 355, 368f, 373f
Circuit components
    capacitor, 364–365, 364f
    inductor, 365, 365f
    resistors, 361–363, 362f
Civil engineer, 7, 64, 108, 294
Clarification, serial discussion for, 92
Clark, Art, 178, 180
Classical style, 475, 479
Classic Etch A Sketch, 159
Class 1 lever, 331
Class 2 lever, 331
Class 3 lever, 331
Clearance, 409–410, 410f
Clearance angle, 277, 277f
Clearance measurement, 199
Clip art, 515, 515f
Closed-loop system, 20, 21f
Closed questions, 188
Cloud computing networks, defined, 68
CNC. *See* Computer numerical code (CNC); Computer numerical control (CNC)
Coach(es), 70, 76
    to team leader, 76t
Coca-Cola Company, 192f, 193, 496, 496f
Code of ethics, 27
Code of Hammurabi, 434
Codes, 473
Coding, 377–378, 378f
Cold forging, 272
Collaboration, defined, 66
Collaborative learning, 69
Colonial style, 484, 485, 485f
Colored pencil techniques, 144, 144f
Color marker techniques, 144, 144f, 145f–146f, 146
Color(s)
    additive color principle, 512, 512f, 513, 513f
    analogous, 125, 125f
    appearance, 471
    chroma, 124, 124f
    complementary, 125, 125f, 519
    as design element, 99, 124–126, 124f, 125f
    and design emphasis, 519
    hue, 124, 124f
    primary, 124–125, 125f
    secondary, 125, 125f
    subtractive color principle, 512, 513, 513f
    tertiary, 125
    value, 124, 124f
Color science and technology, 126
Color wheel, 124, 125, 125f, 513f
Columns, Egyptian, 475, 475f
Combined handset, 497, 497f
Combining materials
    chemical fastening, 279–280, 279f
    mechanical fastening, 278–279, 278f, 279f
COMDEX trade show, 219
Common forces and structural system, 301–304, 302f–304f
Communicating process, in design process, 52, 52f
Communication
    positive, 81, 81t
    team. *See* Team communication
    through drawing, 114–119, 114f–118f
    toolbox, 82, 83f
Commutator, 371
Compatibility, 411, 411f
Competition, first, 36
Complementary colors, 125, 125f, 519
Composition, photographs and, 528, 528f
Composition process, 521, 521f
Comprehensive drawing, 506
Comprehensive or "comp" design, 509, 510f
Compression, 302, 302f, 303f
Computer-aided design (CAD), 108, 206, 237–239, 281
    computer-integrated manufacturing, 280
    defined, 108, 158
    drafting, 219–220, 219f–220f
    ideas development and, 96–97, 96f, 97f, 98f
    program aspects, 208, 235
    technical drawing and, 206, 206f
Computer-aided drafting (CAD), 108, 158
Computer-aided manufacturing (CAM), 146, 160, 246, 280, 281,
Computer-integrated manufacturing (CIM), 280–282, 280f, 281f
Computer numerical code (CNC), 174
Computer numerical control (CNC), 160, 207, 342, 372
Computers, technologists and, 20
Computer spreadsheets and structured programming, 437–448, 438f–447f
Computer work station, 422–424, 422f, 424f
Condiment, 186
Conductivity, 353, 353f
Conductor, electrical, 351, 353, 353f
Cones, 124
Configurations, 97
Consensus
    defined, 72
    reaching, 79
Conservation law, 358
Console computer games, 31, 56
Constable, George, 17
Constraint
    defined, 41
    in design process, 41–43
Construction kits, 95–96, 96f
Construction lines, 228
Consumer, 6, 185
    technical drawing and, 216, 217f
Consumer-based information, 190
Consumer Reports magazine, 246, 246f
Consumers Union, 189, 246
Continuous improvement, defined, 159
Contour drawing, 117–118, 117f, 118f
Contrast, as design principle, 518–520, 519f, 519f
Convention, 216
Convergence, 33
Cook, David, 282
Cooke, Morris Llewellyn, 17
Copernicus, Nicholaus, 348
Copyright, 192, 514
Corinthian columns, 479, 480f
Cornucopian philosophy, 27
Cost, in testing and designing design work, 254, 254f
Cost calculator, 440
Coulomb, Charles, 350
CPU. *See* Central processing unit (CPU)
Cracking Creativity: The Secrets of Creative Genius (Michalko), 87
Craftsman style, 488–489, 488f, 489f
Crating, 142, 142f
Creating/making, in design process, 51–52, 51f
    documenting, 150
Creative thinking, 87, 90
Creativity
    defined, 87
    in design process, 54–56, 54f, 55f
Criteria
    and constraints identification, documenting, 150
    defined, 41
    design process and identifying, 41–43, 42f, 43f
    innovation and, 54–56, 54f, 55f
    selection of, 103
Cross-training, 67
Crystal Palace Park, 11, 11f
CTDs. *See* Cumulative trauma disorders (CTDs)
Cultural differences, as virtual team challenge, 69
Cumulative trauma disorders (CTDs), 406, 413
Current, 355–357, 356f–357f, 372, 373, 374, 374f
Current law, Kirchhoff, 358–360, 360f
Custom products, 406–407, 407f

Cutaway drawing, 141, 141f
Cutting plane line, 230
Cylinders
   double-acting, 394, 395f, 397f, 398, 400f
   pneumatic, 392–396, 393f–396f
   single-acting, 393, 393f, 394, 394f, 396, 396f, 397f, 399

D

DARPA. *See* Defense Advanced Research Projects Agency (DARPA)
Databases, 190, 195
Data organizer form, 166, 166f
Datum dimensioning, 233, 233f
da Vinci, Leonardo, 45f, 89, 207, 207f, 321, 340
DC. *See* Direct current (DC)
Dead loads, 299, 308f
De Bono, Edward, 87, 88
Decimal number, 379, 380
Decision making, as virtual team challenge, 69
Decision matrix, 47, 47f
Defense Advanced Research Projects Agency (DARPA), 35
De la Fuente, Reina, 67
DeLuca, Tom, 417
Demographics, 185
Density, 433
DePuy Orthopaedics, Inc., 342
DePuy Spine, 191
Descartes, René, 434
Description, in test results, 256
Descriptive statistics, 446, 446f, 447f
Design. *See also* Design brief; Design process; Electrical system design; Mechanical system design; Structural system design
   architectural, 473–484, 473f–484f
   careers in, 173
   case study, 53
   defined, 3, 31, 107
   documentation of, 150
   elements of, 99–102, 99f, 101f, 102f
   emphasis, 518–520, 518f–520f
   ethics and, 27
   human factors in, 422–424, 422f, 424f
   ideas, generating, 87–94, 88f–91f, 93f–94f
   and industrial revolution, 8, 8f, 10–12, 10f
   limitations, 56–57
   for manufacture, 282–283, 282f–283f
   principles, 515–521, 516f–520f
   process of. *See* Design process
   styles. *See* Design styles
   sustainable, 24
   universal, 414, 415–418, 415f–417f
Design brief
   in design process, 42, 43f
   documenting specification and, 150
Design communication, 203
Design continuum, 5–7, 5f–7f

Design elements, defined, 101, 101f
Designer
   defined, 108
   fashion, 6, 109, 125
   graphic, 6, 31, 506
   industrial, 6, 471, 494–499
   laser, 53
Designing for manufacture, 282–283, 282f–283f
Designing for People (Dreyfuss), 405, 498
Designing with anthropometric data, 418–420, 418f–420f
The Design of Everyday Things (Norman), 256
Design portfolio, 109–110, 110f
   student, 111–113, 111f–113f
Design principles, defined, 97
Design process
   approach selection in, 46–47, 47f
   brainstorming in. *See* Brainstorming
   case study, 53, 53f
   communicating process in, 52, 52f
   concept of iteration in, 33–34, 34f
   constraints/criteria in, 41–43, 42f, 43f, 44f
   creating/making in, 51–52, 51f
   creativity and innovation in, 54–56, 54f, 55f
   defined, 31
   design brief in, 42, 43f
   design proposal in, 47–48
   developing tests, 246–248, 246f, 248f
   drawing and, 108, 108f, 109f
   exploring possibility in, 44–46, 44f, 45f
   five-step, 32, 32f
   introduction to, 31
   limitations, 56–57
   managing project in, 34–36
   ordering in, 32f, 33
   planning in, 32–33, 33f
   refining in, 50–51
   researching and generating ideas in, 40–41, 40f, 41f
   teams and, 61, 62, 61f
   testing/evaluating in, 49–50, 50f
   twelve-step, 36–52, 36f–45f, 47f–48f, 50f–52f
Design professionals, 4–8, 5f–7f
Design proposal
   defined, 47
   in design process, 47–48
Design skill, evaluating, 259
Design styles, 470–501
   architectural design, 473–484, 473f–481f, 483f–484f
   architectural styles, 484–491, 485f–491f
   industrial design, 494–501, 495f–501f
   introduction, 471–472, 471f, 472f
   postmodern architects, 491–494, 491f–494f
Design teams, 189, 193
Deutsches Institut fur Normung (DIN), 15f
Developmental sketch, 148–149, 148f
Development work, ideas and, 95–102, 96f–99f, 101f–102f

Dial caliper, 168, 169f
Die, 271
Die casting mold, 274, 274f
Die gut gasket, 276
Digital caliper, 168, 169f
Digital processor(s)
   binary numbers and arithmetic, 378–380, 380f
   coding, 377–378, 378f
   logic operation, 378, 378f
   microprocessors, 380–382, 381f–382f
Dimensioning
   features, 235–236, 235f, 236f
   methods of, 233
   in technical drawing, 232–236, 233f–236f
Dimension(s)
   lines, 229–230, 230f
   precision, 234
   on technical drawings, 218, 218f
   tolerances, 234–235, 234f, 235f
DIN. *See* Deutsches Institut fur Normung (DIN)
DIP. *See* Dual in-line pin (DIP)
Direct current (DC), 16, 17, 357, 357f, 371
Direct quote, 193
Disassembly
   in reverse engineering, 165–168, 166f, 167f
Dissatisfaction (storming), in team developing, 75
Divergent thinking, 200
Division of labor, 61, 61f
Document, final design, 52
Documentation, 108
   defined, 108
   of design, 150, 151–154, 151f–154f
   for developing reverse engineering, 176f
Dominant idea, 88
Doodling, 93–94
Doors, 481–482
Doric columns, 479, 480f
Dorsiflexion, 413, 413f, 424
Double-acting cylinders, 394, 395f
Double-acting steam engines, 265
Double-hung window, 482, 483f
Douglas, Stephen, 295
Drafting and CAD, 219–220, 219f, 220f
Draftspersons, 108
Drawing(s), 95, 271. *See also* Technical drawing
   angles, 208
   arcs, 208
   assembly, 216, 217f
   basics, 119–128, 119f–128f
   blind contour, 117–118, 117f, 118f
   communicating through, 114–119, 114f–118f
   comprehensive, 506
   contour, 117–118, , 117f–118f
   conventions, 142–147, 142f–146f, 216
   cutaway, 141, 141f

developing ideas, 107–108
developmental sketches and, 148–149, 148f
documenting the process, 108–113, 108f–113f
exploded, 141, 141f
exploring visible world, 107
freehand, 107, 129
introduction to, 107
isometric, 139–140, 140f
of materials, 273
production, 149, 149f
sectional, 141, 141f
sketching and techniques, 129–141, 129f–141f
types of, 141, 141f
uses of, 107–113, 108f–113f
on vellum, 219
warm-up exercises, 114–119, 114f–118f
Dreyfuss, Henry, 6, 405, 495, 497–498, 497f
Dryer, 392
Dual in-line pin (DIP), 375
Durability, in testing and evaluating design work, 253–254
Durable goods, 296
Dynamic loads, 299
Dynamic microphone, 369
Dynamics, 324, 452–456, 455f, 456f
Dzurko, Ken, 53

E

Earth's resources, technology and, 26
Eaves, 479
  flared, 485
E-Book, 218f
Eccentric cams, 339, 339f
ECN. See Engineering change notice (ECN)
Eco-design. See Sustainable design
Edison, Thomas, 16
Egyptian architects, 475, 475f
Ehrlich, Paul, 27
Eiffel Tower, 292, 292f
Einstein, A., 88
Elasticity, modulus of, 250
Elastic limit, 249
Elastic stage, 301, 301f
Elderly/physically challenged, 416, 416f
Electrical conductor, 351, 353–354
Electrical engineer, 7, 55, 63, 63f, 77
Electrical force, 350, 351, 351f
Electrical method, 277
Electrical resistance, 354–355, 355f
Electrical system design
  basic definitions, 366–383
  careers in, 349
  electronic systems design, 366–383, 366f–382 fintroduction to, 347–348
  major circuit components, 361–365, 362f, 364f, 365f
  science of electricity, 350–361, 351f–361f

Electricity
  science of, 350–361, 351f–361f
  in the twentieth century, 16–18
Electric solenoid, 373, 373f
Electromagnet(s), 360, 361, 371, 372f, 373f, 374
Electronic relays, 373–374, 373f
Electronic system design
  analog processors, 374–376
  system inputs, 367–370, 367f–370f
  system outputs, 370–374, 371f–373f
  system processors, 374, 374f
Electrons, 350, 351, 353, 354, 354f, 355, 356f
Elements, 324, 325
  analyzing, in reverse engineering, 168
  atoms of, 350
Elements of design, 119
Elevations, 479, 479f
Emphasis, design, 518–520, 518f–520f
Encyclopedias, 195–196
Energy, 328
EnergyGuide label, 254
Engineering, 265
  achievements in the twentieth century, 16–18, 16f
  careers, 19
  color, 100
  defined, 139
  design, measurement for, 208–209
  education, 65
  of material testing, technology and, 251
  process, technical drawings and, 218
  professionals, 62, 62f
  rules, 209
  societies, 12, 12f
  solution, testing, 248–252, 249f, 250f, 252f
Engineering change notice (ECN), 218
Engineering design analysis, 298, 298f
Engineers, 7
  and environment, 64–65
Engineer's notebook, 45, 45f, 109, 110–111, 150–151, 165, 170f
Engineer Your Destiny, 62, 62f
Entablature, 479, 480f
Entrepreneur, 10
Environmental engineers, 64
Environmental impacts, 254
EPA. See U. S. Environmental Protection Agency (EPA)
Equilibrium, 299–2300, 299, 300ff
Ergonomics, 199, 253, 253f. See also Human factors (HF)
Estée Lauder, 270
Etch A Sketch
  assembling, 175, 175f
  developing hypothesis, 164–165
  disassembling, 167, 167f
  documentation, 176, 176f
  functional analysis of, 170, 170f
  largest, 179f

  manufacturing analysis of, 175, 175f
  materials analysis of, 173, 173f
  Ohio Art Company and, 159, 159f, 164
  at present, 177–178, 178f
  price point of, 164
  SIGGRAPH, 179, 179f
  structural analysis of, 171, 171f
Ethics
  defined, 26
  and design, 27
Evaluation
  of design for human factors, 422–424, 422f, 424f
  of design skills, 259
  testing and, 244–261
Exploded assembly, 216
Exploded drawing, 141, 141f
Exploration phase, in design process, 45
Exponential function, 462–464
Extension lines, 228, 229, 229f
Exterior walls, 483–484, 484f
Extruding, 273
Eye implant, 39f

F

Fabrello-Streufert, Anne, 77
Fabrication, 278
Faces and vases puzzle, 114, 114f, 115
Failure, team, 71
Failure point, 249
Fallingwater, 474, 474f
Faraday, Michael, 364
Fashion designer, 6, 109, 125
Fastener, threaded, 279, 279f
Fastening
  chemical, 279–280
  mechanical, 278–279, 278f, 279f
Fatigue, 251–252, 252f
FDM. See Fused deposition modeling (FDM)
FEA. See Finite element analysis (FEA)
Feature-based modeling, 237
Features, 239
Federal Highway Administration, 294
Federal style, 486, 486f
Federal Trade Commission of the U. S., 254
Feedback. See Team feedback
Fillet, 235, 236f
Final design document, 52
Finite element analysis (FEA), 160, 170, 171, 171f
FIRST. See For Inspiration and Recognition of Science and Technology (FIRST), 37, 37f
FIRST Robotics Competition (FRC), 37, 37f, 42
FIRST Tech Challenge (FTC), 37
Fischertechnik, 95, 341
555-Timer Chips, 375–376, 375f
Five-port valve, 394–395, 395f

Fixtures, 277, 278
Flared eave, 485
Flexible manufacturing system (FMS), 281
Flint knapping, 269
Floor plan, 477, 478f
Flow regulator, 395–396, 396f
Fluidics, 387
Fluidic systems
    calculating forces in, 396–399, 398f
    safety in, 401
Fluid(s)
    characteristics of, 388–389, 388f
    defined, 388
FMS. See Flexible manufacturing system (FMS)
Fold lines, 225f
Force(s). See also Structural loads and forces
    calculation in fluidic systems, 396–399
    common, 301–304, 302f–304f
    defined, 328
Ford, Henry, 267
Foreshortened, 232
Forging
    cold, 272
    impression-die, 272, 272f
    open-die, 272
For Inspiration and Recognition of Science and Technology (FIRST), 37, 37f
Form, 120
    appearance, 471
    defined, 100
    as design element, 100–101, 101f
Format, 525
Form-cutting tool, 277
Forming, 174t, 269
Fossil fuels, 17, 22
Fox, Charles, 11
Fractions, 209
Franklin, Benjamin, 16, 359, 364
FRC. See FIRST Robotics Competition (FRC)
Free-falling objects, 453–454
Freeform surface modeling, 237–238, 237f
Freehand, 107
Freehand drawing, 93, 107, 147
Freezing cell, in spreadsheets, 440–441, 441f
French Academy of Sciences, 208
French doors, 485
Front face, 130, 131f
FTC. See FIRST Tech Challenge (FTC)
Function, 471, 495
Functional analysis
    of Etch A Sketch, 170, 170f
    in reverse engineering, 168–169, 169f
Fused deposition modeling (FDM), 283, 283f

G

Gable, 479
Galileo, 448
Galley proof, 525
Gantt charts, 206
Gates, 274
Gauge, radius and feeler, 169, 169f
Gaussian curve, 466f
Gaussian distribution, 406, 406f
Geddes, Norman Bel, 405, 496, 496f
Geometric shapes, 120, 120f
Georgian style, 485–486, 485f
Glass box principle, 225, 225f
Global community, 263
Global positioning system (GPS), 367
Goal(s), team, 70
Gold, density of, 433
Golden rectangle, 520f, 521
Golden section. See Golden Rectangle
Goods, durable and nondurable, 296
Google Docs, 198
Google Scholar, 196, 197f
GPS. See Global positioning system (GPS)
Granjean, Arthur, 159
Graphical representation, Excel and, 443–445, 443f–445f
Graphic artists, 108–109
Graphic design
    defined, 505
    directory of schools for studying, 506
    process. See Graphic design process
Graphic designer, 6, 31, 505
Graphic design process, 505–515
    collecting information for design, 506–507, 507f
    creating solutions, 507–510, 508f, 509f, 510f
    producing graphic image, 511–515, 512f–515f
Graphical analysis, calculating loads with, 313, 313f
Graphic image, producing, 511–515, 512f–515f
Graphics and presentation
    design principles and, 515–521, 516f–520f
    graphic design process. See Graphic design process
    illustrations/symbols/logos and, 529, 529f
    introduction to, 505, 505f
    photographs and, 527–529, 528f, 529f
    typography and, 521–526, 521f–526f
    web design, 530–531, 530f, 531f
Graphs
    absolute vs. relative, 434–436, 435f
    polar, 436–437, 436f, 437f
Graves, Michael, 471, 492–494, 492f
Gravitational force, 350, 351f
The Great Exhibition, 11, 11f
Great Pyramid, 475, 475f
Greek architecture, 475, 475f
Greek Revival style, 486, 486f
Green design. See Sustainable design
Greene, Charles, 488, 489f

Greene, Henry, 488, 489f
Grinding machines, 277
Gropius, Walter, 490, 499
Grouping related elements, 516, 516f
Group norms, 71–74
    establishing, 73–74
    establishing team values, 72, 73f
    identifying team mission, 72
    norms defined, 71
Group technology, 281
Gutenberg, Johannes, 89, 521

H

H. J. Heinz Company, 186, 190
HABS. See Historic American Buildings Survey (HABS)
Hammurabi, Code of, 434
Hardness testing, 252, 252f
Hawthorne, Molly, 270
Health and safety engineering, 64
Heat, 368–369, 368f
Helicopter, 322, 322f
Henry, Joseph, 365
Hero's steam turbine, 264, 264f
HF. See Human factors (HF)
Hidden line, 228, 228f
High Country Engineering, 295
High-impact polystyrene (HIPS), 173
High probability thinking, 87
HIPS. See High-impact polystyrene (HIPS)
Histogram, 446–448, 447f
Historic American Buildings Survey (HABS), 484
Hobby models, 227
Hooke's law, 250–251
Hoover Dam, 17, 17f
Hopper, casement window, 482
Hopper, Grace, 110f
Horizon line, 130, 131f, 133f, 134f, 135–136
Howe, Elias, 13
Howe, William, 309
Howe Truss, 309, 309f
How to Solve It (Polya), 32
HTML. See Hypertext markup language (HTML)
Hubbard, Elbert, 488
Hue, 124, 124f, 519, 519f
Human behavior, 410–411, 411f
Human design world
    achievements in, 16–18, 16f, 17f
    design and industrial revolution, 8–12
    design professionals, 4–8
    engineering careers, 19
    engineering societies, 12, 12f
    ethics and design, 27
    impacts of technology, 20–25, 21f, 22f, 24f
    introduction to, 3
    science and technology, 19
    standardization, 14–15

Human factors (HF)
  defined, 405
  designing with anthropometric data, 422–424, 422f, 424f
  engineering, 199
  ergonomics, 199
  human behavior and, 410–411
  human scale, 406–410, 406f–410f
  introduction to, 405
  posture/movement, 412–415, 412f–413f
  universal design, 415–417, 415f–416f
Human factors information, 199
Human scale, 199, 406–410, 406f–410f
Hydraulic system, 387, 388, 397
  defined, 389
  pneumatic system vs., 389–390, 389f–390f
Hypertext markup language (HTML), 530, 530f
Hypothesis, 164–165

I

IAC. See Inventors Assistance Center (IAC)
ID. See Industrial design (ID)
Ideas
  clarification, 92
  dominant, 88
  evaluation, 92
  round-robin recording of, 91–92
Ideas generation and development
  choosing the best solution, 102–103
  creative thinking, 87
  design and, 87–94, 88f
  development work, 95–102
  drawing and, 107
  introduction to, 87
  researching and, 40–41
IDSA. See Industrial Designers Society of America (IDSA)
IEC. See International Electrotechnical Commission (IEC)
Ieoh Ming Pei, 491
Illustration, in graphics, 529
Image, graphic, producing, 511–515, 512f–515f
Impacts, technology and, 20–25, 21f, 22f, 24f
Imperial inch, division of, 209
Imperial system, division of inch, 208
Impression-die forging, 272, 272f
Incubation period, 94, 94f
Independent inventor resources, 193
Inductors, 365, 365f
Industrial Age, 8, 8f
Industrial designer, 6, 218, 471, 494–499
  defined, 494
  postmodern, 499
Industrial Designers Society of America (IDSA), 7, 405, 495
Industrial design (ID), 6, 495

Industrial engineers, 64
Industrial process, 174t
Industrial revolution
  agrarian revolution, 264, 264f
  design and, 8–12, 8f–10f
  interchangeable parts, 266–267
  period of, 8
  steam engine, 264–265, 264f, 265f
  steam engine principles, 265, 266f
Industry, utilization of teams in, 61–76, 61f–64f, 66f, 68f–71f, 73f
Information
  consumer-based, 190
  patent, 192
  saving, 198
  teams and sharing of, 78–83, 79f, 80t, 81t, 83f
Information Age, 8, 10
Infrastructure, as virtual team challenge, 69
Innovation
  creativity and, 90
  defined, 90
  in design process, 54–56, 54f, 55f
  product, 157
  start of, 38
  technological, 157
Input/output, in microprocessor, 381, 381f
Inputs, in technological system, 20
Input section, in electronic system, 366, 366f
Instron® Model 3342 testing system, 194f
Insulator(s), 351, 353–354, 353f, 354f
Integration, team
  as virtual team challenge, 69
Intellectual property (IP), 161, 192, 515
Interchangeable parts, 12, 266–267
Interior views, one-point perspective and, 132–134, 132f–133f, 134f
International Electrotechnical Commission (IEC), 15
International Exposition for Decorative Arts and Modern Industry, 489
International Organization for Standardization (ISO), 15, 216–217, 217f, 520
International style, 490, 490f
Internet, 196–197, 197f
Internet Public Library, 195
Internet Service Provider (ISP), 506, 530
Invention, defined, 157
Inventor resources, 193
Inventors Assistance Center (IAC), 193
Inventor's log, 110, 110f
Inverse square law, 351
Investigation, sketching as, 139
Investigation and research for design development
  asking questions, 188–189, 188f
  average person, myth of, 199, 199f

  careers in, 191
  case study, 186–187, 186f, 187f
  consumer-based information for, 190
  encyclopedias and, 195–196
  human scale, 199
  internet, 196–197, 197f
  introduction to, 185
  laboratory studies, 194, 194f
  library homepages, 194–195, 195f
  magazines, trade journals, and newspapers, 197, 197f
  patent information, 192
  patent searches and independent inventor resources, 193
  saving information and citing sources, 198
  secondary sources, 194–199, 195f–197, 199f
  trademark and copyright protection, 192–193, 192f
  using human factors information and, 199, 199f
  using market research, 189
  using primary sources, 189–190, 190f
  visiting stores, 193, 193f
  web portals, 195, 196f
Investment casting, 274, 274f
Investor-owned utilities (IOUs), 17
Ionic columns, 479, 480f
IOUs. See Investor-owned utilities (IOUs)
IP. See Intellectual property (IP)
ISO. See International Organization for Standardization (ISO)
Isometric drawing(s), 129, 139–140, 140f, 220–222, 221f, 222f
Isometric grid paper, 221–222, 222f, 223f
Isometric planes, 221, 222f
Isometric view drawing, 203
ISP. See Internet Service Provider (ISP)
Iteration, concept of
  defined, 33
  in design process, 33–34
Iterative calculations and formula entry, in spreadsheets, 439–440, 440f
Ive, Jonathan, 252, 499, 500–501, 500f

J

Jalousie, casement window, 482
Jefferson, Thomas, 476
Jigs, 277, 278
JIT. See Just-in-time (JIT)
Johnson, Bill, 186
Johnson, Lyndon, 18
Johnson, Philip, 490
Johnson & Johnson company, 191, 342
Joining process, 174t
Joint extension, 412
Joint flexion, 412
Journals, trade, 195, 197, 197f
Just-in-time (JIT), 281

K

Kamen, Dean, 37
Kerning, 523, 523f
Ketchup, history of, 186–187
Keyboard, ergonomically designed, 424, 424f
Key edge, 134
Keywords, 194
Kinematic chain, 326, 327f
Kinematic diagrams, 326, 326f
Kinematic members, 327–328
Kinematics, 324–328
Kinetic energy, 328
Kirchhoff's Laws, 358, 360
KPFF Consulting Engineers, 77
Kranzberg, Melvin, 26

L

Laboratory study, in investigation and research, 194, 194f
Lambeth, Justin, 186
Landscape, in portfolio pages, 151, 154
Lang, Robert, 430
Language, as virtual team challenge, 69
Laser designer, 53
Laser-electrical discharge machining, 213
Lateral thinking, 87, 88–89
Lateral Thinking: Creativity Step by Step (De Bono), 87
Layout, 109
Lazarus, Charles, 417
Le Corbusier, 490, 490f
Leaders, 230
Leadership, and team success, 71
LEDs. *See* Light-emitting diodes (LEDs)
Left brain, 115
Legal impact, 254
LEGO Corporation, 220
LEGO Digital Designer, 220f
Letter spacing, 523, 523f
Levers and linkages, 331–335, 331f
Leyden jar, 364, 364f
Library homepages, in investigation and research, 194–195, 195f
Libraryspot.com, 195
Life-cycle cost, 254
Lifting, 414–415
Light, 367–368, 367f
Light-dependent resistor (LDR), 367–368, 368f, 374f, 375
Light-emitting diodes (LEDs), 55, 359, 368, 370–371, 371f, 372f
  anode, 371
Light source and shading, 122–123
Limitations, design, 56–57
Limit dimensions, 235, 235f
Limit tolerance dimensioning, 235, 235f
Lincoln, Abraham, 14

Line
  defined, 119, 119f
  as design element, 99, 99f
  horizon, 131–132, 131f, 132f, 134f, 135–136, 135f
Linear motion, 330
Linear perspective, 130, 136
Line conventions, 228–232
Line length, 524–525, 525f
Line spacing, 523, 523f
Line weight, 228
Linkages, 333–335
  bell crank, 333, 333f
  defined, 333
  motion-reversing, 333, 333f
  parallel, 334, 334f
  reversing, 333, 333f
  toggle, 334, 335f
  tradle, 334, 334f
Links, 326
Liquid, 388
Liquid-based hydraulic system, 389
Listening, 81
  active. *See* Active listening
  skills, 81
Live loads, 299
Lockheed Martin, 349
Lockstitch, 13
Loewy, Raymond, 6, 495–496, 495f
Logic operation, 378, 378f
Logistic curve (S-curve), 463, 464f
Logo, 192
  in graphics, 529, 529f
  in portfolio, 151
Lost-wax casting, 274
Louvre Pyramid, Paris, 491, 491f
Low probability thinking, 87
Lubricators, 392
Ludd, Ned, 23
Luddites, 23

M

Machine, 323
Machine tools, 12
Machining information, drawing with, 206f
Magazines, 197
Magnetism, 360–361, 360f, 361f
Managed production systems, 263
Manufacturing. *See also* Manufacturing analysis
  assembly line, 267, 267f
  careers in, 270
  defined, 172
  designing for manufacture, 282–283
  future impacts, 286
  industrial revolution, 263–267
  introduction to, 263
  material processing, 267–282
  process, 268

Manufacturing analysis
  of Etch A Sketch, 175, 175f
  in reverse engineering, 172, 174
Manufacturing engineers, 172
Mark, 192
Markers, 129
Marketplace, 157
Market research
  defined, 185
  using, 189
Market testing, 246
Mass-produced products, 13, 407
Mass production, 51–52
Material choices, 98, 98f
Material failure or breaking point, 301, 301f
Material forming, 269–275
  bending, 269, 271, 269f, 271f
  casting, 273, 273f
  drawing, 273
  extruding, 273, 273f
  forging, 272, 272f
  pressing, 271–272, 271f, 272f
  types of molds, 274–275
Material production cycle, 268
Materials. *See also* Combining materials; Material analysis; Material forming; Materials processing; Materials testing
  combining, 278–280
  forming, 269–275
  importance of, 269
  production cycle, 268
  separating, 275–277
Materials analysis
  of Etch A Sketch, 173, 173f
  in reverse engineering, 171–172, 171f
Materials processing, 267–282
  characteristics of material, 267
  combining materials, 278–280
  computer-integrated manufacturing, 280–282
  forming materials, 269–275
  importance of materials, 269
  organizing for production, 280
  production cycle, 268
  separating materials, 275–277
Materials testing
  in engineering, 249–252, 249f, 250f
  mathematics of, 251
  science of, 251
  technology and engineering of, 251
Math and science applications
  introduction to, 429
  math and the art of Origami, 430–431, 430f, 431f
  measurement and, 429–456
  swinging doors, 456–467
Mathematics
  of color, 100
  of material testing, 251
  perspective and, 136

technical drawings and, 208
  use in calculating, 312
Mattel, 417
Matte surface, 144
McPherson, 421
Mean, 446
Measurements, results of, 212
Measure of Man (Dreyfuss), 405
Mechanical advantage (MA), 332, 332f
Mechanical engineer, 55, 63, 64f
Mechanical fasteners, 278
Mechanical fastening, 278–279, 278f
Mechanical separating techniques, 276–277
Mechanical system design, 320–344
  careers in, 342
  creating motion in mechanism, 329–330
  helicopter, 322, 322f
  introduction to, 321, 321f, 323, 323f
  kinematics, 324–328
  levers and linkages, 331–335, 331f, 333f, 334f, 335f
  mechanisms and machines, 323
  modeling mechanical designs, 341
  rotary mechanisms, 335–341
Mechanism, 323, 324
Memory, in microprocessor, 381, 381f
Mergenthaler, Ottmar, 521
Metric system, 208, 217, 350
Michalko, Michael, 87
Micrometers, 168, 169f
Microphone(s)
  types of, 437
  wireless Bluetooth, 369, 369f
Microprocessor, 348, 366, 376, 377f, 380–382, 381f, 382f, 383f
Microwave cell towers, 42f
Mindmapping, 90, 91f
Minimalism, 490
Mirror images, drawing, 115, 115f
Miscommunication, as virtual team challenge, 69
Mission of team, identifying, 72
MLA. *See* Modern Language Association (MLA)
Model
  defined, 47
  making, 48–49, 48f
Modeling, 146, 147
Modeling or prototyping, documenting, 150
Modern Language Association (MLA), 198
Modern style, 490, 491f
Modulus of elasticity, 250, 301, 301f
Molds, types of, 274–275
Moment calculation, 312, 312f
Moment of inertia, 456, 457, 457f
Momentum, 327
Monostable timer, 375
Monzoni, Willoughby, 308
Moore's law, 376, 377, 462
The Morrill Act of 1862, 14

Morris, William, 488, 499
Mortise joints, 278
Motion, 327, 329–330, 330f
Motion of projectiles, 454, 455f
Motion-reversing linkages, 333f
Motorola company, 158
Multivibrator, 375
Multiview drawings
  defined, 201
  views arrangement in, 224, 224f
Muntins, 482

N

NAE. *See* National Academy of Engineering (NAE)
NASA. *See* National Aeronautics and Space Administration (NASA)
National Academy of Engineering (NAE), 16, 16f
National Aeronautics and Space Administration (NASA), 26
National Bridge Inventory (NBI), 294
National Institute of Neurological Disorders and Stroke (NINDS), 414
National Institute of Occupational Safety and Health (NIOSH), 414
National Institute of Standards and Technology (NIST), 15f
National Society of Professional Engineers (NSPE), 27, 161
Natural exponential function, 462–463
NBI. *See* National Bridge Inventory (NBI)
Neat left, 517
Negative shapes
  defined, 118
  drawing, 118, 118f
Neutrons, 350
Newcomen, Thomas, 264, 265f
New products, learning of, 193
Newspapers, 197
Newton, Isaac, 19, 88, 126, 265, 297–298, 300, 327, 328, 348, 350, 358, 448–449
Newtonian mechanics, 297–298
Newton's third law, 388
Newton's three laws of motion, 448–449
New York City's pneumatic subway, 391
NGT. *See* Nominal group technique (NGT)
Niepce, Nicephore, 130
NINDS. *See* National Institute of Neurological Disorders and Stroke (NINDS)
NIOSH. *See* National Institute of Occupational Safety and Health (NIOSH)
NIST. *See* National Institute of Standards and Technology (NIST)
Node, 358, 360, 360f
Nominal group technique (NGT), 91–92
Nondurable goods, 253, 254, 296
Nonisometric planes, 222f

Nonsequential process, 33
Nonstandard unit, 143
Nonuniform rational basis splines (NURBS), 238
Nonverbal communication, as virtual team challenge, 69
NoodleBib Express, 198
Norman, Donald, 256
Norming and storming, 75
Norms
  defined, 71
  group. *See* Group norms
  team communication, 78
Notebooks, 207, 207f
NSPE. *See* National Society of Professional Engineers (NSPE)
Numbers, in test results, 256
NURBS. *See* Nonuniform rational basis splines (NURBS)

O

Object, envisioning view in, 224–225, 225f
Object lines, 228
Oblique pictorial drawings, 220–223
Oblique projections drawing, 222, 223, 223f
Oblique views, 222–223
Obsolescence, planned, 296
Occupational Safety and Health Administration (OSHA), 406, 414, 422
Ohio Art Company, 159, 164, 167, 173, 174, 178, 180
Ohm, Georg, 355
Ohm's law, 17, 355–358, 360, 363
Olds, Ransome Eli, 267
Omnidirectional microphone, 437
One-point perspective, 130, 131–132, 131f, 132f, 133f, 134, 134f
One-shot molds, 274
Opamp. *See* Operational amplifier
Open-die forging, 272
Open-ended questions, 188
Open-loop system, 20
Operational amplifier, 376, 376f
Optical center, 517, 520
Opticks (Newton), 126
Optimal, 102
Optimal match, 409
Ordering
  defined, 33
  in design process, 33
Organic shape, 120, 120f
Orientation (forming), in team developing, 75
Origami, art of, 430–431, 430f, 431f
Orthographic drawing(s), 206, 207, 214, 216f
  first-angle projection, 224f
  and sketching, 224–227
  third-angle projection, 224f
Orthographic multiview drawings, 203, 204f

Index

Orthographic projection, 141, 141f, 224
Oscillating motion, 330
Oscillators, 460
OSHA. *See* Occupational Safety and Health Administration (OSHA)
Outage, 18
Outlining, 143, 143f
Output
   in electronic system, 366, 366f, 367f, 370–374
   in technological system, 21–23
Output devices, 366, 370, 371

P

Page content, in portfolio, 153
Page layout, in portfolio, 154
Page numbering, in portfolio, 151
Page orientation, in portfolio, 151
Pagination, 525–526
Pairs, 326
Palladian window, 476, 476f, 482
Palladio, Andrea, 476
Palmar flexion, 413, 413f
Paper, sketching and texture of, 129
Parallel circuit, 362
Parallel linkages, 334, 334f
Parameters, 237
Parametric modeling, 237, 239
Parametric Technologies Corporation (PTC), 237
Parapet, 485
Paraphrases, 193
Parthenon in Athens, 475, 475f
Part tolerance, 266, 266f
Pasadena architect, 488, 488f, 489f
Pascal's law, 388
Patent information, 192
Patent infringement, 160
Patent(s)
   defined, 109, 192
   law, 192
   reverse engineering and, 160–161
   searches for, 193
Pauling, Linus, 87
Paxton, Joseph, 11
Pediment, 479
Pencil
   colored, techniques of, 144, 144f
   for sketching, 129
Percentiles, 407–408, 407f, 409, 410f
Performance, 253
Permanent magnet, 360, 371, 372f
Permanent molds, 274, 274f
Personal impact, 254
Perspective, 108
   linear, 130
   mathematics and, 136
   one-point, 130, 131–132, 131f, 132f
   two-point, 134–136, 134f, 136f, 140f, 142f, 221f

Perspective drawing, 129–141
PET. *See* Polyethylene terephthalate (PET)
Photographs in graphics, 527–529
   composition of photographs, 528, 528f
   editing photographs, 529, 529f
   technical qualities of photographs, 527, 528f
   using PowerPoint for presentation, 532–534, 533f, 534f
Photoreceptors, 124
Photoresistor, 367, 368f
Pictorial drawing, 108, 221
Piercing operations, 276
Piezoelectric effect, 369, 369f
Pink, Daniel, 139
Pitch, 479
Plagiarism, 193
   avoiding, 198
Planetary gear chain, 324f
Planned obsolescence, 296
Planning
   defined, 32
   in design process, 32–33
Plastics Additives & Compoundings, 197
Plastic stage, 249, 301
Pneumatic cylinder, 392, 393f, 394f
Pneumatic system. *See also* Pneumatic system design
   automated, 400f
   basic components of, 392–396, 393f, 394f, 395f, 396f
   defined, 389
   hydraulic *vs.*, 389–390, 390f
Pneumatic system design
   basic pneumatic circuits, 399–400, 399f, 400f
   calculating forces in fluidic systems, 396–399
   characteristics of fluids, 388–389
   introduction to, 387
   pneumatics *vs.* hydraulics, 389–390, 390f
   pneumatic-system components, 392–396, 393f, 394f, 395f, 396f
   principles of pneumatics, 390–392
   safety in fluidic systems, 401
Point perspective, 130–138
Pocket Etch A Sketch, 159
Polarity, 357, 359
Polar (non-Cartesian) graphing, 436–437, 436f, 437f
Polya, George, 32
Polya's steps to problem solving, 32–36
Polyethylene terephthalate (PET), 173, 187
Polygon, 208, 313
Polyvinyl chloride (PVC), 354, 354f
Pont du Gard aqueduct, 476f
Popliteal height, 419, 420
Population, growth of, 463–464
Portfolio(s), 109
   developing, 150–151
   online, 110f

   pages, 151–154
   student design, 111, 111f–113f
Portico, 479
Position, 370, 370f
Possibility, exploring, in design process, 44–46
Postmodern architects, 491–494, 491f–494f
Postmodernism, 491, 491f, 494
Posture/movement, 412–415
Potential energy, 328, 358
Potentiometer, 370, 370f
Power, 327
PowerPoint, making presentation using, 532–534, 533f, 534f
Powers of 10, 352, 352f
Powers of 2, 377, 377f
Prairie style, 473, 488, 488f
Pratt, Caleb, 308
Pratt, Thomas, 308
Pratt truss, 308, 308f, 309
Precision, 212
Preconsumer waste, 275
Pre-engineering courses, 65
Preliminary sketch, 147, 147f
Pressing, 271–272, 271f, 272f
Pressure, 388
Pressure-operated five-port valve, 395, 395f, 400f
Pressure regulators, 392
Price point, 164
Primary colors, 124–125, 513, 519
Primary source, 189–190, 192–194
Principles, 5, 5f, 6
Printing, 506
Prismatic, 121, 121f
Probability/statistics applications, 464–467, 466f
Problem-defining phase
   in design process, 38
   documenting, 150
Problem solving
   defined, 31
   Polya's four-step design process, 32–36, 32f, 33f, 34f
   process, 108, 108f
Process(es)
   defined, 374
   in electronic systems, 366
   industrial, 174t
   nonsequential, 33
   separation, 174t
   sequential, 33
   in technological system, 20
Processor(s)
   analog, 374–376
   digital, 376–383
   system, 374
Product innovation, 157
Production, organizing for, 280
Production drawing, 149, 149f

Production (performing), in team developing, 75
Product life cycle, 23–25, 24f
Product redesign, in reverse engineering, 179
Product(s), 503
    custom, 406, 407, 407f
    defined, 505
    mass-produced, 407
    style, 471
Pro/ENGINEER, 239
Professional design societies, 12f
Profit, 254
Project, managing, 34–36
Projectile motion, 454, 455f
Projection, orthographic, 141, 141f, 224
Projection lines, 232
Project risk, 35
Promotion research, 190
Pronation, 412f, 413
Proofreaders' marks, 525, 526f
Proportion, 116, 225, 476, 520
Protons, 350
Prototype
    defined, 47
    making, 48–49, 48f
Proximity, 516, 516f
PTC. *See* Parametric Technologies Corporation (PTC)
PTC's Pro/DESKTOP, 239, 239f
Purpose identification, in reverse engineering, 163
PVC. *See* Polyvinyl chloride (PVC)

Q

Quadrille-ruled graph paper, 223
Questions, 188–189
    closed, 188
    open-ended, 188
Quoins, 484

R

Radial deviation, 413, 413f
Radius, defined, 436
Radius gauges, 169f
Rake angle, 277, 277f
Range of motion, 412–413, 412f
Rapid prototyping, 282–283
Rapid prototyping machines, 207
Ratchford, Patty, 258
Ratio, 94
Rausch, Randy, 9
Raw materials, 268
REA. *See* Rural Electrification Administration (REA)
Reach, 409, 410f
Reaching consensus, 79
Reagan, Ronald, 26
Receiver, 392
Reciprocal motion, 330
Reciprocating-piston compressor, 392, 392f
Rectilinear shapes, 100, 120, 120f
Recycled materials, 268
Refdesk.com, 195
Reference style, 198
Refining
    in design process, 50–51
    documenting, 150
Regenerative braking systems, 340
Relative graphs
    absolute graphs *vs.*, 434–436, 435f
Relay, 373, 373f, 374
Reliability, 18
Repeatability. *See* Precision
Repetition, 518, 518f
Repetitive stress injuries (RSIs), 406
Report preparation, in reverse engineering, 174
Reproducibility. *See* Precision
Researching and generating ideas
    in design process, 40–41
    documenting, 150
Resistance, 17, 354, 355, 355f, 357
Resistivity, 353, 354
Resistors, 355, 355f, 358, 358f, 359, 360, 361, 363–364
Resolution (norming), in team developing, 75
Reuleaux, Franz, 324, 325
Reuleaux's six mechanical elements, 324, 325, 325f
Reuleaux triangle, 324
Reverse engineering, 156–183
    careers in, 180
    case study, 159, 164, 167, 170, 171, 173, 175, 176–178
    defined, 157
    developing hypothesis, 164, 165
    disassembly in, 165–168
    elements analysis in, 168
    functional analysis in, 168–169, 169f
    identifying purpose in, 163
    manufacturing analysis in, 172, 174–179
    materials analysis in, 171–172, 171f
    patents and, 160–161
    process flowchart, 162f
    product redesign in, 179
    products and, 158, 160
    reasons for, 162–163
    report preparation in, 174, 175
    structural analysis in, 169, 170–171
Reversing linkages, 333
Revision, 218
Reynolds, John, 67
Rhythm, 518
Ride, Sally, 199
Ridge, 479
Right brain, 115
Robot Building for Beginners (Cook), 282
Robotics, 35, 37, 280, 282
Robots, 282
Rockwell hardness test, 252
Rods, 124
Roebling, John Augustus, 306
Roman architecture, 476, 476f, 477f
Roof, style of, 479, 480f
Roosevelt, Franklin Delano, 17
Rose, Richard Douglas, 284–285
Rotary mechanisms, 335–341
Rotary motion, 330
Rotational casting, 275
Rotational motion, 454–456
Rotational velocity, 458
Rough layout, 509, 509f
Round, 235, 236f
Round-robin recording, of ideas, 91–92
Rules, 518
Rupture point, 249
RSIs. *See* Repetitive stress injuries (RSIs)
Rural Electrification Administration (REA), 17
Russell, Scott, 11

S

SA. *See* Situational awareness (SA)
Safety
    compressed air, 401
    factor, 297
    in fluidic systems, 401
Savery, Thomas, 264
Scale, 227, 227f
Scaled drawings, 209
Schematic symbols, 399
Schmidt, L. Stephen, 421
Science
    of color, 100
    defined, 19, 265
    of electricity, 350–361, 351f–361f
    of material testing, 249–252
    and technology, 19, 164
Scientific notation, 352
Scrubbers, 22
SD. *See* Standard deviation (SD)
Search engine, 196–197
Secondary colors, 125
Secondary source, 194–199
Sectional drawing, 141, 141f
Sectional view, 141, 230–231, 231f
Section lines, 231
Section planes, 231f
Segway, 37, 37f
Self-directed team, 71
Semiconductors, 354
Semi-permanent magnet, 360
Separating materials
    defined, 275
    mechanical techniques, 276–277
Separation process, 174t
Sequential process, 33
Serial discussion
    for clarification/evaluation, 92
Series circuit, 362
Sewing machine, 13, 13f

Shading
  defined, 119
  light source and, 122–123
  sphere with colored pencil, 144f
Shape
  appearance, 471
  defined, 120
  as design element, 100–101, 101f
  drawing positive and negative, 118–119, 118f
  geometric, 120, 120f
  organic, 120, 120f
  prismatic, 121, 121f
  rectilinear, 120, 120f
  types of, 120f
Shear force(s), 304, 304f, 310
Shearing, 276, 276f
Shotgun microphone, 437
Shuttle valve, 396, 396f
SIGGRAPH, 179, 179f
Sighting for proportion, 143, 143f
Sikorsky, Igor, 322, 322f
Simon, Julian, 27
Simple testing, of loads, 310–312, 311f
Simulate, 127, 127f
Singer, Isaac, 13
Single-acting cylinders, 393, 393f, 396, 396f, 397f, 399f
Situational awareness (SA), 411
Sketch(es)/sketching
  analyzing simple mechanism through, 102
  annotated, 148, 148f
  casting, 273
  creating in solid modeling software, 239
  defined, 107
  developing ideas and, 93–94, 93f, 94f
  developmental, 148–149, 148f
  drawing techniques and, 129–141
  hypothesis in engineering notebook, 165f
  as investigation, 139
  preliminary, 147, 147f
  technical, 94f
  in solid modeling software, 239
  techniques, 127f
Sliding-vane compressor, 391, 392f
Smith, Adam, 61
Social impacts, 254
Socialization, team
  as virtual team challenge, 69
Society(ies)
  engineering, 12, 12f, 161
  professional design, 12f
  technologically dependent, 3
Solenoids, 373, 373f
Solenoid valve, 396, 396f
Solid modeling, 220, 237, 238
  sketches creation in, 239, 239f
Solution
  best, choosing, 102–103
  engineering testing, 248–252

Somerville, Bob, 17
Sonic welding, 167
Sound, 369–370
Sources, citing, in investigation and research, 198
Space, 116, 128
  as design element, 101
Space Shuttle Challenger, 26, 26f, 27, 421
Spacing between views, 227
Spatial ability, 224
Spatial relations, 117, 224
Spinning, forming by, 272f
Sport utility vehicle (SUV), 41, 45, 263
Spreadsheet, 146
Stage designers, 6, 6f
Standage, Tom, 256
Standard deviation (SD), 409, 466
Standard industry materials, 268
Standardization, 14–15
Standards, 143
  American National Standards Institute, 15, 216
  technical drawing, 216–218
"Stanley," autonomous vehicle, 35f
Static loads, 299
Statics, 450–452, 450f, 451f, 452f
Statistics, Excel and, 445–448
Steam engine, 264–265
  atmospheric, 265, 266f
  double-acting, 265
  idea of, 264–265, 264f
  James Watt's, 323
  principles of, 265
Steinmetz, Charles, 16
STEM connection, 265
Stephenson, Robert, 11
Stepper motor, 372, 373f
Stereolithography, 282, 282f
Stickley, Gustav, 488, 499
Sticky notes concept, 39f
Store owners, 193, 193f
Stores visiting, 193
Strain, 249, 300–301, 300f, 301f
Streamlining, 496, 496f
Strength
  team member, identifying, 74, 74t
  ultimate, 249
Stress, 249, 300, 300f
Structural analysis
  of Etch A Sketch, 171, 171f
  in reverse engineering, 169, 170–171
Structural components, 305
Structural failure, 294, 296–297
Structural loads and forces
  bending force, 302–303, 307f
  calculating loads, 310–316
  compression forces, 302
  dead loads, 299
  dynamic loads, 299

  equilibrium, 299–300
  graphical analysis use in calculating, 313
  live loads, 299
  shear forces, 304
  simple testing of loads, 310–312
  static loads, 299
  strain, 300
  stress, 300
  of structural system design, 299–301
  tension force, 302, 302f
  torsion force, 304
Structural system design
  basics of, 292
  bending force, 301, 302–303, 307f
  bridge structures, 305–310, 307f–310f
  calculating loads on structures, 310–316
  careers in, 295
  common forces and, 301–304
  introduction to, 291, 291f
  newtonian mechanics, 297–298
  overview of, 292–294
  structural components, 305
  structural loads and forces, 299–301
  technological structures, 294, 296–297
Structure, 292
Student design portfolio, 111, 111f–113f
Styles
  building, 477–484
  defined, 471
  design. See Design styles
Subtracting fraction, 209
Success, team, 70–71, 71f
Supination, 412f, 413,
Suspension bridge, 309–310, 310f
Sustainable design, 24
SUV. See Sport utility vehicle (SUV)
Swinging doors, 456–467, 456f–462f, 464f–466f
Switches, types of, 370f
Symbols in graphics, 399, 529, 529f
Symmetry, 479, 517
Synectics, 92–93
Synergy, defined, 70
Syringe, 393, 393f
System processors, 374
System outputs, 370–374

T

Table of contents, in portfolio, 151
Tabloid, 217
Tacoma Narrows Bridge, failure of, 296, 297f
Taipei 101 Tower, 291, 291f
Target population, 409
Task team, 67
Team-building, 67
Team communication
  format, 81
  level of effective, 78–79
  norms, 78
  reaching consensus, 79

Team control, 76
Team development
    careers in, 77
    case study, 67
    information sharing and, 78–83
    introduction to, 61
    team leadership and team control, 76
    utilization of teams in industry, 61–76, 61f–64f, 66f, 68f–71f
Team failure, 71
Team feedback, 79–80, 79f
    giving/receiving, 80, 80t
Team integration/socialization, as virtual team challenge, 69
Team leadership, from coach to, 76t
Team maturity, cycles of, 75
Team(s)
    defined, 61
    design process and, 65–66, 66f
    failure of, 71
    formation/structure of, 70
    goals, 70
    and information sharing, 78–83
    self-directed, 71
    setback of, 67
    stages of development, 75
    success of, 70–71, 71f
    task, 67
    virtual, 67–70
    work, 67
Teamwork, 67
Teardown, 165
Technical drawing
    architectural scale, 209–210, 209f, 210f
    computer-aided design, 237–239, 237f, 238f, 239f
    consumer and, 216
    defined, 95, 129, 146
    dimensioning, 232–236–236f
    dimensions on, 218
    drafting and CAD, 219–220, 219f, 220f
    engineering design, measurement for, 208–209, 208f, 209f
    engineering process and, 218
    introduction to, 203
    isometric, 220–223, 221f–223f
    line conventions, 228–232, 228f–232f
    math and, 208
    oblique pictorial drawings, 220–223, 221f–223f
    orthographic drawing and sketching, 224–227, 224f–227f
    process of, 207
    standards, 216–218, 217f
    types of, 203–207, 204f–207f
    use of, 213–214, 214f–216f
Technical qualities, photographs and, 527, 528f
Techniques, 114
Technological Age, 8

Technological development, historical perspectives on, 8f
Technological literacy, defined, 3
Technological structure, 294, 296–297, 296f, 297f
Technological system, components of, 20–23
Technologist, 19
    computer and, 20
    defined, 107
Technology
    assistive/adaptive, 415–416, 416f
    of color, 100
    color science and, 126
    defined, 3, 19
    and earth's resources, 26
    engineering of material testing, 251
    impacts of, 20–25
    resources of, 174
    science and, 19
Technology and engineering, of material testing, 251
Technology curve, 463, 464f
Technology professionals, 3
Temperature sensors, 369
Tenon joints, 278
Tensile test, 249–250, 249f, 250f
Tension, 302, 302f, 307f
Tension force, 300, 302, 304f
Termination (adjourning), in team developing, 75
Tertiary colors, 125
Tesla, Nikola, 16
Testimonials, in test results, 259, 259f
Testing/evaluating
    careers in, 258
    in design process, 49–50, 50f
    design work, 252–256
    developing appropriate tests, 246–248
    documenting, 150
    engineering solution, 248–252, 249f, 250f, 252f
    evaluating design skills, 259
    hardness testing, 252
    introduction to, 245
    presenting test results, 256–257, 256f, 257f
    testing and evaluating own design work, 252–256, 253f–255f
    testing materials, 249–252
    testing of loads, 310–312, 311f
Test result(s)
    checklists, 256, 257f
    descriptions, 256
    numbers, 256
    testimonials, 256
Texture, 126–127
    appearance, 471
    defined, 119
    as design element, 102, 102f
    simulating, 127, 127f

Thermal bonding, 279
Thermal method, 277
Thermistor, 368, 368f, 383f
Thermocouple, 368
Thinking
    analytical, 103
    creative, 87
    lateral, 87, 88–89
    verbal, 114
    vertical, 87
    visual, 114
Third-angle projection, 224, 224f
Threaded fastener, 279, 279f
Three-dimensional (3-D) forms, 122
3-D LEGO model, 220, 220f
Three-port valve, 394, 394f
Three-view drawings. See Multiview drawings
Three views, envisioning an object in, 224–226, 224f, 225f, 226f
Thumbnail sketch, 507, 508f
Thurston, Robert Henry, 12
Tiffany, Louis Comfort, 488, 498
Time, 327
Time zone, as virtual team challenge, 69, 69f
Tint, 124, 124f
Title block, 151, 218
Title page, in portfolio, 151
Toggle linkages, 334, 334f, 335f
Tolerance(s), 213, 213f
    bilateral, 234
    defined, 234, 266
    dimension, 232–233
    part, 266, 267
    unilateral, 234, 234f
Tone, 520, 520f
Torque, 304, 327, 328, 328f, 456, 456f, 457
Torque amplifier from Meccano, 96f
Torsion force, 304, 305
Total quality control (TQC), 280
Toy Guide for Differently Abled Children, 417
Toys "R" Us, Inc., 417
Toys testing, 246–248, 248f
TQC. See Total quality control (TQC)
Track Changes tool, Microsoft Word, 526
Trade journals, 195, 197
Trademark, 192–193
Trade-off, 23
Training, cross, 67
Transistors, 374–375, 374f, 377f
Transition, 144, 144f
Transparent grid, 137, 138, 138f
Travel Etch A Sketch, 159
Treadle linkages, 334
Trigonometric functions, 460–462, 460f, 461f
    determining force vectors using, 312f
Truss bridge, 307, 308–309, 308f

Truss designs, 309
Twentieth century
  engineering achievements of, 16–18, 16f
  microprocessors in, 348
Twentieth-Century styles, 487
Two-dimensional shape, 120
Two-point perspective, 134
Two-view drawing, 205f
Typeface or font, 523–524, 524f
Type IV technological impact, 296
Type size, 522, 522f
Type specifications
  line length, 524–525, 525f
  line spacing, 523, 523f
  typeface or font, 523–524, 524f
  type format, 525
  type pagination, 525–526, 526f
  type size, 522, 522f
Typography, use of, 521–526, 521f–526f

U

UDL. *See* Uniform distributed load (UDL)
UL. *See* Underwriters Laboratories (UL)
Ulnar deviation, 413, 413f, 424
Ultimate Resource, 27
Ultimate strength, 249
Ultrasonic machines, 277
Underwriters Laboratories (UL), 246
Uniform distributed load (UDL), 312, 312f
Uniform Resource Locator (URL), 530
Unilateral tolerances, 234, 234f
United States Patent and Trademark Office (USPTO), 192
Units analysis, 433–434
Unity, principle of, 516–517
Universal design
  defined, 414
  human factors and, 415–418, 415f, 416f
URL. *See* Uniform Resource Locator (URL)
U.S. Department of Commerce, 296
U.S. Department of Labor
  design professions career information, 4
  number of engineers working in U.S., 7, 8
U.S. Department of Transportation, 294
U.S. Environmental Protection Agency (EPA), 286
U.S. National Committee (USNC), 15
USNC. *See* U.S. National Committee (USNC)
USPTO. *See* United States Patent and Trademark Office (USPTO)

V

Valley, 479
Value(s), 122, 471, 495
  defined, 72
  as design element, 102, 102f
  scale, 122, 122f
  team, establishing, 72, 73f
Valve
  five-port, 394–395, 395f
  pressure-operated five-port, 395, 395f
  shuttle, 396, 396f
  solenoid, 396, 396f
  three-port, 394, 394f
van der Rohe, Ludwig Mies, 490
Vanishing points, 129, 130, 133f, 134, 135–136, 135f
Variability. *See* Variation (variability)
Variation (variability), 210
  concept of, 429–433, 429f–432f
  defined, 429
  rules for measurement, 213
Vector polygons, 313
Vector(s), 313, 328, 450–452, 450f
Vellum, 219
Velocity, 327, 441
Velocity ratio (VR), 332
Verbal thinking, 114
Vernier, Pierre, 209
Vertical thinking, 87
Vex gear, 220f
Vex Robotics Design system, 220
Victorian Internet: The Remarkable Story of the Telegraph and the Nineteenth Century's On-Line Pioneers, The, (Standage) 256
Victorian style, 486–487, 487f
Views
  auxiliary, 232, 232f
  envisioning an object, 224–225, 225f, 226f
  section, 230–231, 231f
  spacing between, 227
  three, 228f
Virtual Reference Desk, 195
Virtual teams, 67–70, 68f, 69f
  benefits of, 68–69
  building, 69–70
  challenges, 69
  defined, 67
  growth of, 68
Viscosity, 389
Visible world, drawing and, 107
Vision, 70
Visual brainstorming, 147, 148, 148f
Visualize, ideas, 114
Visual measurement, 143
Visual thinking, 114
Voltage, 355, 357
Voltage law, Kirchhoff, 360
Voting, 92
VR. *See* Velocity ratio (VR)

W

Walls, exterior, 483–484, 484f
Warm-up exercises, 114–119
Warren, James, 308
Warren truss, 308, 308f
Water resistance, 356
Watt, James, 323
Wavelength, 124
Wax patterns, 274
Weaknesses of team members, identifying, 74, 74t
Weary, Kristin, 349
Web design, 530–531, 530f
Web page development, 530–531
Web portals, 195, 196f
Web sites
  design professions, career information, 4
  on engineering careers, 19
  on greatest engineering achievements, 16
  hydraulic and pneumatics principles, 401
  for researching and generating ideas, 40
  standards organization information, 16
Wedge-shaped cutting, 276–277, 277f
Weight in garbage, 25
Westinghouse, George, 16
Whitney, Eli, 160–161, 160f, 161f
Whole-brain drawing, 114
A Whole New Mind (Pink), 139
Wikipedia, 195
Windows, 481–482, 481f, 483f
  anatomy of, 481f
Wireless Bluetooth, 369, 369f
Wood, variability in, 431–432, 431f
Work, defined, 327
Working styles, historical view of, 61f
Work team, 67
Worthington, Henry Rossiter, 12
Wright, Frank Lloyd, 473–474, 473f

Y

Yield Point, 250, 251
Yield strength, 249
Young, Thomas, 301
Young's modulus of elasticity, 250, 301, 301f